MW01201344

Introduction to Radar Using Python and MATLAB®

For a complete listing of titles in the
Artech House Radar Series,
turn to the back of this book.

Introduction to Radar Using Python and MATLAB®

Andy Harrison

ARTECH HOUSE
BOSTON | LONDON
artechhouse.com

Library of Congress Cataloging-in-Publication Data
A catalog record for this book is available from the U.S. Library of Congress.

British Library Cataloguing in Publication Data
A catalogue record for this book is available from the British Library.

Cover design by John Gomes

ISBN 13: 978-1-63081-597-4

685 Canton Street
Norwood, MA 02062

Accompanying software is located at:
https://radarbook.github.io/, and the password is: $PythonRadarBook!

10 9 8 7 6 5 4 3 2 1

Contents

Preface

I firmly believe even the most difficult topics can and should be presented in a clear and concise manner such that the student can fully comprehend the material. Education is not the aggregation of facts; it is the accumulation of understanding and knowledge about particular subjects. That said, there are many radar engineering handbooks and online references containing vast amounts of very useful and detailed material with little or no explanation or analysis. These are valuable resources and have their place as references for students and engineering professionals. However, there is a tendency to indiscriminately use equations, generalities, and approximations in situations where they are not applicable. Thus, the intent of this book is to provide the reader with a straightforward and succinct introduction to, and explanation of, a number of radar fundamentals. To reinforce these fundamentals, a tool suite consisting of Python® scripts and corresponding MATLAB® scripts is included with this book. This gives the user a mechanism to analyze and predict radar performance for various scenarios and applications. The tool suite includes a user-friendly GUI that produces many graphical outputs and visualizations of the concepts being covered. The user has full access to the Python and MATLAB scripts to modify for their particular application. Several examples using the tool suite are given at the end of each chapter. These examples, combined with the end of chapter problem sets, make this an excellent choice for a senior or first-year graduate level course in engineering, as well as professionals seeking to enter the field of radar engineering.

While there are numerous topics in the field of radar engineering, this book focuses on the more crucial concepts. The book begins with a brief history of radar and description of the various categories of radar systems. Included in this chapter are detailed instructions for installing and executing the accompanying software. Chapter 2 begins with a review of the theory of electromagnetic fields and waves necessary for understanding the interaction of energy transmitted by the radar with the environment and targets of interest. Specific environmental effects such as refraction, diffraction, rain attenuation, fog attenuation, and atmospheric attenuation are also covered. Chapter 3 introduces various types of antennas used by radar systems and the parameters needed to describe their performance and characteristics. These devices transmit and receive electromagnetic energy generated by the radar, and include linear wire, loop, aperture, horn, and array antennas. Chapter 4 derives various forms of the radar range equation to predict performance measures such as maximum detection range, noise effects, output signal to noise ratio, and others. Special versions of the radar range equation for specific missions and bistatic radar configurations are derived. Chapter 5 discusses the processes used by radar receivers to take the signals received by the antenna and then amplify, filter, downconvert, digitize, and deliver them to the radar signal and data processors. An overview of digital receivers is also presented. The detection process is a key part of any radar system, and Chapter 6 deals with the detection of targets in the presence of noise. This chapter addresses coherent and noncoherent receivers, pulse integration, fluctuating targets, and constant false alarm rate processing. Radar cross section plays an important role in received signal power, probability of detection, probability of false alarm, tracking accuracy, classification, and discrimination. Therefore, the study of radar cross section is fundamental to the understanding of radar system analysis and design. Chapter 7 is dedicated to radar cross section theory, prediction, and measurement. Chapter 8 begins with an overview of range resolution for simple waveforms and considers the detection of closely spaced targets through the use of pulse compression techniques. This is followed by the study of intrapulse and interpulse approaches to pulse compression. Finally, the ambiguity function is examined for various waveforms, including pulse trains and phase-coded waveforms. Chapter 9 first considers simple alpha-beta and alpha-beta-gamma tracking filters. Then, the Kalman filter is introduced in one-dimension and subsequently extended to the multivariate formulation. Next, adaptive filtering techniques for maneuvering targets are presented. Multiple target tracking techniques are also covered. Chapter 10 introduces tomographic synthetic aperture radar (SAR) imaging and its various applications. The Radon transform is presented, which leads to a set of line integrals and projection functions. Image reconstruction techniques, including the filtered backprojection algorithm, are then used to recreate the target's reflectivity function from the spatial frequency domain data. A generalized three-dimensional framework is also given. Chapter 11 presents a compact yet thorough review of countermeasures, including active and passive jamming techniques. Methods for mitigating jamming techniques, such as space-time adaptive processing (STAP) and moving target indication (MTI) are presented. Finally, a sophisticated jamming technique known as digital radio frequency memory (DRFM) is discussed.

This book is based on the broad range of experience I have gained as a student, instructor, and professional engineer. My journey in the field of radar engineering began with an undergraduate course in microwave engineering and continued with more advanced coursework in radar-related areas. My professional career began as a radar test engineer and has advanced into areas such as distributed aperture radar, SAR imaging, and advanced signal processing.

There have been many individuals along the way from whom I have learned much and who were influential in my career. However, there are a few who deserve special mention. First, my gratitude to Caleb Stewart for our valuable discussions about Qt and GUI development, which led to an interactive and educational set of tools accompanying this book. I must also thank Vince Rodriguez for his review of the material, in particular his expertise in antenna systems and radar cross section measurement techniques. Next, I want to express my sincere appreciation to Will Halcomb. He very generously gave of his time and effort reviewing the material in this book, which notably improved the overall quality. I would like to also thank my good friend Bassem Mahafza who urged me to pursue this book, gave me invaluable opportunities and advice, and without whom this book would not have been possible.

Most importantly, I would like to thank my wife, Lacon, and my son, Conner, for their love, support, patience, understanding, and sacrifice during the countless hours committed to this endeavor.

Chapter 1

Introduction

While some may argue that the development of radar did very little to contribute to theoretical science, its significance in areas of engineering and technology cannot be disputed. The emergence of radar systems capable of detecting enemy aircraft played a vital role in the Allied victory in World War II [1]. Since that time, radar has been the precursor to a wide range of modern technology and has numerous practical applications. For example, medical imaging, radar astronomy, meteorological prediction, safe air travel, communication systems, modern computers, television, and microwave ovens arose from radar technologies. This chapter begins with a brief history of radar, and then describes various categories of radar systems. This is followed by an overview of the software tool suite accompanying this book. Instructions for downloading and installing Python® [2] and the necessary libraries, along with the tool suite are presented. The chapter concludes with a brief description of MATLAB® [3] and the scripts associated with this book.

1.1 HISTORY OF RADAR

The term radar was first used in the 1940s by the United States Signal Corps working on these systems for the United States Navy, the origin of these systems dates back much further. In the late 1880s, German physicist Heinrich Hertz definitively showed the existence of electromagnetic waves [4] as formulated by James Clerk Maxwell in his profound work on the theory of electromagnetism [5]. Using electromagnetic energy at a frequency of about 455 MHz, Hertz demonstrated that electromagnetic waves are reflected by metallic structures and refracted by dielectric materials. The experiments performed by Hertz provided the basis for the detection of targets, and in 1904 German physicist Christian Hülsmeyer was issued a patent for an obstacle detector and ship navigation device named the *telemobiloscope* [6]. While Hülsmeyer demonstrated the telemobiloscope to the German navy, there was little interest until the 1930s when the urgency to detect enemy aircraft arose due to the advancement of long-range, large-capacity bombers. During this time, development of systems to use short pulses of electromagnetic energy to detect aircraft took place independently in the United States, Great Britain, Germany, France, the Soviet Union, Italy, the Netherlands, and Japan. In 1939, a system with a single antenna for both transmitting and receiving was used on the battleship USS *New York* for detecting and tracking aircraft and ships. Immediately following World War II, advancement in radar technology slowed significantly [7]. However, the 1950s saw the emergence of highly accurate tracking radars and the use of the klystron amplifier [8] for high-power, long-range systems. The statistical theory of detection of signals in noise and the matched filter theory were published during this time [9, 10]. Digital technology advancements in the 1970s gave rise to many signal and data processing techniques including target discrimination. During the 1980s, improvements in phased array systems, including solid-state and microwave circuit technology, made remote sensing of environmental effects such as wind shear possible. In 1990, an operational prototype of the WSR-88D (Weather Surveillance Radar, 1988, Doppler) was completed, and the first installation for daily weather forecasting was completed in Sterling, Virginia in 1992 [11]. These were the first of the next-generation radar (NEXRAD) systems, which is a network of 159 high-resolution Doppler weather radars [12]. Over the years, many enhancements have been made to weather radars, including super resolution, dual polarization, and the automated volume scan evaluation and termination (AVSET) algorithm [12]. Current digital technology and new signal and data processing techniques have given rise to multiple-input multiple-output (MIMO) systems with digital beamforming and diverse waveforms, pushing radar systems toward the goal of an all-digital design [13].

1.2 RADAR CLASSIFICATION

The early driving force behind the development of radar systems was the military need to detect enemy aircraft. Since then, radar systems have seen application in many areas such as cancer detection, autonomous vehicle navigation, weather forecasting, terrain mapping, and through-wall detection. Therefore, it is often convenient to delineate systems into various categories. Radar types may be classified based on the operating frequency band, type of waveform used, application, and configuration. While there may be other possible categories, the following sections focus on these four areas.

1.2.1 Frequency Band

The first method for classifying radar systems is by frequency band, which emerged during World War II (e.g., L-band radars are used for air turbulence studies). Table

Table 1.1
Radar Classification by Operating Frequency

Band	*Frequency Range*	*Applications*
HF (high frequency)	3 – 30 MHz	Coastal radar and over the horizon systems
VHF (very high frequency)	30 – 300 MHz	Very long-range and ground-penetrating systems
UHF (ultrahigh frequency)	300 – 1000 MHz	Very long-range and foliage-penetrating systems
L (long)	1 – 2 GHz	Long-range air traffic control and surveillance
S (short)	2 – 4 GHz	Medium-range surveillance, long-range weather, and marine systems
C (compromise)	4 – 8 GHz	Long-range tracking, medium-range surveillance, and wind profiles
X (secret during WW II)	8 – 12 GHz	Missile guidance, marine radar, and medium-resolution mapping
Ku (under K)	12 – 18 GHz	High-resolution imaging and marine applications
K (*kurz*)	18 – 24 GHz	Cloud detection and police radar guns
Ka (above K)	24 – 40 GHz	High-resolution mapping and short-range surveillance

1.1 gives the IEEE radar designation [14] for frequency bands commonly used in radar systems and a description of the types of applications for each band.

1.2.2 Waveform

The next method for classifying radar systems is by the type of waveform used. Typical radar waveforms include continuous wave (CW), where the signal is transmitted continuously at a given frequency; pulsed waveforms, where bursts of energy are transmitted at various pulse repetition frequencies (PRF), and more complex waveforms such as phase-coded waveforms. Table 1.2 gives a description of commonly used waveform categories and their applications.

1.2.3 Application

Another useful method for classifying radar systems is based on the particular application of the system. For example, the WSR-88D may be categorized as a weather radar as it was

Table 1.2
Radar Classification by Waveform Type

Waveform Type	*Note*	*Applications*
Continuous wave	Unmodulated	Police radars and human gait recognition
	Modulated	Automotive collision avoidance, intelligent housing, ground clutter, and wind speed/direction
Pulsed CW	Low PRF	Airborne and ground-based moving target indication (MTI) and long-range surveillance
	Medium PRF	Ground-based surveillance and airborne interceptors
	High PRF	Imaging, missile seekers, and short-range weapon control
Pulsed CW	Noncoherent	Surveillance radar and some MTI applications
	Coherent	Imaging, clutter suppression, and weather phenomena
Coded waveforms	Phase, frequency, and time codes	Reduce peak power requirements, improve maximum range, improve range/Doppler ambiguities, and for better jamming immunity

designed, manufactured, and emplaced for the specific purpose of daily meteorological prediction. Table 1.3 gives a few of the common radar application categories and example systems in each category.

1.2.4 Configuration

The final method of classifying radar systems is by the configuration; in particular, monostatic, quasi-monostatic, bistatic, and multistatic configurations. Referring to Figures 1.1 and 1.2, the different radar configurations are defined as follows:

- Monostatic systems use the same antenna for transmitting and receiving ($\alpha = 0$).
- Quasi-monostatic systems have the transmitting and receiving antennas at slightly different locations ($\alpha \approx 0$).
- Bistatic systems use a separate transmitting and receiving antenna ($\alpha \neq 0$).
- Multistatic systems use multiple transmitting and receiving antennas ($\alpha_1, \alpha_2, \cdots, \alpha_n \neq 0$).

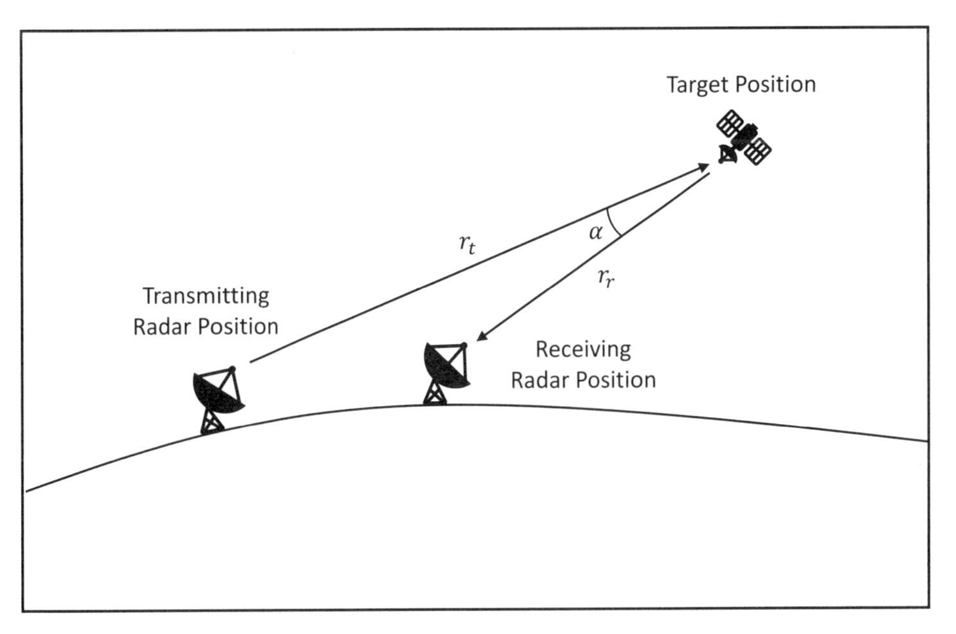

Figure 1.1 Bistatic radar transmitting and receiving configuration.

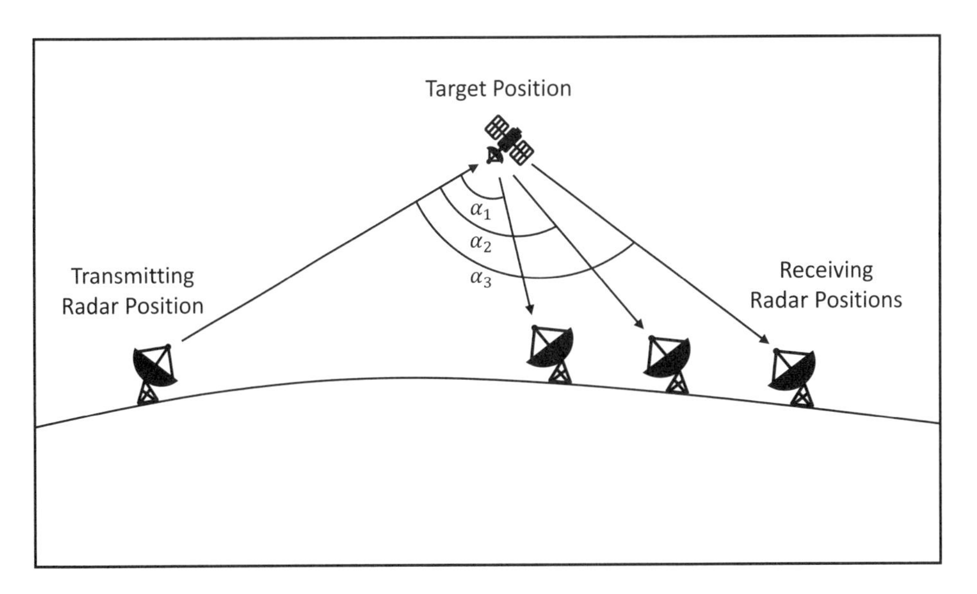

Figure 1.2 Multistatic radar transmitting and receiving configuration.

Table 1.3
Radar Classification by Application

Application	*Systems*
Weather forecasting	WSR-88D, WF44, OU-PRIME, Baron Gen3
Marine navigation	Raymarine Quantum 2, Garmin GMR, Furuno 1st Watch Wireless
Automotive collision avoidance	Audi Pre Sense, BMW Active Protection, Ford Forward Alert, GM Forward Collision Alert
Airborne military	AN/APG-79, AN/APG-82, AN/APY-12
Air traffic control	Monopulse Secondary Surveillance Radar, SKLYER, Enterprise Air Surveillance Radar (EASR)
Air and missile defense	AN/MPQ-53, AN/MPQ-64, AN/TPQ-53
Ballistic missile defense	AN/TPY-2, Upgraded Early Warning Radar (UEWR), Long Range Discrimination Radar (LRDR)
Synthetic aperture radar	RADARSAT, UAVSAR, ASARS, TRACER

1.3 ACCOMPANYING SOFTWARE

To aid reader comprehension of the concepts covered in the subsequent chapters, a radar software tool suite is included with this book. The tool suite is written in the Python programming language [2] and gives the reader a mechanism to analyze and predict radar performance for various scenarios and applications. The tool suite includes a user-friendly GUI developed with Qt [15] that updates plots and images automatically as the user changes parameters. This results in a very interactive tool, helping the user gain valuable insight about the topic of interest. Also included are scripts written in MATLAB corresponding to each of the scripts written in Python. Full access to both the Python and MATLAB scripts gives the user the ability to customize and extend each tool for their particular application or include as part of a larger simulation. Several examples using the tool suite are given at the end of each chapter.

The Python and MATLAB sets of code are obtained from the GitHub repository at `https://radarbook.github.io`. The user may clone the repository or download as a zip file, as illustrated in Figure 1.3. Once the software has been obtained, the folder structure is shown in Figure 1.4. Note that the subfolders are identical, as the MATLAB scripts correspond to each of the Python scripts. More information about the installation and execution of the software is given in the following sections.

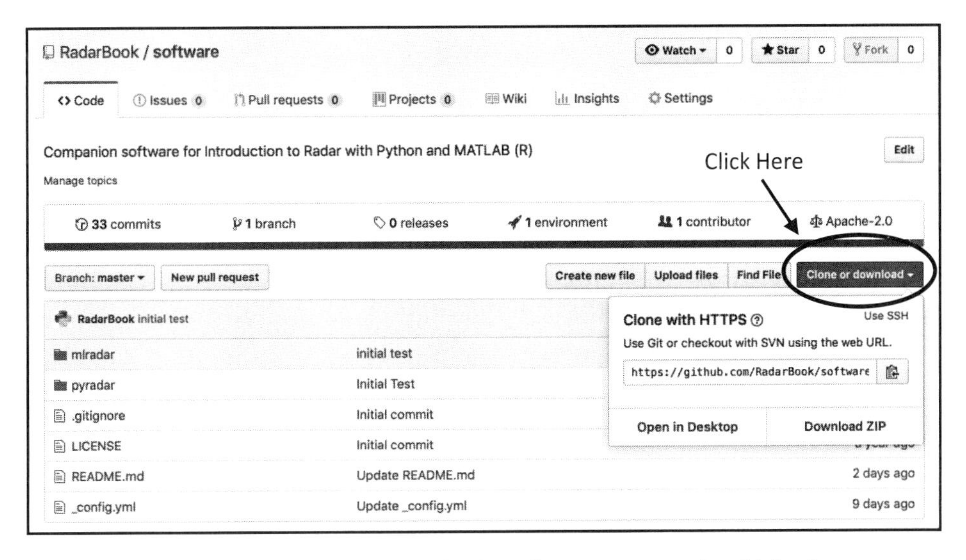

Figure 1.3 GitHub repository for the software accompanying this book.

1.3.1 Python

The Python tools were developed with version 3.6.4, and the GUIs were developed with Qt 5.10.0. The tool suite makes use of packages, including SciPy, which is a Python-based system of open-source software for mathematics, science, and engineering [16]. It also makes use of the core package, NumPy, which is a fundamental package for scientific computing, and includes a powerful N-dimensional array object, useful linear algebra, Fourier transform, and random number capabilities. [17]. For plotting, the tool suite takes advantage of Matplotlib, a Python plotting library that produces publication-quality figures in a variety of formats and interactive environments across platforms [18]. To install Python and the necessary packages, begin by navigating to `https://www.python.org` and downloading the installer for the operating system of choice, as shown in Figure 1.5. Launch the Python installer and follow the onscreen instructions. Once the installation is complete, ensure the following packages are installed by typing the commands given in Listing 1.1 at the command prompt.

Listing 1.1 Installation of Necessary Python Packages

```
>> pip install numpy
>> pip install scipy
>> pip install matplotlib
>> pip install pyqt5
```

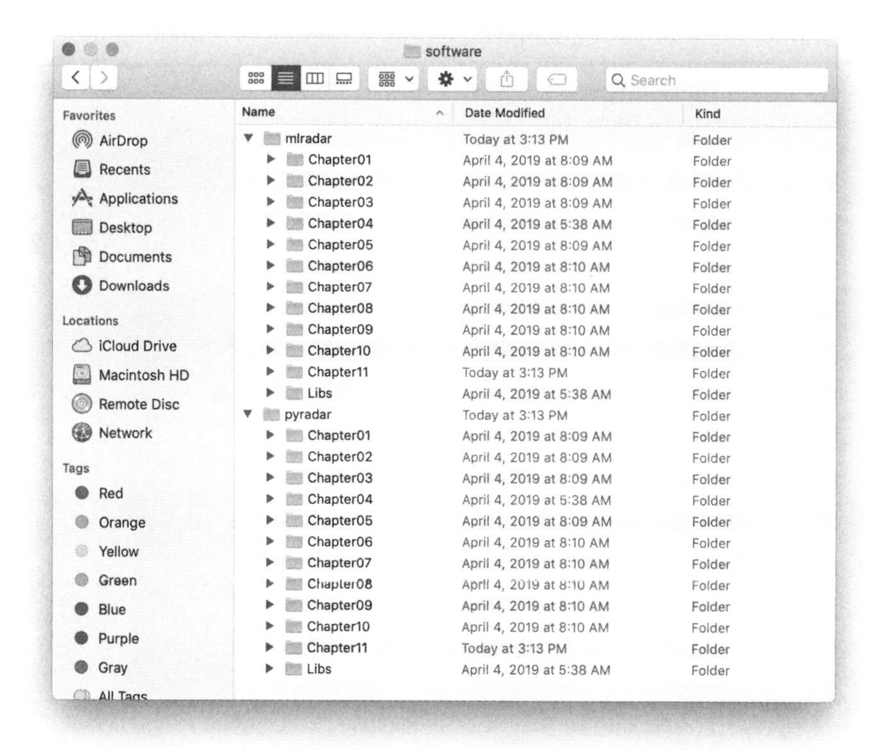

Figure 1.4 Folder structure for the software accompanying this book.

Then install the libraries and examples associated with this book by typing "pip install -e ***mypath***/pyradar", where ***mypath*** is the path to the folder "pyradar". To test if all packages were installed correctly, run the main GUI given in the file *pyradar\Chapter01\RadarBook.py*. This displays the GUI shown in Figure 1.6, which allows the user to execute any of the examples from the chapters by selecting the desired chapter and double-clicking on the example from the tree widget. There are different methods for running a Python script, such as typing "python pyradar\Chapter01\RadarBook.py" at the command prompt, or from an integrated development environment (IDE). It is recommended to use an IDE as these tools are dedicated to software development and provide a mechanism for the user to modify the code for their specific application. As the name implies, IDEs integrate several tools that usually include an editor with syntax highlighting and auto completion; build, execution, and debugging tools; and some form of source code control. There are several excellent IDEs for use with Python, including PyCharm, Spyder, and PyDev [19–21].

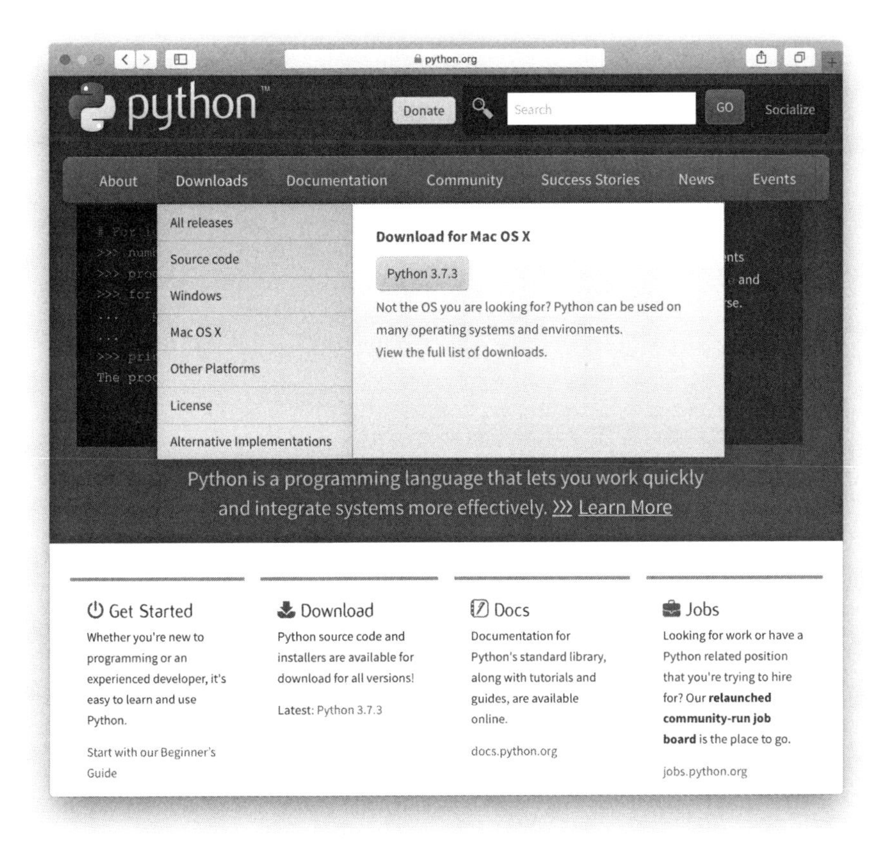

Figure 1.5 Download Python for the particular operating system.

1.3.2 MATLAB

MATLAB combines a desktop environment tuned for iterative analysis and design processes with a programming language that expresses matrix and array mathematics directly. It includes the Live Editor for creating scripts that combine code, output, and formatted text in an executable notebook [3]. The MATLAB scripts associated with this book were developed with R2018a. In order to run the MATLAB scripts provided with this book, the library folder and subfolders must be added to the MATLAB path, as shown in Figure 1.7. Once these folders have been added to the path, the MATLAB scripts may be executed from the editor, as shown in Figure 1.8 or from the command window by typing the name of the script. An example output from one of the MATLAB scripts is shown in Figure 1.8.

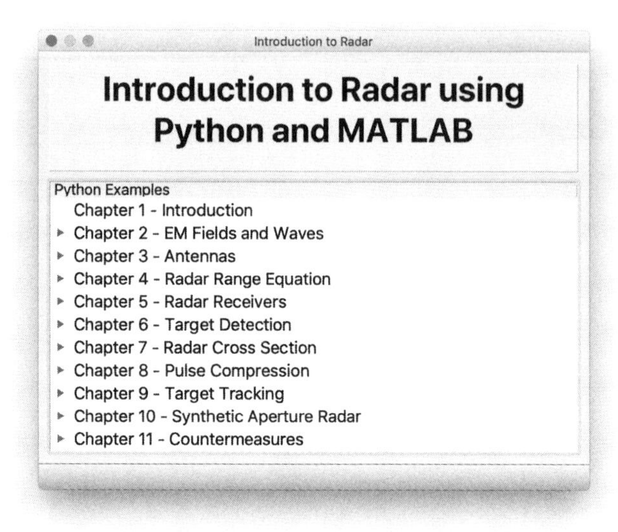

Figure 1.6 Main GUI for the Python radar tool suite.

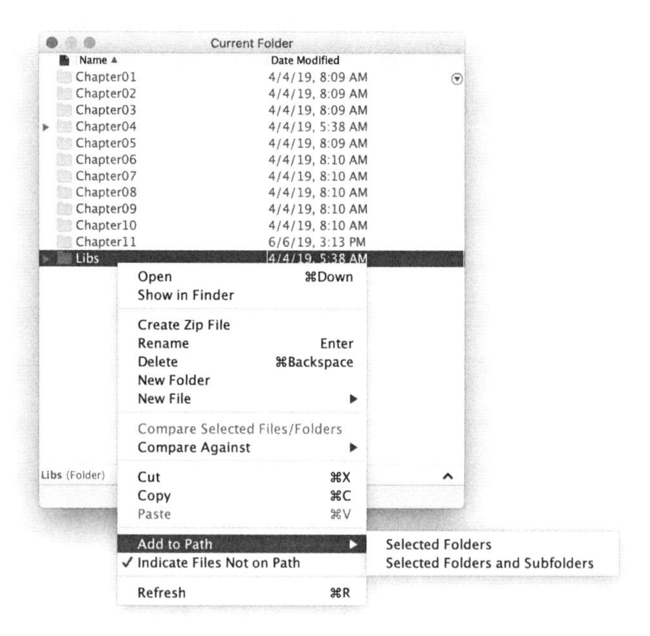

Figure 1.7 Add library folder and subfolders to the MATLAB path.

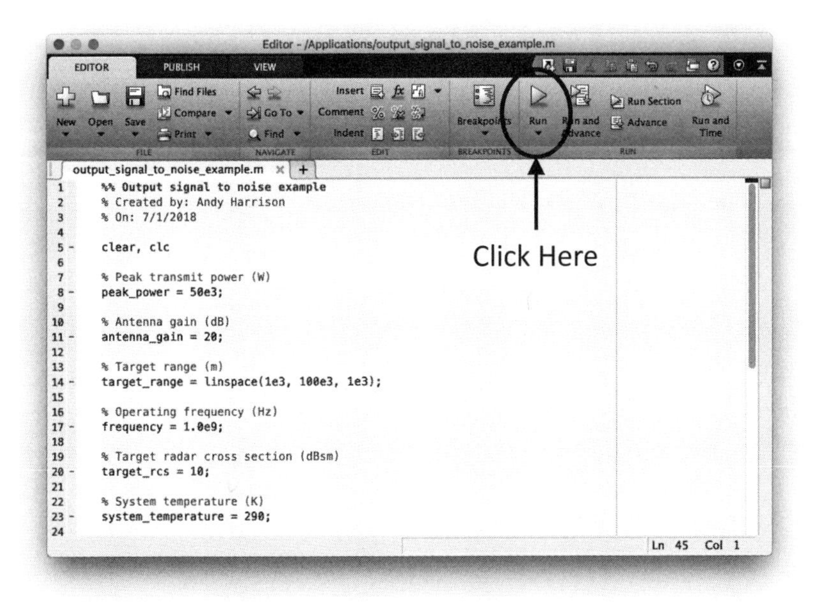

Figure 1.8 Execute a MATLAB script from the editor.

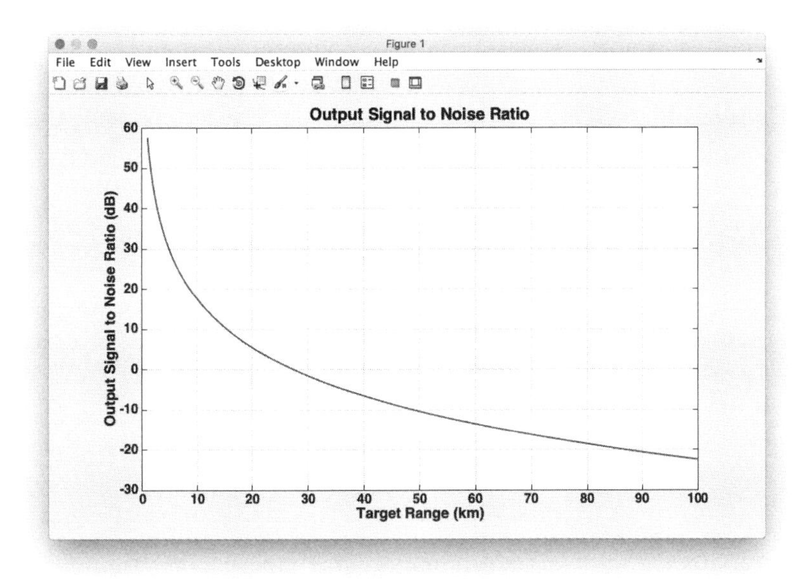

Figure 1.9 Example plot generated by the MATLAB script.

PROBLEMS

1.1 Describe the significance of radar development to the Allied victory in WWII. Identify the major advancements and the timeline leading to the deployment of these systems.

1.2 Using $\lambda = c/f$, where c is the speed of light, and f is the operating frequency, calculate the wavelength, λ, of radar systems operating at L, S, and X-band, and compare to the dimension of common radar targets such as aircraft, ships, and automobiles.

1.3 Applying $t = 2\,r/c$, where r is the range from the radar to the target, and c is the speed of light, calculate the round-trip time delay, t, for targets located at ranges of 100m, 1 km, and 100 km.

1.4 Describe the major differences between continuous wave and pulsed radar waveforms. Give examples of practical applications of each.

1.5 Explain the difference between monostatic and bistatic radar systems. Give the advantages and disadvantages of each type of system.

1.6 Determine the operating frequency band of a weather forecasting radar, such as the WSR-88D, and explain the choice of the frequency band.

1.7 Obtain the Python code associated with this book, as outlined in Section 1.3. Then install Python as described in Section 1.3.1, and verify successful installation by executing *RadarBook.py*, which produces the window shown in Figure 1.6.

References

[1] D. K. Keuren. Science goes to war: The radiation laboratory, radar, and their technological consequences. *Reviews in American History*, 25(4):643–647, 1997.

[2] G. van Rossum. *Python Software Foundation*. `https://www.python.org/`.

[3] MATLAB. *version 9.4.0 (R2018a)*. The MathWorks Inc., Natick, Massachusetts, 2018.

[4] H. R. Hertz. Ueber sehr schnelle electrische schwingungen. *Annalen der Physik*, 267(7).

[5] J. C. Maxwell. *A Treatise on Electricity and Magnetism*. Oxford University Press, London, UK, 1873.

[6] The telemobiloscope. *Electrical Magazine (London)*, 2:388, 1904.

[7] M. Skolnik. *Radar Handbook*, 3rd ed. McGraw-Hill Education, 2008.

[8] A. S. Gilmour. *Klystrons, Traveling Wave Tubes, Magnetrons, Cross-Field Amplifiers, and Gyrotrons*. Artech House, Norwood, MA, 2011.

[9] P. Swerling. Probability of detection for fluctuating targets. *IRE Transaction on Information Theory*, pages 269–308, April 1960.

[10] P. M. Woodward. *Probability and Information Theory with Applications to Radar*. Pergamon Press, London, 1953.

[11] The National Weather Service. *Everything You Ever Wanted to Know about the NWS WSR-88D*. https://www.weather.gov/iwx/wsr_88d.

[12] Wikipedia Contributors. *NEXRAD*. https://en.wikipedia.org/wiki/NEXRAD.

[13] W. Wiesbeck, et al. Radar 2020: The future of radar systems. *2014 International Radar Conference*, March 2014.

[14] IEEE standard for radar definitions. *IEEE Std 686-2017 (Revision of IEEE Std 686-2008)*, pages 1–54, September 2017.

[15] Qt Group. This is Qt, 1995–. http://www.qt.io.

[16] E. Jones, et al. SciPy: Open source scientific tools for Python, 2001–. http://www.scipy.org/.

[17] T. Oliphant. NumPy: A guide to NumPy. USA: Trelgol Publishing, 2006–. http://www.numpy.org.

[18] J. D. Hunter. Matplotlib: A 2D graphics environment. *Computing in Science & Engineering*, 9(3):90–95, 2007.

[19] JetBrains. PyCharm: the python IDE for professional developers, 2010–. https://www.jetbrains.com/pycharm.

[20] P. Raybaut. SPYDER: the scientific python development environment, 2009–. https://www.spyder-ide.org.

[21] Appcelerator. Pydev, 2003–. https://www.pydev.org.

Chapter 2

Electromagnetic Fields and Waves

Radar systems make use of the transmission, reflection, and reception of electromagnetic energy to determine the presence, direction, distance, velocity, and other features of various types of objects. The electromagnetic energy transmitted by a radar system interacts with the environment as well as the objects of interest. To properly account for these interactions requires an understanding of radiation and scattering mechanisms. This chapter begins with a presentation of Maxwell's equations, which fully describe electromagnetic radiation and scattering. Next, the boundary conditions associated with electric and magnetic fields are studied and solutions to Maxwell's equations are developed. This allows for investigation of the reflection and transmission of electromagnetic energy at interfaces between various types of media. Specific environmental effects such as refraction, diffraction, rain attenuation, fog attenuation, and atmospheric attenuation are then covered. The chapter concludes with several Python examples to reinforce the concepts of electromagnetic fields and waves.

2.1 MAXWELL'S EQUATIONS

James Clerk Maxwell unified the theories of electricity and magnetism in his profound work, *A Treatise on Electricity and Magnetism*, which was first published in 1873 [1]. He presented a set of linear, consistent (although not independent) equations eloquently showing the relationship between electricity and magnetism. These equations, which are now known as *Maxwell's equations*, were formulated through experiments by other scientists, but it took the genius of Maxwell to modify them and put them into a consistent set implying new physical phenomena. The differential form lends itself to the solution of boundary value electromagnetic problems, such as those arising in the propagation of radar signals, and is given as

$$\nabla \times \vec{E} + \frac{\partial \vec{B}}{\partial t} = 0 \qquad \text{Faraday's law,} \tag{2.1}$$

$$\nabla \times \vec{H} - \frac{\partial \vec{D}}{\partial t} = \vec{J} \qquad \text{Ampere's law,} \tag{2.2}$$

$$\nabla \cdot \vec{D} = \rho \qquad \text{Gauss's law,} \tag{2.3}$$

$$\nabla \cdot \vec{B} = 0 \qquad \text{no free magnetic charges,} \tag{2.4}$$

where

$\vec{E}$ = electric field intensity (V/m),
$\vec{H}$ = magnetic field intensity (A/m),
$\vec{D}$ = electric flux density (C/m^2),
$\vec{B}$ = magnetic flux density (Wb/m^2),
$\vec{J}$ = electric current density (A/m^2),
ρ = electric charge density (C/m^3).

As can be seen from (2.1)–(2.4), the source of electric and magnetic fields is the electric current density, $\vec{J}$, and the electric charge density, ρ. While it is mathematically useful to include a fictitious magnetic current, $\vec{M}$, and magnetic charge, ρ_m, in Maxwell's equations to aid in the solution of radiation and scattering problems, these will not be included as it is beyond the scope of this book. For further detailed information on this subject, the reader is referred to several excellent texts on the matter [2–5].

One of the major contributions from Maxwell is the generalization of *Ampere's law*, which states $\nabla \cdot \vec{J} = 0$. While this is true for steady-state currents, the general relationship between charge and current is given by the *continuity equation*, which is derived by taking the divergence of (2.2). This gives

$$\nabla \cdot \left(\nabla \times \vec{H}\right) = \nabla \cdot \vec{J} + \frac{\partial(\nabla \cdot \vec{D})}{\partial t}. \tag{2.5}$$

Since the divergence of the curl of a vector is zero, then

$$\nabla \cdot \vec{J} + \frac{\partial(\nabla \cdot \vec{D})}{\partial t} = 0, \tag{2.6}$$

and substituting (2.3) into (2.6) results in

$$\nabla \cdot \vec{J} + \frac{\partial \rho}{\partial t} = 0, \tag{2.7}$$

which states that if charge is moving out of a differential volume, then the amount of charge within the volume is going to decrease, and so the rate of change of charge density is negative. Therefore, the continuity equation is a conservation of charge. However, this modification to Ampere's law (2.2) is of crucial importance as it means a time-varying electric field causes a magnetic field even in the absence of a current. It is the converse of Faraday's law (2.1), and without it there would be no electromagnetic radiation.

2.2 TIME HARMONIC ELECTROMAGNETICS

In (2.1)–(2.4), the field quantities written in vector notation are functions of space and time (x, y, z, t). While these equations are valid for arbitrary time dependence, this book is concerned with electric and magnetic fields having sinusoidal, or harmonic, time dependence. Therefore, it is convenient to use phasor notation such that all field quantities will be complex vectors with an implied $e^{j\omega t}$ time dependence. Using this convention, the electric and magnetic fields are written as

$$\vec{E}(x, y, z, t) = Real\Big\{\mathbf{E}(x, y, z)e^{j\omega t}\Big\}, \tag{2.8}$$

$$\vec{H}(x, y, z, t) = Real\Big\{\mathbf{H}(x, y, z)e^{j\omega t}\Big\}. \tag{2.9}$$

With a time dependence of $e^{j\omega t}$, time derivatives are replaced with $j\omega$. Considering a homogeneous medium with the constitutive parameters, μ and ϵ, (2.1)–(2.4) are written as

$$\nabla \times \mathbf{E} + j\omega\mu\mathbf{H} = 0, \tag{2.10}$$

$$\nabla \times \mathbf{H} - j\omega\epsilon\mathbf{E} = \mathbf{J}, \tag{2.11}$$

$$\nabla \cdot \mathbf{D} = \rho, \tag{2.12}$$

$$\nabla \cdot \mathbf{B} = 0. \tag{2.13}$$

The *permittivity*, which is denoted by ϵ, is the amount of resistance to the formation of an electric field in a medium. The SI unit for permittivity is Farad per meter (F/m). The permittivity of vacuum is defined as

$$\epsilon_0 = \frac{1}{c^2 \mu_0} \approx 8.854 \times 10^{-12} \qquad \text{(F/m)}, \tag{2.14}$$

where μ_0 is the *permeability* of free space. The permittivity of a material is often specified by its relative permittivity, which is its value relative to free space. The absolute permittivity is given as $\epsilon = \epsilon_r \, \epsilon_0$. For linear, homogeneous, isotropic materials, $\mathbf{D} = \epsilon \mathbf{E}$, where ϵ is a scalar.

The permeability is a measure of the ability of a material to support the formation of a magnetic field. The SI unit for permeability is henries per meter (H/m). The permeability of vacuum is given as

$$\mu_0 = 4\pi \times 10^{-7} \qquad \text{(H/m)}. \tag{2.15}$$

The permeability of a material is often specified by its relative permeability, which is its value relative to free space. The absolute permeability is given as $\mu = \mu_r \, \mu_0$. For linear, homogeneous, isotropic materials, $\mathbf{B} = \mu \mathbf{H}$, where μ is a scalar.

2.3 ELECTROMAGNETIC BOUNDARY CONDITIONS

In practical radar applications, objects of various material composition will exist. To solve electromagnetic problems involving regions of differing constitutive parameters, μ and ϵ, it is necessary to know the boundary conditions the fields must satisfy at the interfaces between the materials. Consider the general plane material interface between two regions of differing parameters illustrated in Figure 2.1. The integral form of Maxwell's equations can be used to deduce the boundary conditions on the normal and tangential components of the fields at the interface [4]. These results are summarized in the next sections for various special cases.

2.3.1 General Material Interface

At a general material interface, as shown in Figure 2.1, the generalized boundary conditions for the electric and magnetic fields are written as

$$\hat{\mathbf{n}} \times (\mathbf{E}_1 - \mathbf{E}_2) = 0, \tag{2.16}$$

$$\hat{\mathbf{n}} \times (\mathbf{H}_1 - \mathbf{H}_2) = \mathbf{J}_s, \tag{2.17}$$

$$\hat{\mathbf{n}} \cdot (\mathbf{D}_1 - \mathbf{D}_2) = \rho_s, \tag{2.18}$$

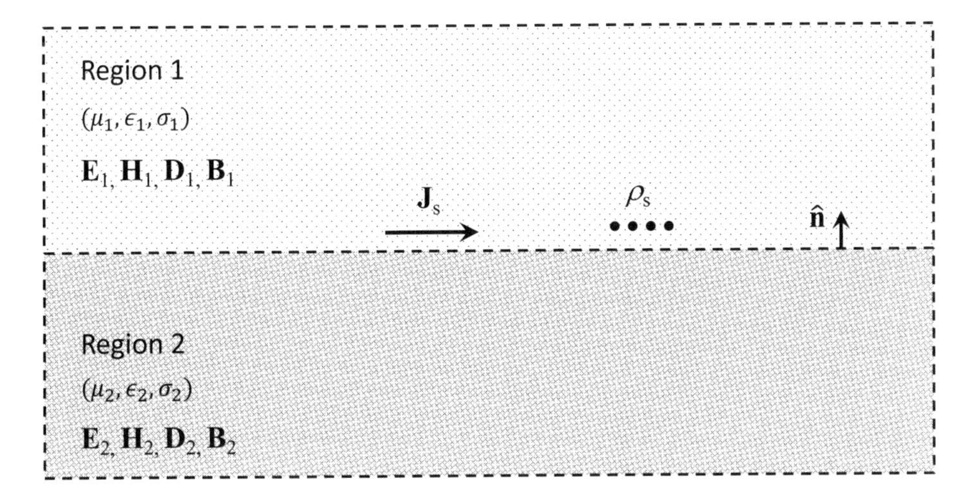

Figure 2.1 A general interface between two regions of differing constitutive parameters.

$$\hat{\mathbf{n}} \cdot (\mathbf{B}_1 - \mathbf{B}_2) = 0. \tag{2.19}$$

From (2.16)–(2.19), the tangential component of the electric field is continuous across a boundary. The tangential component of the magnetic field is discontinuous across a boundary. The amount of discontinuity is the surface current $\mathbf{J}_s$. The normal component of the electric flux density is discontinuous across a boundary. The amount of discontinuity is the surface charge density ρ_s. The normal component of the magnetic flux density is continuous across a boundary.

2.3.2 Dielectric Interface

At a dielectric interface, there are ordinarily no surface currents or charges. Therefore, the boundary conditions become

$$\hat{\mathbf{n}} \times (\mathbf{E}_1 - \mathbf{E}_2) = 0, \tag{2.20}$$

$$\hat{\mathbf{n}} \times (\mathbf{H}_1 - \mathbf{H}_2) = 0, \tag{2.21}$$

$$\hat{\mathbf{n}} \cdot (\mathbf{D}_1 - \mathbf{D}_2) = 0, \tag{2.22}$$

$$\hat{\mathbf{n}} \cdot (\mathbf{B}_1 - \mathbf{B}_2) = 0. \tag{2.23}$$

2.3.3 Perfect Electric Conductor Interface

Many problems of interest to radar engineers involve boundaries with good conductors (e.g., ships, aircraft, automobiles). Most metals are categorized as good conductors, where $\sigma >> \omega\epsilon$. The electric and magnetic fields inside good conductors approach zero as the conductivity increases, $\sigma \rightarrow \infty$. Therefore, the expressions for the boundary conditions for a perfect conducting interface become

$$\hat{\mathbf{n}} \times \mathbf{E}_1 = 0, \tag{2.24}$$

$$\hat{\mathbf{n}} \times \mathbf{H}_1 = \mathbf{J}_s, \tag{2.25}$$

$$\hat{\mathbf{n}} \cdot \mathbf{D}_1 = \rho_s, \tag{2.26}$$

$$\hat{\mathbf{n}} \cdot \mathbf{B}_1 = 0. \tag{2.27}$$

2.3.4 Perfect Magnetic Conductor Interface

The perfect magnetic conductor boundary is a dual to the perfect electric conductor boundary. For such a boundary, the tangential components of the magnetic field must vanish. That is, $\hat{\mathbf{n}} \times \mathbf{H}_1 = 0$. This type of boundary does not exist in practice, and may only be approximated by other boundaries, such as corrugated surfaces. However, this idealistic boundary condition is convenient for solving certain types of planar transmission line problems and is presented here for completeness. The fields at a perfect magnetic conductor interface satisfy the following

$$\hat{\mathbf{n}} \times \mathbf{E}_1 = 0, \tag{2.28}$$

$$\hat{\mathbf{n}} \times \mathbf{H}_1 = 0, \tag{2.29}$$

$$\hat{\mathbf{n}} \cdot \mathbf{D}_1 = 0, \tag{2.30}$$

$$\hat{\mathbf{n}} \cdot \mathbf{B}_1 = 0. \tag{2.31}$$

2.3.5 Radiation Condition

For radiation problems, a boundary condition must also be enforced on the fields at infinity. This condition is known as the *radiation condition*, and is a statement of the conservation of energy. As defined by Sommerfeld [6], the fields infinitely far from a radiating source must be vanishingly small or be outward radiating. No energy may be radiated from infinity into the field. This is intuitive, since a field coming from infinity and having a finite amplitude would require an infinite source.

2.4 WAVE EQUATIONS AND SOLUTIONS

The solutions to Maxwell's equations give the answers to all classical electromagnetic problems. This is true because Maxwell's equations completely describe the relationship between electric fields, magnetic fields, charges, and current distributions. While this is true, the solutions to many problems are extremely difficult to obtain. Even with special analytical and numerical methods, many problems remain intractable. Special techniques aid in obtaining solutions, but they do not add to nor refine the fundamental structure. This illustrates the significance of Maxwell's equations.

2.4.1 Scalar and Vector Potentials

Maxwell's equations can be solved directly as they are written for simple situations. For more practical problems, it is often convenient to introduce potential functions [2, 3] to obtain a smaller number of second-order partial differential equations that satisfy some of Maxwell's equations identically. Since the magnetic field is always solenoidal ($\nabla \cdot \mathbf{B} = 0$), define $\mathbf{B}$ in terms of a vector potential $\mathbf{A}$. The magnetic field will be represented as the curl of another vector, since the divergence of the curl is always zero, $\nabla \cdot (\nabla \times \mathbf{A}) = 0$. Therefore

$$\mathbf{B} = \nabla \times \mathbf{A}. \tag{2.32}$$

Substituting (2.32) into (2.10) results in

$$\nabla \times (\mathbf{E} + j\omega\mathbf{A}) = 0. \tag{2.33}$$

A function of vanishing curl may be written as the gradient of a scalar function. To be consistent with the electrostatic definition of the scalar electric potential, this is expressed as

$$\nabla \times (-\nabla V) = 0. \tag{2.34}$$

The electric field in terms of scalar and vector potentials is then

$$\mathbf{E} = -\nabla V - j\omega\mathbf{A}. \tag{2.35}$$

These two expressions identically satisfy the two homogeneous Maxwell's equations, (2.10) and (2.13). The dynamic behavior of the potentials, $\mathbf{A}$ and V, will be determined by the two inhomogeneous equations, (2.11) and (2.12). Substituting into (2.11) and (2.12), and making use of the constitutive relations $\mathbf{B} = \mu\mathbf{H}$ and $\mathbf{D} = \epsilon\mathbf{E}$ gives

$$\nabla \times (\nabla \times \mathbf{A}) = \mu\mathbf{J} + j\omega\mu\epsilon(-\nabla V - j\omega\mathbf{A}). \tag{2.36}$$

Using the vector identity

$$\nabla \times (\nabla \times \mathbf{A}) = \nabla(\nabla \cdot \mathbf{A}) - \nabla^2\mathbf{A}, \tag{2.37}$$

results in

$$\nabla^2\mathbf{A} + \omega^2\mu\epsilon\mathbf{A} = -\mu\mathbf{J} - \nabla(\nabla \cdot \mathbf{A} + j\omega\mu\epsilon V). \tag{2.38}$$

The vector potential, $\mathbf{A}$, is arbitrary to the extent that the gradient of some scalar function can be added. This means $\nabla \cdot \mathbf{A}$ may be specified in any manner and still obtain the same physical results. This is expressed as

$$\mathbf{A} \rightarrow \mathbf{A} + \nabla\psi. \tag{2.39}$$

The magnetic field, $\mathbf{B}$, remains unchanged by this transform. However, for the electric field to remain unchanged, V must be simultaneously transformed as

$$V \rightarrow V - j\omega\psi. \tag{2.40}$$

From (2.39) and (2.40), there is freedom to choose a set of potentials, $(\mathbf{A}, V)$, satisfying the *Lorenz condition* [7],

$$\nabla \cdot \mathbf{A} + j\omega\mu\epsilon V = 0. \tag{2.41}$$

This uncouples the pair of equations and leaves two inhomogeneous equations, one for V and one for $\mathbf{A}$. This is expressed as

$$\nabla^2\mathbf{A} + k^2\mathbf{A} = -\mu\mathbf{J}, \tag{2.42}$$

$$\nabla^2 V + k^2 V = -\rho/\epsilon, \tag{2.43}$$

where

$$k = \omega\sqrt{\mu\epsilon} = \frac{\omega}{v_p} = \frac{2\pi}{\lambda}, \tag{2.44}$$

is the *wavenumber*, and v_p is the velocity of propagation in the medium. This transform is known as a *gauge transformation*, and the invariance of the fields under such a transform is called the *gauge invariance*. For potentials satisfying the Lorenz condition, there is still arbitrariness. The so-called *restricted gauge transform*, expressed as

$$\mathbf{A} \rightarrow \mathbf{A} + \nabla \psi, \tag{2.45}$$

$$V \rightarrow V - j\omega\psi, \tag{2.46}$$

where

$$\nabla^2 \psi + \omega^2 \mu \epsilon \psi = 0, \tag{2.47}$$

preserves the Lorenz condition, provided the scalar and vector potentials, ($\mathbf{A}$, V), initially satisfy the Lorenz condition. All potentials in this restricted class belong to the *Lorenz gauge*. This gauge, which is independent of coordinate system, is commonly used because it leads to the wave equations in (2.42) and (2.43). There are other useful gauges, such as

- The *Coulomb gauge*, which is typically used when there are no sources present, is defined as $\nabla \cdot \mathbf{A} = 0$, $V = 0$. The fields are then given as

$$\mathbf{E} = -j\omega\mathbf{A}, \quad \mathbf{B} = \nabla \times \mathbf{A}. \tag{2.48}$$

- The {*radiation gauge*, also known as the *transverse gauge*, is defined as $\nabla \cdot \mathbf{A} = 0$.
- The *temporal gauge*, also known as the *Weyl gauge* or *Hamiltonian gauge*, is defined as $V = 0$.
- The *axial gauge* is defined as $\hat{\mathbf{n}} \cdot \mathbf{A} = 0$.

There are also a number of less commonly used gauges, such as *Maximal Abelian gauge*, *Poincare gauge*, *FockSchwinger gauge*, and *Dirac gauge*. The reader is referred to [8, 9] for more information on these and other gauges.

2.4.2 Fields Due to Sources

Following the treatment in [4], the solution to the inhomogeneous wave equations given in (2.42) and (2.43) is

$$V(r) = \frac{1}{4\pi\epsilon} \int_{V'} \rho \frac{e^{-jkr}}{r} \, dv', \tag{2.49}$$

$$\mathbf{A}(r) = \frac{\mu}{4\pi} \int_{V'} \mathbf{J} \frac{e^{-jkr}}{r} \, dv'. \tag{2.50}$$

The equations above represent the scalar and vector retarded potentials for time harmonic sources. The general procedure for calculating the electric and magnetic fields due to time harmonic charge and current sources is

1. Calculate the solutions for the scalar and vector potentials, $(V, \mathbf{A})$, from (2.49) and (2.50).
2. Calculate the solution for the phasor representation of the electric and magnetic fields, $\mathbf{E}(r)$ and $\mathbf{B}(r)$ from (2.32) and (2.35).
3. Calculate the instantaneous electric and magnetic fields, $\vec{E}(r,t)$ and $\vec{B}(r,t)$ from (2.8).

2.4.3 Source Free Fields

In a source free, linear, isotropic, homogeneous medium, Maxwell's curl equations are given by

$$\nabla \times \mathbf{E} + j\omega\mu\mathbf{H} = 0, \tag{2.51}$$

$$\nabla \times \mathbf{H} - j\omega\epsilon\mathbf{E} = 0. \tag{2.52}$$

Using (2.51) and (2.52), and following the procedure for obtaining the wave equations for the scalar and vector potentials, as shown in Section 2.4.1, results in

$$\nabla^2\mathbf{E} + k^2\mathbf{E} = 0, \tag{2.53}$$

$$\nabla^2\mathbf{H} + k^2\mathbf{H} = 0. \tag{2.54}$$

The equations above are of the form of an elliptic partial differential equation known as *Helmholtz's equation*, named for Hermann von Helmholtz [10], and ∇^2 is the vector Laplacian [11]. Equations (2.53) and (2.54) form the basis for introducing wave behavior.

2.5 PLANE WAVES

The solutions to (2.53) and (2.54) represent waves. To simplify the analysis, begin with waves having one-dimensional spatial dependence. Waves of this type are known as *plane waves*. A particular form of this solution, known as *uniform plane waves*, has an electric field with uniform direction, magnitude, and phase in infinite planes perpendicular to the direction of propagation. The same is also true of the magnetic field. Uniform plane waves cannot exist in practice, as it requires a source of infinite extent to create such electric and magnetic fields. However, if the observer is far enough away from the source, the surfaces of constant phase *wavefronts* become nearly spherical. On a very small portion of a very large sphere, the wavefront then becomes nearly planar. While the properties of uniform plane waves are simple, their study is of importance both theoretically and practically.

2.5.1 Plane Waves in Lossless Media

A lossless medium is described by real valued μ and ϵ. Therefore, the wavenumber, k, is also real valued. The wave equation for the electric field, $\mathbf{E}$, in Cartesian coordinates is equivalent to three scalar Helmholtz equations; one for each component E_x, E_y, and E_z. To find the basic plane wave solution, consider an electric field possessing only an E_x component and propagating in the positive z direction. $E_y = E_z = 0$, and the expression for E_x is written as

$$\left(\frac{\partial^2}{\partial x^2} + \frac{\partial^2}{\partial y^2} + \frac{\partial^2}{\partial z^2} + k^2\right) E_x = 0. \tag{2.55}$$

Recall, a uniform plane wave is described by uniform magnitude and phase over planar surface perpendicular to the direction of propagation. This is expressed as

$$\frac{\partial^2 E_x}{\partial x^2} = 0 \quad \text{and} \quad \frac{\partial^2 E_x}{\partial y^2} = 0, \tag{2.56}$$

which results in

$$\frac{d^2 E_x}{dz^2} + k^2 E_x = 0. \tag{2.57}$$

This is an ordinary differential equation as E_x is a function of only z. The solution to this equation is easily obtained by substitution, and is of the form

$$E_x(z) = E^+ e^{-jkz} + E^- e^{jkz}. \tag{2.58}$$

Since (2.57) is a second-order equation, its solution contains two constants of integration, E^+ and E^-. These are arbitrary constants that are determined by enforcing the boundary conditions. Using (2.8), the time domain result is then

$$E_x(z,t) = E^+ \cos(\omega t - kz) + E^- \cos(\omega t + kz). \tag{2.59}$$

The first term on the right-hand side of (2.59) represents a wave traveling in the positive z direction. To illustrate this, consider Figure 2.2, where $E^+ cos(\omega t - kz)$ has been plotted for various values of time, t. At each successive time, the curve has effectively traveled in the positive z direction. Considering a particular phase on the wave, set $cos(\omega t - kz)$ equal to a constant. Alternatively, set $\omega t - kz = C$, where C is a constant. As time, t, increases, z must also increase to keep the phase term constant.

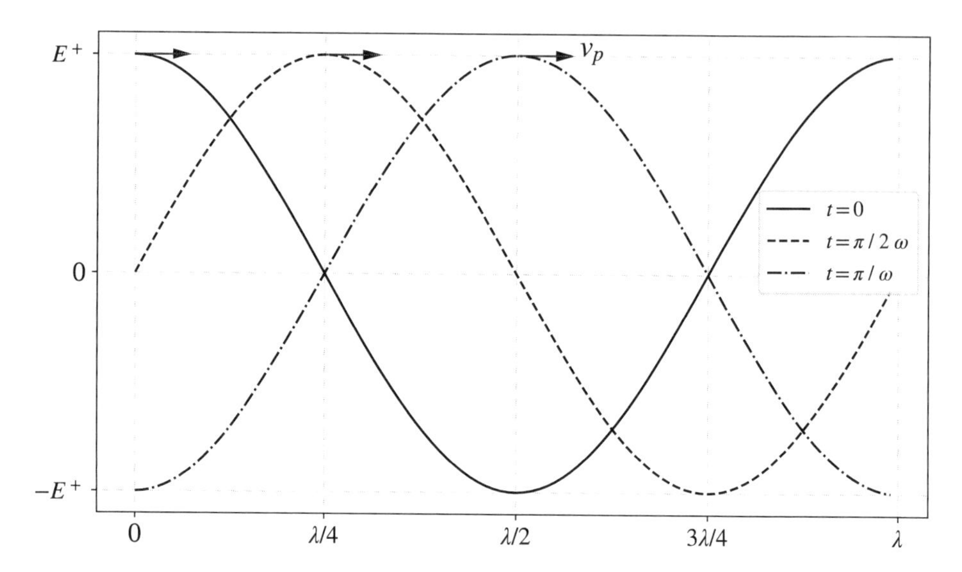

Figure 2.2 A wave traveling in the positive z direction for varying values of t.

The velocity of propagation of the wavefront is referred to as the *phase velocity*, and is given by

$$v_p = \frac{dz}{dt} = \frac{d}{dt}\left(\frac{\omega t - C}{k}\right) = \frac{\omega}{k} \qquad \text{(m/s)}. \tag{2.60}$$

In free space this becomes

$$v_p = \frac{\omega}{k} = \frac{1}{\sqrt{\mu_0 \epsilon_0}} = c \qquad \text{(m/s)}. \tag{2.61}$$

The speed of light in free space, c, is a universal physical constant, and its exact value is 299,792,458 (m/s). Referring again to Figure 2.2, the *wavelength*, λ, is the distance between two successive reference points, such as a maximum or minimum, at a particular instance of time. This represents the spatial period of the wave. Therefore,

$$(\omega t - kz) - \Big[\omega t - k(z + \lambda)\Big] = 2\pi \qquad \text{(rad)}, \tag{2.62}$$

$$\lambda = \frac{2\pi}{k} = \frac{2\pi v_p}{\omega} = \frac{v_p}{f} \qquad \text{(m)}. \tag{2.63}$$

The second term on the right-hand side of (2.59) represents a wave traveling in the negative z direction. For now, only consider waves traveling in the positive z direction.

As will be shown in later sections, if there are discontinuities in the medium, there will be reflected waves traveling in the opposite direction, and those must be considered. To complete the definition of the plane wave, the magnetic field, **H**, is found from (2.51) as

$$\mathbf{H} = \frac{1}{-j\omega\mu}\nabla \times \mathbf{E} = \frac{1}{-j\omega\mu}\begin{vmatrix} \hat{\mathbf{x}} & \hat{\mathbf{y}} & \hat{\mathbf{z}} \\ 0 & 0 & \frac{\partial}{\partial z} \\ E_x^+(z) & 0 & 0 \end{vmatrix}, \tag{2.64}$$

where $E_x^+(z) = E^+ e^{-jkz}$. This gives

$$\mathbf{H} = \frac{1}{\eta} E^+ e^{-jkz}\,\hat{\mathbf{y}}, \tag{2.65}$$

where $\eta = \sqrt{\mu/\epsilon}$, is the *wave impedance* for a plane wave and is the ratio of the electric and magnetic fields. Since η is a real number, **E** and **H** are in phase, and the instantaneous expression for **H** is

$$H_y(z,t) = \frac{E^+}{\eta}\cos(\omega t - kz). \tag{2.66}$$

E and **H** are orthogonal to each other, and both are orthogonal to the direction of propagation, as shown in Figure 2.3. This is a characteristic of transverse electromagnetic (TEM) waves.

2.5.2 Plane Waves in Lossy Media

In a lossy material with conductivity, σ, a conduction current density will be present. This current is given by

$$\mathbf{J} = \sigma\mathbf{E}. \tag{2.67}$$

This represents *Ohm's law* from an electromagnetic field point of view. Substituting (2.67) into (2.10) and (2.11) results in

$$\nabla \times \mathbf{E} + j\omega\mu\mathbf{H} = 0, \tag{2.68}$$

$$\nabla \times \mathbf{H} - j\omega\epsilon\mathbf{E} - \sigma\mathbf{E} = 0. \tag{2.69}$$

Following the approach in Section 2.4.1, the wave equation for the electric field is now

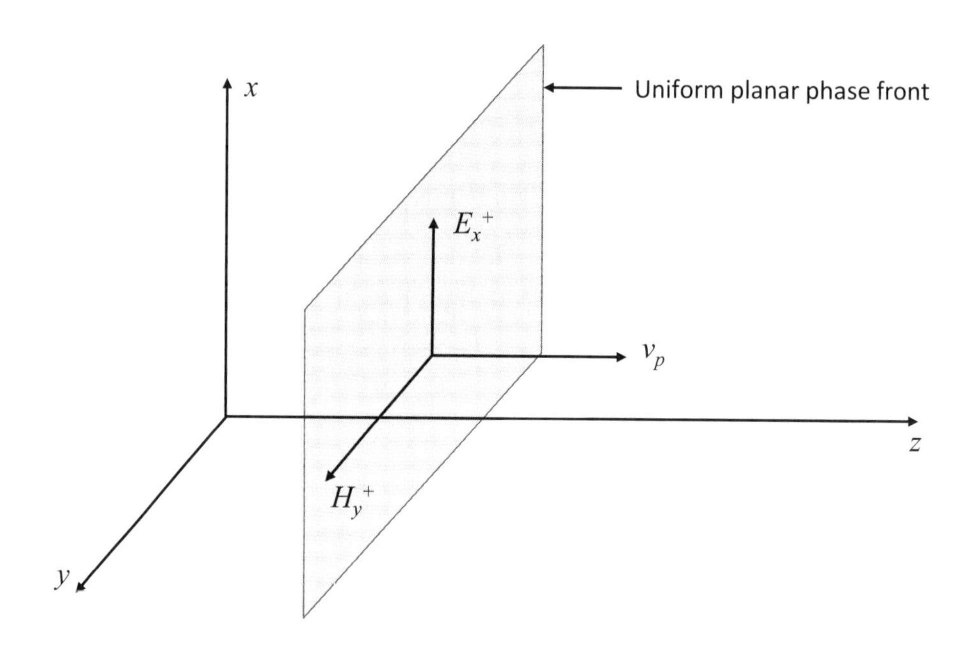

Figure 2.3 The electric and magnetic field vectors for a uniform plane wave.

$$\nabla^2 \mathbf{E} + \omega^2 \mu \epsilon \left(1 - j\frac{\sigma}{\omega \epsilon}\right) \mathbf{E} = 0. \tag{2.70}$$

Comparing (2.70) to the results in (2.53), $k^2 = \omega^2 \mu \epsilon$ has been replaced by

$$\omega^2 \mu \epsilon \left[1 - j\left(\frac{\sigma}{\omega \epsilon}\right)\right]. \tag{2.71}$$

Therefore, a complex propagation constant is defined as

$$\gamma = \alpha + j\beta = j\omega\sqrt{\mu\epsilon}\sqrt{1 - j\frac{\sigma}{\omega\epsilon}}, \tag{2.72}$$

where α is referred to as the *attenuation constant*, and β is the *phase constant*. Following Section 2.5.1, consider an electric field with only an x component. Also, this field is uniform in the x and y directions. The wave equation (2.70) becomes

$$\frac{\partial^2 E_x}{\partial z^2} - \gamma^2 E_x = 0. \tag{2.73}$$

As before, the solution has waves traveling in both the positive and negative z directions,

$$E_x(z) = E^+ e^{-\gamma z} + E^- e^{\gamma z}. \tag{2.74}$$

Considering for now only the positive z traveling wave,

$$E_x(z) = E^+ e^{-\gamma z} = E^+ e^{-\alpha z} e^{-j\beta z}. \tag{2.75}$$

Using (2.8), the time domain representation is given by

$$E_x(z,t) = E^+ e^{-\alpha z} \cos(\omega t - \beta z). \tag{2.76}$$

The first term, $e^{-\alpha z}$, decreases as z increases, and is therefore an attenuation that is expressed in units of nepers per meter. The second term, $cos(\omega t - \beta z)$, indicates the wave is traveling in the positive z direction with a phase velocity of ω/β and wavelength $\lambda = 2\pi/\beta$. As $\sigma \to 0$, the expressions for this case reduce to the lossless case. Namely, $\gamma = jk$ and $\alpha \to 0$, $\beta \to k$. The magnetic field is found from (2.69) as

$$H_y(z) = \frac{j}{\omega\mu}\frac{\partial E_x}{\partial z} = \frac{-j\gamma}{\omega\mu}\left(E^+ e^{-\gamma z} - E^- e^{\gamma z}\right). \tag{2.77}$$

As in the previous case, the wave impedance is the ratio of the electric and magnetic fields. For the lossy case, this becomes

$$H_y(z) = \frac{1}{\eta}\left(E^+ e^{-\gamma z} - E^- e^{\gamma z}\right), \tag{2.78}$$

where

$$\eta = \frac{j\omega\mu}{\gamma} \qquad \text{(ohms)}. \tag{2.79}$$

2.5.3 Plane Waves in Low-Loss Dielectrics

A low-loss dielectric medium is defined as a good insulator, but with nonvanishing conductivity. Specifically, a low-loss dielectric has a conductivity meeting the requirement

$$\frac{\sigma}{\omega\epsilon} \ll 1. \tag{2.80}$$

The binomial expansion is used to write the propagation constant as

$$\gamma = \alpha + j\beta \approx j\omega\sqrt{\mu\epsilon}\left[1 - j\frac{\sigma}{2\omega\epsilon} + \frac{1}{8}\left(\frac{\sigma}{\omega\epsilon}\right)^2\right]. \tag{2.81}$$

The attenuation constant is

$$\alpha = \frac{\sigma}{2}\sqrt{\frac{\mu}{\epsilon}} \qquad \text{(Np/m)}, \tag{2.82}$$

and the phase constant is

$$\beta = j\omega\sqrt{\mu\epsilon}\left[1 + \frac{1}{8}\left(\frac{\sigma}{\omega\epsilon}\right)^2\right] \qquad \text{(rad/m)}. \tag{2.83}$$

The attenuation constant is a positive value directly related to the conductivity. The propagation constant varies only slightly from the lossless case. The wave impedance is found to be

$$\eta = \sqrt{\frac{\mu}{\epsilon}}\left(1 - j\frac{\sigma}{\omega\epsilon}\right)^{-1/2} \approx \sqrt{\frac{\mu}{\epsilon}}\left(1 + j\frac{\sigma}{2\omega\epsilon}\right). \tag{2.84}$$

Recall that the wave impedance is the ratio of the electric and magnetic fields, E_x and H_y, for a uniform plane wave. Since η is a complex value, this means the electric and magnetic fields are not in time phase as they are in the case of a lossless material. The phase velocity is calculated from ω/β as

$$v_p = \frac{\omega}{\beta} \approx \frac{1}{\sqrt{\mu\epsilon}}\left[1 - \frac{1}{8}\left(\frac{\sigma}{\omega\epsilon}\right)^2\right] \qquad \text{(m/s)}. \tag{2.85}$$

2.5.4 Plane Waves in Good Conductors

Many problems of interest in radar applications involve good, but not perfect, conductors. A good conductor is defined as a material in which the conductive current is much larger than the displacement current (i.e., $\sigma \gg \omega\epsilon$). Most metals are classified as good conductors. Table 2.1 shows the conductivity of a few common materials [12]. The propagation constant is approximated by ignoring the displacement current, which results in

$$\gamma = \alpha + j\beta \approx j\omega\sqrt{\mu\epsilon}\sqrt{\frac{\sigma}{j\omega\epsilon}} = (1 + j)\sqrt{\frac{\omega\mu\sigma}{2}}. \tag{2.86}$$

This result indicates α and β for a good conductor are approximately equal. The wave impedance is found to be

$$\eta = \frac{j\omega\mu}{\gamma} \approx (1 + j)\sqrt{\frac{\omega\mu}{2\sigma}}. \tag{2.87}$$

One characteristic of good conductors is the phase angle of the wave impedance is 45°.

The *skin depth*, or *depth of penetration*, is defined as the distance through which the amplitude of the fields decreases by a factor of e^{-1} or 36.8%. This distance is given mathematically as

$$\delta_s = \frac{1}{\alpha} = \sqrt{\frac{2}{\omega\mu\sigma}} \qquad \text{(m)}. \tag{2.88}$$

At microwave frequencies, the skin depth is small, so the fields and currents are confined to a very thin layer on the surface of the conductor. Tables 2.2 and 2.3 summarize the parameters for plane wave propagation in various media.

2.6 PLANE WAVE REFLECTION AND TRANSMISSION

The previous sections examined uniform plane waves propagating in unbounded homogeneous media. Practical radar problems involve waves propagating in bounded regions

Table 2.1

Resistivity and Conductivity of Various Materials at 20° C

Material	*Resistivity (Ohm · m)*	*Conductivity (S/m)*
Silver	1.59×10^{-8}	6.30×10^{7}
Copper	1.68×10^{-8}	5.96×10^{7}
Gold	2.44×10^{-8}	4.10×10^{7}
Aluminum	2.65×10^{-8}	3.77×10^{7}
Iron	9.71×10^{-8}	1.00×10^{7}
Lead	2.20×10^{-7}	4.55×10^{6}
Seawater	2.00×10^{-1}	4.8
Wood	10^{3} to 10^{4}	10^{-4} to 10^{-3}
Glass	10^{11} to 10^{15}	10^{-15} to 10^{-11}
Rubber	10^{13}	10^{-14}

Table 2.2

Plane Wave Propagation Parameters in Various Media

Parameter	*Lossless*	*Lossy*
Propagation constant (γ)	$j\omega\sqrt{\mu\epsilon}$	$j\omega\sqrt{\mu\epsilon}\sqrt{1 - j\sigma/\omega\epsilon}$
Phase constant (β)	$\omega\sqrt{\mu\epsilon}$	$\mathrm{Im}(\gamma)$
Attenuation constant (α)	0	$\mathrm{Re}(\gamma)$
Wave impedance (η)	$\sqrt{\mu/\epsilon}$	$j\omega\mu/\gamma$
Skin depth (δ_s)	∞	$1/\alpha$
Wavelength (λ)	$2\pi/\beta$	$2\pi/\beta$
Phase velocity (v_p)	ω/β	ω/β

Table 2.3

Plane Wave Propagation Parameters in Various Media

Parameter	*Low-Loss Dielectric*	*Good Conductor*
Propagation constant (γ)	$j\omega\sqrt{\mu\epsilon}\left[1 - j\sigma/2\omega\epsilon + (1/8)\left(\sigma/\omega\epsilon\right)^2\right]$	$(1+j)\sqrt{\omega\mu\sigma/2}$
Phase constant (β)	$j\omega\sqrt{\mu\epsilon}\left[1 + (1/8)\left(\sigma/\omega\epsilon\right)^2\right]$	$\sqrt{\omega\mu\sigma/2}$
Attenuation constant (α)	$(\sigma/2)\sqrt{\mu/\epsilon}$	$\sqrt{\omega\mu\sigma/2}$
Wave impedance (η)	$\sqrt{\mu/\epsilon}\left(1 + j\sigma/2\omega\epsilon\right)$	$(1+j)\sqrt{\omega\mu/2\sigma}$
Skin depth (δ_s)	$1/\alpha$	$\sqrt{2/\omega\mu\sigma}$
Wavelength (λ)	$2\pi/\beta$	$2\pi/\beta$
Phase velocity (v_p)	ω/β	ω/β

and interacting with media of differing constitutive parameters. These media may include the targets of interest to the radar system as well as other regions such as rain, buildings, trees, and birds. Therefore, it is beneficial to study the reflection and refraction of electromagnetic waves occurring when a wave traveling in a given medium impinges on another medium with a different set of constitutive parameters. The general case of oblique incidence upon a lossy dielectric region will be shown. Other special cases, such as normal incidence, lossless dielectrics and perfect conductors may then be obtained from these results.

Consider the geometry shown in Figure 2.4, where the region $z > 0$ is characterized by the constitutive parameters μ_2, ϵ_2, σ_2, and the region $z < 0$ is characterized by μ_1, ϵ_1, σ_1. There are two canonical cases for the incident fields. The first case is when the incident electric field is in the x-z plane (parallel polarization). The second case is when the incident electric field is normal to the x-z plane (perpendicular polarization). Any arbitrary incident plane wave may then be expressed as a linear combination of these two cases.

2.6.1 Perpendicular Polarization

The perpendicular polarization case is defined by an electric field perpendicular to the x-z plane. This means the electric field only has a component along the y axis. Waves of this type are also known as transverse electric (TE) waves, as the electric field is transverse to the plane of incidence. Referring to Figure 2.5, the incident electric field is written as

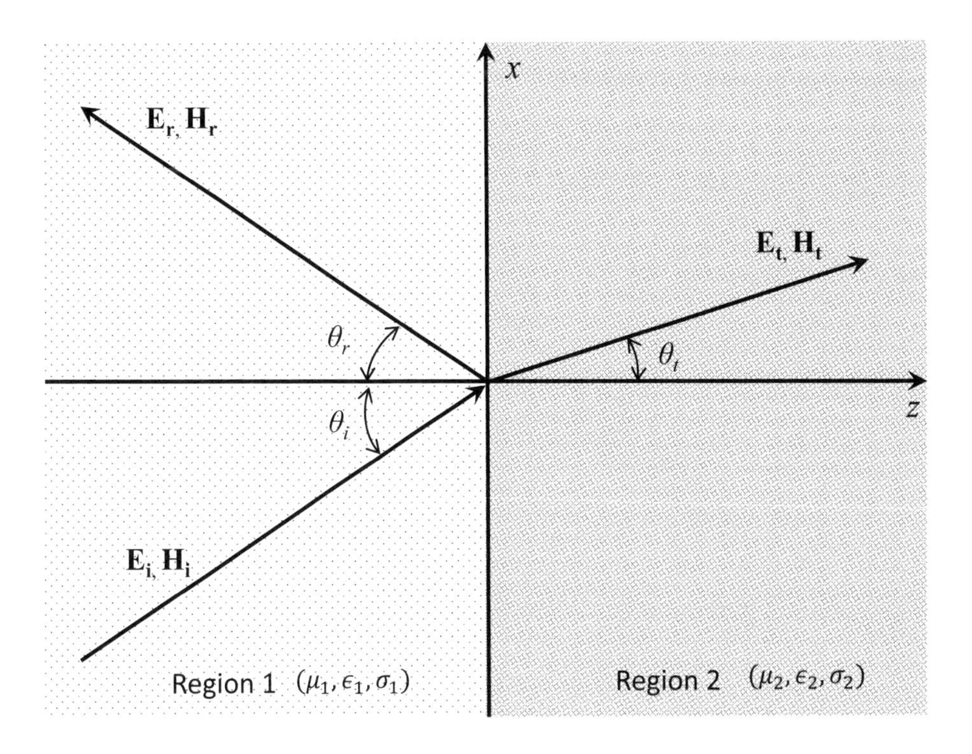

Figure 2.4 Plane wave obliquely incident on the interface between two lossy dielectrics.

$$\mathbf{E}_i = E_0\, e^{-\gamma_1(x \sin\theta_i + z\cos\theta_i)}\ \hat{\mathbf{y}}, \tag{2.89}$$

where the direction of propagation is now $(x\sin\theta_i + z\cos\theta_i)$. This reduces to a wave propagating in the z direction for $\theta_i = 0$. Using (2.51), the incident magnetic field is

$$\mathbf{H}_i = \frac{1}{-j\omega\mu}\nabla\times\mathbf{E}_i = \frac{1}{-j\omega\mu}\begin{vmatrix} \hat{\mathbf{x}} & \hat{\mathbf{y}} & \hat{\mathbf{z}} \\ \dfrac{\partial}{\partial x} & \dfrac{\partial}{\partial y} & \dfrac{\partial}{\partial z} \\ 0 & E_y(x,z) & 0 \end{vmatrix}, \tag{2.90}$$

where $E_y(x,z) = E_0\, e^{-\gamma_1(x\sin\theta_i + z\cos\theta_i)}$. Expanding the curl results in

$$\mathbf{H}_i = \frac{E_0}{\eta_1}\left[e^{-\gamma_1(x\sin\theta_i + z\cos\theta_i)}\right]\ (-\cos\theta_i\ \hat{\mathbf{x}} + \sin\theta_i\ \hat{\mathbf{z}})\,. \tag{2.91}$$

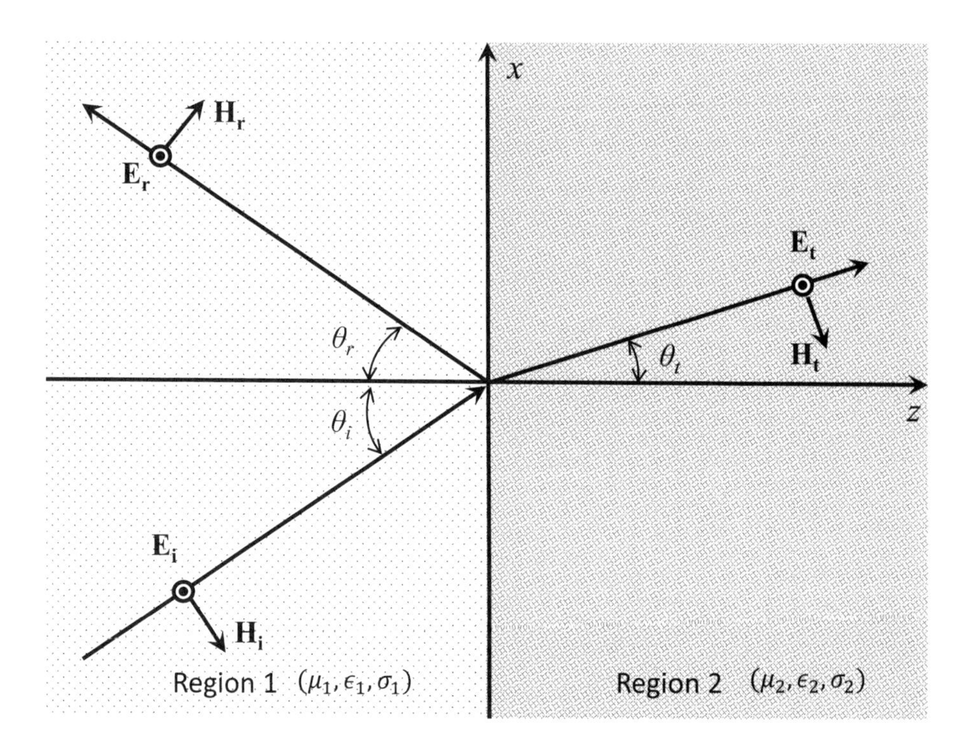

Figure 2.5 Plane wave obliquely incident on the interface between two lossy dielectrics for perpendicular polarization.

Referring to Figure 2.4, the fields reflected back into Region 1 are seen to propagate along the direction $(x \sin\theta_i - z\cos\theta_i)$. The reflected electric field is then expressed as

$$\mathbf{E}_r = E_0\,\Gamma_{TE}\,e^{-\gamma_1(x\sin\theta_i - z\cos\theta_i)}\;\hat{\mathbf{y}}. \tag{2.92}$$

Γ_{TE} is known as the *reflection coefficient* and is the ratio of the amplitude of the reflected electric field to the incident electric field. It is therefore a measure of how much of the incident field is reflected back into Region 1. The reflection coefficient, Γ_{TE}, is often written as $\Gamma_\perp$ to explicitly indicate perpendicular polarization. Just as in the case of the incident fields, the reflected magnetic field is found from (2.51), and is given as

$$\mathbf{H}_r = \frac{E_0\,\Gamma_{TE}}{\eta_1}\left[e^{-\gamma_1(x\sin\theta_r - z\cos\theta_r)}\right](\cos\theta_r\,\hat{\mathbf{x}} + \sin\theta_r\,\hat{\mathbf{z}})\,. \tag{2.93}$$

The fields transmitted into Region 2 follow the same form as the incident fields, with the incident angle, θ_i, being replaced by transmitted angle, θ_t. Also, the propagation constant for Region 1, γ_1, is replaced by the propagation constant for Region 2, γ_2. The *transmission coefficient*, T_{TE}, is the ratio of the amplitude of the transmitted electric field to the incident electric field. The transmission coefficient is a measure of how much of the incident field is transmitted into Region 2. The transmission coefficient, T_{TE}, is often written as $\mathrm{T}_{\perp}$. Therefore, the transmitted electric and magnetic fields are written as

$$\mathbf{E}_t = E_0\, \mathrm{T}_{TE}\, e^{-\gamma_2 (x \sin\theta_t + z \cos\theta_t)}\, \hat{\mathbf{y}}, \tag{2.94}$$

$$\mathbf{H}_t = \frac{E_0\, \mathrm{T}_{TE}}{\eta_2} \left[e^{-\gamma_2 (x \sin\theta_t + z \cos\theta_t)} \right] \left(-\cos\theta_i\, \hat{\mathbf{x}} + \sin\theta_i\, \hat{\mathbf{z}} \right). \tag{2.95}$$

At this point there are four unknowns, Γ_{TE}, T_{TE}, θ_r, and θ_t. In order to find these unknowns, the boundary conditions from Section 2.3 are used, specifically (2.16) and (2.17), which relate the tangential components of the electric and magnetic fields at the interface. Enforcing the continuity of E_y and H_x at the interface results in

$$e^{-\gamma_1 x \sin\theta_i} + \Gamma_{TE}\, e^{-\gamma_1 x \sin\theta_r} = \mathrm{T}_{TE}\, e^{-\gamma_2 x \sin\theta_t}, \tag{2.96}$$

$$-\frac{1}{\eta_1} \cos\theta_i\, e^{-\gamma_1 x \sin\theta_i} + \frac{\Gamma_{TE}}{\eta_1} \cos\theta_r\, e^{-\gamma_1 x \sin\theta_r} = -\frac{\mathrm{T}_{TE}}{\eta_2} \cos\theta_t\, e^{-\gamma_2 x \sin\theta_t}. \tag{2.97}$$

For the tangential components of the electric and magnetic fields to be continuous at the interface, $z = 0$, for all x, then the phase terms must have the same x variation on both sides of (2.96) and (2.97). This is often referred to as the *phase matching condition* as it ensures the phase varies with the same rate in x on both sides of the interface. Therefore

$$\gamma_1 \sin\theta_i = \gamma_1 \sin\theta_r = \gamma_2 \sin\theta_t, \tag{2.98}$$

which leads to

$$\theta_i = \theta_r, \qquad \gamma_1 \sin\theta_i = \gamma_2 \sin\theta_t. \tag{2.99}$$

This is the well-known Snell's laws of reflection and refraction [13]. Using these results in (2.96) and (2.97) gives

$$1 + \Gamma_{TE} = \mathrm{T}_{TE}, \tag{2.100}$$

$$\frac{\cos\theta_i}{\eta_1} - \Gamma_{TE} \frac{\cos\theta_r}{\eta_1} = \mathrm{T}_{TE} \frac{\cos\theta_t}{\eta_2}. \tag{2.101}$$

Solving these equations for the reflection and transmission coefficients gives

$$\Gamma_{TE} = \frac{\eta_2 \cos\theta_i - \eta_1 \cos\theta_t}{\eta_2 \cos\theta_i + \eta_1 \cos\theta_t}, \tag{2.102}$$

$$\mathrm{T}_{TE} = \frac{2\,\eta_2 \cos\theta_i}{\eta_2 \cos\theta_i + \eta_1 \cos\theta_t}. \tag{2.103}$$

2.6.2 Parallel Polarization

The parallel polarization case is described by an electric field in the x-z plane. The magnetic field is now transverse to the plane of incidence, as shown in Figure 2.6, so these waves are also known as transverse magnetic (TM) waves. The magnetic field will only have a y component and is written as

$$\mathbf{H}_i = \frac{E_0}{\eta_1} e^{-\gamma_1(x\sin\theta_i + z\cos\theta_i)}\,\hat{\mathbf{y}}. \tag{2.104}$$

The electric field is found from (2.52) as

$$\nabla \times \mathbf{H}_i - j\omega\epsilon_1 \mathbf{E}_i = 0. \tag{2.105}$$

$$\mathbf{E}_i = \frac{1}{j\omega\epsilon_1}\nabla \times \mathbf{H}_i = \frac{1}{j\omega\epsilon_1}\begin{vmatrix} \hat{\mathbf{x}} & \hat{\mathbf{y}} & \hat{\mathbf{z}} \\ \frac{\partial}{\partial x} & \frac{\partial}{\partial y} & \frac{\partial}{\partial z} \\ 0 & H_y(x,z) & 0 \end{vmatrix}, \tag{2.106}$$

where $H_y(x,z) = \dfrac{H_0}{\eta_1}\, e^{-\gamma_1(x\sin\theta_i + z\cos\theta_i)}$. Expanding the curl gives

$$\mathbf{E}_i = E_0\left[e^{-\gamma_1(x\sin\theta_i + z\cos\theta_i)}\right]\left(\cos\theta_i\,\hat{\mathbf{x}} - \sin\theta_i\,\hat{\mathbf{z}}\right). \tag{2.107}$$

Following a procedure similar to the one given in Section 2.6.1, the reflected magnetic field is written as

$$\mathbf{H}_r = -\frac{E_0\,\Gamma_{TM}}{\eta_1}\, e^{-\gamma_1(x\sin\theta_r - z\cos\theta_r)}\,\hat{\mathbf{y}}, \tag{2.108}$$

where Γ_{TM} is the reflection coefficient and is sometimes written as $\Gamma_{\|}$ to explicitly indicate parallel polarization. The reflected electric field is found from (2.52) as

$$\mathbf{E}_r = E_0\,\Gamma_{TM}\, e^{-\gamma_1(x\sin\theta_r - z\cos\theta_r)}\left(\cos\theta_r\,\hat{\mathbf{x}} + \sin\theta_r\,\hat{\mathbf{z}}\right). \tag{2.109}$$

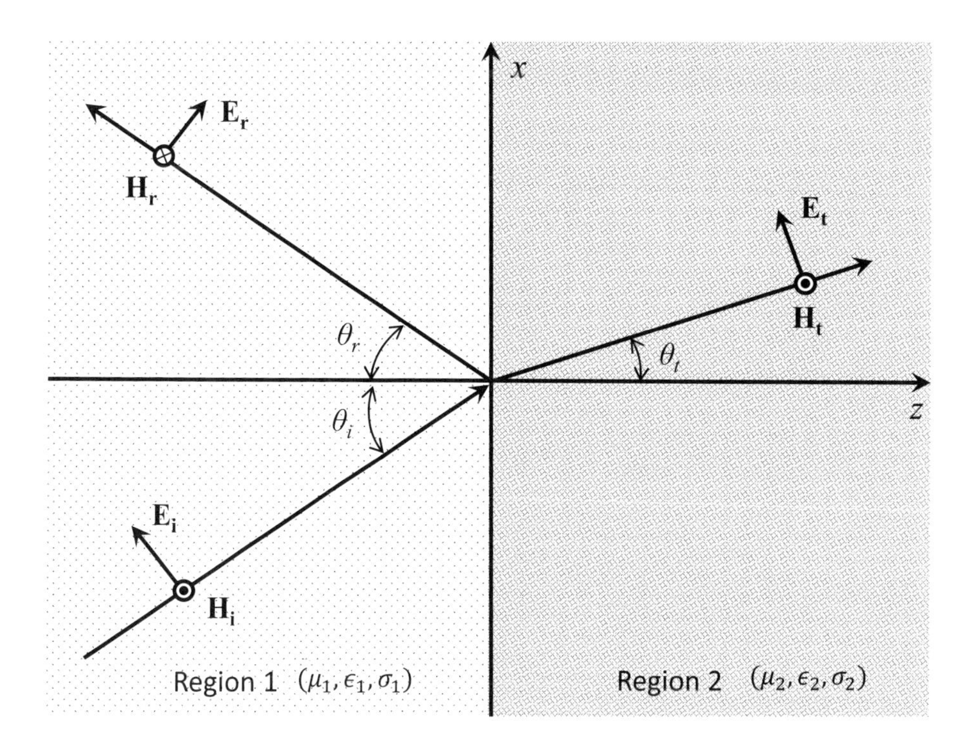

Figure 2.6 Plane wave obliquely incident on the interface between two lossy dielectrics for parallel polarization.

The transmitted fields also follow a similar approach to Section 2.6.1 and are given here as

$$\mathbf{E}_t = E_0 \, \mathrm{T}_{TM} \, e^{-\gamma_2(x \sin\theta_t + z \cos\theta_t)} \left(\cos\theta_t \, \hat{\mathbf{x}} - \sin\theta_t \, \hat{\mathbf{z}}\right), \tag{2.110}$$

$$\mathbf{H}_t = \frac{E_0}{\eta_2} \, \mathrm{T}_{TM} \, e^{-\gamma_2(x \sin\theta_t + z \cos\theta_t)} \, \hat{\mathbf{y}}. \tag{2.111}$$

Using Snell's law results in

$$\cos\theta_i + \Gamma_{TM} \cos\theta_r = \mathrm{T}_{TM} \cos\theta_t, \tag{2.112}$$

$$\frac{1}{\eta_1}\left(1 - \Gamma_{TM}\right) = \frac{\mathrm{T}_{TM}}{\eta_2}. \tag{2.113}$$

Solving these equations for the reflection and transmission coefficients gives

$$\Gamma_{TM} = \frac{\eta_2 \cos\theta_t - \eta_1 \cos\theta_i}{\eta_2 \cos\theta_t + \eta_1 \cos\theta_i}, \tag{2.114}$$

$$\mathrm{T}_{TM} = \frac{2\,\eta_2 \cos\theta_i}{\eta_2 \cos\theta_t + \eta_1 \cos\theta_i}. \tag{2.115}$$

2.6.3 Brewster Angle

For parallel polarization, an incident angle exists where there is no reflection from Region 2 and all waves are absorbed. This angle is known as the Brewster angle and is denoted as θ_B. The Brewster angle is found by setting $\Gamma_{TM} = 0$ in (2.114) as

$$\Gamma_{TM} = \frac{\eta_2 \cos\theta_t - \eta_1 \cos\theta_i}{\eta_2 \cos\theta_t + \eta_1 \cos\theta_i} = 0. \tag{2.116}$$

This leads to

$$\eta_2 \cos\theta_t = \eta_1 \cos\theta_i = \eta_1 \cos\theta_B. \tag{2.117}$$

Again using Snell's law, (2.117) becomes

$$\sin^2\theta_B = \left[\frac{\eta_1^2/\eta_2^2 - 1}{\eta_1^2/\eta_2^2 - \gamma_1^2/\gamma_2^2}\right]. \tag{2.118}$$

Figure 2.7 shows the magnitude of the reflection coefficient for both TE and TM waves as a function of incident angle. The relative permittivity for Region 1 is 1.3 and the relative permittivity for Region 2 is 2.8. It is seen in this figure the Brewster angle is located at 55.7°. A Brewster angle does not exist for perpendicular polarization and nonmagnetic material. For optics and lasers, the Brewster angle is referred to as the *polarizing angle*. The reader is referred to [13] for more detailed information.

2.6.4 Critical Angle

The critical angle is the angle of incidence corresponding to the threshold of total reflection and is commonly denoted as θ_c. When the incident angle is greater than or equal to the critical angle, then the transmission angle, θ_t, is equal to 90°, as shown in Figure 2.8. The critical is found from Snell's law of refraction as

$$\gamma_1 \sin\theta_i = \gamma_2 \sin\theta_t. \tag{2.119}$$

Setting $\theta_t = 90°$ results in

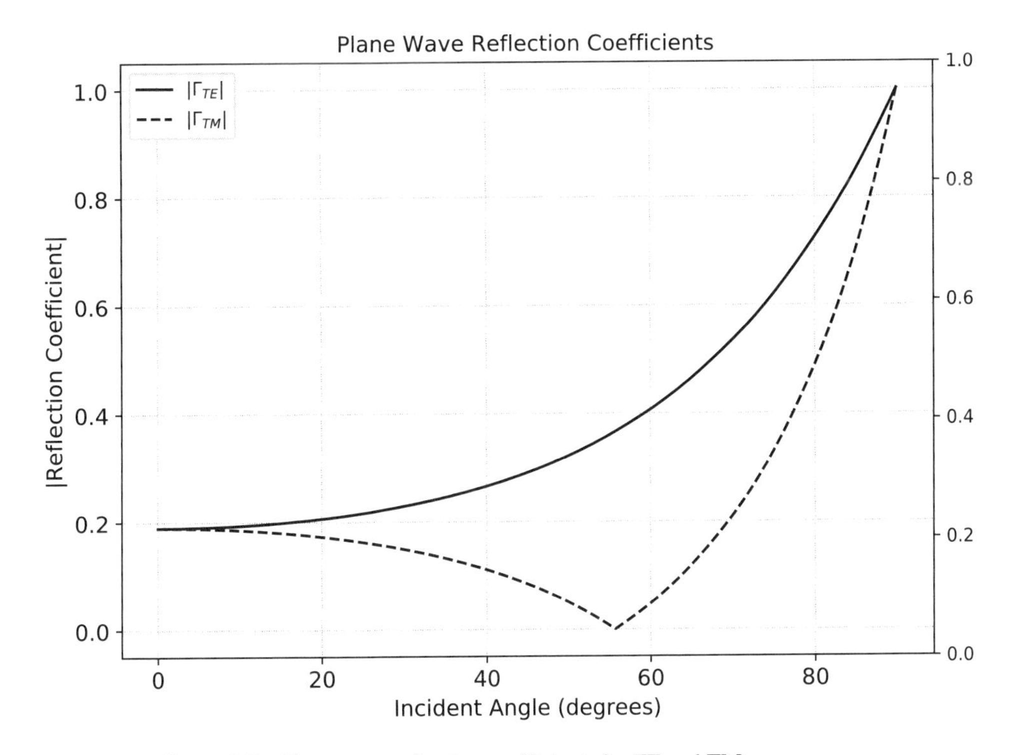

Figure 2.7 Plane wave reflection coefficients for TE and TM waves.

$$\theta_c = \sin^{-1}\left(\frac{\gamma_2}{\gamma_1}\right). \tag{2.120}$$

For nonmagnetic dielectrics, the critical angle is written as

$$\theta_c = \sin^{-1}\left(\sqrt{\frac{\epsilon_2}{\epsilon_1}}\right). \tag{2.121}$$

Therefore, to have a real valued critical angle, $\epsilon_2 < \epsilon_1$. If this condition is not true, then the Poynting vector is purely imaginary, and no real power flows along the z axis in Region 2 [14].

2.7 TROPOSPHERIC REFRACTION

As illustrated in Section 2.6, a uniform plane wave incident upon the interface of a medium with different constitutive parameters experiences a change in the direction

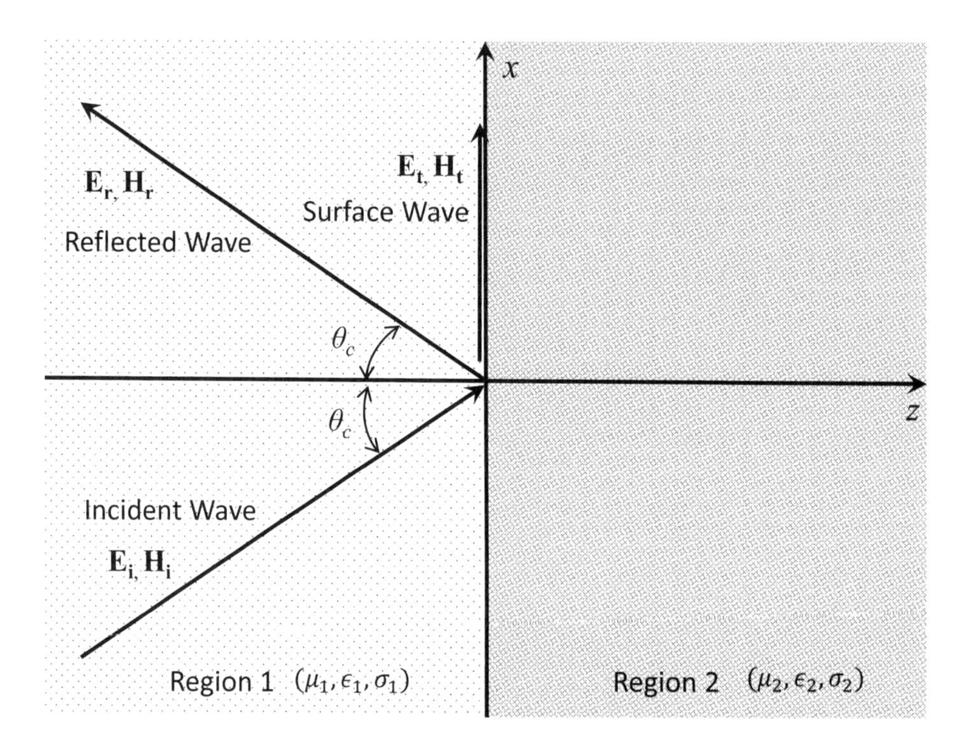

Figure 2.8 Critical angle for total reflection.

of propagation. This change of direction or *refraction* was studied for various types of media and incident angles. Since the constitutive parameters of the nonionized lower layers of the atmosphere vary with altitude, a transmitted radar signal also experiences a change in the direction of propagation through the atmosphere. The refraction is now continuous as the constitutive parameters are continuous functions of altitude, although some approaches treat the atmosphere as stratified. This continuous refraction is often referred to as *ray bending*. Generally, only the vertical gradient of the constitutive parameters is considered, which allows the curvature at a point to be written as

$$\frac{1}{r} = -\frac{\cos\phi}{n}\frac{dn}{dh}, \tag{2.122}$$

where

r = radius of curvature of the path of the wave (m),
ϕ = angle of the direction of propagation with the horizontal (rad),
n = refractive index of the atmosphere,
h = height above Earth's surface (m).

Therefore, dn/dh represents the vertical gradient of the refractive index of the atmosphere. Convention is to define the ray bending as positive when the path bends towards the surface of the earth. This refraction is virtually independent of frequency as long as the gradient does not vary significantly over a wavelength.

2.7.1 Apparent Elevation

As described above, a signal transmitted through the atmosphere bends toward the earth. In many radar applications, this would require transmission of the signal at a higher elevation angle in order to have the energy intercept the target, as shown in Figure 2.9. This angle is referred to as the apparent elevation angle, as the target appears to be at a different angle than its true position. Following the treatment in [15], the amount of correction to the true target elevation is

$$\Delta\theta = -\int_h^\infty \frac{n'(z)}{n(z)\tan\phi}\,dz \qquad \text{(deg)}, \tag{2.123}$$

where

$$\cos\phi = \frac{c}{(r_e + z)\,n(z)}, \qquad c = (r_e + h)\,n(h)\cos\theta. \tag{2.124}$$

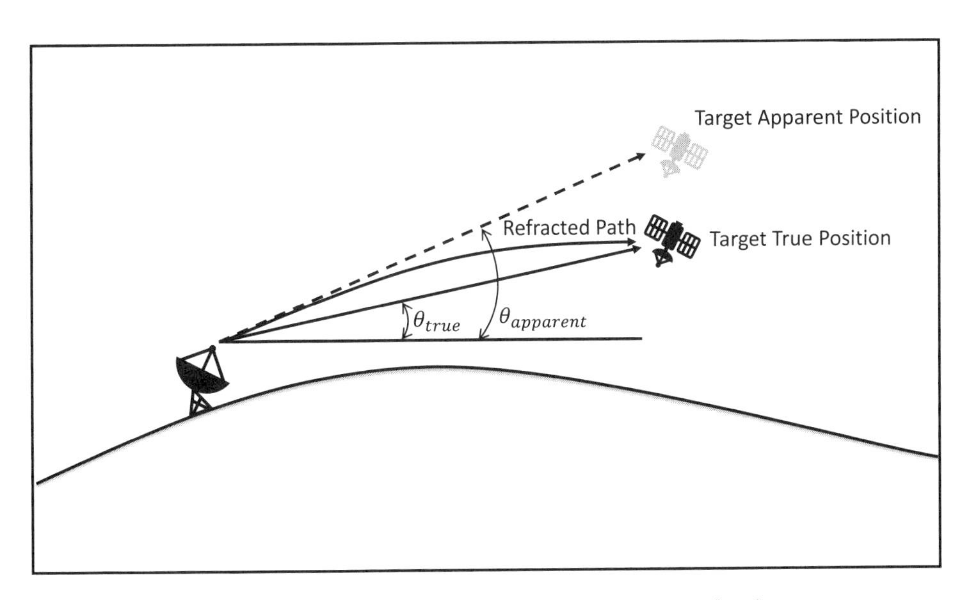

Figure 2.9 True and apparent target positions due to refraction.

r_e is the radius of Earth, and h is the height of the target. Since the refraction in the atmosphere is largely determined by the lower layers, a model based on the exponential atmosphere for terrestrial propagation allows the index of refraction at some altitude, z, to be expressed as [16]

$$n(z) = 1 + \alpha\, e^{-\beta z}, \tag{2.125}$$

where $\alpha = 0.000315$, and $\beta = 0.1361$. The relationship between the true elevation angle, θ_t, and apparent elevation angle, θ_a, is given by

$$\theta_a = \theta_t + \Delta\theta. \tag{2.126}$$

2.7.2 Apparent Range

Since a wave transmitted through the atmosphere experiences a refractive index that is a function of altitude, the path length from the transmitter to the target exceeds the geometrical path length, as shown in Figure 2.9. This difference in range is described by the integral [15]

$$\Delta R = \int_A^B (n - 1)\, dl, \tag{2.127}$$

where l is the length along the path, and (A, B) is the starting and ending points of the path. This expression is used when the variation of the constitutive parameters along the integration path are known. A semiempirical method was developed in order to calculate the apparent range when the temperature, atmospheric pressure, and relative humidity are known at ground level [15]. This method was derived in 1979 using atmospheric radio profiles at 500 meteorological stations over the course of one year. The expression for the difference in range for this method is

$$\Delta R = \frac{\Delta R_V}{\sin\theta\sqrt{(1 + k\cot^2\theta)}} + \delta(\theta, \Delta R_V), \tag{2.128}$$

where

θ	=	elevation angle at the target (rad),
ΔR_V	=	difference in vertical path length (m),
k	=	corrects for variations of the elevation angle along the path,
$\delta(\theta, \Delta R_V)$	=	correction to the effects of refraction (m).

The reader is referred to [15] for further details on this semiempirical method.

2.7.3 Beam Spreading

Since the refractive index is a function of altitude, the antenna beam will spread in the vertical direction [15]. This is due to the refractive index causing energy at the bottom of the beam to travel along a slightly different path than energy at the top of the beam. This beam spreading results in a loss in the peak gain, which is insignificant and may be ignored for elevation angles greater than 5°. In addition, there is no spreading in the horizontal direction. If the signal travels through the total atmosphere, then the total loss is

$$A = -10\log(B) \qquad \text{(dB)}, \tag{2.129}$$

where

$$\begin{aligned} B = 1- &\left[0.5411 + 0.07446\,\theta + (0.06272 + 0.0276\,\theta)\,h + 0.008288\,h^2\right] / \\ &\left[1.728 + 0.5411\,\theta + (0.1815 + 0.06272\,\theta + 0.0138\,\theta^2)\,h \right. \\ &\left. + (0.01727 + 0.008288\,\theta)\,h^2\right]^2, \end{aligned} \tag{2.130}$$

and θ is the elevation angle in degrees. The expression above is only valid when $\theta < 10^\circ$ and $h \leq 5$ km.

2.7.4 Ducting

An atmospheric duct is a region in the lower layers of the atmosphere where the vertical gradient of the refractive index is such that the propagating waves are ducted or guided and tend to follow the curvature of the earth, as shown in Figure 2.10. Ducts provide a mechanism for waves to propagate beyond the line of sight and limit the spread of the wavefront to only the horizontal dimension [15]. A duct is possible whenever the vertical refractivity gradient, (dN/dh), is less than about -157 N/km. A first order approximation of the refractivity is [17]

$$N = (n-1)10^6 = \frac{77.6}{T}\left(p + \frac{4810e}{T}\right), \tag{2.131}$$

where

T = air temperature (K),
p = atmospheric pressure (hPa),
e = the water vapor pressure (hPa).

The critical angle for waves to become trapped in a duct is

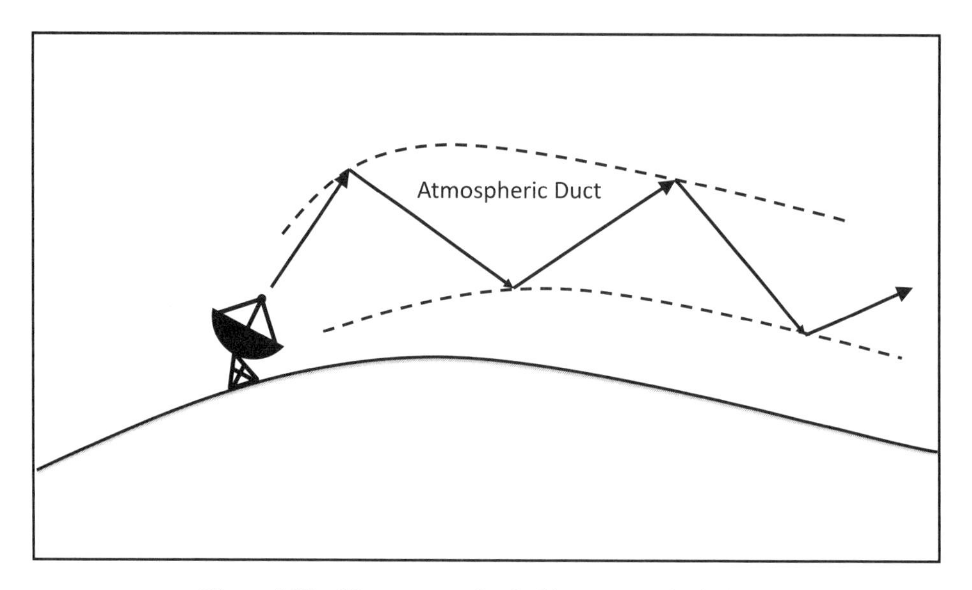

Figure 2.10 Wave propagation inside an atmospheric duct.

$$\alpha = \sqrt{2 \times 10^{-6} \left| \frac{dM}{dh} \right| \Delta h} \qquad \text{(rad)}. \tag{2.132}$$

M is the modified refractivity defined in [18] as

$$M = N + 10^6 \left(\frac{h}{r_e} \right), \tag{2.133}$$

and Δh is the height of the duct above the transmitting antenna.

2.8 EARTH DIFFRACTION

Diffraction is the mechanism by which waves spread out when passing through an aperture or around the corner of an object and into the *shadow region*. While diffraction may be caused by many objects and geometries in a given scenario, those discussions are covered in Chapter 7, and here the focus is on the diffraction caused by the earth. Following a procedure similar to the one given in [19], the diffraction loss associated with wave propagation above a spherical Earth at frequencies above 10 MHz is determined. Referring to Figure 2.11, begin by calculating a marginal line of sight distance as

$$d_{los} = \sqrt{2\,r_e}\left(\sqrt{h_1} + \sqrt{h_2}\right) \qquad \text{(m)}. \tag{2.134}$$

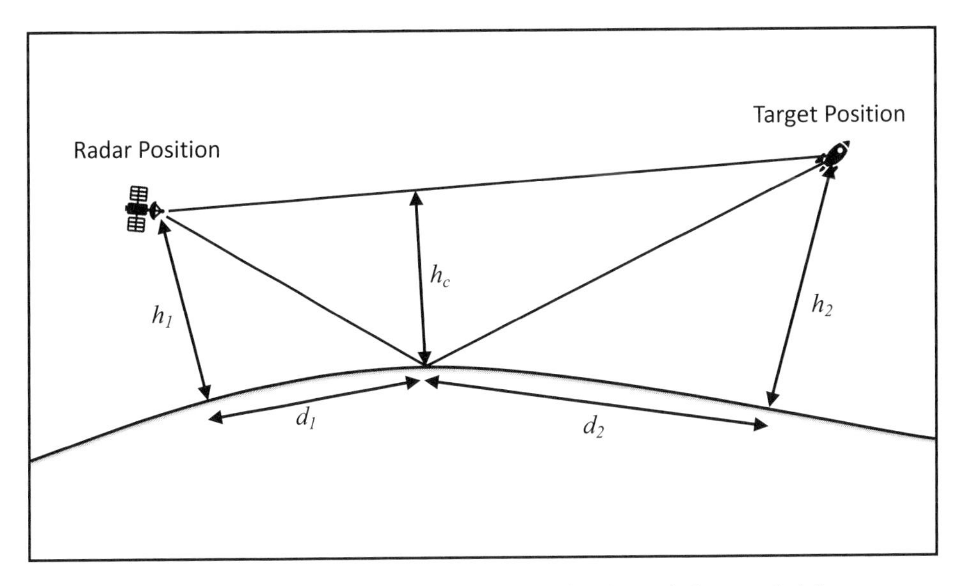

Figure 2.11 Geometry for spherical Earth diffraction and clearance height.

This leads to two possible cases described in the following sections.

2.8.1 Case 1: $d \geq d_{los}$

For this case, the diffraction is greatly influenced by the electrical characteristics of the surface of the earth. This influence is determined by calculating a normalized surface admittance, Y, for both horizontal and vertical polarization given by

$$Y_H = \left(\frac{2\pi\, r_e}{\lambda}\right)^{-1/3} \left[(\epsilon_r - 1)^2 + (60\,\lambda\,\sigma)^2\right]^{-1/4}, \tag{2.135}$$

$$Y_V = Y_H \left[\epsilon_r^2 + (60\,\lambda\,\sigma)^2\right]^{1/2}, \tag{2.136}$$

where

ϵ_r = relative permittivity,
σ = conductivity (S/m),
λ = wavelength (m).

The field strength including diffraction, relative to the field strength without diffraction is given by

$$20\log_{10}\left(\frac{E}{E_0}\right) = F(d_n) + G(h_{1n}) + G(h_{2n}) \qquad \text{(dB)}, \tag{2.137}$$

where

F = gain term for the distance,
G = gain term for the height,
d_n = normalized path length between the transmitter and the target,
h_{1n} = normalized height of the transmitter,
h_{2n} = normalized height of the target.

These normalized distances are found from

$$d_n = \beta\, d \left(\frac{\pi}{\lambda\, r_e^2}\right)^{1/3}, \tag{2.138}$$

$$h_{1n} = 2\,\beta\, h_1 \left(\frac{\pi^2}{\lambda^2\, r_e}\right)^{1/3}, \tag{2.139}$$

$$h_{2n} = 2\,\beta\, h_2 \left(\frac{\pi^2}{\lambda^2\, r_e}\right)^{1/3}, \tag{2.140}$$

where β accounts for the electrical characteristics of the earth. It is related to the admittance given in (2.135) and (2.136) by the semiempirical formula

$$\beta = \frac{1 + 1.6\,Y^2 + 0.67\,Y^4}{1 + 4.5\,Y^2 + 1.53\,Y^4}, \tag{2.141}$$

where Y is either Y_H or Y_V. The gain terms in (2.137) are expressed as

$$F(d_n) = \begin{cases} 11 + 10\log_{10}(d_n) - 17.6\,d_n & \text{for } d_n \geq 1.6 \\ -20\log_{10}(d_n) - 5.6488\,d_n^{1.425} & \text{for } d_n < 1.6, \end{cases} \tag{2.142}$$

and

$$G(h) = \begin{cases} 17.6(B - 1.1)^{1/2} - 5\log_{10}(B - 1.1) - 8 & \text{for } B > 2 \\ 20\log_{10}(B + 0.1B^3) & \text{for } B \leq 2, \end{cases} \tag{2.143}$$

where $B = \beta h$. The diffracted field strength in (2.137) is an approximation to the classical residue series [20, 21], and is accurate to better than 2 dB for long path lengths [19].

2.8.2 Case 2: $d < d_{los}$

For this case, begin by calculating the smallest clearance height, h_c, between the line of sight path and the curved earth, as shown in Figure 2.11. The clearance height is given by

$$h_c = \frac{\left(h_1 - \frac{d_1^2}{2\,r_e}\right) d_2 + \left(h_2 - \frac{d_2^2}{2\,r_e}\right) d_1}{d}, \tag{2.144}$$

where

$$d_1 = 0.5\,d\,(1+b), \qquad d_2 = d - d_1, \tag{2.145}$$

$$b = 2\sqrt{\frac{m+1}{3m}}\cos\left(\frac{\pi}{3} + \frac{1}{3}\cos^{-1}\left(\frac{3c}{2}\sqrt{\frac{3m}{(m+1)^3}}\right)\right), \tag{2.146}$$

$$c = \frac{h_1 - h_2}{h_1 + h_2}, \qquad m = \frac{d^2}{4\,r_e\,(h_1 + h_2)}. \tag{2.147}$$

Next calculate the required clearance height for no diffraction loss, h_0, as

$$h_0 = 0.552\sqrt{\frac{d_1\,d_2\,\lambda}{d}}. \tag{2.148}$$

If the calculated clearance height, h_c, is greater than the required clearance height, h_0, then there is no diffraction loss for this path. Otherwise, calculate a modified effective Earth radius by

$$r_{em} = 0.5\left(\frac{d}{\sqrt{h_1} + \sqrt{h_2}}\right) \qquad \text{(m)}. \tag{2.149}$$

This modified effective Earth radius is then used in the procedure described in Section 2.8.1, and the loss is designated as A_{em}. If the calculated loss using the modified effective Earth radius is negative, then the diffraction loss for this path is zero. Otherwise an interpolated diffraction loss is used, and is calculated by

$$A = \left[1 - h/h_0\right] A_{em} \qquad \text{(dB)}. \tag{2.150}$$

2.9 PLANE WAVE ATTENUATION

There are several mechanisms contributing to the attenuation of propagating waves. These can include rain, clouds, fog, the atmosphere, and vegetation. Several of these mechanisms are examined in the following sections.

2.9.1 Atmospheric Attenuation

Electromagnetic waves are attenuated in the atmosphere primarily due to oxygen (O_2) and water vapor (H_2O). The first peak in attenuation occurs at approximately 22 GHz and is due to water vapor. The second peak is located at 63 GHz due to oxygen. For frequencies up to 1 THz, the attenuation due to dry air and water vapor is most accurately evaluated by means of a summation of the individual resonance lines from oxygen and water vapor [22]. Additional factors are added to this result to account for the non-resonant Debye spectrum of oxygen, pressure-induced nitrogen attenuation, and excess water vapor absorption. The specific atmospheric attenuation is written as

$$A_s = A_o + A_w = 0.1820\, f\, \big(N_o(f) + N_w(f)\big) \qquad \text{(dB/km)}, \tag{2.151}$$

where

A_o = attenuation due to oxygen and nitrogen (dB/km),
A_w = attenuation due to water vapor (dB/km),
f = the operating frequency (GHz).

The terms $N_o(f)$ and $N_w(f)$ are the imaginary parts of the complex refractivity and are expressed as

$$N_o(f) = \sum_i S_i^o\, F_i^o + N_d(f), \qquad N_w(f) = \sum_i S_i^w\, F_i^w, \tag{2.152}$$

where

S_i = strength of the ith oxygen or water vapor line,
F_i = shape factor for the ith oxygen or water vapor line,
$N_d(f)$ = pressure induced nitrogen absorption.

The reader is referred to [22] for the tables of coefficients for these terms. For terrestrial paths, the total path attenuation is then found from

$$A = A_s\, r = (A_o + A_w)\, r \qquad \text{(dB)}. \tag{2.153}$$

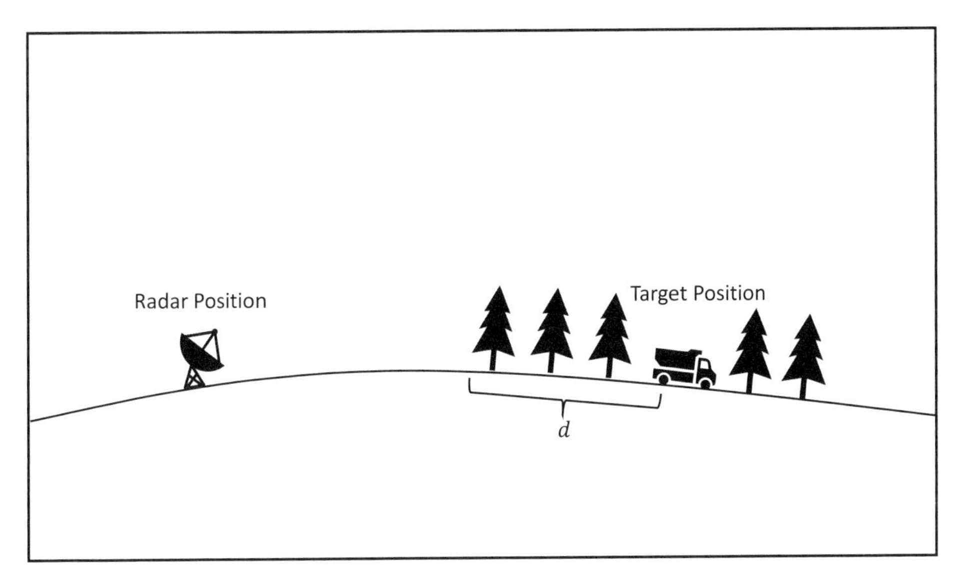

Figure 2.12 Geometry for a target located at a distance, d, within vegetation.

2.9.2 Attenuation in Vegetation

For some radar applications, the attenuation in vegetation can be important. However, there is a wide range of conditions and foliage types making it extremely difficult to develop generalized prediction methods. Furthermore, there is a lack of experimental data collected in a single database for use. However, models have been developed for extensive vegetation such as woodland for particular frequency ranges and different types of geometry [23]. Two geometries are considered in the following sections. The first is when the transmitter or the target is located within the vegetation, and the other is when both the transmitter and target are both located outside of the vegetation.

2.9.2.1 Target within Vegetation

Consider a signal path where the target is located within extensive vegetation, such as woodland, as shown in Figure 2.12. The attenuation is characterized by two main quantities. The first is the specific attenuation rate due to the scattering of energy in directions away from the target. The second is the maximum total attenuation, which is limited by mechanisms such as surface wave propagation over the top of the vegetation and the forward scatter within the vegetation [23]. The attenuation due to the vegetation is expressed as

$$A_v = A_m \left[1 - e^{-d\gamma/A_m}\right] \qquad \text{(dB)}, \qquad (2.154)$$

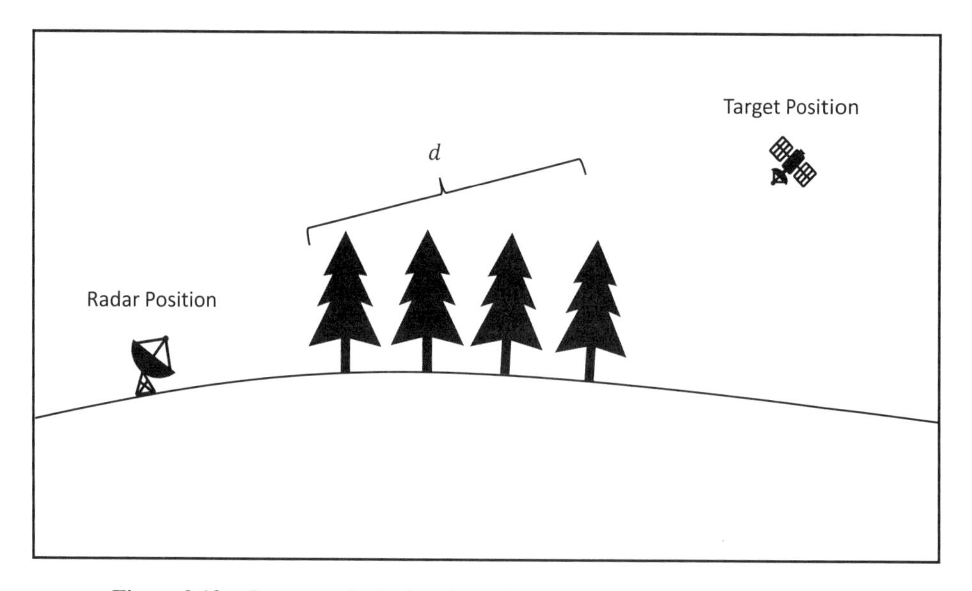

Figure 2.13 Geometry for both radar and target located outside the vegetation.

where

d	=	depth within the vegetation (m),
γ	=	specific attenuation for short vegetation paths (dB/km),
A_m	=	maximum attenuation for the specific type of vegetation (dB).

A_m is the clutter loss for targets obstructed by ground clutter. The maximum attenuation and the specific attenuation depend on the species and density of the vegetation as well as the polarization of the transmitted signal. Experiments have been conducted on various species of trees, and those results are given in [23].

2.9.2.2 Target Outside Vegetation

Consider a signal path where the target is outside the vegetation and the signal must pass through the vegetation to reach the target, as shown in Figure 2.13. A model has been proposed to describe the attenuation along both horizontal and slant foliage paths [23], and is given here as

$$A_v = A\, f^B\, d^C\, (\theta + E)^G \qquad \text{(dB)}, \tag{2.155}$$

where

f = operating frequency (MHz),
d = depth of the vegetation (m),
θ = elevation angle (deg).

The parameters A, B, C, E, and G are found empirically by fitting to measurements. For example, a fit to measurements in pine woodland in Austria [23] is

$$L = 0.25\, f^{0.39}\, d^{0.25}\, \theta^{0.05} \qquad \text{(dB)}. \tag{2.156}$$

The reader is referred to [23] to find measurements for other types of trees, including the silver maple, London plane, horse chestnut, ginkgo, and dawn redwood.

2.9.3 Rain Attenuation

When operating radar systems in regions with precipitation, it is necessary to account for the attenuation of the propagating wave due to the precipitation. Following the recommendation from the International Telecommunication Union Radiocommunication Sector (ITU-R) [24], the specific attenuation due to rain is given as

$$\gamma = kR^{\alpha} \qquad \text{(dB/km)}, \tag{2.157}$$

where R is the rain rate in millimeters per hour. The factors k and α are functions of frequency and their expressions are derived by curve fitting scattering calculations to the form in (2.157). From [24], the factors k and α are given by

$$k = 0.5 \times \big[k_H + k_V + (k_H - k_V)\cos^2\theta\cos(2\tau)\big], \tag{2.158}$$

$$\alpha = 0.5 \times \big[k_H\alpha_H + k_V\alpha_V + (k_H\alpha_H - k_V\alpha_V)\cos^2\theta\cos(2\tau)\big]/k, \tag{2.159}$$

where θ is the elevation angle and τ is the polarization tilt angle. The expressions in (2.158) and (2.159) are seen to be a combination of horizontal and vertical polarization components and are valid for both linear and circular polarization and for all path geometries. The terms k_H and k_V are given in [24] as

$$\log_{10} k = \sum_{i=1}^{4} a_i\, e^{-\beta^2} + d_k \log_{10} f + e_k, \tag{2.160}$$

$$\alpha = \sum_{i=1}^{5} a_i\, e^{-\beta^2} + d_\alpha \log_{10} f + e_\alpha, \tag{2.161}$$

where

$$\beta = \frac{\log_{10} f - b_i}{c_i}, \tag{2.162}$$

Table 2.4

Coefficients for Calculating k_H

a_i	b_i	c_i	d_k	e_k
−5.33980	−0.10008	1.13098	−0.18961	0.71147
−0.35351	1.26970	0.45400		
−0.23789	0.86036	0.15354		
−0.94158	0.64552	0.16817		

Table 2.5

Coefficients for Calculating k_V

a_i	b_i	c_i	d_k	e_k
−3.80595	0.56934	0.81061	−0.16398	0.63297
−3.44965	−0.22911	0.51059		
−0.39902	0.73042	0.11899		
0.50167	1.07319	0.27195		

and f is the frequency in GHz, k is either k_H or k_V, and α is either α_H or α_V. The values of the coefficients, a, b, c, d, and e are given in [24], and are listed here in Tables 2.4–2.7 for reference.

2.9.4 Cloud and Fog Attenuation

The attenuation of electromagnetic waves due to clouds and fog is of particular importance not only to airborne and ground-based upward-looking radars, but also to vehicle based radars dealing with automatic emergency braking, radar systems for mining, agriculture, and search and rescue. Many of these sensors operate in higher frequency regimes where the attenuation due to fog and cloud becomes significant. Following the recommendation in [25], and considering clouds and fog made up of small droplets less than 0.01 cm, the Rayleigh approximation allows the attenuation within the cloud or fog to be written as

$$\gamma = k_l(f, T)\, M \qquad \text{(dB/km)}, \tag{2.163}$$

where

Table 2.6
Coefficients for Calculating α_H

a_i	b_i	c_i	d_α	e_α
−0.14318	1.82442	−0.55187	0.67849	−1.95537
0.29591	0.77564	0.19822		
0.32177	0.63773	0.13164		
−5.37610	−0.96230	1.47828		
16.1721	−3.29980	3.43990		

Table 2.7
Coefficients for Calculating α_V

a_i	b_i	c_i	d_α	e_α
−0.07771	2.33840	−0.76284	−0.053739	0.83433
0.56727	0.95545	0.54039		
−0.20238	1.14520	0.26809		
−48.2991	0.791669	0.116226		
48.5833	0.791459	0.116479		

f = operating frequency (GHz),
T = cloud temperature (K),
k_l = liquid water specific attenuation coefficient ((dB/km)/(g/m^3)),
M = liquid water density (g/m^3).

This expression is valid for frequencies up to about 200 GHz. Typical values for the liquid water density in fog is 0.05 g/m^3 for medium fog with a visibility of about 300 meters, and 0.5 g/m^3 for heavy fog with a visibility of about 50 meters [25]. The model for the liquid water specific attenuation coefficient, k_l, is based on Rayleigh scattering. The permittivity is modeled as double Debye [26], and is used to calculate k_l as

$$k_l = \frac{0.819\, f}{\epsilon'' \left(1 + \eta^2\right)} \qquad \text{(dB/km)/(g/m}^3\text{)}, \tag{2.164}$$

where f is the frequency in GHz, and

$$\eta = \frac{2 + \epsilon'}{\epsilon''}. \tag{2.165}$$

The complex permittivity of water is

$$\epsilon'' = \frac{f(\epsilon_0 - \epsilon_1)}{f_p\left[1 + (f/f_p)^2\right]} + \frac{f(\epsilon_1 - \epsilon_2)}{f_s\left[1 + (f/f_s)^2\right]}, \tag{2.166}$$

$$\epsilon' = \frac{\epsilon_0 - \epsilon_1}{\left[1 + (f/f_p)^2\right]} + \frac{\epsilon_1 - \epsilon_2}{\left[1 + (f/f_s)^2\right]} + \epsilon_2, \tag{2.167}$$

where

$$\epsilon_0 = 77.66 + 103.3\,(\theta - 1), \quad \epsilon_1 = 0.0671\,\epsilon_0, \quad \epsilon_2 = 3.52, \quad \theta = 300/T, \tag{2.168}$$

and T is the liquid water temperature in degrees Kelvin [25]. In (2.166) and (2.167), the frequencies (f_p, f_s), represent the principal and secondary relaxation frequencies, respectively. These are given in GHz as

$$f_p = 20.20 - 146\,(\theta - 1) + 316\,(\theta - 1)^2, \qquad f_s = 39.8\,f_p. \tag{2.169}$$

The total attenuation along the path is then written as

$$A = \gamma\, r \qquad \text{(dB)}, \tag{2.170}$$

where r is the path length in the cloud or fog in km.

2.10 EXAMPLES

The sections below illustrate the concepts of this chapter with several Python examples. The examples for this chapter are in the directory *pyradar\Chapter02* and the matching MATLAB examples are in the directory *mlradar\Chapter02*. The reader should consult Chapter 1 for information on how to execute the Python code associated with this book.

2.10.1 Plane Wave Propagation

For this example, calculate the plane wave propagation parameters listed in Table 2.2 for a medium with the following constitutive parameters: $\epsilon_r = 4.3$, $\mu_r = 1.0$, $\sigma = 0.05$ S/m at a frequency of 300 MHz. Also, plot the electric and magnetic fields for this case.

Solution: The solution to the above example is given in the Python code *plane_waves_example.py* and in the MATLAB code *plane_waves_example.m*. Running the Python

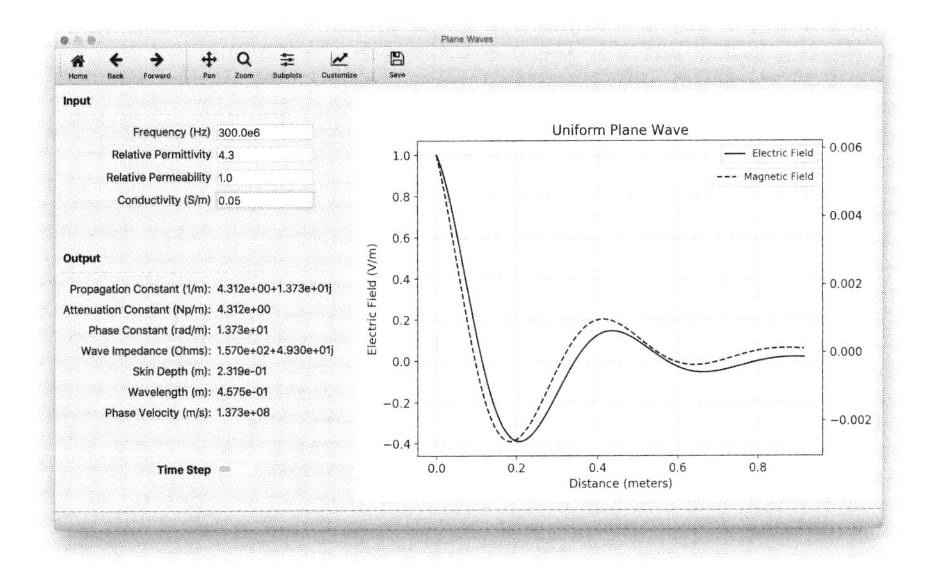

Figure 2.14 Plane wave parameters calculated by *plane_waves_example.py*.

example code displays a GUI, allowing the user to enter the constitutive parameters, and calculates the plane wave parameters as well as creating a plot of the electric and magnetic fields. Figure 2.14 shows the solution to this example obtained from *plane_waves_example.py*.

2.10.2 Reflection and Transmission

Consider an interface between two media where media 1 has the following constitutive parameters: $\epsilon_r = 1.3$, $\mu_r = 1.0$, $\sigma = 0.01$ S/m. Media 2 has the following constitutive parameters: $\epsilon_r = 2.8$, $\mu_r = 1.0$, $\sigma = 0.01$ S/m. The operating frequency is 300 MHz. Plot the reflection and transmission coefficients for both parallel and perpendicular polarization. Also calculate the critical and Brewster angles if they exist.

Solution: The solution to the above example is given in the Python code *reflection_transmission_example.py* and in the MATLAB code *reflection_transmission_example.m*. Running the example Python code displays a GUI, allowing the user to enter the constitutive parameters. The code then calculates the critical and Brewster angles, as shown in Figure 2.15. As seen in this figure, both values are complex, indicating neither of these angles exist for this interface. Also, *reflection_transmission_example.py* generates

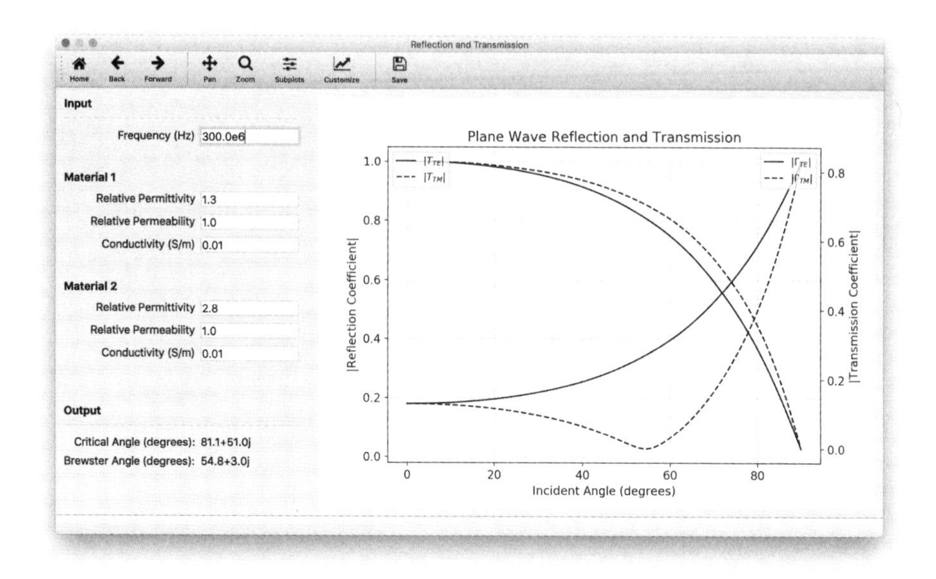

Figure 2.15 Critical angle, Brewster angle, transmission and reflection coefficients calculated by *reflection_transmission_example.py*.

the plots of the reflection and transmission coefficients for both parallel and perpendicular polarization, illustrated in Figure 2.15.

2.10.3 Tropospheric Refraction

Tropospheric refraction can play a significant role in the manner in which plane waves propagate, as was shown in Section 2.7. The following sections give examples for apparent elevation and range, beam spreading loss, and ducting.

2.10.3.1 Apparent Elevation

For this example, plot the apparent elevation angle due to refraction when the true elevation in 20°, and the height varies from 0 to 5 km.

Solution: The solution to the above example is given in the Python code *apparent_elevation_example.py* and in the MATLAB code *apparent_elevation_example.m*. Running the Python example code displays a GUI, allowing the user to enter the true elevation angle and the height. The code then produces a plot of the apparent elevation angle as a function of height, as shown in Figure 2.16.

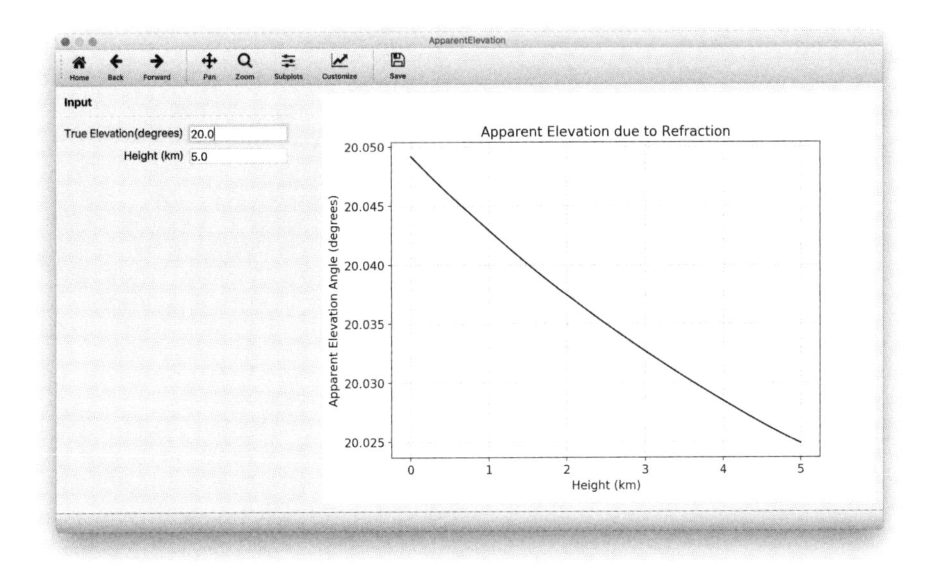

Figure 2.16 Apparent elevation angle calculated by *apparent_elevation_example.py*.

2.10.3.2 Apparent Range

In this example, calculate the apparent range due to refraction when the radar is located at 34° latitude, 84° longitude, and has an altitude of 120 meters. The target is located at 34° latitude, 80° longitude, and has an altitude of 12 km.

Solution: The solution to the above example is given in the Python code *apparent_range_example.py* and in the MATLAB code *apparent_range_example.m*. Running the Python example code displays a GUI, allowing the user to enter the radar location and the target location as latitude, longitude and altitude. The code then calculates both the true and apparent range, as shown in Figure 2.17.

2.10.3.3 Beam Spreading

For this example, calculate and plot the beam spreading loss due to refraction as the elevation angle varies from 0 to 5°, and the height varies from 0 to 5 km.

Solution: The solution to the above example is given in the Python code *beam_spreading_example.py* and in the MATLAB code *beam_spreading_example.m*. Running the Python

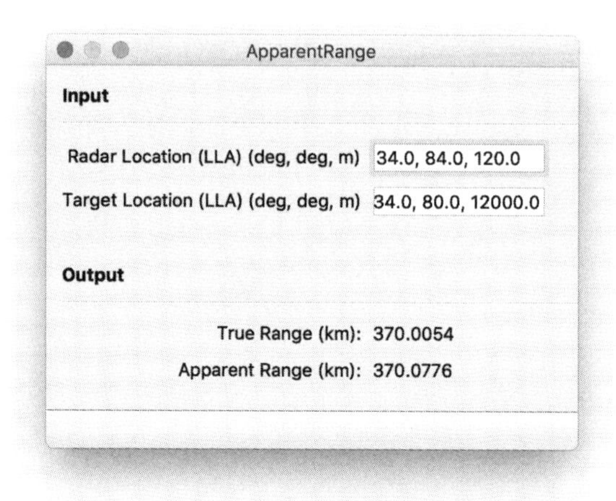

Figure 2.17 True and apparent range calculated by *apparent_range_example.py*.

example code displays a GUI, allowing the user to enter the elevation angle and height and produces a plot of the beam spreading loss, as shown in Figure 2.18.

2.10.3.4 Ducting

In this example, calculate and plot the critical angle for ducting as a function of refractivity gradient for duct thicknesses of 10, 20, and 50 m. Allow the refractivity gradient to vary from -150 to -500 N/km.

Solution: The solution to the above example is given in the Python code *ducting_example.py* and in the MATLAB code *ducting_example.m*. Running the Python example code displays a GUI, allowing the user to enter the duct thicknesses and creates a plot of the critical angle for ducting as a function of refractivity gradient, as shown in Figure 2.19.

2.10.4 Earth Diffraction

The example for Earth diffraction is to calculate and plot the diffraction loss as a function of frequency for horizontal polarization when the radar is located at 26.5° latitude, 97.0° longitude and has an altitude of 1.0 km, and the target is located at 31.0° latitude, 96.0° longitude and has an altitude of 13.0 km. The constitutive parameters are $\epsilon_r = 1.3$ and $\sigma = 0.01$ S/m. The frequency range is from 1 MHz to 300 MHz.

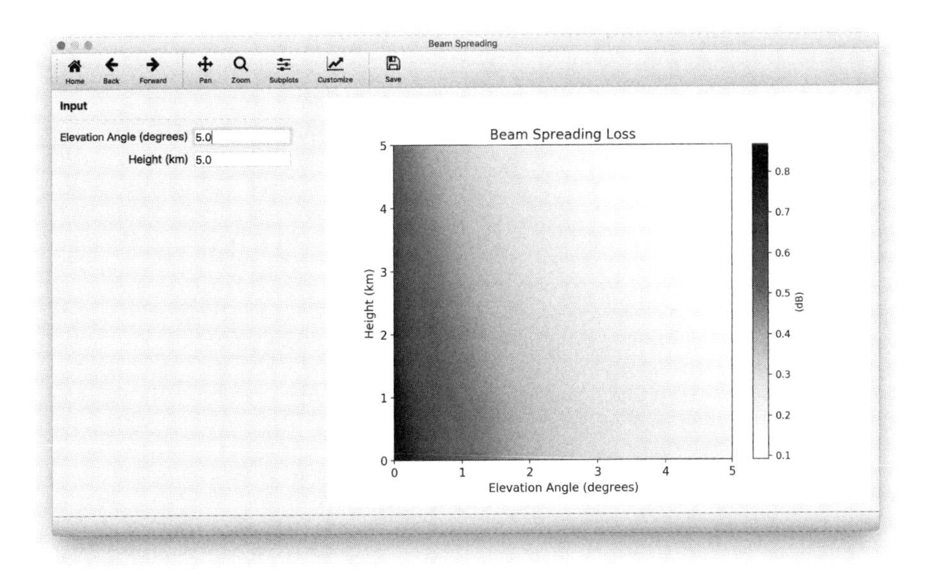

Figure 2.18 Beam spreading loss calculated by *beam_spreading_example.py*.

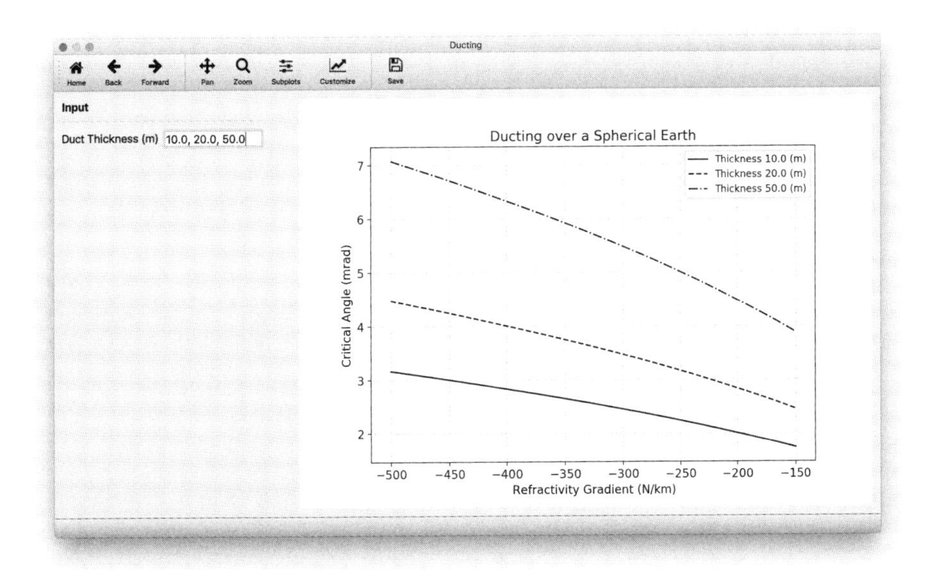

Figure 2.19 Critical angle calculated by *ducting_example.py*.

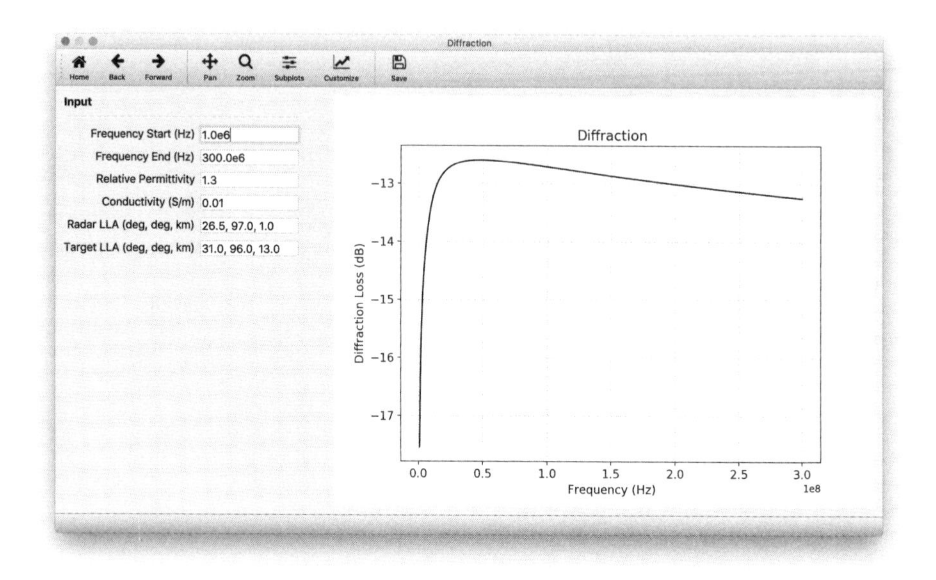

Figure 2.20 Diffraction loss calculated by *diffraction_example.py*.

Solution: The solution to the above example is given in the Python code *diffraction_example.py* and in the MATLAB code *diffraction_example.m*. Running the Python example code displays a GUI, allowing the user to enter the frequency range, radar location, target location, and constitutive parameters. The code then creates a plot of the diffraction loss as a function of frequency, as shown in Figure 2.20.

2.10.5 Attenuation

From Section 2.9, there are several mechanisms contributing to the attenuation of propagating waves. These can include rain, clouds, fog, the atmosphere, and vegetation. The following sections give examples for each of these attenuation mechanisms.

2.10.5.1 Atmospheric

As illustrated in Section 2.9.1, the attenuation due to the atmosphere is primarily due to water vapor, oxygen, and pressure-induced nitrogen attenuation. For this example, plot the atmospheric attenuation in the frequency range from 0 to 1000 GHz. Use a temperature of 290° K, a dry air pressure of 1013.25 hPa, and a water vapor density of 7.5 g/m^3.

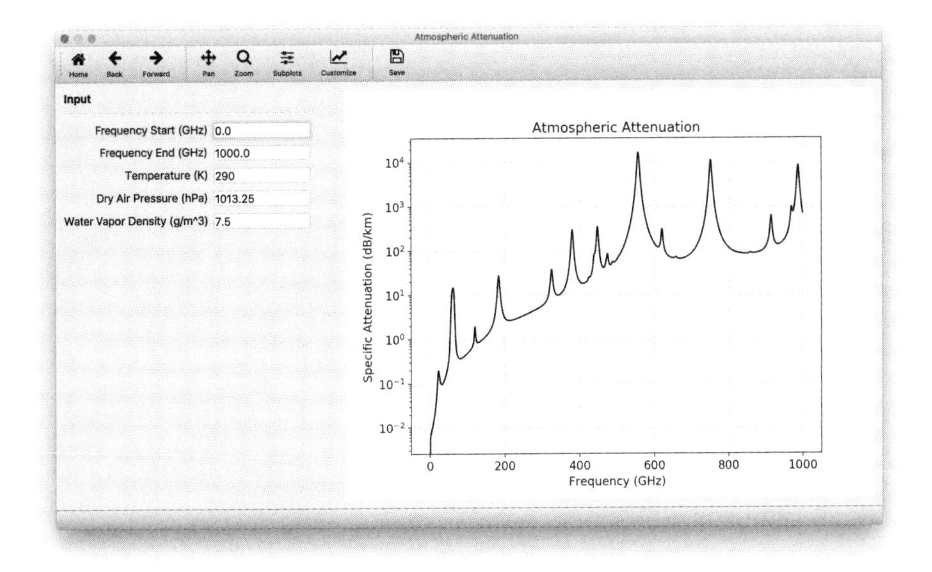

Figure 2.21 Atmospheric attenuation calculated by *atmosphere_example.py*.

Solution: The solution to the above example is given in the Python code *atmosphere_example.py* and in the MATLAB code *atmosphere_example.m*. Running the Python example code displays a GUI, allowing the user to enter the frequency range, temperature, dry air pressure, and water vapor density. The code then calculates and creates a plot of the atmospheric attenuation as a function of frequency, as shown in Figure 2.21.

2.10.5.2 Vegetation

In this example, consider the target to be located within the vegetation. Calculate and plot the total attenuation as a function of target distance in the vegetation. Consider vegetation with a specific attenuation of 0.39 dB/m and a maximum attenuation of 34.10 dB. Allow the range to vary from 0 to 100 meters.

Solution: The solution to the above example is given in the Python code *vegetation_example.py* and in the MATLAB code *vegetation_example.m*. Running the Python example code displays a GUI, allowing the user to enter the specific and maximum attenuation of the vegetation and distance within the vegetation. The code then calculates and creates a plot of the vegetation attenuation as a function of distance within the vegetation, as shown in Figure 2.22.

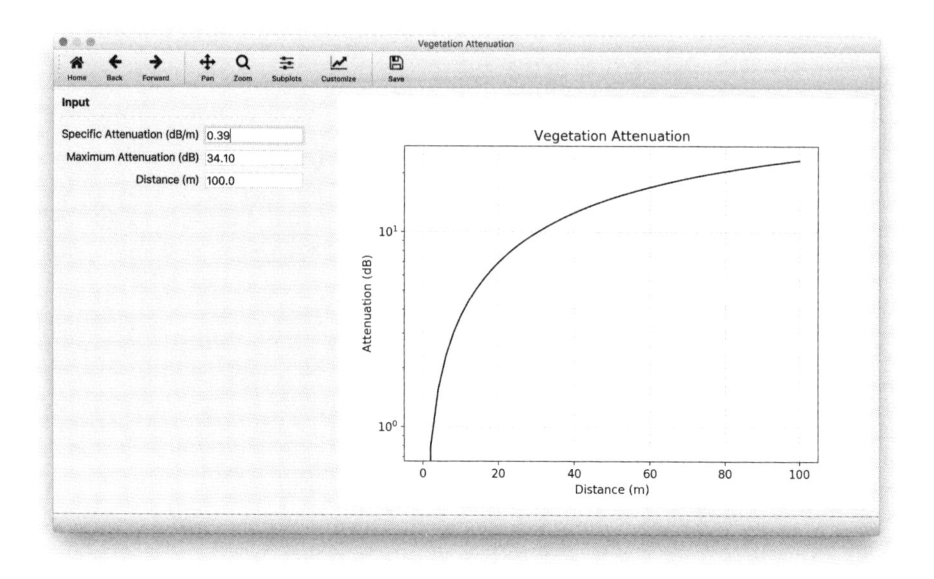

Figure 2.22 Vegetation attenuation calculated by *vegetation_example.py*.

2.10.5.3 Rain

For this example, calculate and display the specific attenuation due to rain in the frequency range from 1 to 1000 GHz. Assume the rain rate is 5 mm/hr, the elevation angle is 10°, and the polarization tilt angle is 45° (circular polarization).

Solution: The solution to the above example is given in the Python code *rain_example.py* and in the MATLAB code *rain_example.m*. Running the Python example code displays a GUI, allowing the user to enter the frequency range, rain rate, elevation angle, and polarization tilt angle. The code then calculates and creates a plot of the rain attenuation as a function of frequency, as shown in Figure 2.23.

2.10.5.4 Cloud and Fog

In this example, calculate and create a plot of the specific attenuation due to fog as a function of frequency. Allow the frequency to vary from 1 to 200 GHz. Assume a heavy fog with a visibility of about 50 meters. This corresponds to a fog density of 0.5 g/m^3 at a temperature of 290° K.

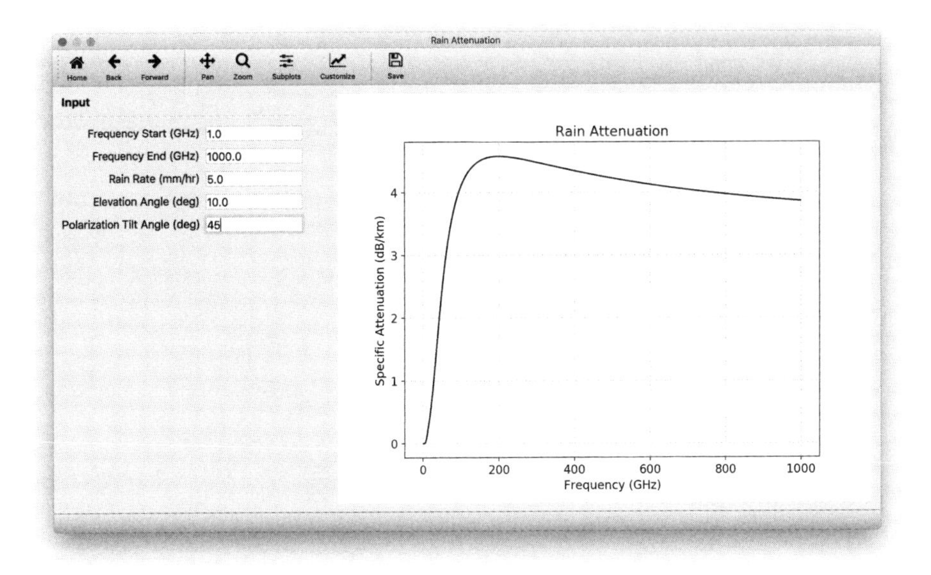

Figure 2.23 Rain attenuation calculated by *rain_example.py*.

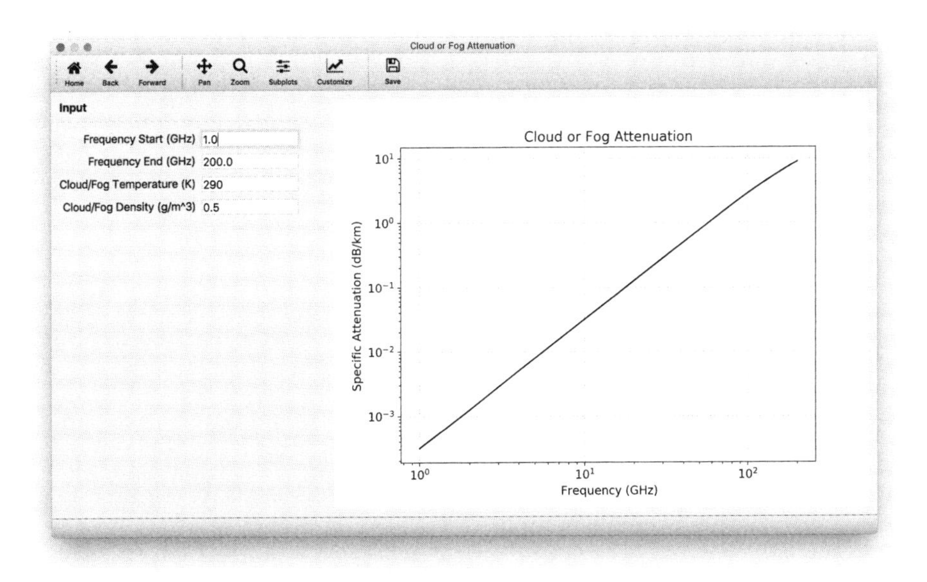

Figure 2.24 Fog attenuation calculated by *cloud_fog_example.py*.

Solution: The solution to the above example is given in the Python code *cloud_fog_ example.py* and in the MATLAB code *cloud_fog_example.m*. Running the Python example code displays a GUI, allowing the user to enter the frequency range, cloud or fog temperature, and cloud or fog density. The code then calculates and creates a plot of the specific attenuation as a function of frequency, as shown in Figure 2.24.

PROBLEMS

2.1 Describe Maxwell's equations and their importance in the development of radar systems.

2.2 Write Maxwell's equations, given in (2.1)–(2.4), as eight first-order scalar equations.

2.3 For the electric field given by $\mathbf{E} = 2.3\,\cos(3\pi x)\sin(6\pi 10^9 t - kz)$ V/m, find the magnetic field, **H**, and the propagation constant, k, in air.

2.4 Describe the difference between perfect electric conductors (PEC) and perfect magnetic conductors (PMC), and the boundary conditions associated with each.

2.5 Using Table 2.3, calculate the wave propagation parameters for seawater at a temperature of $T = 25^\circ$ C, a salinity of 25 ppt, and a frequency of 1.7 GHz. The relative permeability is $\mu_r = 1.0$, the relative permittivity is $\epsilon_r \approx 77.0 + j6.5$, and the conductivity is given by $\sigma = 2.903602 + 8.607 \times 10^{-2}\,T + 4.738817 \times 10^{-4}\,T^2 - 2.991 \times 10^{-6}\,T^3 + 4.3047 \times 10^{-9}\,T^4$ S/m.

2.6 Describe Snell's laws of reflection and refraction, and how these are related to the phase matching condition.

2.7 For a uniform plane wave in air incident on a material interface, calculate the reflection and transmission coefficients for both parallel and perpendicular polarizations. The material is defined by $\epsilon_r = 4.0$, $\mu_r = 2.3$, $\sigma = 4.0$ S/m, and the incident angle is 60°. Discuss the significance of the resulting quantities.

2.8 Find the Brewster angle ($\gamma = 0$) when the material in the first region has a relative permittivity of $\epsilon_r = 1.3$, a relative permeability of $\mu_r = 1.0$, and conductivity of $\sigma = 0.0$ S/m. The material in the second region has a relative permittivity of $\epsilon_r = 2.8$, a relative permeability of $\mu_r = 1.0$, and conductivity of $\sigma = 0.0$ S/m.

2.9 Determine the critical angle (total reflection) when the material in the first region has a relative permittivity of $\epsilon_r = 1.3$, a relative permeability of $\mu_r = 1.0$, and conductivity of $\sigma = 0.0$ S/m. The material in the second region is free space.

2.10 Calculate the total loss due to beam spreading as a plane wave travels through the atmosphere at an elevation angle of 3.0° and a height of 4.5 km.

2.11 For an atmospheric duct with a thickness of 15 m, calculate the critical angle when the refractive gradient is 300 N/km.

2.12 For vegetation with a specific attenuation of 0.4 dB/m and a maximum attenuation of 31.2 dB, calculate the total attenuation when a target is located at a distance of 40 m.

2.13 Calculate the attenuation due to rain when the rain rate is 4.3 mm/hr, the elevation angle is 15°, the polarization tilt angle is 0°, and the operating frequency is 10 GHz.

2.14 Calculate the attenuation due to clouds when the cloud density is 0.7 g/m^3, the cloud temperature is 300° K, and the operating frequency is 75 GHz.

References

[1] J. C. Maxwell. *A Treatise on Electricity and Magnetism.* Oxford University Press, London, UK, 1873.

[2] C. A. Balanis. *Advanced Engineering Electromagnetics,* 2nd ed. John Wiley and Sons, New York, 2012.

[3] J. D. Jackson. *Classical Electrodynamics,* 3rd ed. John Wiley and Sons, New York, 1998.

[4] D. K. Cheng. *Field and Wave Electromagnetics,* 2nd ed. Addison-Wesley, Reading, MA, 1989.

[5] W. Panofsky and M. Phillips. *Classical Electricity and Magnetism,* 2nd ed. Addison-Wesley, Reading, MA, 1962.

[6] A. Sommerfeld. Uber die beugung der rontgenstrahlen. *Annalen der Physik*, 343(8):473–506, 1912.

[7] L. Lorenz. On the identity of the vibrations of light with electrical currents. *Philosophical Magazine*, 4(34):287–301, 1867.

[8] J. D. Jackson and L. B. Okun. Historical roots of gauge invariance. *Reviews of Modern Physics*, 73:663–680, 2001.

[9] J. D. Jackson. From Lorenz to Coulomb and other explicit gauge transformations. *American Journal of Physics*, 70:917, 2002.

[10] Wikipedia Contributors. *Hermann von Helmholtz.* https://en.wikipedia.org/wiki/Hermann_von_Helmholtz.

[11] P. Moon and D.E. Spencer. *Field Theory Handbook, Including Coordinate Systems, Differential Equations, and Their Solutions,* 2nd ed. Springer Verlag, New York, 1988.

[12] Wikipedia Contributors. *Electrical Resistivity and Conductivity.* https://en.wikipedia.org/wiki/Electrical_resistivity_and_conductivity.

[13] M. Born and E. Wolf. *Principles of Optics: Electromagnetic Theory of Propagation, Interference and Diffraction of Light,* 7th ed. Cambridge University Press, Cambridge, UK, 1999.

[14] D. Pozar. *Microwave Engineering,* 4th ed. John Wiley and Sons, New York, 2012.

[15] ITU-R. Effects of tropospheric refraction on radiowave propagation. Recommendation P.834-9, International Telecommunication Union, Geneva, 2017.

[16] ITU-R. The radio refractive index: Its formula and refractivity data. Recommendation P.453-13, International Telecommunication Union, Geneva, 2017.

[17] P. Alberoni, et al. Use of the vertical reflectivity profile for identification of anomalous propagation. *Meteorological Applications*, 8:257–266, 2001.

[18] ITU-R. Definition of terms relating to propagation in non-ionized media. Recommendation P.310-9, International Telecommunication Union, Geneva, 1994.

[19] ITU-R. Propagation by diffraction. Recommendation P.526-14, International Telecommunication Union, Geneva, 2018.

[20] B. van der Pol and H. Bremmer. XIII. The diffraction of electromagnetic waves from an electrical point source round a finitely conducting sphere, with applications to radiotelegraphy and the theory of the rainbow. Part I. *The London, Edinburgh, and Dublin Philosophical Magazine and Journal of Science*, 24(159):141–176, 1937.

[21] B. van der Pol and H. Bremmer. LXXVI. The diffraction of electromagnetic waves from an electrical point source round a finitely conducting sphere, with applications to radiotelegraphy and the theory of the rainbow. Part II. *The London, Edinburgh, and Dublin Philosophical Magazine and Journal of Science*, 24(164):825–864, 1937.

[22] ITU-R. Attenuation by atmospheric gasses. Recommendation P.676-11, International Telecommunication Union, Geneva, 2016.

[23] ITU-R. Attenuation in vegetation. Recommendation P.833-9, International Telecommunication Union, Geneva, 2016.

[24] ITU-R. Specific attenuation model for rain for use in prediction methods. Recommendation P.838-3, International Telecommunication Union, Geneva, 2005.

[25] ITU-R. Attenuation due to clouds and fog. Recommendation P.840-7, International Telecommunication Union, Geneva, 2017.

[26] H. J. Liebe, G. A. Hufford and T. Manabe. A model for the complex permittivity of water at frequencies below 1 THz. *International Journal of Infrared and Millimeter Waves*, 12(7):659–675, 1991.

Chapter 3

Antenna Systems

Chapter 2 discussed how radar systems make use of the transmission, reflection, and reception of electromagnetic energy to determine the presence, direction, distance, velocity, and other features about targets of interest to the radar. The antenna is the device for transmitting and receiving electromagnetic energy. Depending on the application, radar systems use various types of antennas, including single elements, dishes, lenses, and arrays. This chapter introduces the antenna parameters needed to describe the performance and characteristics of an antenna system. Various types of radiating elements such as linear wire, loop, and aperture antennas are then analyzed. Combining radiating elements into antenna arrays is covered, and the chapter concludes with several Python examples to demonstrate antenna theory and analysis.

3.1 ANTENNA PARAMETERS

This section deals with various parameters describing the performance and characteristics of an antenna system. Not all of the parameters are independent and may not need to be specified. In addition, certain parameters will be more important than others depending on the type of antenna system and the requirements on the system. This section will cover the more important parameters common across various types of antenna systems.

3.1.1 Radiation Pattern

The antenna pattern or radiation pattern of an antenna is defined as "The spatial distribution of a quantity that characterizes the electromagnetic field generated by an antenna" [1]. The distribution may be given as an equation, a set of measured data, or a graphical representation. The quantities most often considered are power density, radiation intensity, gain, and electric field strength. For radar applications, the antenna pattern is most often found in the far field of the antenna, as discussed in Section 3.1.1.1. The

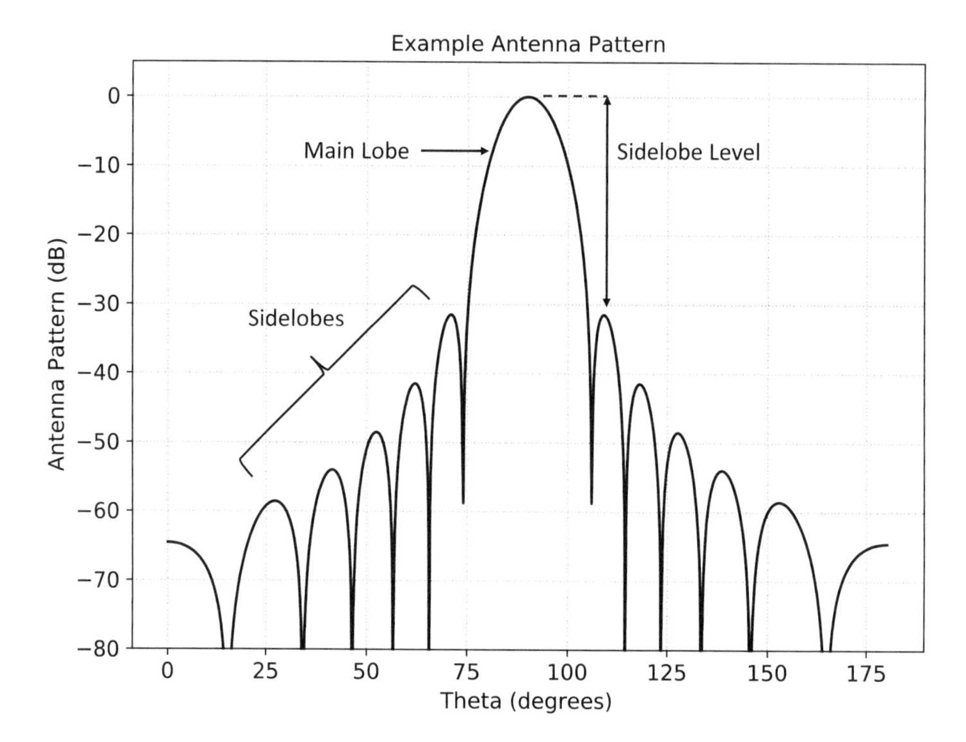

Figure 3.1 Example antenna pattern cut.

antenna pattern is usually given in terms of the angular direction coordinates (θ, ϕ). While the full three-dimensional antenna pattern may be measured or calculated, a few patterns as a function of θ, for a few values of ϕ, often provide the most useful information. These are referred to as *pattern cuts*, and an example pattern cut is given in Figure 3.1. The most important directions, also known as principal directions, are chosen for the pattern cuts. For linearly polarized antennas, the *E-plane* and *H-plane* directions are used. The E-plane is defined as "the plane containing the electric-field vector and the direction of maximum radiation," and the H-plane is "the plane containing the magnetic-field vector and the direction of maximum radiation" [1].

Referring to Figure 3.1, different angular regions of the antenna pattern are referred to as *lobes*. The IEEE defines lobes as "the portion of the antenna pattern bounded by regions of relatively weak radiation" [1]. While these regions or lobes may be divided into several subclasses, this book will use the convention of main lobe and sidelobe, as shown in Figure 3.1. The sidelobes represent energy radiated in an undesired direction. The relative strength of the sidelobes is expressed as a ratio of the power density or directivity and is called the *sidelobe level*. In most radar applications, it is desirable to reduce the sidelobe level in order to minimize jammers, clutter, and false targets from

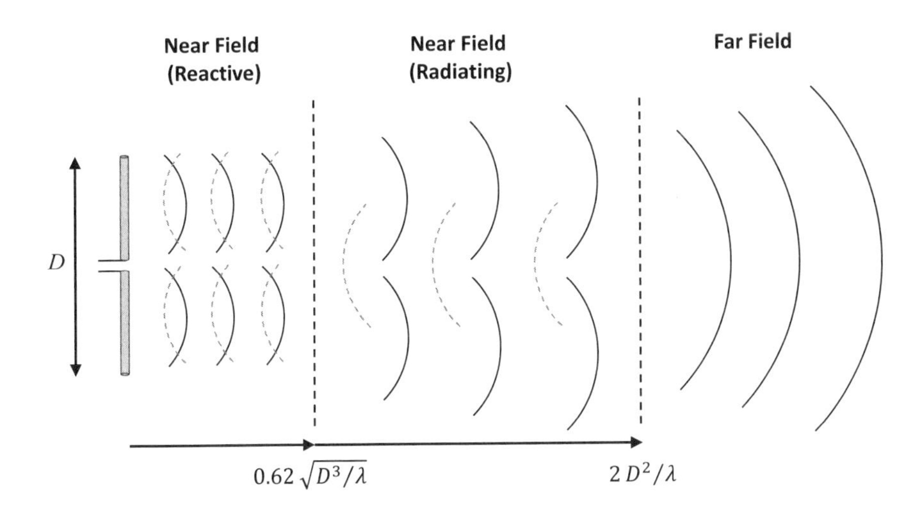

Figure 3.2 Radiation zones for an antenna with dimension D.

being introduced through these sidelobes. While it is desirable to have low sidelobe levels and a narrow beamwidth, lower sidelobes lead to a wider beamwidth and vice versa. This will be shown in Section 3.2.5.5.

3.1.1.1 Radiation Zones

The area surrounding the antenna is broken into three main regions based on the field behavior in each of the regions. These three regions are the *reactive near field*, *radiating near field*, and *far field*, as illustrated in Figure 3.2.

Reactive Near Field

The reactive near field region is the region immediately surrounding the antenna. In this region, $kr \ll 1$, and the terms varying as $1/kr^3$ are dominant. The fields in this region are similar to those of a static electric dipole and to those of a static current element. The electric and magnetic fields are out of phase and there is no time average power flow, no radiated power, and all the energy is stored in the near field around the antenna.

Radiating Near Field

The radiating near field is an intermediate region between the near-field and far-field regions. In this region, $kr > 1$, and the dominant terms are the ones constant with respect to kr. The angular field distribution in this region is dependent on the distance from the antenna. The electric and magnetic fields start to become in phase, and there is a formation of time average power flow in the outward direction. The distance to the radiating near field is given by

$$R_1 = 0.62\sqrt{D^3/\lambda}, \qquad \text{(m)}, \tag{3.1}$$

where D is the largest dimension of the antenna, as shown in Figure 3.2. If the antenna has a maximum dimension that is not large compared to the wavelength, this region may not exist.

Far Field

In the far field region, $kr \gg 1$, and the terms involving $1/r$ are dominant. In this region, there are no radial field components and the angular field distribution does not vary with the distance from the antenna. Furthermore, the radiating fields are transverse electromagnetic and have properties described in Section 2.5. In many radar applications, the target is in the far field of the radiating antenna. For antennas larger than a half wavelength, the distance to the far field is defined by the Fraunhofer distance [2], which is given by

$$R_2 = \frac{2D^2}{\lambda} \qquad \text{(m)}. \tag{3.2}$$

3.1.2 Beamwidth

With respect to the main lobe, an important figure of merit is the beamwidth, which is illustrated in Figure 3.3. The beamwidth gives a measure of the angular resolution of the antenna. This beamwidth provides information about the radar system's capability to distinguish between two adjacent targets. A smaller beamwidth provides more angular resolution between targets. In other words, a radar system with a smaller beamwidths may resolve targets that are more closely spaced in angle than a system with a larger beamwidth. The two quantities most often given are the half-power beamwidth and the first-null beamwidth, shown in Figure 3.3. The half-power beamwidth is found from the angle between the two directions where the antenna pattern is one half the maximum value. Similarly, the first-null beamwidth is the angle between the two directions where the first nulls from the main lobe of the antenna pattern occur. Other definitions are sometimes used, including the 10-dB beamwidth, which is the angular separation between the two locations that are 10 dB lower than the peak value. In practice, if the term beamwidth is used with no other qualifier, it is taken to be the half-power or 3-dB beamwidth.

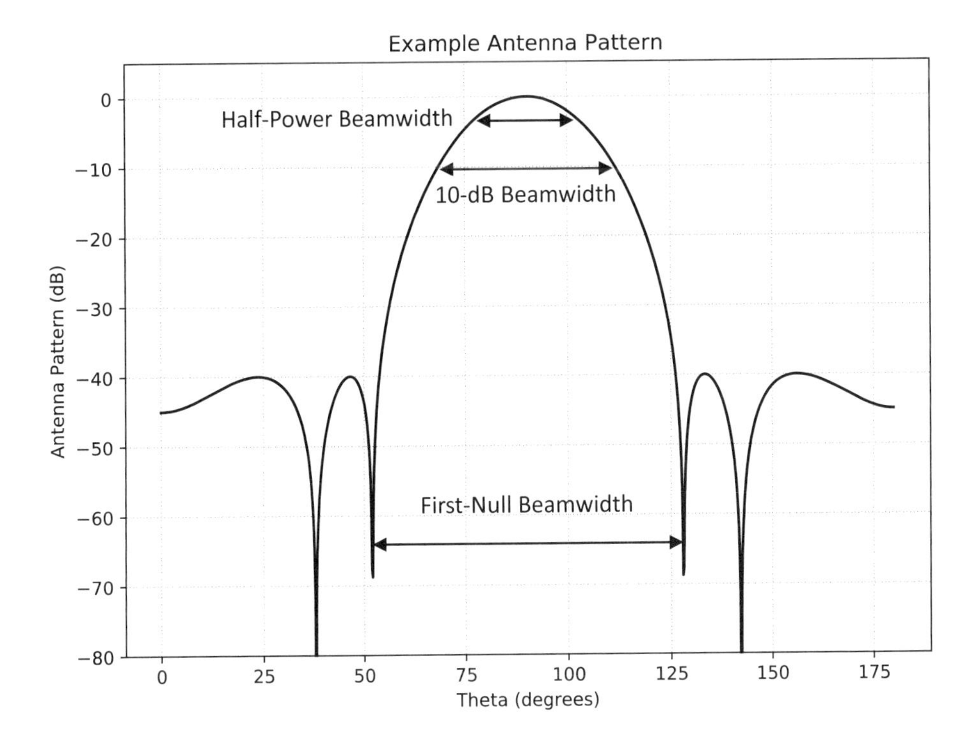

Figure 3.3 Illustration of antenna beamwidth.

3.1.3 Power Density

In radar applications, the antenna is used to focus the transmitted power on the target of interest. The *Poynting vector* is used to study the power associated with an electromagnetic wave, and is given here as

$$\vec{W} = \vec{E} \times \vec{H} \qquad \text{(W/m}^2\text{)}, \tag{3.3}$$

where

$\vec{E}$ = instantaneous electric field intensity (V/m),
$\vec{H}$ = instantaneous magnetic field intensity (A/m),
$\vec{W}$ = instantaneous Poynting vector (W/m^2).

Since radar applications involve time-varying fields, it is desirable to calculate the average power density incident on a target. Considering time harmonic electromagnetic fields, and using (2.8) and (2.9), the time average Poynting vector is written as

$$\mathbf{W} = \vec{W}_{average} = \frac{1}{2} Real\left[\mathbf{E} \times \mathbf{H}^*\right] \qquad \text{(W/m}^2\text{)}. \tag{3.4}$$

The factor of $1/2$ is necessary when **E** and **H** represent peak values. If these are RMS values, the factor of $1/2$ is omitted. Since the Poynting vector represents a power density, the average power radiated by the antenna is found by integrating the normal component of the Poynting vector over a closed surface surrounding the antenna. This is expressed as

$$P_{rad} = \oiint_S \mathbf{W} \cdot d\mathbf{s} = \frac{1}{2} \oiint_S Real\left[\mathbf{E} \times \mathbf{H}^*\right] \cdot d\mathbf{s} \qquad \text{(W)}. \tag{3.5}$$

Typically, this surface is taken to be a sphere with a large radius, such that **E** and **H** are far-field values.

3.1.4 Radiation Intensity

The radiation intensity associated with an antenna is defined by the IEEE as, "In a given direction, the power radiated from an antenna per unit solid angle" [1]. The radiation intensity is closely related to the power density and is found by multiplying the power density by the distance squared. This is written as

$$U(\theta, \phi) = r^2 \left|\mathbf{W}\right| \qquad \text{(W/sr)}. \tag{3.6}$$

The radiation intensity is a far-field quantity and may be expressed in terms of the electric field of the antenna as

$$U(\theta, \phi) = \frac{r^2}{2\eta} \left|\mathbf{E}(\theta, \phi)\right|^2 \qquad \text{(W/sr)}, \tag{3.7}$$

where

$\mathbf{E}(\theta, \phi)$ = far zone electric field of the antenna (V/m),
η = intrinsic impedance of the medium (ohms),
$U(\theta, \phi)$ = radiation intensity (W/sr).

Since the radiation intensity is the power per unit solid angle, the total power radiated by the antenna is found by integrating the radiation intensity over the entire solid angle of 4π,

$$P_{rad} = \oiint_\Omega U(\theta, \phi)\, d\Omega = \int_0^{2\pi} \int_0^{\pi} U(\theta, \phi)\, \sin\theta\, d\theta\, d\phi \qquad \text{(W)}. \tag{3.8}$$

3.1.5 Directivity

Radar systems utilize antennas to focus the transmitted energy on the target of interest. The directivity is a measure of the degree to which an antenna focuses radiation in a particular direction. Therefore, it is a direct measure of how well a radar system can focus energy onto a specific target. The IEEE defines directivity as "The ratio of the radiation intensity in a given direction from the antenna to the radiation intensity averaged over all directions" [1]. In other words, it is the ratio of the radiation intensity in a given direction to that of an isotropic radiator. If the direction is not given, it is taken to be the direction of maximum radiation. The average radiation intensity, or intensity of an isotropic radiator, is equal to the total power divided by 4π. This is expressed as

$$D(\theta,\phi) = 4\pi\,\frac{U(\theta,\phi)}{P_{rad}} = 4\pi\,\frac{U(\theta,\phi)}{\displaystyle\int_0^{2\pi}\int_0^{\pi} U(\theta,\phi)\,\sin\theta\,d\theta\,d\phi}. \tag{3.9}$$

From Section 3.1.2, it is desirable for the antenna beamwidth to be as small as practical in order to have good angular resolution. Small beamwidths indicate the energy is focused well in a particular direction and the directivity is large. This is illustrated in Figure 3.4, where the directivity of an isotropic antenna is compared to the directivity of dipoles of length $\lambda/2$ and λ. For first cut design and analysis purposes, it is convenient to use approximate expressions for the directivity rather than the exact expression given in (3.9). The directivity in the maximum direction is [2]

$$D_0 = \frac{4\pi}{\Omega}, \tag{3.10}$$

where

D_0 = maximum directivity,
Ω = beam solid angle (sr).

The beam solid angle is defined by the IEEE as "The solid angle through which all the radiated power would stream if the power per unit solid angle were constant throughout this solid angle and at the maximum value of the radiation intensity" [1]. For antenna patterns with one main lobe and relatively small sidelobes, the beam solid angle may be approximated by multiplying the half-power beamwidths in any two perpendicular planes,

$$\Omega \approx \Theta_1\,\Theta_2 \qquad \text{(sr)}, \tag{3.11}$$

where

Θ_1 = half-power beamwidth in one plane (rad),
Θ_2 = half-power beamwidth in the perpendicular plane (rad),
Ω = beam solid angle (sr).

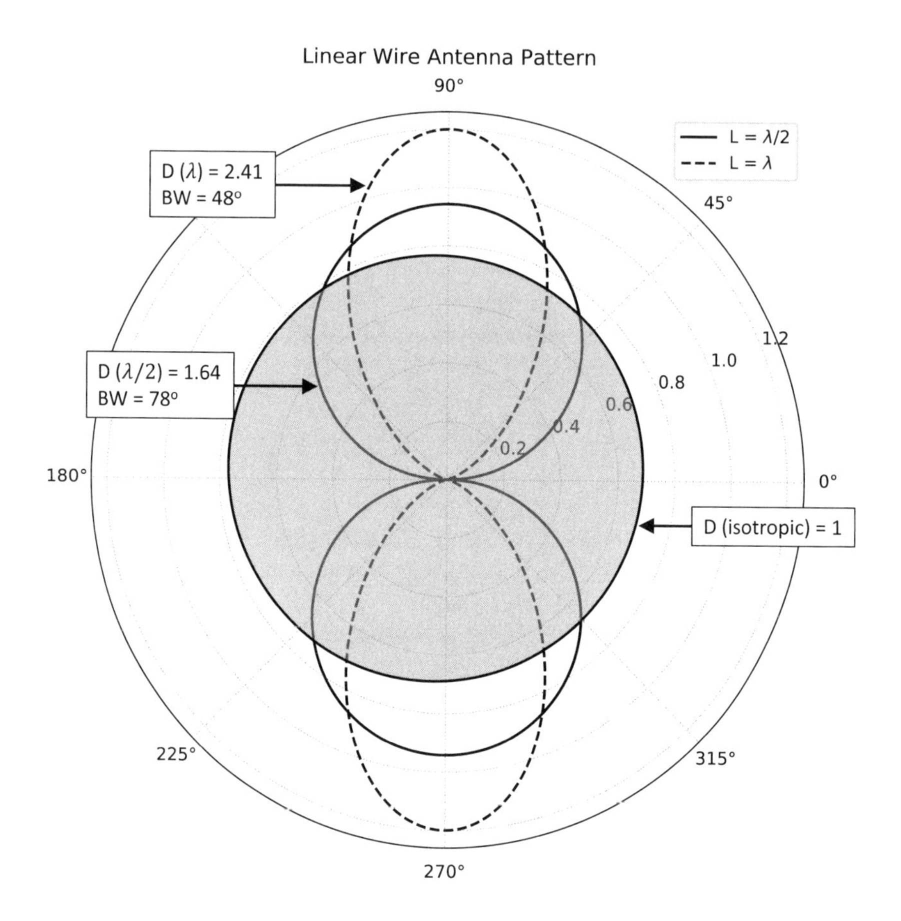

Figure 3.4 Illustration of directivity and beamwidth.

The expression for the directivity then becomes

$$D_0 \approx \frac{4\pi}{\Theta_1 \, \Theta_2}. \tag{3.12}$$

If the half-power beamwidths, $(\Theta_{1d}, \Theta_{2d})$, are known in degrees this becomes

$$D_0 \approx \frac{4\pi \, (180/\pi)^2}{\Theta_{1d} \, \Theta_{2d}} \approx \frac{41,253}{\Theta_{1d} \, \Theta_{2d}}. \tag{3.13}$$

For planar arrays, a better approximation is given by [2]

$$D_0 \approx \frac{32,400}{\Theta_{1d} \, \Theta_{2d}}. \tag{3.14}$$

3.1.6 Gain

For radar applications, a closely related but perhaps more important parameter than directivity is gain. From Section 3.1.5, the directivity is a measure of how well an antenna focuses energy in a particular direction. The gain takes into account both the directionality and the efficiency of the antenna. The antenna efficiency considers losses associated with the antenna itself. These losses are typically due to mismatch between the feed network and the antenna, thermal loss due to conductive materials, and dielectric losses. Often, these are grouped into a single term simply called efficiency. The efficiency relates the power input to the antenna to the power radiated by the antenna as

$$P_{rad} = e \cdot P_{in} \qquad \text{(W)}, \tag{3.15}$$

where

P_{in} = power input to the antenna (W),
e = total efficiency of the antenna,
P_{rad} = power radiated by the antenna (W).

The IEEE defines gain as "The ratio of the radiation intensity in a given direction to the radiation intensity that would be produced if the power accepted by the antenna were isotropically radiated" [1]. This is written mathematically as

$$G(\theta, \phi) = 4\pi \frac{U(\theta, \phi)}{P_{in}}, \tag{3.16}$$

where

$U(\theta, \phi)$ = radiation intensity in the (θ, ϕ) direction (W/sr),
P_{in} = power input to the antenna (W),
$G(\theta, \phi)$ = antenna gain in the (θ, ϕ) direction.

Gain is a function of direction, and as with directivity, if the direction is not specified, the gain is taken to be in the direction of maximum radiation. Using the antenna efficiency in (3.15) allows the gain to be written in terms of the directivity as

$$G(\theta, \phi) = e \cdot D(\theta, \phi). \tag{3.17}$$

For many practical antennas, an efficiency between 0.6 and 0.7 is reasonable and a good approximation for first cut design and analysis. Using a value in this range for the efficiency along with (3.13), an approximation for the gain is

$$G \approx \frac{30,000}{\Theta_{1d}\,\Theta_{2d}}, \tag{3.18}$$

where

Θ_{1d} = half-power beamwidth in one plane (deg),
Θ_{2d} = half-power beamwidth in the perpendicular plane (deg),
G = antenna gain in the maximum direction.

3.1.7 Bandwidth

The bandwidth of an antenna is considered to be the range of frequencies over which the antenna parameters fall within the bounds of some performance metrics. Since antenna parameters do not vary the same with frequency, there is no single characterization of bandwidth. For example, the antenna pattern of an electrically short linear dipole varies far less with frequency than the input impedance. Therefore, the input impedance is the limiting parameter when considering the antenna operating bandwidth. For narrowband antennas, the bandwidth is expressed as a percentage of the center operating frequency as

$$BW = \frac{F_{upper} - F_{lower}}{F_{center}} \times 100 \qquad (\%), \tag{3.19}$$

where

F_{center} = center operating frequency (Hz),
F_{upper} = upper limit of operation (Hz),
F_{lower} = lower limit of operation (Hz),
BW = antenna operating bandwidth (%).

For broadband antennas, the bandwidth is expressed as a ratio of the upper operating frequency to the lower operating frequency. The definition of broadband is somewhat ambiguous and greatly depends on the antenna type. If the antenna pattern and the antenna input impedance fall within performance limits over a bandwidth of 2:1, the antenna is typically considered to be broadband.

3.1.8 Polarization

Polarization refers to the orientation of the electric field of an electromagnetic wave. Since the antenna is the mechanism by which the radar radiates energy, it is the antenna design that determines the polarization of the transmitted electromagnetic wave. Analogous to the antenna pattern, the polarization is also a function of (θ, ϕ) and thus varies with direction. In general, the electric field traces out an elliptical path as a function of time. For radar applications, there are two important special cases of elliptical polarization, namely linear and circular polarization.

3.1.8.1 Linear

For linearly polarized waves, the electric field vector is always in the same plane. Within linear polarization, consider vertical and horizontal polarization, as shown in Figure 3.5.

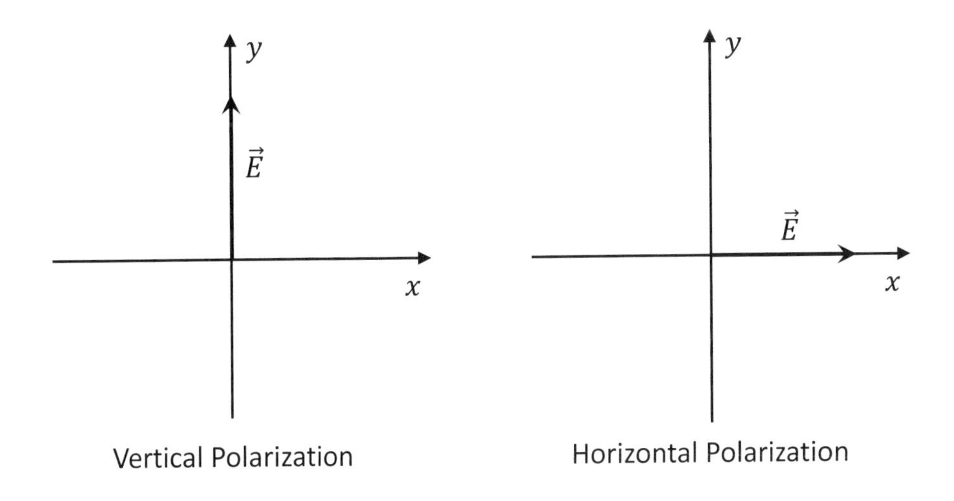

Figure 3.5 Linear polarization of the electric field.

Vertically polarized waves tend to be less affected by reflections over the transmission path, while horizontal polarization tends to be less affected by man-made objects. Linear polarization may be achieved by an electric field that either has a single component or two orthogonal components either in phase or 180 degrees out of phase.

3.1.8.2 Circular

With circularly polarized waves, the electric field vector traces a circle at a point as a function of time, completing one full turn per cycle. To achieve circular polarization the electric field must have two orthogonal components of the same amplitude and a phase difference that is an odd multiple of $\pi/2$. The sense of rotation may be either clockwise (right hand) or counterclockwise (left hand), as shown in Figure 3.6, and is determined by the phase leading component of the electric field.

3.1.8.3 Polarization Mismatch

Polarization mismatch occurs when the transmitting antenna and the receiving antenna do not have the same polarization or they are not aligned. Polarization mismatch results in a loss in received power and degradation in overall system performance. This is commonly referred to as the *polarization loss factor*. The incoming electric field may be written as

$$\mathbf{E}_i = E_i\,\hat{\mathbf{r}}_i \qquad \text{(V/m)}, \tag{3.20}$$

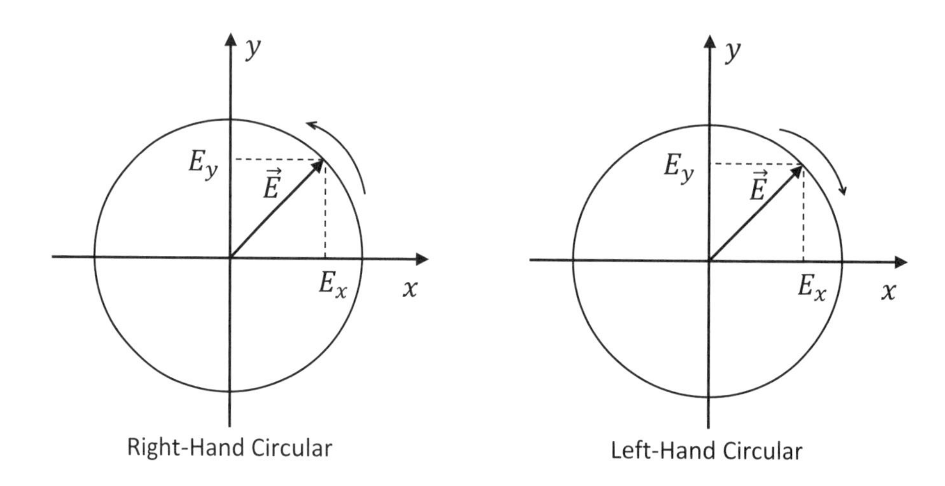

Figure 3.6 Circular polarization of the electric field.

and the polarization of the electric field of the receiving antenna is represented by

$$\mathbf{E}_a = E_a \, \hat{\mathbf{r}}_a \qquad \text{(V/m)}, \tag{3.21}$$

where $\hat{\mathbf{r}}_i$ and $\hat{\mathbf{r}}_a$ are unit vectors representing the polarization of the electric field. The polarization loss factor is then defined as

$$PLF = \left|\hat{\mathbf{r}}_i \cdot \hat{\mathbf{r}}_a\right|^2 = \cos^2(\psi). \tag{3.22}$$

As an example, consider an incoming linearly polarized wave whose electric field vector is written as $\mathbf{E}_i = E_i\,(\hat{\mathbf{x}} + \hat{\mathbf{y}})$, and is incident upon a receiving antenna whose electric field vector is given by $\mathbf{E}_a = E_a\,\hat{\mathbf{y}}$. Therefore,

$$\hat{\mathbf{r}}_i = \frac{1}{\sqrt{2}}(\hat{\mathbf{x}} + \hat{\mathbf{y}}), \qquad \hat{\mathbf{r}}_a = \hat{\mathbf{y}}. \tag{3.23}$$

The polarization loss factor is then

$$PLF = \left|\hat{\mathbf{y}} \cdot \frac{1}{\sqrt{2}}(\hat{\mathbf{x}} + \hat{\mathbf{y}})\right|^2 = \frac{1}{2}. \tag{3.24}$$

This indicates there would be a reduction in the received voltage by a factor of $1/2$.

3.2 ANTENNA TYPES

In radar systems, there are many different types of antennas used for different applications. These range from simple wire antennas to complex arrays with active elements capable of digital beamforming and pattern synthesis. The following sections cover some of the more common types of antennas used for radar applications.

3.2.1 Linear Wire Antennas

While these types of antennas are some of the oldest, they are still in use in radar systems today. They have the advantage of being cheap, easy to design, and easy to fabricate. The treatment of the infinitesimal dipole is saved for Chapter 4, and this analysis begins with small finite length dipoles.

3.2.1.1 Small Dipole

Consider a small dipole whose length is on the order of $\lambda/50 < l < \lambda/10$. For dipole antennas of this size, a good approximation of the current distribution is triangular, as shown in Figure 3.7. The triangular current distribution is written as

$$\mathbf{I}(x', y', z') = \begin{cases} I_0 \left(1 - \dfrac{2}{l} z'\right) \hat{\mathbf{z}} & \text{for } 0 \le z' \le l/2 \\ \\ I_0 \left(1 + \dfrac{2}{l} z'\right) \hat{\mathbf{z}} & \text{for } -l/2 \le z' < 0, \end{cases} \tag{3.25}$$

where

l = length of the dipole (m),
I_0 = peak current at the center of the dipole (A),
z' = source coordinate along the z direction (m),
$\mathbf{I}$ = current distribution on the dipole (A).

Following the procedure of computing the fields from sources given in Section 2.4.2, the far zone electric and magnetic fields are

$$E_r \approx 0,\ E_\phi \approx 0,\ E_\theta = j\eta \frac{k\, I_0\, l\, e^{-jkr}}{8\pi r} \sin\theta \qquad \text{(V/m)}, \tag{3.26}$$

$$H_r \approx 0,\ H_\theta \approx 0,\ H_\phi = j \frac{k\, I_0\, l\, e^{-jkr}}{8\pi r} \sin\theta \qquad \text{(A/m)}. \tag{3.27}$$

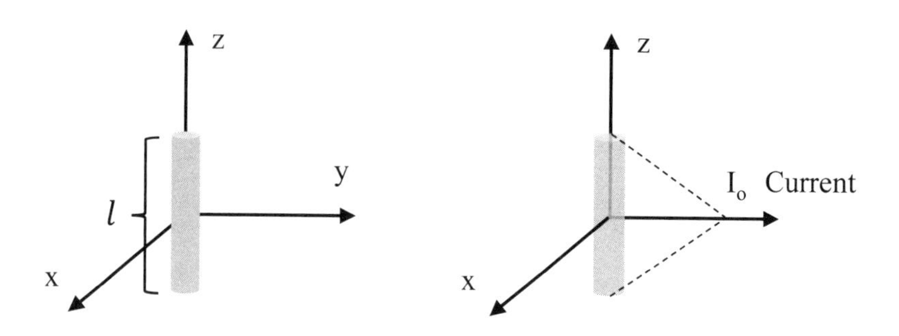

Figure 3.7 Current distribution on a small dipole of length $\lambda/50 < l < \lambda/10$.

From the fields in (3.26) and (3.27), and using the definitions of Section 3.1, the directivity is found to be 1.5, and the half-power beamwidth is 90°. The maximum effective aperture is

$$A_e = \frac{3\lambda^2}{8\pi} \qquad (\text{m}^2), \tag{3.28}$$

and the radiation resistance is

$$R_r = 20\,\pi^2 \left(\frac{l}{\lambda}\right)^2 \qquad (\text{ohms}). \tag{3.29}$$

3.2.1.2 Finite Length Dipole

The analysis of finite length dipole antennas follows the same procedure as the previous section. However, the current distribution needs to be changed to a sinusoidal form. For these types of antennas, the length is considered to be much greater than the radius of the wire. This allows the current distribution on the wire to be written as

$$\mathbf{I}(x', y', z') = \begin{cases} I_0 \left[k\left(\frac{l}{2} - z'\right)\right] \hat{\mathbf{z}} & \text{for } 0 \le z' \le l/2 \\ \\ I_0 \left[k\left(\frac{l}{2} + z'\right)\right] \hat{\mathbf{z}} & \text{for } -l/2 \le z' < 0, \end{cases} \tag{3.30}$$

Using the current distribution of (3.30) and following the procedure given in Section 2.4.2, the far zone electric and magnetic fields are

$$E_\theta = j\,\eta\, I_0\, \frac{e^{-jkr}}{2\pi r}\left[\frac{\cos\left(\frac{kl}{2}\cos\theta\right)-\cos\left(\frac{kl}{2}\right)}{\sin\theta}\right] \qquad \text{(V/m).} \qquad (3.31)$$

$$H_\phi = j\, I_0\, \frac{e^{-jkr}}{2\pi r}\left[\frac{\cos\left(\frac{kl}{2}\cos\theta\right)-\cos\left(\frac{kl}{2}\right)}{\sin\theta}\right] \qquad \text{(A/m).} \qquad (3.32)$$

Using the electric field given in (3.31) and the definition in (3.7), the radiation intensity for the finite length dipole is

$$U = \eta\frac{|I_0|^2}{8\pi^2}\left[\frac{\cos\left(\frac{kl}{2}\cos\theta\right)-\cos\left(\frac{kl}{2}\right)}{\sin\theta}\right]^2 \qquad \text{(W/sr).} \qquad (3.33)$$

Using (3.33) and (3.8), the power radiated is

$$P_{rad} = \eta\frac{|I_0|^2}{8\pi^2}\int_0^{2\pi}\int_0^{\pi}\left[\frac{\cos\left(\frac{kl}{2}\cos\theta\right)-\cos\left(\frac{kl}{2}\right)}{\sin\theta}\right]^2 \sin\theta\, d\theta\, d\phi \qquad \text{(W).} \qquad (3.34)$$

The evaluation of the integrals in (3.34) results in [2]

$$P_{rad} = \frac{\eta|I_0|^2}{4\pi}\Big[\gamma + \ln(kl) - C_i(kl) + \frac{1}{2}\sin(kl)\big(S_i(2kl) - 2S_i(kl)\big) + \frac{1}{2}\cos(kl)\big(\gamma + \ln(kl/2) + C_i(2kl) - 2C_i(kl)\big)\Big] \qquad \text{(W),} \qquad (3.35)$$

where γ is the Euler-Mascheroni constant (≈ 0.5772) and S_i, C_i are the sin and cos integrals given by

$$S_i = \int_0^x \frac{\sin z}{z}\,dz, \qquad C_i = \int_\infty^x \frac{\cos z}{z}\,dz. \qquad (3.36)$$

The radiation resistance is then found as

$$R_r = \frac{2\,P_{rad}}{|I_0|^2} = \frac{\eta}{2\pi}\Big[\gamma + \ln(kl) - C_i(kl) + \frac{1}{2}\sin(kl)\big(S_i(2kl) - 2S_i(kl)\big) + \frac{1}{2}\cos(kl)\big(\gamma + \ln(kl/2) + C_i(2kl) - 2C_i(kl)\big)\Big] \qquad \text{(ohms).} \qquad (3.37)$$

The directivity is found from (3.9), (3.33), and (3.35), and the maximum effective aperture is

$$A_e = \frac{\lambda^2 D_0}{4\pi} \qquad (\text{m}^2). \tag{3.38}$$

3.2.2 Loop Antennas

Similar to the linear wire antenna, the loop antenna is also easy to fabricate, easy to analyze, and inexpensive to build. Loops are generally classified as either electrically small, where the circumference is less than a tenth of a wavelength, or electrically large, where the circumference is about a wavelength. Small loops are not effective radiators and are usually used as probes or other receiving type elements, whereas the large circular loops are typically used in array configurations. This section covers both small circular loops and constant current loops.

3.2.2.1 Small Circular Loop

Assume a small circular loop in the x-y plane, as shown in Figure 3.8. The wire is very thin such that the current distribution is $I_\phi = I_0$. To determine the fields radiated by a small circular loop, follow the same procedure as for the linear dipole. The vector potential for this current distribution is

$$\mathbf{A} = \frac{\mu}{4\pi} \int_C \mathbf{I} \frac{e^{-jkr}}{r} \, dl' \tag{3.39}$$

where

C	=	the contour around the loop,
dl'	=	differential length on the loop (m),
$\mathbf{I}$	=	current distribution on the loop (A),
$\mathbf{A}$	=	vector potential.

Writing the current in terms of its rectangular components gives

$$\mathbf{I} = -I_0 \sin(\phi') \, \hat{\mathbf{x}} + I_0 \cos(\phi') \, \hat{\mathbf{y}} \qquad (\text{A}). \tag{3.40}$$

However, the radiated fields are usually represented in spherical coordinates. Therefore, the unit vectors in (3.40) are converted to spherical unit vectors and the current distribution is now written as

$$\mathbf{I} = I_0 \sin\theta \sin(\phi - \phi') \, \hat{\mathbf{r}} + I_0 \cos\theta \sin(\phi - \phi') \, \hat{\boldsymbol{\theta}} + I_0 \cos(\phi - \phi') \hat{\boldsymbol{\phi}} \qquad (\text{A}). \tag{3.41}$$

Using (3.39) and (3.41) allows the electric and magnetic fields to be written as

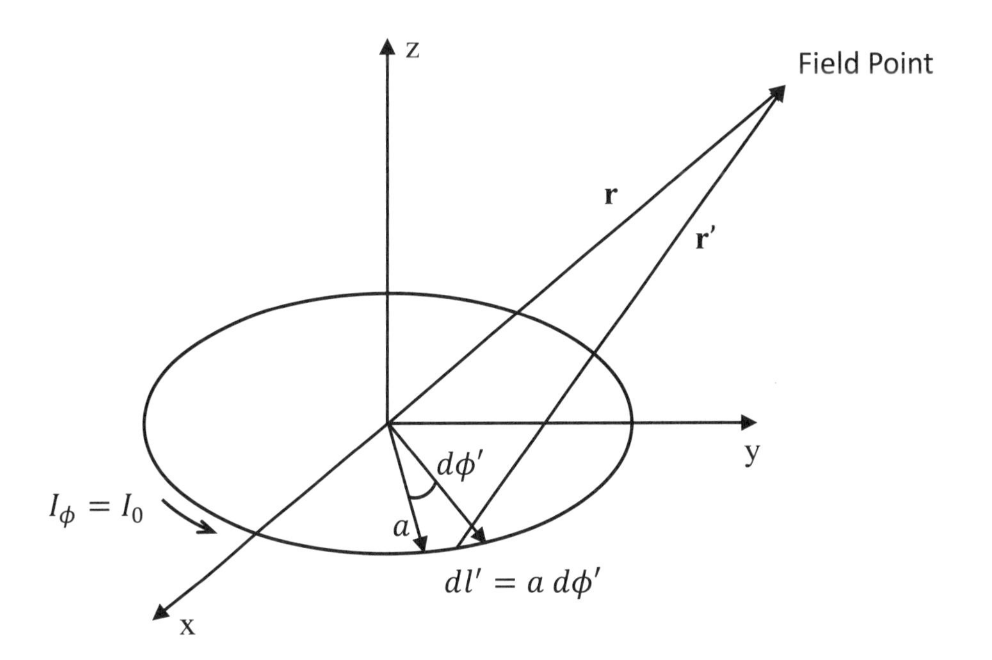

Figure 3.8 Current distribution on a small circular loop.

$$E_\phi = \eta \sin\theta \, I_0 \, (ka)^2 \, \frac{e^{-jkr}}{4r} \qquad \text{(V/m)}, \tag{3.42}$$

$$H_\theta = -\sin\theta \, I_0 \, (ka)^2 \, \frac{e^{-jkr}}{4r} \qquad \text{(A/m)}, \tag{3.43}$$

where

a = radius of the loop (m),
k = wavenumber at the operating frequency (rad/m),
η = wave impedance (ohms),
θ = angle to the field point (rad),
r = distance to the field point (m),
I_0 = current on the loop (A),
E_ϕ = electric field intensity (V/m),
H_θ = magnetic field intensity (A/m).

Using (3.42), (3.43), and (3.7), the radiation intensity is

$$U = \eta\,|I_0|^2 \sin^2\theta\,\frac{(ka)^4}{32} \qquad \text{(W/sr)}. \tag{3.44}$$

Integrating the radiation intensity over a closed sphere gives the total radiated power as

$$P_{rad} = \eta\,(ka)^4\,|I_0|^2 \frac{\pi}{12} \qquad \text{(W)}. \tag{3.45}$$

The radiation resistance is found as

$$R_r = \frac{2\,P_{rad}}{|I_0|^2} = \eta\,(ka)^4\,\frac{\pi}{6} \qquad \text{(ohms)}. \tag{3.46}$$

The radiation resistance in (3.46) is small and makes the small loop inefficient as a radiating element. The resistance can be increased by adding more turns. For a loop comprised of N turns, the radiation resistance becomes

$$R_r = \eta\,(ka)^4\,\frac{\pi}{6}\,N^2 \qquad \text{(ohms)}. \tag{3.47}$$

For example, a loop with a radius equal to $\lambda/30$ has a radiation resistance of $\approx 0.38\,\Omega$. A loop with the same radius and 10 turns has a radiation resistance of $\approx 38\,\Omega$. Using (3.44) and (3.45) the directivity of a small circular loop is

$$D_0 = 4\pi\frac{U_{max}}{P_{rad}} = 1.5, \tag{3.48}$$

and its maximum effective aperture is

$$A_e = \left(\frac{\lambda^2}{4\pi}\right) D_0 = \frac{3\lambda^2}{8\pi} \qquad \text{(m}^2\text{)}. \tag{3.49}$$

Since the antenna pattern of the small circular loop is the same as the short linear dipole, their directivity and maximum effective aperture are identical.

3.2.2.2 Constant Current Loop

Consider a circular loop antenna with a radius that is not considered small but does have a constant current distribution. Following the same procedure as for the small circular loop antenna in Section 3.2.2.1 allows the fields for a constant current loop antenna to be written as

$$E_\phi = a\,k\,\eta\,I_0\,\frac{e^{-jkr}}{2r}\,J_1(ka\sin\theta) \qquad \text{(V/m)}, \tag{3.50}$$

$$H_\theta = -a\,k\,I_0\,\frac{e^{-jkr}}{2r}\,J_1(ka\sin\theta) \qquad \text{(A/m)}, \tag{3.51}$$

where

a	=	radius of the loop (m),
k	=	wavenumber at the operating frequency (rad/m),
η	=	wave impedance (ohms),
θ	=	angle to the field point (rad),
r	=	distance to the field point (m),
I_0	=	current on the loop (A),
J_1	=	Bessel function of the first kind of order 1,
E_ϕ	=	electric field intensity (V/m),
H_θ	=	magnetic field intensity (A/m).

Using the electric field from (3.50) along with (3.7), the radiation intensity is

$$U = (a\,\omega\,\mu)^2 \,\frac{|I_0|^2}{8\,\eta}\, J_1^2(ka\sin\theta) \qquad \text{(W/sr)}, \tag{3.52}$$

where $\omega = 2\pi f$, and f is the operating frequency in Hz. The total power radiated is found by integrating (3.52) over a closed sphere, and is

$$P_{rad} = \pi\,(a\,\omega\,\mu)^2 \,\frac{|I_0|^2}{4\,\eta} \int_0^\pi J_1^2(ka\sin\theta)\,\sin\theta\,d\theta \qquad \text{(W)}. \tag{3.53}$$

Assuming the radius of the loop is larger than half a wavelength, the integral in (3.53) may be approximated, and the total power radiated is given by

$$P_{rad} \approx \pi\,(a\,\omega\,\mu)^2 \,\frac{|I_0|^2}{4\,\eta\,k\,a} \qquad \text{(W)}. \tag{3.54}$$

Using the expression in (3.54) the radiation resistance is

$$R_r = \frac{2\,P_{rad}}{|I_0|^2} = \eta\,\pi\,\frac{ka}{2} \qquad \text{(ohms)}. \tag{3.55}$$

The directivity is found to be

$$D_0 = 4\pi\frac{U_{max}}{P_{rad}} = 2\,ka\,(0.58)^2. \tag{3.56}$$

The maximum of the radiation intensity, U_{max}, is when $ka\sin\theta = 1.84$, and $J_1^2(1.84) \approx 0.58$. The maximum effective aperture is

$$A_e = \left(\frac{\lambda^2}{4\pi}\right) D_0 = \frac{\lambda^2\,ka}{4\pi}\,(0.58)^2 \qquad (\text{m}^2). \tag{3.57}$$

3.2.3 Aperture Antennas

While some aperture antennas radiate through an opening, the term *aperture* has more to do with the method by which these types of antennas are analyzed rather than their physical characteristics. For example, open-ended waveguides, horn antennas, microstrip antennas, reflector antennas, and lens antennas are all considered aperture antennas. In the previous sections, the radiation from antennas was found from the current distribution. For aperture antennas however, the radiated fields are not determined from a current distribution but rather the fields on or in the vicinity of the antenna structure. One method for accomplishing this is through the use of the *equivalence principle*, in which the actual sources are replaced by equivalent sources. These equivalent sources are mathematical and are equivalent in the sense they produce the same fields as the actual sources within a defined region. While this is outside the scope of this book, the reader is referred to a few excellent texts covering the subject [3–5]. Due to size constraints, aperture antennas are usually found at microwave and millimeter frequencies. In radar applications, the aperture opening is typically covered to protect from environmental conditions such as moisture and dirt. The following sections consider rectangular and circular apertures with various field distributions on the aperture.

3.2.3.1 Rectangular Apertures

Uniform Distribution in a Ground Plane

The rectangular aperture is one of the most common microwave antennas. Consider a rectangular aperture with a uniform field distribution on an infinite ground plane, as illustrated in Figure 3.9. The uniform field distribution in the aperture is given by

$$\mathbf{E}_a = \begin{cases} E_0\,\hat{\mathbf{y}} & \text{for} \quad -a/2 \le x' \le a/2, -b/2 \le y' \le b/2 \\ 0 & \text{otherwise} \end{cases} \qquad \text{(V/m)}, \tag{3.58}$$

where E_0 is a constant, a is the aperture width, and b is the aperture height. The far-zone radiated fields from this aperture are found by using the equivalence principle and are given here as [2]

$$E_r = 0, \quad H_r = 0, \tag{3.59}$$

$$E_\theta = j\,a\,b\,k\,E_0\,\frac{e^{-jkr}}{2\pi r}\left[\sin\phi\left(\frac{\sin X}{X}\right)\left(\frac{\sin Y}{Y}\right)\right], \tag{3.60}$$

$$E_\phi = j\,a\,b\,k\,E_0\,\frac{e^{-jkr}}{2\pi r}\left[\cos\theta\cos\phi\left(\frac{\sin X}{X}\right)\left(\frac{\sin Y}{Y}\right)\right], \tag{3.61}$$

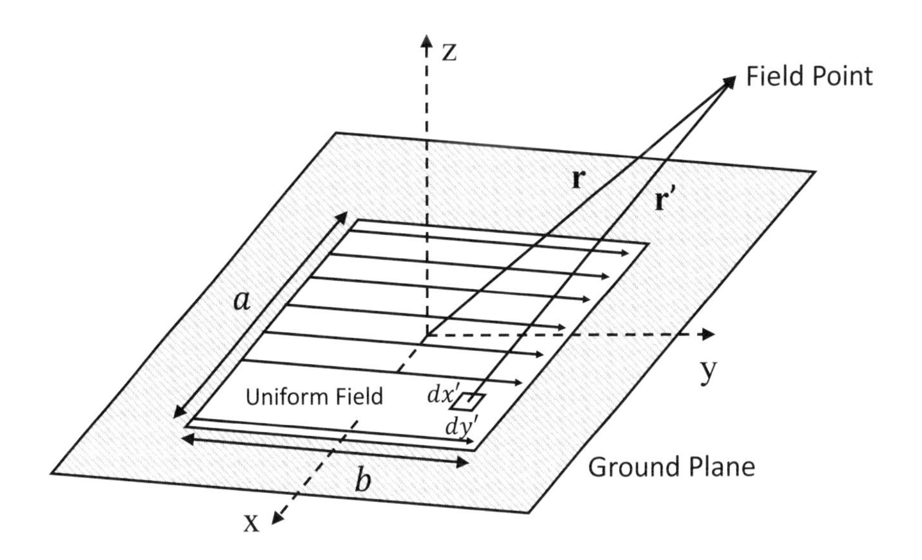

Figure 3.9 Rectangular aperture with a uniform distribution on an infinite ground plane.

$$H_\theta = -j\,a\,b\,k\,E_0\,\frac{e^{-jkr}}{2\pi\eta r}\left[\cos\theta\cos\phi\left(\frac{\sin X}{X}\right)\left(\frac{\sin Y}{Y}\right)\right], \tag{3.62}$$

$$H_\phi = j\,a\,b\,k\,E_0\,\frac{e^{-jkr}}{2\pi\eta r}\left[\sin\phi\left(\frac{\sin X}{X}\right)\left(\frac{\sin Y}{Y}\right)\right], \tag{3.63}$$

where

$$X = \frac{ka}{2}\sin\theta\cos\phi,$$
$$Y = \frac{kb}{2}\sin\theta\sin\phi,$$

θ, ϕ = angular direction to the field point (rad),
a = aperture width (m),
b = aperture height (m),
k = wavenumber (rad/m),
η = wave impedance (ohms).

From Section 3.1.1, the principal directions contain the most useful information. For this case, the E-plane is when $\phi = \pi/2$, and the H-plane corresponds to $\phi = 0$. As can be seen in (3.59), the antenna pattern in the E-plane is a function of dimension b, and the H-plane pattern is a function of the dimension a. These patterns are plotted in Figure 3.10. The first sidelobe in both the E- and H-plane is -13.26 dB below the peak of the main lobe.

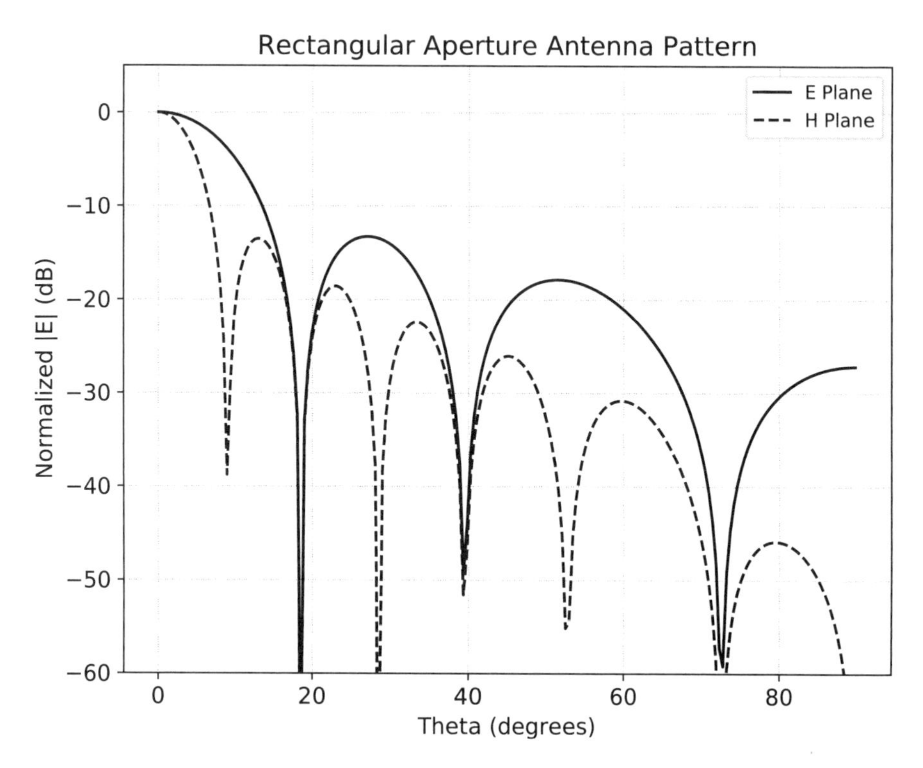

Figure 3.10 E- and H-plane antenna patterns for a rectangular aperture with a uniform distribution on an infinite ground plane, with dimensions $a = 2\lambda$ and $b = 1\lambda$.

The maximum effective aperture for this case is simply the physical area of the aperture and the maximum directivity is

$$D_0 = 4\pi \left(\frac{ab}{\lambda^2}\right). \tag{3.64}$$

Uniform Distribution in Free Space

The next rectangular aperture studied has a uniform field distribution but is now in free space rather than on an infinite ground plane, as illustrated in Figure 3.11. The field distribution for this case is given by

$$\mathbf{E}_a = \begin{cases} E_0\,\hat{\mathbf{y}} & \text{for} \quad -a/2 \leq x' \leq a/2, -b/2 \leq y' \leq b/2 \\ 0 & \text{otherwise} \end{cases} \qquad \text{(V/m)}, \tag{3.65}$$

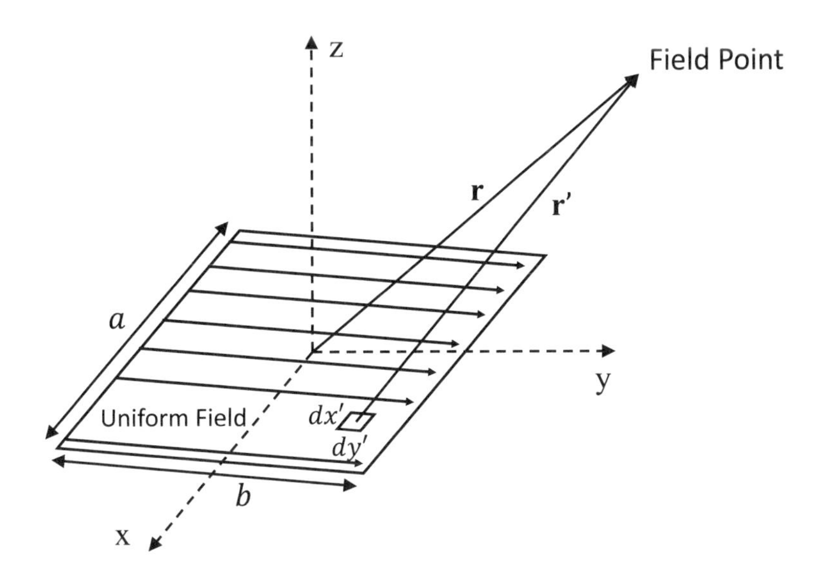

Figure 3.11 Rectangular aperture with a uniform distribution in free space.

$$\mathbf{H}_a = \begin{cases} -\dfrac{E_0}{\eta}\,\hat{\mathrm{x}} & \text{for} \quad -a/2 \le x' \le a/2, -b/2 \le y' \le b/2 \\ 0 & \text{otherwise} \end{cases} \quad \text{(A/m)}, \tag{3.66}$$

where E_0 is a constant, a is the aperture width, and b is aperture height. The far zone electric and magnetic fields are found to be

$$E_r = 0, \quad H_r = 0, \tag{3.67}$$

$$E_\theta = j\,a\,b\,k\,E_0\,\frac{e^{-jkr}}{4\pi r}\left[\sin\phi\,(1+\cos\theta)\left(\frac{\sin X}{X}\right)\left(\frac{\sin Y}{Y}\right)\right], \tag{3.68}$$

$$E_\phi = j\,a\,b\,k\,E_0\,\frac{e^{-jkr}}{4\pi r}\left[\cos\phi\,(1+\cos\theta)\left(\frac{\sin X}{X}\right)\left(\frac{\sin Y}{Y}\right)\right], \tag{3.69}$$

$$H_\theta = -j\,a\,b\,k\,E_0\,\frac{e^{-jkr}}{4\pi\eta r}\left[\cos\phi\,(1+\cos\theta)\left(\frac{\sin X}{X}\right)\left(\frac{\sin Y}{Y}\right)\right], \tag{3.70}$$

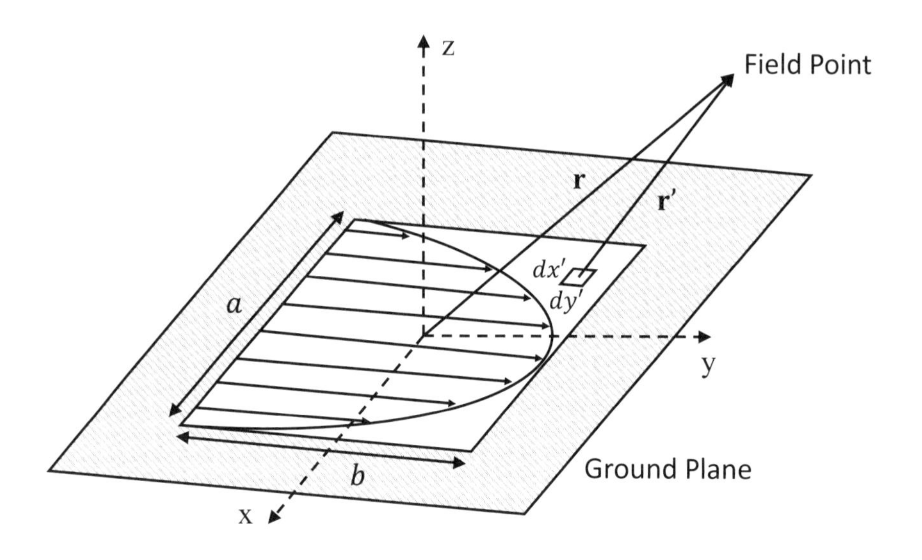

Figure 3.12 Rectangular aperture with a TE_{10} distribution in an infinite ground plane.

$$H_\phi = j\,a\,b\,k\,E_0\,\frac{e^{-jkr}}{2\pi\eta r}\left[\sin\phi\,(1+\cos\theta)\left(\frac{\sin X}{X}\right)\left(\frac{\sin Y}{Y}\right)\right]. \tag{3.71}$$

The first sidelobe, in both the E- and H-planes, and the maximum directivity are identical to that of the aperture in the ground plane given in (3.64).

TE_{10} Distribution in a Ground Plane

The final rectangular aperture analyzed has a field distribution of the TE_{10} waveguide mode and is in an infinite ground plane, as shown in Figure 3.12. The field distribution for the TE_{10} mode is given by

$$\mathbf{E}_a = \begin{cases} E_0\cos\left(\dfrac{\pi x}{a}\right)\hat{\mathbf{y}} & \text{for} \quad -a/2 \le x' \le a/2, -b/2 \le y' \le b/2 \\ 0 & \text{otherwise} \end{cases} \quad \text{(V/m)}, \tag{3.72}$$

where E_0 is a constant. The far zone electric and magnetic fields are given by

$$E_r = 0, \quad H_r = 0, \tag{3.73}$$

$$E_\theta = -j\,a\,b\,k\,E_0\,\frac{e^{-jkr}}{4r}\left[\sin\phi\left(\frac{\cos X}{X^2-(\pi/2)^2}\right)\left(\frac{\sin Y}{Y}\right)\right], \tag{3.74}$$

$$E_\phi = -j\,a\,b\,k\,E_0\,\frac{e^{-jkr}}{4r}\left[\cos\phi\,\cos\theta\left(\frac{\cos X}{X^2-(\pi/2)^2}\right)\left(\frac{\sin Y}{Y}\right)\right], \tag{3.75}$$

$$H_\theta = j\,a\,b\,k\,E_0\,\frac{e^{-jkr}}{4\eta r}\left[\cos\phi\,\cos\theta\left(\frac{\cos X}{X^2-(\pi/2)^2}\right)\left(\frac{\sin Y}{Y}\right)\right], \tag{3.76}$$

$$H_\phi = -j\,a\,b\,k\,E_0\,\frac{e^{-jkr}}{4\eta r}\left[\sin\phi\left(\frac{\cos X}{X^2-(\pi/2)^2}\right)\left(\frac{\sin Y}{Y}\right)\right]. \tag{3.77}$$

The first sidelobe in the E-plane is -13.26 dB below the peak of the main lobe, and in the H-plane is -23 dB below the peak. The maximum directivity is

$$D_0 = \frac{8}{\pi^2}\left[4\pi\left(\frac{ab}{\lambda^2}\right)\right]. \tag{3.78}$$

The maximum effective aperture for this case is

$$A_e = \frac{8ab}{\pi^2} \qquad (\text{m}^2). \tag{3.79}$$

3.2.3.2 Circular Apertures

Circular apertures are quite common as microwave antennas because they are relatively easy to construct, and closed form expressions exist for all the modes of field distributions existing on such an aperture. The analysis procedure is the same as for rectangular apertures, with the field distribution being the main difference.

Uniform Distribution in an Infinite Ground Plane

For this case, consider a circular aperture in an infinite ground plane that has a uniform field distribution on the aperture, as illustrated in Figure 3.13. The uniform field distribution in the aperture is given by

$$\mathbf{E}_a = \begin{cases} E_0\,\hat{\mathbf{y}} & \text{for} \quad \rho' \le a \\ 0 & \text{otherwise} \end{cases} \qquad (\text{V/m}), \tag{3.80}$$

where E_0 is a constant, and a is the radius of the aperture. The far zone electric and magnetic fields are found to be

$$E_r = 0, \quad H_r = 0, \tag{3.81}$$

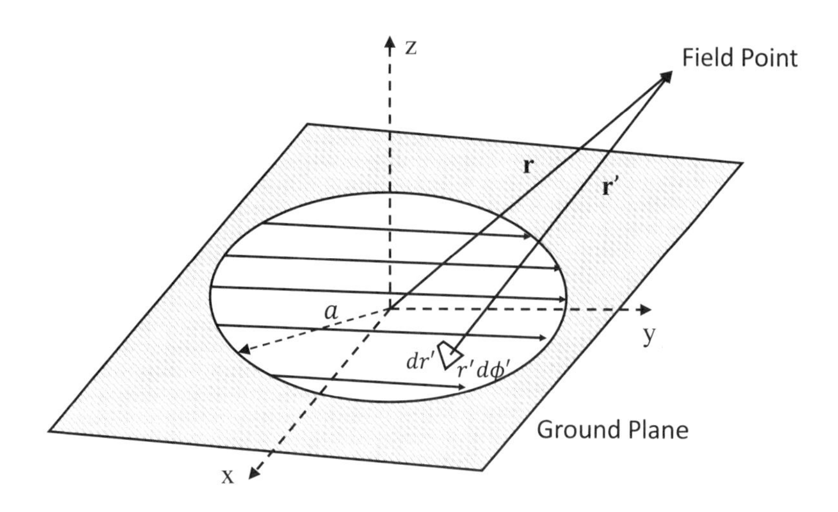

Figure 3.13 Circular aperture with a uniform distribution in a ground plane.

$$E_\theta = j\,k\,a^2\,E_0\,\frac{e^{-jkr}}{r}\left[\sin\phi\,\frac{J_1(ka\sin\theta)}{ka\sin\theta}\right], \tag{3.82}$$

$$E_\phi = j\,k\,a^2\,E_0\,\frac{e^{-jkr}}{r}\left[\cos\theta\cos\phi\,\frac{J_1(ka\sin\theta)}{ka\sin\theta}\right], \tag{3.83}$$

$$H_\theta = -j\,k\,a^2\,E_0\,\frac{e^{-jkr}}{\eta r}\left[\cos\theta\cos\phi\,\frac{J_1(ka\sin\theta)}{ka\sin\theta}\right], \tag{3.84}$$

$$H_\phi = j\,k\,a^2\,E_0\,\frac{e^{-jkr}}{\eta r}\left[\sin\phi\,\frac{J_1(ka\sin\theta)}{ka\sin\theta}\right]. \tag{3.85}$$

The principal E- and H-planes for this case occur at $\phi = \pi/2$ and $\phi = 0$, respectively. The first sidelobe in the E- and H-planes is -17.6 dB below the peak of the main lobe. The maximum effective aperture for this case is equal to the physical area of the aperture and the maximum directivity is

$$D_0 = \left(\frac{2\pi\,a}{\lambda}\right)^2. \tag{3.86}$$

TE_{11} Distribution in an Infinite Ground Plane

A common application of aperture antennas is a circular aperture on a large ground plane with the aperture fields being those of the dominant TE_{11} waveguide mode. The field distribution for this case is

$$\mathbf{E}_a = \begin{cases} E_\rho \, \hat{\boldsymbol{\rho}} + E_\phi \, \hat{\boldsymbol{\phi}} & \text{for} \quad \rho' \leq a \\ 0 & \text{otherwise} \end{cases} \qquad \text{(V/m)}, \tag{3.87}$$

where

$$\begin{aligned} E_\rho &= E_0 \, J_1(\chi'_{11} \, \rho'/a) \, \sin\phi'/\rho' \text{ (V/m)}, \\ E_\phi &= E_0 \, J_1(\chi'_{11} \, \rho'/a) \, \cos\phi' \text{ (V/m)}, \\ E_0 &= \text{constant (V/m)}, \\ \chi'_{11} &= \text{first root of the Bessel function of the first kind } (\approx 1.841). \end{aligned}$$

The far zone electric and magnetic fields are found to be

$$E_r = 0, \quad H_r = 0, \tag{3.88}$$

$$E_\theta = j \, k \, a \, E_0 \, J_1(\chi'_{11}) \, \frac{e^{-jkr}}{r} \left[\sin\phi \, \frac{J_1(ka \sin\theta)}{ka \sin\theta} \right], \tag{3.89}$$

$$E_\phi = j \, k \, a \, E_0 \, J_1(\chi'_{11}) \, \frac{e^{-jkr}}{r} \left[\cos\theta \cos\phi \, \frac{J'_1(ka \sin\theta)}{1 - (ka \sin\theta/\chi'_{11})^2} \right], \tag{3.90}$$

$$H_\theta = -j \, k \, a \, E_0 \, J_1(\chi'_{11}) \, \frac{e^{-jkr}}{\eta r} \left[\cos\theta \cos\phi \, \frac{J'_1(ka \sin\theta)}{1 - (ka \sin\theta/\chi'_{11})^2} \right], \tag{3.91}$$

$$H_\phi = j \, k \, a \, E_0 \, J_1(\chi'_{11}) \, \frac{e^{-jkr}}{\eta r} \left[\sin\phi \, \frac{J_1(ka \sin\theta)}{ka \sin\theta} \right]. \tag{3.92}$$

The first sidelobe in the E-plane is -17.6 dB below the peak of the main lobe and the first sidelobe in the H-plane is -26.2 dB below the peak. The maximum directivity is

$$D_0 = 0.836 \left(\frac{2\pi a}{\lambda} \right)^2. \tag{3.93}$$

The maximum effective aperture for this case is

$$A_e = 0.836 \, \pi a^2 \qquad (\text{m}^2). \tag{3.94}$$

3.2.4 Horn Antennas

Horn antennas are used in many applications as feed horns for reflector and lens antennas, as standard gain calibration antennas, and as elements in larger antenna arrays. The popularity of horn antennas is in part due to their simple design, relatively wide bandwidth, and moderately high gain. Since horn antennas typically have no resonant structures, they can operate over bandwidths of 10:1 and higher. Horn antennas can have many different geometries and are flared out from the waveguide opening, as shown in Figure 3.14. A sectoral horn has only one pair of sides flared, while the other pair of sides are parallel. This type of geometry produces a fan-shaped antenna pattern. The pattern is narrower in the plane of the flared sides. This section covers three of the more basic horn antenna geometries: E-plane sectoral, H-plane sectoral, and pyramidal.

3.2.4.1 E-plane Sectoral

An E-plane sectoral horn has a flared opening in the direction of the electric field, as shown in Figure 3.14. Horn antennas are analyzed in the same fashion as the aperture antennas of Section 3.2.3. In this case, the aperture is taken to be the plane of the horn

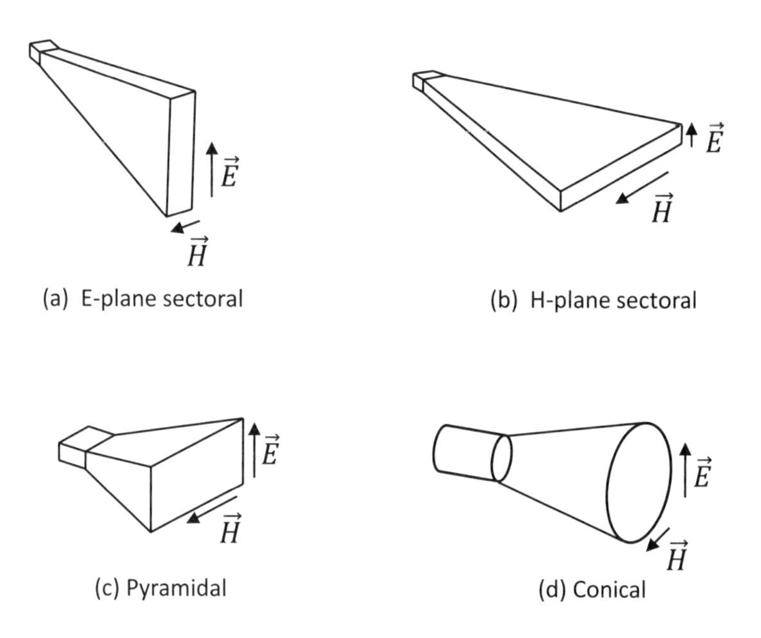

Figure 3.14 Various horn antenna geometries.

opening, and a common approximation is to assume the fields outside the aperture to be zero. If the waveguide is operating in the dominant TE_{10} mode and the length of the horn is large compared to the waveguide dimensions, then the fields at the aperture of the horn can be shown to be [2]

$$E_x = 0, \quad E_z = 0, \quad H_y = 0, \tag{3.95}$$

$$E_y = E_0 \cos\left(\frac{\pi x'}{a}\right) e^{-jky'^2/(2l)}, \tag{3.96}$$

$$H_x = -\frac{E_0}{\eta} \cos\left(\frac{\pi x'}{a}\right) e^{-jky'^2/(2l)}, \tag{3.97}$$

$$H_z = j\frac{\pi\, E_0}{\eta\, k\, a} \sin\left(\frac{\pi x'}{a}\right) e^{-jky'^2/(2l)} \tag{3.98}$$

where

E_0 = constant (V/m),
l = horn effective length (m),
a = waveguide width (m),

The far zone electric fields are given by

$$E_r = 0, \tag{3.99}$$

$$\begin{aligned} E_\theta = &-j\, a\sqrt{\pi k\, l}\, E_0 \frac{e^{-jkr}}{8\, r} \\ &\times \left[e^{jk_y^2 l/2k} \sin\phi(1+\cos\theta)\left(\frac{\cos(k_x a/2)}{(k_x a/2)^2 - (\pi/2)^2}\right)\right] \\ &\times \Big[\big(C(t_2) - C(t_1)\big) - j\big(S(t_2) - S(t_1)\big)\Big], \end{aligned} \tag{3.100}$$

$$\begin{aligned} E_\phi = &-j\, a\sqrt{\pi k\, l}\, E_0 \frac{e^{-jkr}}{8\, r} \\ &\times \left[e^{jk_y^2 l/2k} \cos\phi(1+\cos\theta)\left(\frac{\cos(k_x a/2)}{(k_x a/2)^2 - (\pi/2)^2}\right)\right] \\ &\times \Big[\big(C(t_2) - C(t_1)\big) - j\big(S(t_2) - S(t_1)\big)\Big], \end{aligned} \tag{3.101}$$

where

k_x = x component of the wavenumber, $(k \sin\theta \cos\phi)$ (rad/m),
k_y = y component of the wavenumber, $(k \sin\theta \sin\phi)$ (rad/m),
h = horn height in the flare direction (m).

C and S are the Fresnel sin and cos integrals, which are written as

$$C(x) = \int_0^x \cos\left(\frac{\pi\, t^2}{2}\right) dt, \qquad S(x) = \int_0^x \sin\left(\frac{\pi\, t^2}{2}\right) dt, \tag{3.102}$$

and the arguments t_1 and t_2 are given by

$$t_1 = \sqrt{\frac{1}{\pi k\, l}}\left(-\frac{kh}{2} - k_y\, l\right), \qquad t_2 = \sqrt{\frac{1}{\pi k\, l}}\left(\frac{kh}{2} - k_y\, l\right). \tag{3.103}$$

The radiation intensity for the E-plane sectoral horn is found from (3.99) and (3.7) as

$$U_{max} = \frac{4\, a^2\, l\, |E_0|^2}{\eta\, \lambda\, \pi^2}\left[C^2\left(\frac{h}{\sqrt{2\lambda\, l}}\right) + S^2\left(\frac{h}{\sqrt{2\lambda\, l}}\right)\right] \qquad \text{(W/sr)}. \tag{3.104}$$

The total power radiated by the E-plane sectoral horn is then found by integrating (3.104) over a close sphere, which results in

$$P_{rad} = |E_0|^2 \frac{ah}{4\eta} \qquad \text{(W)}. \tag{3.105}$$

Finally, the directivity for the E-plane sectoral horn is determined by

$$D_0 = 4\pi\, \frac{U_{max}}{P_{rad}} = \frac{64\, a\, l}{\pi\, \lambda\, h}\left[C^2\left(\frac{h}{\sqrt{2\lambda\, l}}\right) + S^2\left(\frac{h}{\sqrt{2\lambda\, l}}\right)\right]. \tag{3.106}$$

3.2.4.2 H-plane Sectoral

An H-plane sectoral horn has a flared opening in the direction of the magnetic field, as shown in Figure 3.14. The analysis procedure is very similar to that of the E-plane horn. If the waveguide is operating in the dominant TE_{10} mode and the length of the horn is large compared to the waveguide dimensions, then the fields at the aperture of the horn can be shown to be [2]

$$E_x = 0, \quad H_y = 0, \tag{3.107}$$

$$E_y = E_0 \cos\left(\frac{\pi x'}{w}\right) e^{-jkx'^2/(2l)}, \tag{3.108}$$

$$H_x = -\frac{E_0}{\eta} \cos\left(\frac{\pi x'}{w}\right) e^{-jkx'^2/(2l)}, \tag{3.109}$$

where

E_0 = constant (V/m),
l = horn effective length (m),
w = horn width in the flare direction (m).

From the aperture fields, the far zone electric fields found to be

$$E_r = 0, \tag{3.110}$$

$$E_\theta = j\, E_0 \sqrt{\frac{k\,l}{\pi}}\, \frac{e^{-jkr}}{8r}\, \sin\phi\,(1+\cos\theta) \left[\frac{\sin\, Y}{Y}\right]$$
$$\left[e^{jk_x'^2 l/2k}\Big[(C(t_2') - C(t_1')) - j(S(t_2') - S(t_1'))\Big]\right.$$
$$\left. + \, e^{jk_x''^2 l/2k}\Big[(C(t_2'') - C(t_1'')) - j(S(t_2'') - S(t_1''))\Big]\right], \tag{3.111}$$

$$E_\phi = j\, E_0 \sqrt{\frac{k\,l}{\pi}}\, \frac{e^{-jkr}}{8r}\, \cos\phi\,(1+\cos\theta) \left[\frac{\sin\, Y}{Y}\right]$$
$$\left[e^{jk_x'^2 l/2k}\Big[(C(t_2') - C(t_1')) - j(S(t_2') - S(t_1'))\Big]\right.$$
$$\left. + \, e^{jk_x''^2 l/2k}\Big[(C(t_2'') - C(t_1'')) - j(S(t_2'') - S(t_1''))\Big]\right], \tag{3.112}$$

where

$$Y = \frac{k\,h}{2}\, \sin\theta\, \sin\phi, \tag{3.113}$$

$$t_1' = \sqrt{\frac{1}{\pi k\, l}} \left(-\frac{kw}{2} - k_x'\, l\right), \qquad t_2' = \sqrt{\frac{1}{\pi k\, l}} \left(-\frac{kw}{2} + k_x'\, l\right), \tag{3.114}$$

$$t_1'' = \sqrt{\frac{1}{\pi k\, l}} \left(-\frac{kw}{2} - k_x''\, l\right), \qquad t_2'' = \sqrt{\frac{1}{\pi k\, l}} \left(-\frac{kw}{2} + k_x''\, l\right), \tag{3.115}$$

$$k'_x = k\,\sin\theta\,\cos\phi + \pi/w, \qquad\qquad k''_x = k\,\sin\theta\,\cos\phi - \pi/w. \tag{3.116}$$

The radiation intensity for the H-plane sectoral horn is then found to be

$$U_{max} = |E_0|^2\frac{h^2\,l}{4\,\eta\,\lambda}\left[\Big(C(u) - C(v)\Big)^2 + \Big(S(u) - S(v)\Big)^2\right] \qquad \text{(W/sr)}. \tag{3.117}$$

where

$$u = \frac{1}{\sqrt{2}}\left(\frac{\sqrt{\lambda\,l}}{w} + \frac{w}{\sqrt{\lambda\,l}}\right), \qquad v = \frac{1}{\sqrt{2}}\left(\frac{\sqrt{\lambda\,l}}{w} - \frac{w}{\sqrt{\lambda\,l}}\right). \tag{3.118}$$

As before, the total power radiated is found by integrating the radiation intensity over a closed sphere, which leads to

$$P_{rad} = |E_0|^2\,\frac{wh}{4\eta} \qquad \text{(W)}. \tag{3.119}$$

The directivity is then

$$D_0 = \frac{4\pi\,h\,l}{w\,\lambda}\left[\Big(C(u) - C(v)\Big)^2 + \Big(S(u) - S(v)\Big)^2\right]. \tag{3.120}$$

3.2.4.3 Pyramidal

A pyramidal horn is flared in both the direction of the electric and magnetic fields, as shown in Figure 3.14, and is the most widely used horn antenna. A very good approximate form for the fields in the horn aperture is [2]

$$E_y = E_0\,\cos\left(\frac{\pi x'}{w}\right)\,e^{-jk(x'^2/l_h + y'^2/l_e)/2}, \tag{3.121}$$

$$H_x = -\frac{E_0}{\eta}\,\cos\left(\frac{\pi x'}{w}\right)\,e^{-jk(x'^2/l_h + y'^2/l_e)/2}, \tag{3.122}$$

where

l_e = horn effective length in the E-plane (m),
l_h = horn effective length in the H-plane (m),
w = horn width (m).

The far zone electric field generated by the aperture fields in (3.121) is

$$E_r = 0, \tag{3.123}$$

$$E_\theta = j\,k\,E_0\,\frac{e^{-jkr}}{4\pi r}\,\big[\sin\phi(1+\cos\theta)\,I_1\,I_2\big], \tag{3.124}$$

$$E_\phi - j\,k\,E_0\,\frac{e^{-jkr}}{4\pi r}\,\big[\cos\phi(1+\cos\theta)\,I_1\,I_2\big], \tag{3.125}$$

where

$$I_1 = \frac{1}{2}\sqrt{\frac{\pi\,l_h}{k}}\,\Big[e^{jk_x'^2 l_h/2k}\Big(\big[C(t_2') - C(t_1')\big] - j\big[S(t_2') - S(t_1')\big]\Big)$$
$$+\,e^{jk_x''^2 l_h/2k}\Big(\big[C(t_2'') - C(t_1')\big] - j\big[S(t_2'') - S(t_1'')\big]\Big], \tag{3.126}$$

and

$$I_2 = \sqrt{\frac{\pi\,l_e}{k}}\,e^{jk_y{}^2 l_e/2k}\Big(\big[C(t_2) - C(t_1)\big] - j\big[S(t_2) - S(t_1)\big]\Big). \tag{3.127}$$

The radiation intensity for the pyramidal horn is found from (3.123) and (3.7), and is written as

$$U_{max} = |E_0|^2\,\frac{l_e\,l_h}{2\,\eta}\,\Big[\big(C(u) - C(v)\big)^2 + \big(S(u) - S(v)\big)^2\Big]$$
$$\times\left[C^2\left(\frac{h}{2\,\lambda\,l_e}\right) + S^2\left(\frac{h}{2\,\lambda\,l_e}\right)\right] \quad \text{(W/sr)}. \tag{3.128}$$

The total power radiated by the pyramidal horn antenna is found by integrating (3.128) over a closed sphere, which results in

$$P_{rad} = |E_0|^2\,\frac{wh}{4\eta} \qquad \text{(W)}. \tag{3.129}$$

The directivity for the pyramidal horn antennas is then found as

$$D_0 = \frac{8\pi\,l_e\,l_h}{w\,h}\,\Big[\big(C(u) - C(v)\big)^2 + \big(S(u) - S(v)\big)^2\Big]$$

$$\times \left[C^2 \left(\frac{h}{2\,\lambda\, l_e} \right) + S^2 \left(\frac{h}{2\,\lambda\, l_e} \right) \right]. \tag{3.130}$$

For the pyramidal horn antenna, the directivity may also be written as

$$D_P = \frac{\pi\,\lambda^2}{32\,a\,b} D_E\, D_H, \tag{3.131}$$

where

D_E = directivity of the E-plane sectoral horn,
D_H = directivity of the H-plane sectoral horn,
D_P = directivity of the pyramidal horn.

3.2.5 Antenna Arrays

The antenna patterns associated with single elements typically have low values of directivity and wide beamwidths. In many radar applications it is desirable to have an antenna with a narrow beamwidth (high directivity). This allows the radar to focus energy on the target of interest and to increase angular resolution. To increase the directivity requires an increase in the electrical size, D/λ, of the antenna. While increasing the size of a single element does increase the directivity, another method is to assemble single-element antennas into a geometrical pattern, as shown in Figure 3.15. Such a multi-element antenna is referred to as an *array*. Arrays are usually constructed of identical individual elements, although this is not necessary. The total field radiated by an array is the vector addition of the field radiated by individual elements. It is assumed the current distribution on each element is the same as if the element were isolated. To provide insight into the process of calculating the antenna pattern of arrays, a simple two-element array is first analyzed. This is followed by N-element linear arrays, circular arrays, and finally planar arrays.

3.2.5.1 Two-Element Array

Consider the two-element array shown in Figure 3.16. It consists of two infinitesimal dipoles along the x axis with a horizontal orientation. Assuming there is no coupling between the two elements, the total electrical field is found by the vector summation of the fields from the individual elements given (4.9). This is expressed as

$$\mathbf{E}_{total} = j\,\frac{\eta\, k\, I_0\, l}{4\pi} \left[\frac{e^{-j\,(kr_1 - \alpha/2)}}{r_1} \cos\theta_1 + \frac{e^{-j\,(kr_2 - \alpha/2)}}{r_2} \cos\theta_2 \right] \hat{\boldsymbol{\theta}} \;\; \text{(V/m)}, \tag{3.132}$$

where α is the phase difference between the two dipoles. Considering the observation point to be in the far field of the dipoles allows the angles and ranges for amplitude variations to be approximated as

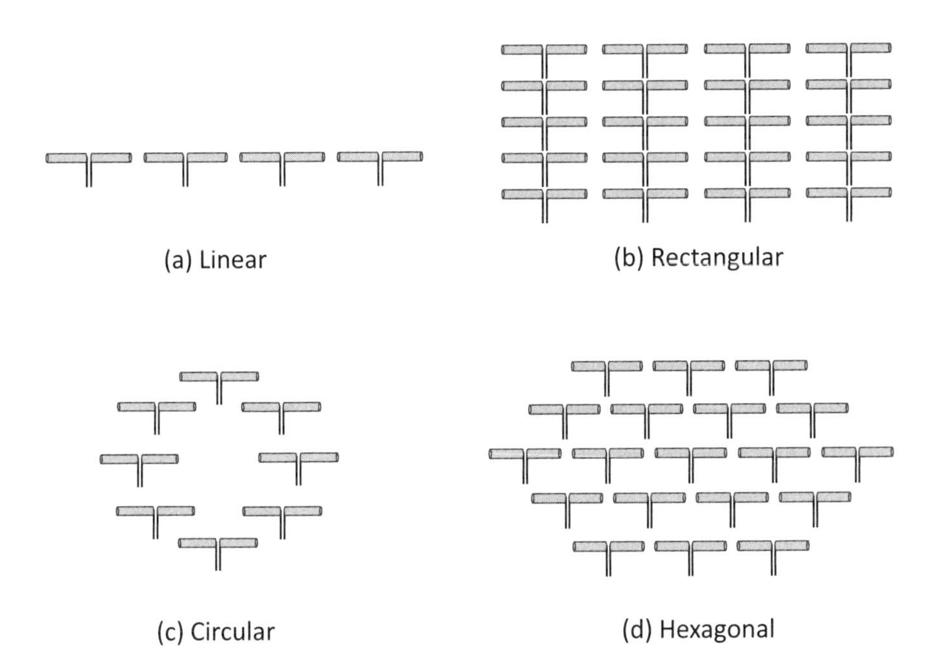

Figure 3.15 Typical array antenna geometries.

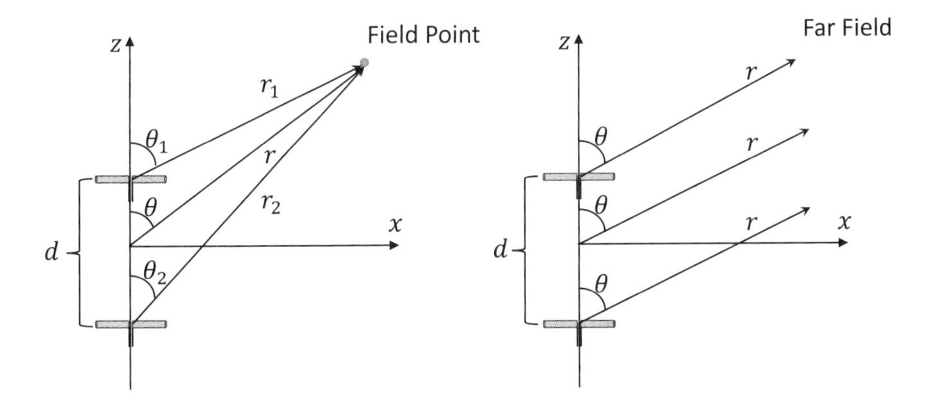

Figure 3.16 Two-element array of infinitesimal dipoles.

$$\theta_1 \approx \theta_2 \approx \theta \qquad \text{(rad)}, \tag{3.133}$$

$$r_1 \approx r_2 \approx r \qquad \text{(m)}, \tag{3.134}$$

and for phase variations

$$r_1 \approx r - \frac{d}{2}\cos\theta, \qquad r_2 \approx r + \frac{d}{2}\cos\theta \qquad \text{(m)}. \tag{3.135}$$

The expression for the electric field in (3.132) is now

$$\begin{aligned}\mathbf{E}_{total} &= j\,\frac{\eta\,k\,I_0\,l}{4\pi}\cos\theta\left[e^{j(kd\cos\theta+\alpha)/2} + e^{-j(kd\cos\theta+\alpha)/2}\right]\hat{\boldsymbol{\theta}} \\ &= \underbrace{\left[j\,\frac{\eta\,k\,I_0\,l}{4\pi}\cos\theta\right]}_{\text{Single Dipole Pattern}}\;\underbrace{\left[2\,\cos\theta\left[(kd\cos\theta+\alpha)/2\right]\right]}_{\text{Array Factor}}\hat{\boldsymbol{\theta}} \qquad \text{(V/m)}.\end{aligned} \tag{3.136}$$

From (3.136), the total field of the two element dipole array is equal to the field of a single dipole at the origin multiplied by a factor. This factor is commonly referred to as the *array factor*. More generically, the antenna pattern of an array antenna is

$$\mathbf{E}_{total} = \mathbf{E}_{element} \times \text{Array Factor}. \tag{3.137}$$

The product in (3.137) is commonly referred to as *pattern multiplication*. Although this illustration consists of only two radiating elements of identical amplitude, pattern multiplication is valid for arrays with varying number of elements, element spacing, and element amplitude and phase. The array factor is a function of many variables, including number of elements, element spacing, element excitation, and geometrical layout.

3.2.5.2 Uniform Linear Array

Extending the two-element array to N elements results in the configuration shown in Figure 3.17. For a uniform linear array, each element has the same amplitude excitation and a progressive phase shift, α. The array factor for this case is given as

$$AF = \sum_{n=1}^{N} e^{j(n-1)(kd\cos\theta+\alpha)}. \tag{3.138}$$

Letting $\psi = kd\cos\theta + \alpha$, the array factor is written as

$$AF = \sum_{n=1}^{N} e^{j(n-1)\psi}. \tag{3.139}$$

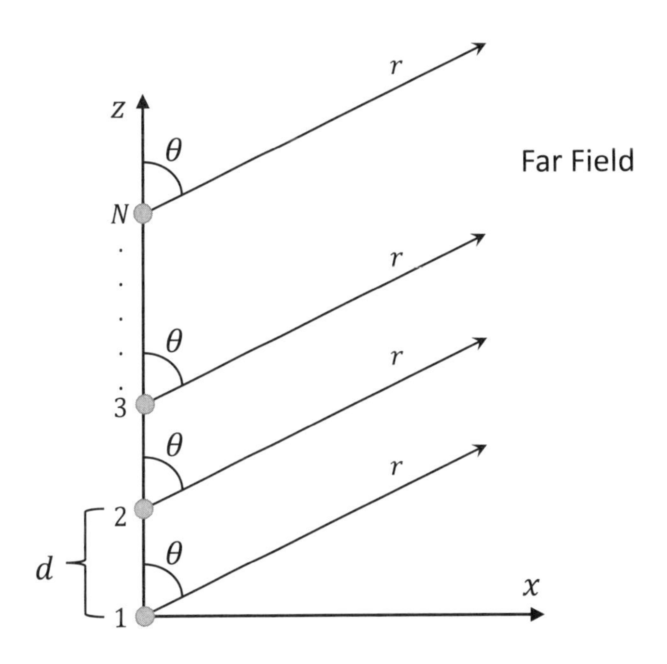

Figure 3.17 N-element uniform linear array.

The array factor in (3.139) can be written in an alternate more compact form as [2]

$$AF = \frac{1}{N}\left[\frac{\sin\left(\frac{N}{2}\psi\right)}{\sin\left(\frac{1}{2}\psi\right)}\right]. \tag{3.140}$$

The factor of $1/N$ is necessary to normalize the array factor such that the maximum always has a value of 1. The array factor in (3.140) has a maximum when

$$\psi = kd\cos\theta + \alpha = n\pi. \tag{3.141}$$

If the maximum is required to be located $0^{\circ} \leq \theta_{max} \leq 180^{\circ}$, this results in

$$\psi = kd\cos\theta_{max} = 0 \rightarrow \alpha = -kd\cos\theta_{max}. \tag{3.142}$$

Therefore, changing the progressive phase shift, α, between the elements points the maximum radiation (main lobe) to any desired direction. This type of array is known as a *phased array*. To steer the beam to any arbitrary angle would require a continuous phase shift. Often the phase shifters in radar systems have discrete values, which in turn causes degradation in the antenna pattern.

3.2.5.3 Binomial Array

To study nonuniform linear arrays, the array factor is written in a different form from (3.140). For a linear array with an even or odd number of elements, the array factor is

$$AF_{even} = \sum_{n=1}^{N} a_n \cos\big((2n-1)\,u\big), \quad AF_{odd} = \sum_{n=1}^{N+1} a_n \cos\big(2(n-1)\,u\big), \tag{3.143}$$

where

$$\begin{aligned} u &= \frac{\pi\, d}{\lambda} \cos\theta, \\ 2N &= \text{total elements for even case}, \\ 2N+1 &= \text{total elements for odd case}, \\ a_n &= \text{amplitude excitation for each element}. \end{aligned}$$

For the binomial array, the excitation coefficient for each element is found using the binomial expansion as [6]

$$\begin{aligned}(1+x)^{m-1} = 1 + (m-1)x + &\frac{(m-1)(m-2)}{2!}x^2 \\ &+ \frac{(m-1)(m-2)(m-3)}{3!}x^3 + \ldots \end{aligned} \tag{3.144}$$

The coefficients for various values of m are given by Pascal's triangle, shown in Table 3.1, which is named after French mathematician Blaise Pascal [7]. The binomial coefficients may be calculated in Python using the SciPy implementation ***scipy.special.binom(n,k)*** [8]. For example, in a six-element binomial array, the amplitude coefficients would be $a_1 = 10$, $a_2 = 5$, $a_3 = 1$. The advantage of the binomial array is very low sidelobes, especially compared to the uniform array. If the element spacing is half a wavelength ($d = \lambda/2$), then there are no sidelobes at all. The disadvantage of the binomial array is poor directivity and larger beamwidths compared to the uniform array.

3.2.5.4 Dolph-Tschebyscheff Array

The Dolph-Tschebyshceff array, sometimes written Chebyshev, is a compromise between the uniform and binomial arrays. It was originally proposed by Dolph [9], and later advanced by others [10, 11]. In this method, the designer specifies the sidelobe level,

Table 3.1
Pascal's Triangle

m						a_n					
1						1					
2					1		1				
3				1		2		1			
4			1		3		3		1		
5		1		4		6		4		1	
6	1		5		10		10		5		1

and the antenna pattern is approximated by a Tschebyscheff polynomial of order m. The order is large enough to meet the requirements for the sidelobe level. If the sidelobe level is specified to be zero, then the Dolph-Tschebyscheff design reduces to the binomial design. The Dolph-Tschebyscheff coefficients are given by

$$a_n = \frac{\cos\left[N\cos^{-1}\left\{\beta\cos\left(\frac{n\pi}{N}\right)\right\}\right]}{\cosh\left[\frac{1}{N}\cosh^{-1}\left(10^{sll/20}\right)\right]}, \tag{3.145}$$

where sll is the desired sidelobe level expressed in dB. The Dolph-Tschebyscheff coefficients may be calculated in Python using the SciPy implementation ***scipy.signal. chebwin(M,A)*** [8], where M is the number of elements in the array and A is the sidelobe level. Consider a six-element array with a sidelobe level of -20 dB, then the amplitude coefficients are found using the code in Listing 3.1. The coefficients have been normalized so the maximum value is one. Allowing the sidelobe level to become very small reduces the coefficients to those of the binomial case, as shown in Listing 3.2.

Listing 3.1 Dolph-Tschebyscheff Coefficients

```
from scipy.signal import chebwin
print(chebwin(6, -20))
...
([0.54057352,  0.77676753, 1.0, 1.0, 0.77676753, 0.54057352])
```

Listing 3.2 Dolph-Tschebyscheff Coefficients for Small Sidelobe Level

```
from scipy.signal import chebwin
print(chebwin(6, -200))
...
([0.10004548, 0.50007579, 1.0, 1.0, 0.50007579, 0.10004548])
```

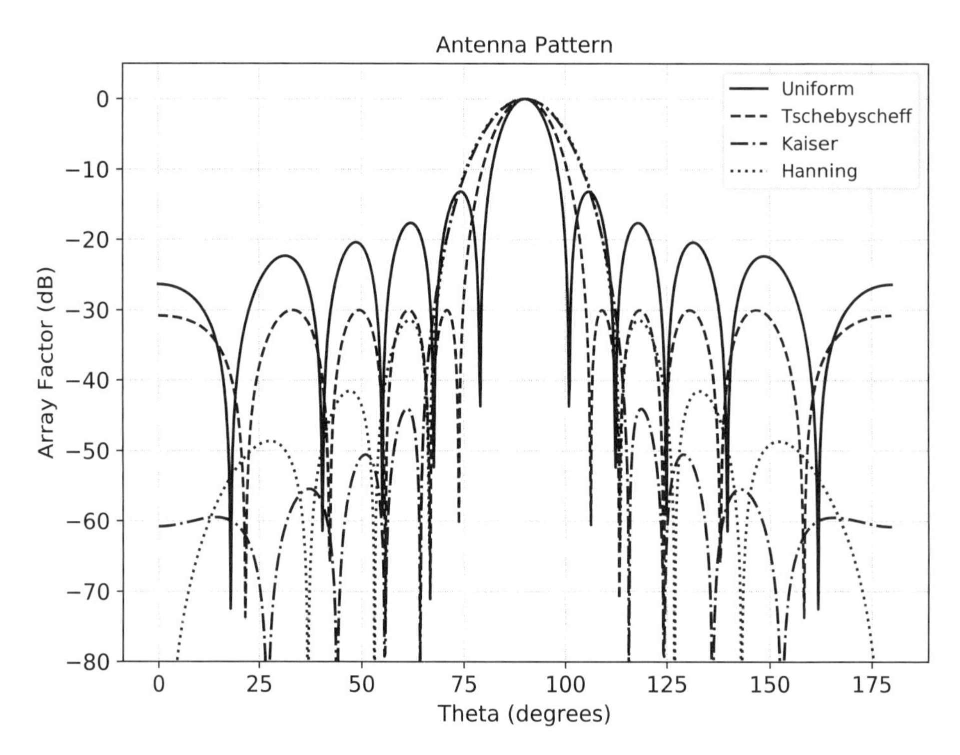

Figure 3.18 Comparison of window functions for antenna sidelobe reduction.

3.2.5.5 Windowing

The Chebyshev coefficients above provide a means of reducing sidelobes in the antenna pattern. Many other sets of coefficients may be generated, each of which has trade-offs between sidelobe level and beamwidth. These are often referred to as *window functions*. SciPy has a suite of window functions for filtering and spectral estimation, many of which may be used for finding antenna coefficients [8]. Some of the more important ones for antenna applications include *Hamming*, *Hanning*, *Kaiser*, and *Blackman-Harris*. Figure 3.18 shows a comparison between these window functions for an eight-element linear array with an element spacing of $d = \lambda/2$. The uniform window is seen to have the highest sidelobes, but also the smallest beamwidth. The Kaiser window has the lowest sidelobes and the largest beamwidth. There is a trade-off between low sidelobes and narrow beamwidths. The chosen window must meet the performance requirements of the system in both sidelobe level and beamwidth.

3.2.5.6 Planar Array

An extension of the linear array antenna is the planar arrays. Planar arrays are formed by placing the radiating elements in a grid and may take on various configurations, as shown in Figure 3.15. Planar arrays are much more versatile than single radiating elements and linear arrays. Planar arrays are used to scan the main lobe to any point in space, have lower sidelobes, may have more symmetrical patterns, and have many applications in search and tracking radars, remote sensing, and communications. Referring to Figure 3.19, the array factor for a rectangular planar array is written as

$$AF = \sum_{m=1}^{M} \sum_{n=1}^{N} a_{mn} \, e^{j(m-1)(k\, d_x \sin\theta \, \cos\phi + \alpha_x)} \, e^{j(n-1)(k\, d_y \sin\theta \, \sin\phi + \alpha_y)}, \tag{3.146}$$

where

d_x = element spacing along the x direction (m),
d_y = element spacing along the y direction (m),
α_x = progressive phase shift along the x direction (rad),
α_y = progressive phase shift along the y direction (rad),
a_{mn} = excitation coefficient for the (m, n)th element.

The main lobe of the planar array may be steered to any point in space. To have a single main beam directed toward (θ_0, ϕ_0), the progressive phase shift between each element in the x and y directions is given by

$$\alpha_x = -k\, d_x \sin\theta_0 \cos\phi_0, \qquad \alpha_y = -k\, d_y \sin\theta_0 \sin\phi_0 \qquad \text{(rad)}. \tag{3.147}$$

Window functions are also applied to planar arrays in a similar fashion as the linear arrays. For the planar case, a window function is applied in the x and y directions. The window function for the two directions may or may not be identical, depending on the requirements. For a rectangular planar array, using window functions in both the x and y directions results in very small coefficients in the corner areas of the array. Therefore, these elements are often not included in the design of the array. Typically, a circular or elliptical boundary is specified and elements outside this boundary are not used, as shown in Figure 3.20. To calculate the antenna pattern for this case, use (3.146) and set the coefficients outside the boundary equal to zero as

$$a_{mn} = \begin{cases} window(x) \times window(y) & \text{for } \; r_{mn} \le r_{boundary} \\ 0 & \text{otherwise,} \end{cases} \tag{3.148}$$

where

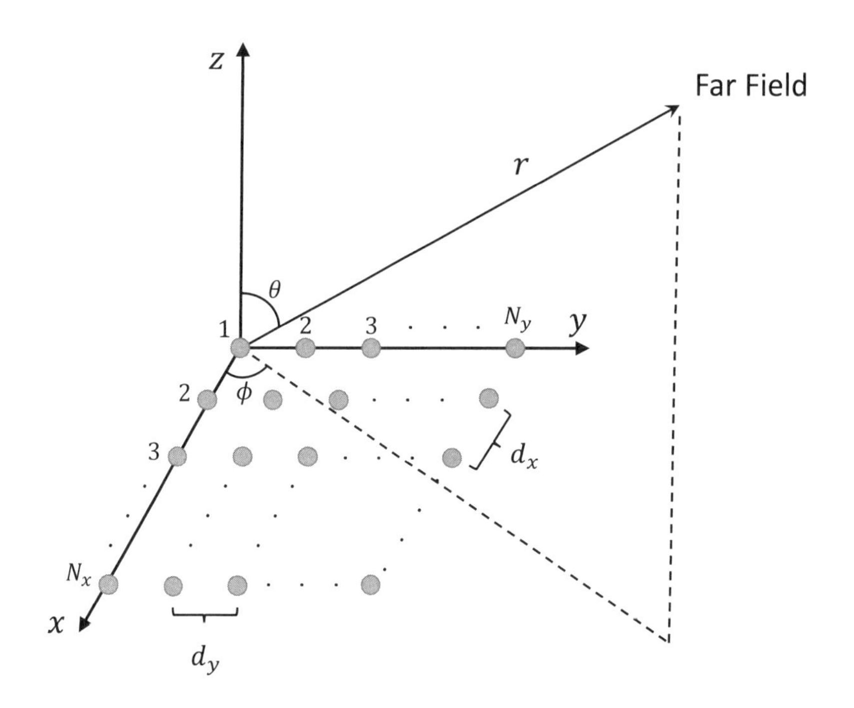

Figure 3.19 Rectangular planar array.

$window(x)$	=	window function in the x direction,
$window(y)$	=	window function in the y direction,
r_{mn}	=	distance to the (m, n)th element (m),
$r_{boundary}$	=	distance to the boundary (m),
a_{mn}	=	coefficient of the (m, n)th element.

Another form of the planar array is the hexagonal array with a circular boundary, as illustrated in Figure 3.21. The array factor for the hexagonal array is written as

$$AF = \sum_{n=1}^{N} \sum_{m=1}^{M} a_{mn} \, e^{j\psi_{mn}}, \tag{3.149}$$

where the phase of the $(m, n)^{th}$ element is given by [12]

$$\psi_{mn} = \frac{2\pi}{\lambda} \sin\theta \left[d_x \left(m + \frac{n}{2} \right) \cos\phi + d_y \, n \, \sin\phi \right] \qquad \text{(rad)}. \tag{3.150}$$

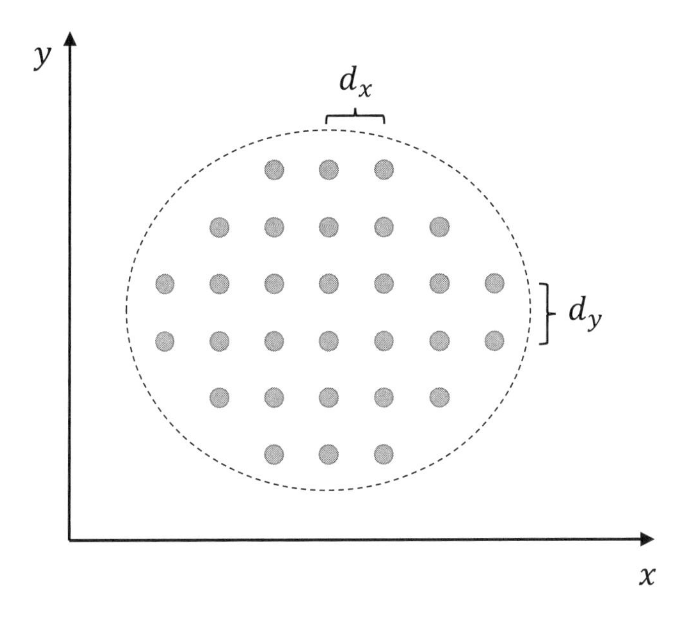

Figure 3.20 Planar array with elliptical boundary.

3.2.5.7 Circular Array

The final antenna array type presented is the circular array, shown in Figure 3.22. For this type of array, the radiating elements are placed on a circular path with spacing of $a\,\Delta\phi$, where a is the radius, and $\Delta\phi$ is the angular difference between the elements. The normalized electric field is written as

$$E(r,\theta,\phi) = \sum_{n=1}^{N} a_n \frac{e^{-jkr_n}}{r_n} \qquad \text{(V/m)}, \tag{3.151}$$

where

$$r_n = \sqrt{r^2 + a^2 - 2\,a\,r\,\cos\psi_n} \qquad \text{(m)}. \tag{3.152}$$

For phase terms, r_n is approximated by

$$r_n \approx r - a\,\sin\theta\,\cos(\phi - \phi_n) \qquad \text{(m)}, \tag{3.153}$$

where ϕ_n is the angular position of the nth element, and is expressed as

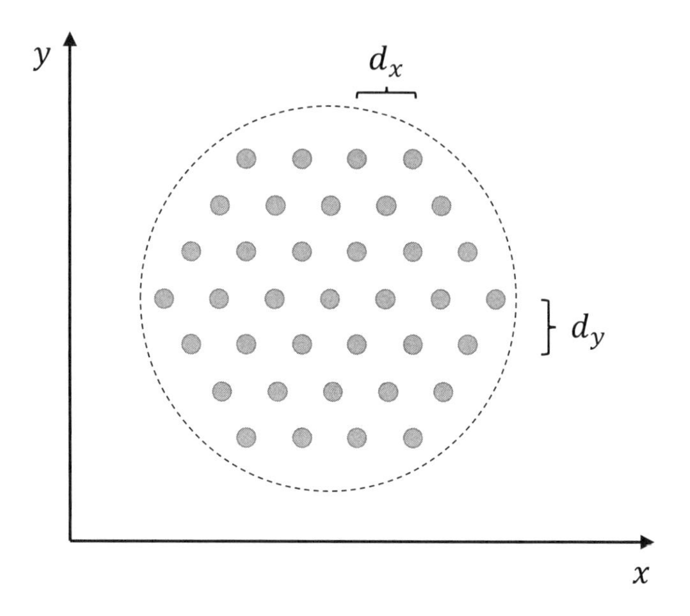

Figure 3.21 Hexagonal array with circular boundary.

$$\phi_n = \frac{2\pi\, n}{N} \qquad \text{(rad)}, \tag{3.154}$$

and for amplitude $r_n \approx r$. The coefficients, a_n, are represented by an amplitude and phase as

$$a_n = A_n\, e^{j\alpha_n}, \tag{3.155}$$

where A_n is the amplitude and α_n is the phase. The array factor for the circular array is now written as

$$AF(\theta, \phi) = \sum_{n=1}^{N} A_n\, e^{j[ka \sin\theta\, \cos(\phi - \phi_n) + \alpha_n]}. \tag{3.156}$$

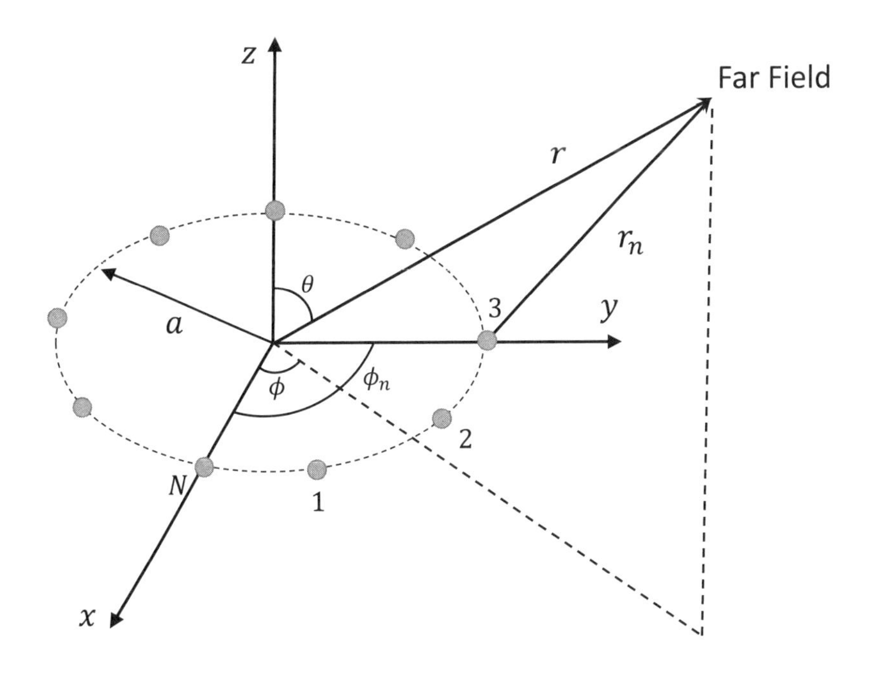

Figure 3.22 Circular array geometry.

3.3 EXAMPLES

The sections below illustrate the concepts of this chapter with several Python examples. The examples for this chapter are in the directory *pyradar\Chapter03* and the matching MATLAB examples are in the directory *mlradar\Chapter03*. The reader should consult Chapter 1 for information on how to execute the Python code associated with this book.

3.3.1 Finite Length Dipole

Consider a dipole of length of 0.4 m, with a current of 1 A operating at a frequency of 1 GHz. Calculate the total power radiated, radiation resistance, beamwidth, directivity, and maximum effective aperture. Also calculate and display the antenna pattern.

Solution: The solution to the above example is given in the Python code *linear_wire_example.py* and in the MATLAB code *linear_wire_example.m*. Running the Python example code displays a GUI, allowing the user to enter the length, current, and operating frequency of the wire antenna. The GUI allows the user to select from infinitesimal, small,

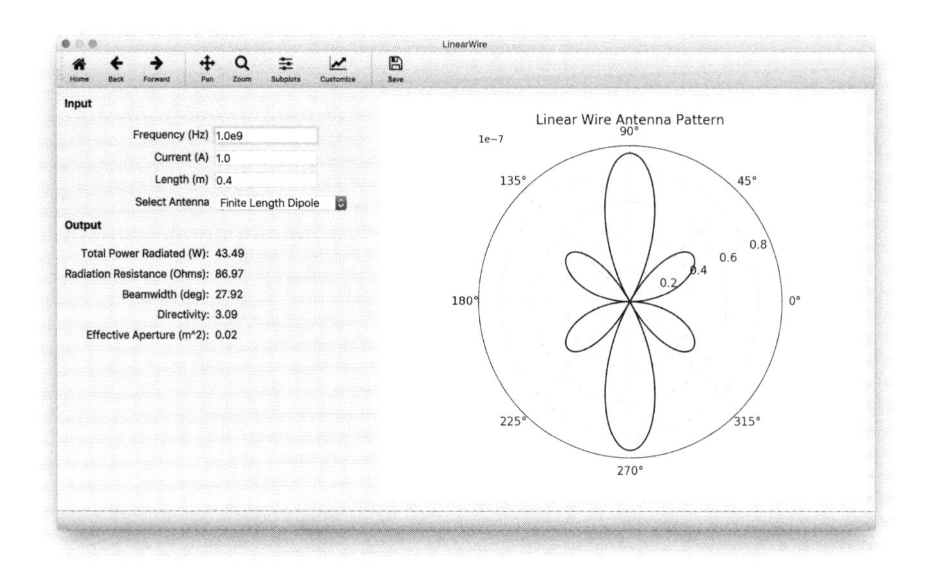

Figure 3.23 The antenna pattern and operating parameters of a finite length dipole calculated by *linear_wire_example.py*.

and finite length dipoles. The code then calculates and displays the total power radiated, radiation resistance, beamwidth, directivity, and maximum effective aperture. The GUI also displays a plot of the antenna pattern in polar format, as shown in Figure 3.23.

3.3.2 Circular Loop

For this example, calculate the total power radiated, radiation resistance, beamwidth, directivity, and maximum effective aperture of a circular loop antenna operating at a frequency of 1 GHz with a radius of 0.1 m and a constant current of 1 A. Also calculate and display the antenna pattern.

Solution: The solution to the above example is given in the Python code *loop_example.py* and in the MATLAB code *loop_example.m*. Running the Python example code displays a GUI, allowing the user to enter the radius, current, and operating frequency of the loop antenna. The GUI allows the user to select from very small loops and circular loops with a constant current. The code then calculates and displays the total power radiated, radiation resistance, beamwidth, directivity, and maximum effective aperture. The GUI also displays a plot of the antenna pattern in polar format, as shown in Figure 3.24.

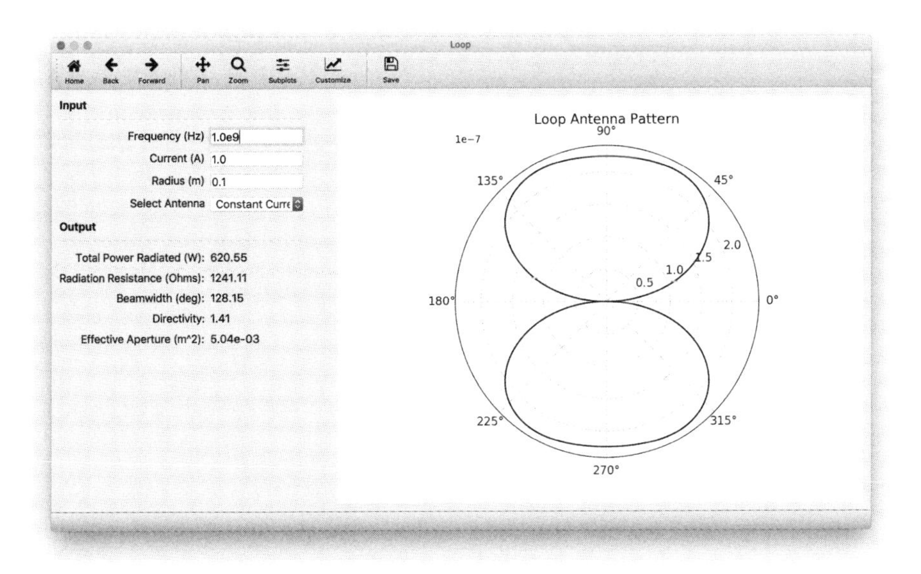

Figure 3.24 The antenna pattern and operating parameters of a circular loop of constant current calculated by *loop_example.py*.

3.3.3 Rectangular Aperture

For this example, calculate the directivity and the E- and H-plane sidelobe levels for a rectangular aperture in an infinite ground plane with a width of 0.5 m and a height of 0.3 m operating in the dominant TE^{10} mode at a frequency of 1 GHz. Also, create a 2D contour plot of the antenna pattern as well as line plots of the principal E- and H-plane patterns.

Solution: The solution to the above example is given in the Python code *rectangular_aperture_example.py* and in the MATLAB code *rectangular_aperture_example.m*. Running the Python example code displays a GUI, allowing the user to enter the width, height, and operating frequency of the aperture as well as the field distribution in the aperture. The code then calculates and displays the directivity and the E- and H-plane sidelobe levels. The GUI also displays the color and contour plots of the antenna pattern and line plots of the principal E- and H-plane patterns. The line and contour plots are shown in Figures 3.25 and 3.26, respectively.

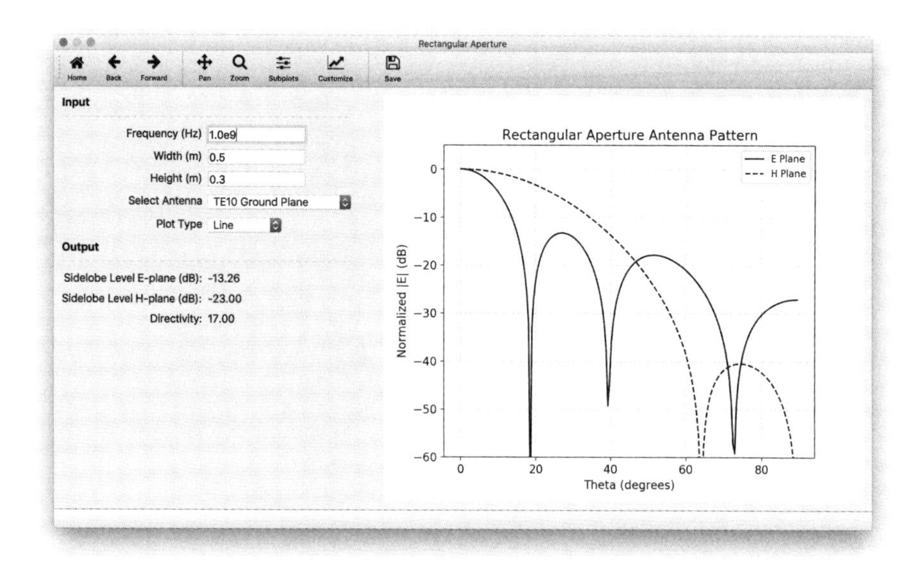

Figure 3.25 The principal E- and H-plane patterns of a rectangular aperture operating in the dominant TE^{10} mode calculated by *rectangular_aperture_example.py*.

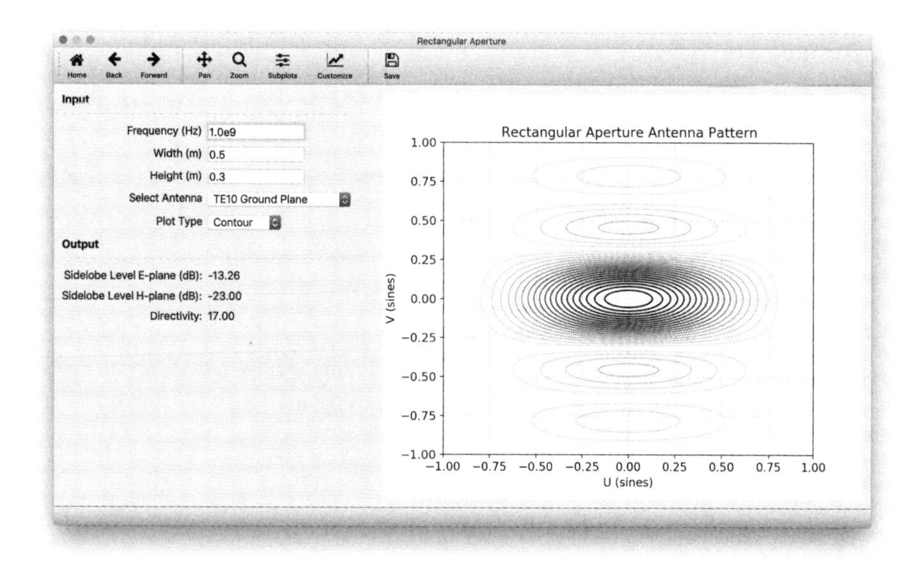

Figure 3.26 The antenna pattern and operating parameters of a rectangular aperture operating in the dominant TE^{10} mode calculated by *rectangular_aperture_example.py*.

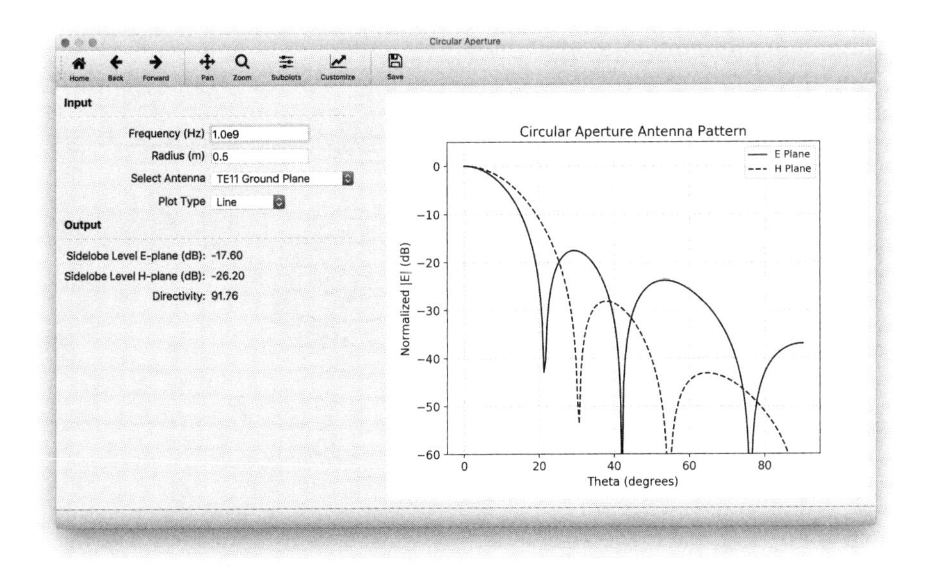

Figure 3.27 The principal E- and H-plane patterns of a circular aperture operating in the dominant TE^{11} mode calculated by *circular_aperture_example.py*.

3.3.4 Circular Aperture

Consider a circular aperture of radius 0.5 m in an infinite ground plane. The field distribution in the aperture is that of the dominant TE^{11} mode, and the operating frequency is 1 GHz. Calculate the directivity and sidelobe levels in the E- and H-planes. Create a 2D contour plot of the antenna pattern as well as line plots of the principal E- and H-plane patterns.

Solution: The solution to the above example is given in the Python code *circular_aperture_example.py* and in the MATLAB code *circular_aperture_example.m*. Running the Python example code displays a GUI, allowing the user to enter the radius and operating frequency of the aperture as well as the field distribution in the aperture. The code then calculates and displays the directivity and the E- and H-plane sidelobe levels. The GUI also displays color and contour plots of the antenna pattern and line plots of the principal E- and H-plane patterns. The line and contour plots are shown in Figures 3.27 and 3.28, respectively.

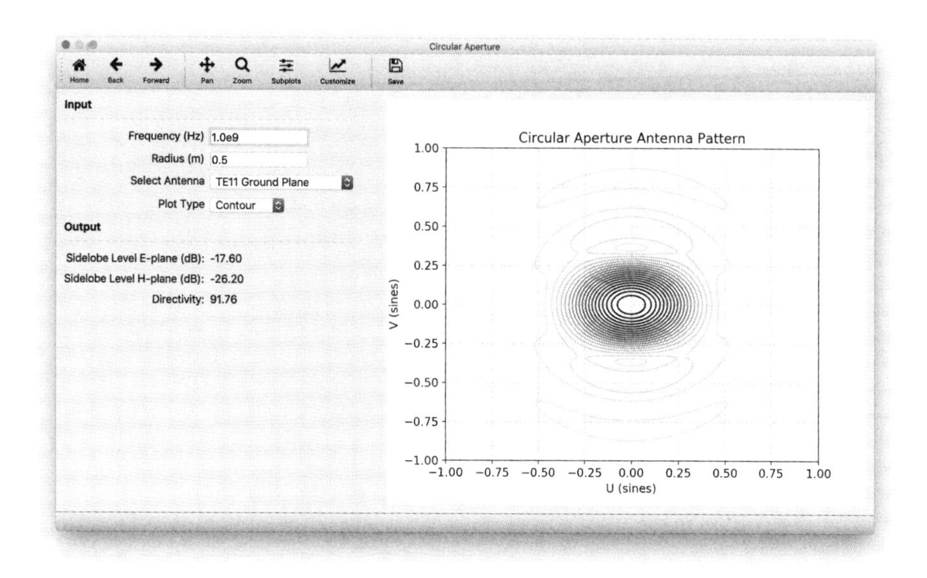

Figure 3.28 The antenna pattern and operating parameters of a circular aperture operating in the dominant TE^{11} mode calculated by *circular_aperture_example.py*.

3.3.5 Pyramidal Horn

In this example, calculate the normalized radiated power and directivity for a pyramidal horn antenna with a waveguide width of 0.2λ, waveguide height of 0.3λ, horn width of 5.5λ, horn height of 2.75λ, E-plane effective length of 6.0λ, and H-plane effective length of 6.0λ. Also, create a 2D contour plot of the antenna pattern as well as line plots of the principal E- and H-plane patterns.

Solution: The solution to the above example is given in the Python code *horn_example.py* and in the MATLAB code *horn_example.m*. Running the Python example code displays a GUI, allowing the user to enter the parameters specified above as well as choosing between E-plane sectoral, H-plane sectoral, and pyramidal horn geometries. The code then calculates and displays the directivity and normalized power radiated. The GUI also displays color and contour plots of the antenna pattern and line plots of the principal E- and H-plane patterns. Figures 3.29 and 3.30 show the lines plots and contour plot.

3.3.6 Tschebyscheff Linear Array

For this example, plot the array factor for a Tschebyscheff linear array consisting of 31 elements with a spacing of $\lambda/4$. The desired sidelobe level is -30 dB, and the main lobe

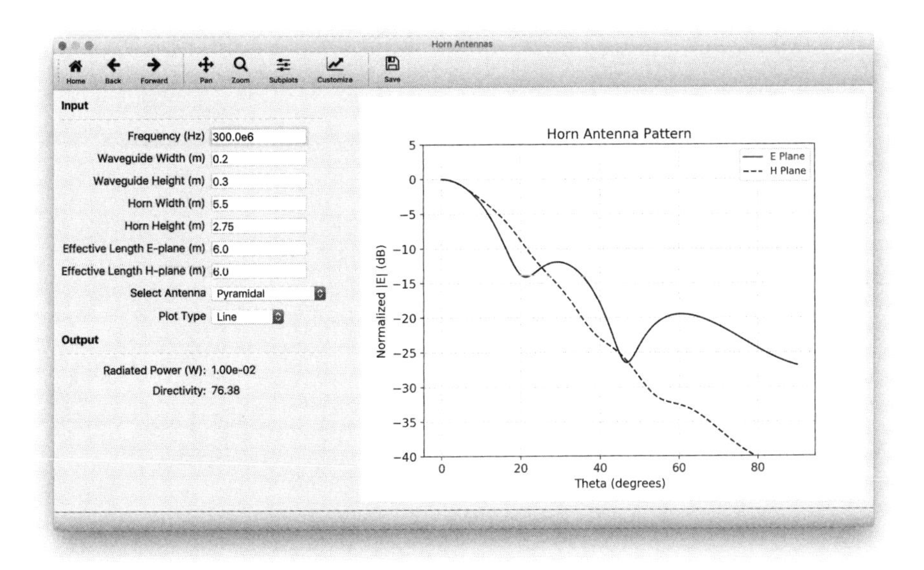

Figure 3.29 The principal E- and H-plane patterns of a pyramidal horn antenna calculated by *horn_example.py*.

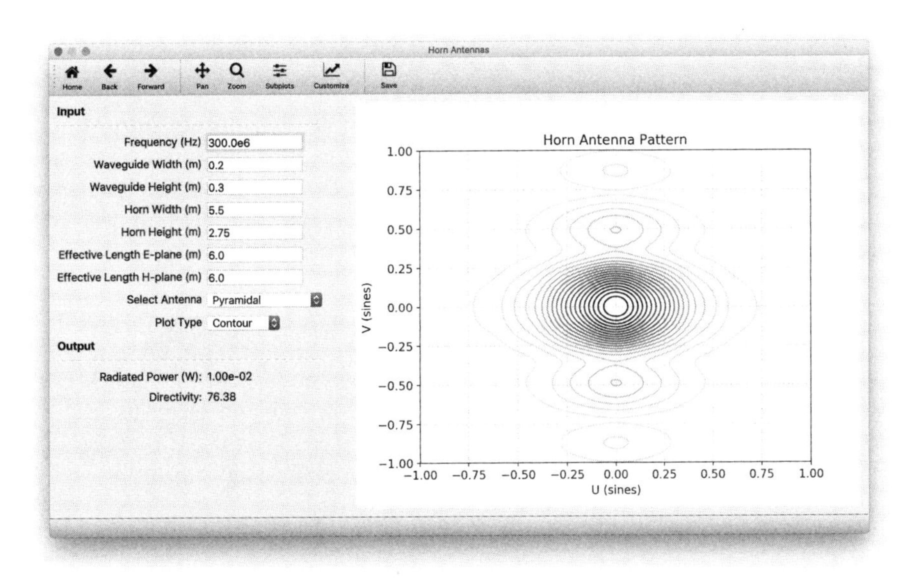

Figure 3.30 The antenna pattern and operating parameters of a pyramidal horn antenna calculated by *horn_example.py*.

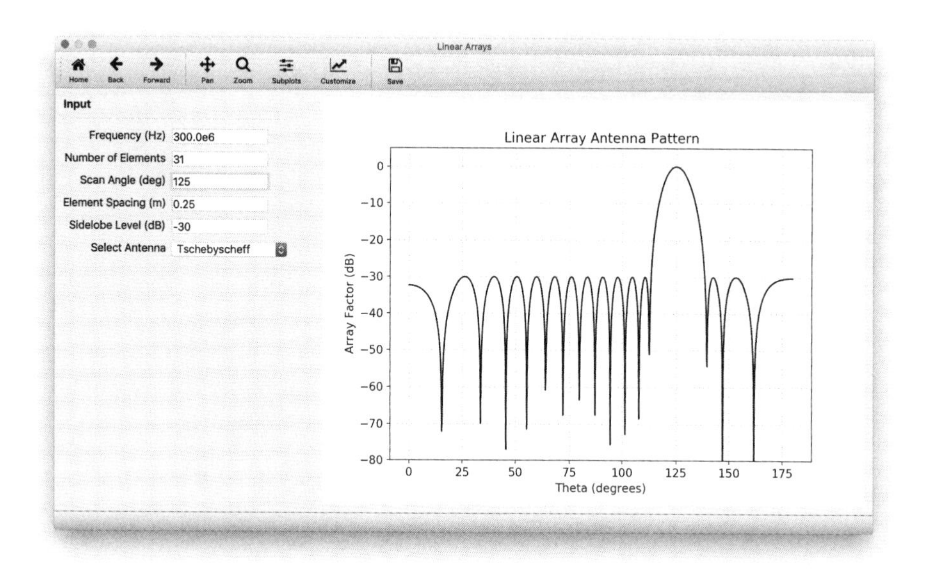

Figure 3.31 The array factor for a linear Tschebyscheff array calculated by *linear_array_example.py*.

is to be scanned to 125°.

Solution: The solution to the above example is given in the Python code *linear_array_example.py* and in the MATLAB code *linear_array_example.m*. Running the Python example code displays a GUI, allowing the user to enter the parameters specified above as well as choosing between several window types, including Tschebyscheff. The code then calculates and displays the array factor for the specified linear array, as shown in Figure 3.31.

3.3.7 Planar Array

Plot the array factor for a rectangular uniform planar array with 11 elements in the x direction, 21 elements in the y direction, element spacing of $\lambda/2$ in the x direction, $\lambda/4$ in the y direction, scan angle of 30° in the θ direction , and 45° in the ϕ direction.

Solution: The solution to the above example is given in the Python code *planar_array_example.py* and in the MATLAB code *planar_array_example.m*. Running the Python example code displays a GUI, allowing the user to enter the parameters specified above as well as choosing plot types. The code then calculates and displays the array factor as

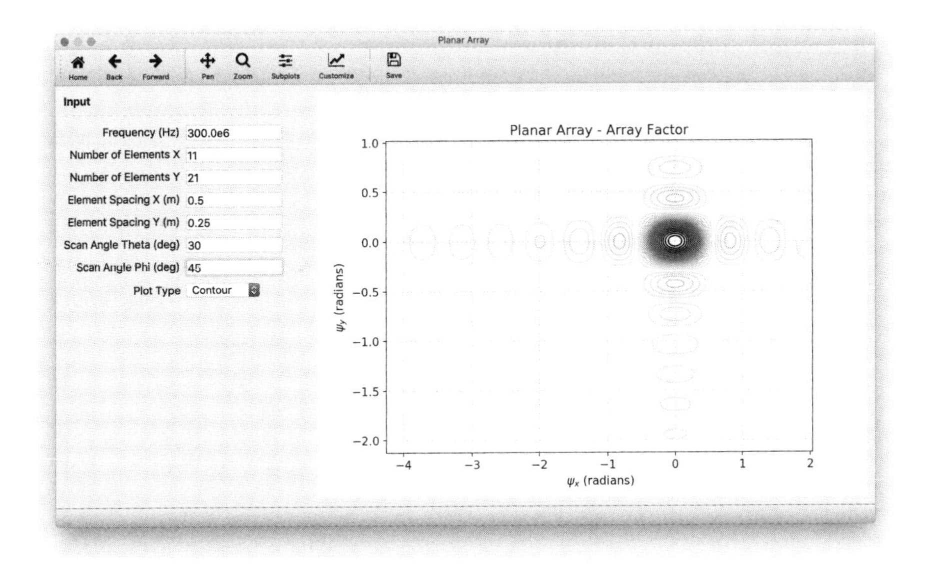

Figure 3.32 The array factor for a rectangular uniform planar array calculated by *planar_array_example.py*.

either a color, contour, or line plot for the specified rectangular uniform planar array. The contour plot is shown in Figure 3.32.

3.3.8 Circular Array

Plot the array factor for a circular array with 40 elements, radius of 1.1λ, scan angle of 30° in the θ direction, and 30° in the ϕ direction.

Solution: The solution to the above example is given in the Python code *circular_array_example.py* and in the MATLAB code *circular_array_example.m*. Running the Python example code displays a GUI, allowing the user to enter the parameters specified above as well as choosing plot types. The code then calculates and displays the array factor as either a color, contour, or line plot for the specified circular array. The contour plot is shown in Figure 3.33.

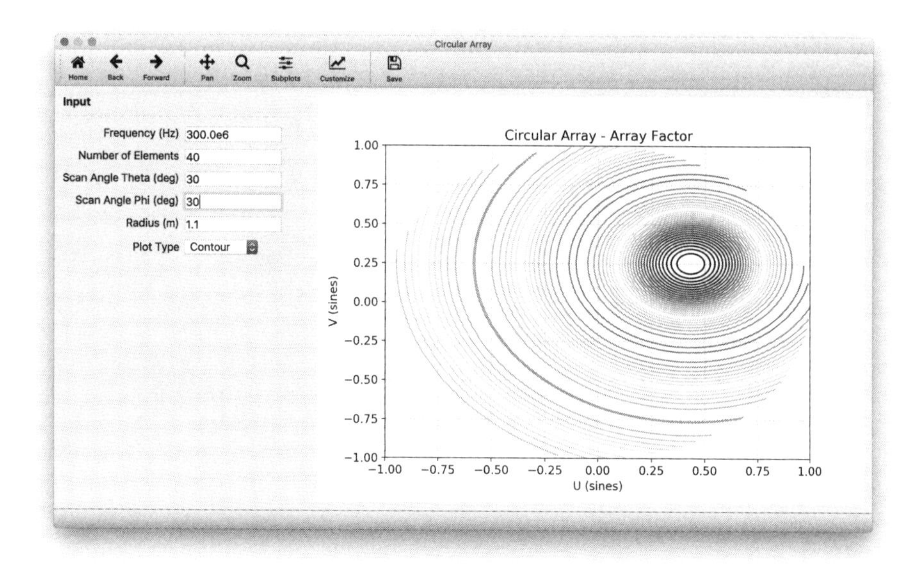

Figure 3.33 The array factor for a circular array calculated by *circular_array_example.py*.

PROBLEMS

3.1 Describe the type of antenna pattern, in terms of beamwidth, sidelobe level, directivity, and gain, that would be suitable for distinguishing closely spaced targets.

3.2 Calculate the distance to the radiating near field and the far field for the WSR-88D weather radar operating at S-band with an antenna diameter of 8.54 m.

3.3 Find the approximate azimuth and elevation beamwidths for the WSR-88D weather radar, which uses a center-fed parabolic dish antenna with a gain of 45.5 dB.

3.4 Determine the polarization loss factor when a linearly polarized incoming electric field, given by $\mathbf{E}_i = E_i\ (\hat{\mathbf{x}} + 2\hat{\mathbf{y}})$ V/m, is incident on a receiving antenna whose electric vector is given by $\mathbf{E}_a = E_a\ (10\hat{\mathbf{x}} + \hat{\mathbf{y}})$ V/m.

3.5 For a small dipole of length 0.25 m operating at a frequency of 100 MHz, calculate the maximum effective aperture and the radiation resistance.

3.6 Given a small circular loop antenna with four turns of radius 0.1 m operating at a frequency of 300 MHz, find the radiation intensity, total power radiated, radiation resistance, and maximum effective aperture.

3.7 Calculate the radiation intensity, total power radiated, radiation resistance, and maximum effective aperture for a constant current circular loop antenna with a radius of 1.5 m operating at a frequency of 900 MHz.

3.8 For a rectangular aperture with a uniform field distribution situated in a ground plane, find the half-power beamwidths, first-null beamwidths, maximum directivity, and maximum effective aperture. The aperture dimensions are 0.03×0.1 m and the operating frequency is 1 GHz.

3.9 Determine the maximum directivity and maximum effective aperture for a rectangular aperture with a TE_{10} field distribution situated in a ground plane. The aperture dimensions are 0.07×0.2 m and the operating frequency is 10 GHz.

3.10 Calculate the half-power beamwidths, first-null beamwidths and maximum directivity for a circular aperture with a uniform field distribution in a ground plane. The aperture radius is 0.15 m and the operating frequency is 2 GHz.

3.11 For a circular aperture with a TE_{11} field distribution in a ground plane, determine the half-power beamwidths, first-null beamwidths, and maximum directivity. The aperture radius is 0.08 m and the operating frequency is 4 GHz.

3.12 Calculate the radiation intensity, total power radiated, and the maximum directivity for an E-plane sectoral horn antenna with dimensions of 5.5×2.75 m fed by a waveguide of dimension 0.2×0.3 m, operating at a frequency of 300 MHz.

3.13 Calculate the radiation intensity, total power radiated, and the maximum directivity for an H-plane sectoral horn antenna with dimensions of 1.3×6.75 m fed by a waveguide of dimension 0.2×0.7 m, operating at a frequency of 600 MHz.

3.14 Calculate the radiation intensity, total power radiated, and the maximum directivity for a pyramidal horn antenna with dimensions of 1.1×1.5 m fed by a waveguide of dimension 0.2×0.1 m, operating at a frequency of 450 MHz.

3.15 Describe the advantages and disadvantages of using an array antenna versus a single-element antenna for radar applications.

3.16 Using the SciPy implementation for calculating binomial coefficients, create Pascal's triangle shown in Table 3.1.

3.17 Calculate the Dolph-Tschebyscheff coefficients for a ten-element array with a sidelobe level of -30 dB.

3.18 List the advantages and disadvantages of Hanning, Hamming, and Blackman-Harris windowing functions for radar antenna applications.

3.19 For a planar array with an element spacing of 0.5λ in the x direction and 0.25λ in the y direction, find the phase shift between each element in the x and y directions required to steer the main beam in the direction $\theta_0 = 30^o$, $\phi_0 = 45^o$.

3.20 Compute the phase shift of each element of a hexagonal planar array consisting of three rows, and the center row contains three elements. The element spacing along the x direction is 0.5λ, and the spacing along the y direction is 0.433λ.

References

[1] IEEE. IEEE standard for definitions of terms for antennas. *IEEE Std 145-2013 (Revision of IEEE Std 145-1993)*, pages 1–50, March 2014.

[2] C. A. Balanis. *Antenna Theory Analysis and Design,* 3rd ed. John Wiley and Sons, New York, 2005.

[3] C. A. Balanis. *Advanced Engineering Electromagnetics,* 2nd ed. John Wiley and Sons, New York, 2012.

[4] J. D. Jackson. *Classical Electrodynamics,* 3rd ed. John Wiley and Sons, New York, 1998.

[5] W. C. Gibson. *The Method of Moments in Electromagnetics,* 2nd ed. CRC Press, Boca Raton, FL, 2015.

[6] J. S. Stone. United States Patents No. 1,643,323 and No 1,715,433.

[7] Wikipedia Contributors. *B. Pascal.* `https://en.wikipedia.org/wiki/Blaise_Pascal`.

[8] E. Jones, et al. SciPy: Open source scientific tools for Python, 2001–. `http://www.scipy.org/`.

[9] C. L. Dolph. A current distribution for broadside arrays which optimizes the relationship between beamwidth and side-lobe level. *Proceedings IRE and Waves and Electrons*, pages 489–492, May 1946.

[10] D. Barbiere. A method for calculating the current distribution of Tschebyscheff arrays. *Proceedings IRE*, pages 78–82, January 1952.

[11] C. J. Drane. Useful approximations for the directivity and beamwidth of large scanning Dolph-Chebyshev arrays. *Proceedings of the IEEE*, pages 1779–1787, November 1968.

[12] B. R. Mahafza and A. Elsherbeni. *MATLAB Simulations for Radar Systems Design.* Chapman and Hall/CRC, New York, 2003.

Chapter 4

The Radar Range Equation

In its basic form, the radar range equation estimates the power at the input to the receiver for a target of a given radar cross section at a specified range. While the equation is simple, it is often misunderstood and used incorrectly. There are numerous versions of the radar range equation as well as various derivations. Different forms of the radar range equation allow for studies of particular values of interest, such as maximum detection range, noise effects, output signal to noise ratio, and others. The derivation of the radar range equation begins by studying the fields radiated by an infinitesimal dipole. While it is difficult to physically realize such a radiating element, it provides insight into phenomena often passed over in other derivations. Next, a basic form of the radar range equation is examined, and the effects of noise and losses are included. Then a special version of the radar range equation for search-specific missions is derived. The final special case considered is for bistatic radar configurations. The chapter concludes with several Python examples to illustrate the application of various forms of the radar range equation.

4.1 HERTZIAN DIPOLE

The study of the radar range equation begins with the *Hertzian dipole*. This is a very simplistic radiating element and is often referred to as an infinitesimal dipole. This type of radiating element is characterized by a length much less than the wavelength and has a uniform current distribution as shown in Figure 4.1. While this is very difficult to implement, there are many advantages to studying this type of radiating element. It is useful as a building block for analyzing larger wire antennas and for studying the radar range equation. To solve for the fields radiated from the Hertzian dipole, begin by calculating the vector potential **A**. Recall from (2.50) that the vector potential for the Hertzian dipole is written as

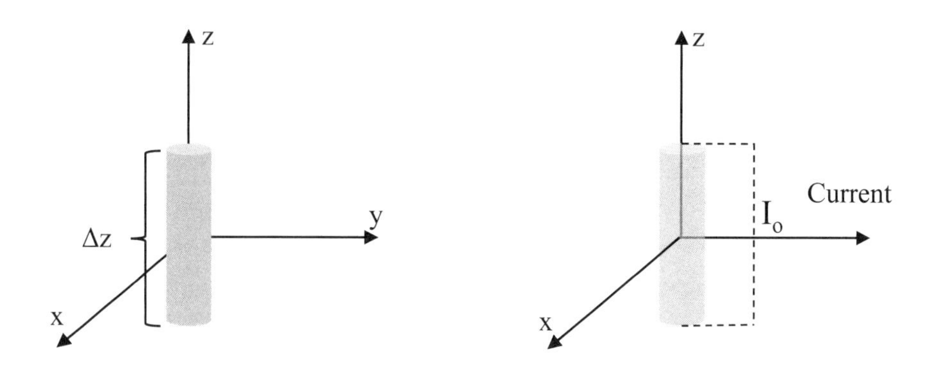

Figure 4.1 Hertzian dipole with a uniform current distribution.

$$\mathbf{A}(r) = \frac{\mu}{4\pi}\int_{V'} \mathbf{J}(r')\,\frac{e^{-jkr}}{r}\,dv' = \frac{\mu}{4\pi}\iiint \mathbf{J}(x', y', z')\,\frac{e^{-jkr}}{r}\,dx'\,dy'\,dz'. \tag{4.1}$$

The current density for the Hertzian dipole is expressed as

$$\mathbf{J}(x', y', z') = \begin{cases} I_o\,\delta(x')\,\delta(y')\,\hat{\mathbf{z}} & \text{for } -\Delta z/2 \le z' \le \Delta z/2 \\ 0 & \text{otherwise.} \end{cases} \tag{4.2}$$

Substituting (4.2) into (4.1), and performing the integration results in

$$\mathbf{A}(r) = \mu I_o \frac{e^{-jkr}}{4\pi r}\,\Delta z\,\hat{\mathbf{z}}. \tag{4.3}$$

The magnetic field radiated by the Hertzian dipole is found from (2.32), and is

$$\mathbf{H} = \frac{\nabla \times \mathbf{A}}{\mu} \qquad \text{(A/m)}. \tag{4.4}$$

Expanding the curl of the vector potential gives

$$\nabla \times \mathbf{A} = \left[\mu I_o \frac{e^{-jkr}}{4\pi}\,\Delta z\,\sin\theta\left(\frac{jk}{r} + \frac{1}{r^2}\right)\right]\hat{\boldsymbol{\phi}}. \tag{4.5}$$

The magnetic field is therefore

$$\mathbf{H} = I_o\,\Delta z\,\sin\theta\,\frac{e^{-jkr}}{4\pi r}\left[jk + \frac{1}{r}\right]\hat{\boldsymbol{\phi}} \qquad \text{(A/m)}. \tag{4.6}$$

The electric field is obtained from (2.11) as

$$\mathbf{E} = \frac{\nabla \times \mathbf{H}}{j\omega\epsilon} \qquad \text{(V/m)}, \tag{4.7}$$

and evaluating the curl results in

$$\mathbf{E} = 2\eta\, I_o\, \Delta z\, \frac{e^{-jkr}}{4\pi r}\, \cos\theta \left[\frac{1}{r} - \frac{j}{kr^2}\right] \hat{\mathbf{r}}$$

$$+ j\eta k\, I_o\, \Delta z\, \frac{e^{-jkr}}{4\pi r}\, \sin\theta \left[1 + \frac{1}{jkr} - \frac{1}{(kr)^2}\right] \hat{\boldsymbol{\theta}} \qquad \text{(V/m)}. \tag{4.8}$$

4.1.1 Radiated Power

To determine the power radiated by the Hertzian dipole, the far-field electric and magnetic fields must first be found. From Section 3.1.1.1, the terms involving $1/r$ are dominant in the far field. Therefore, the electric and magnetic far fields for the Hertzian dipole are written as

$$\mathbf{E} \approx j\,\eta\, k\, I_o\, \Delta z\, \frac{e^{-jkr}}{4\pi r}\, \sin\theta\, \hat{\boldsymbol{\theta}} \qquad \text{(V/m)}, \tag{4.9}$$

$$\mathbf{H} \approx j\, k\, I_o\, \Delta z\, \frac{e^{-jkr}}{4\pi r}\, \sin\theta\, \hat{\boldsymbol{\phi}} \qquad \text{(A/m)}. \tag{4.10}$$

The radiated power density is calculated from the time average complex Poynting vector as [1]

$$\mathbf{P}_d = \frac{\mathbf{E} \times \mathbf{H}^*}{2} = \frac{E_\theta\, H_\phi^*}{2}\, \hat{\mathbf{r}} \qquad \text{(W/m}^2\text{)}. \tag{4.11}$$

Substituting (4.9) and (4.10) into (4.11) gives

$$\mathbf{P}_d = \frac{\eta}{2} \left[\frac{k\, I_o\, \Delta z}{4\pi r}\, \sin\theta\right]^2 \hat{\mathbf{r}} \qquad \text{(W/m}^2\text{)}. \tag{4.12}$$

The power density is real-valued and varies as $1/r^2$. This means if the distance to the target is doubled, then the power loss is quadrupled. To compute the total power radiated, integrate the power density, $\mathbf{P}_d$, over a sphere as

$$P_{rad} = \int_0^{\pi} \int_0^{2\pi} \mathbf{P}_d \cdot r^2 \sin\theta\, d\theta\, d\phi\, \hat{\mathbf{r}} \qquad \text{(W)}. \tag{4.13}$$

Substituting (4.12) in (4.13), and performing the integration results in

$$P_{rad} = \eta \, \frac{\pi}{3} \left[\frac{I_o \, \Delta z}{\lambda} \right]^2 \qquad \text{(W)}. \tag{4.14}$$

4.1.2 Radiation Intensity

Recall from Section 3.1.4, the radiation intensity is a far-field parameter, and is defined as the power radiated per unit solid angle in a given direction. To calculate the radiation intensity, multiply the power density by the square of the distance [2]

$$U = r^2 \, P_d \qquad \text{(W/sr)}. \tag{4.15}$$

The total power radiated is related to the radiation intensity as

$$P_{rad} = \frac{1}{2} \int_S \left(E \times H^* \right) \cdot d\mathbf{S} = \int \int U(\theta, \phi) \, \sin\theta \, d\theta \, d\phi \qquad \text{(W)}. \tag{4.16}$$

From this, the radiation intensity for the Hertzian dipole is given by

$$U(\theta, \phi) = \frac{\eta}{2} \left[\frac{k \, I_o \, \Delta z}{4\pi} \, \sin\theta \right]^2 \qquad \text{(W/sr)}. \tag{4.17}$$

4.1.3 Directivity and Gain

From Section 3.1.5, the directivity of an antenna is a dimensionless unit, and is defined as the ratio of the radiation intensity in a given direction from the antenna to the radiation intensity averaged over all directions. Using the radiation intensity for the Hertzian dipole from the previous section, the directivity is calculated as

$$D(\theta, \phi) = \frac{4\pi \, U(\theta, \phi)}{P_{rad}} = \frac{3}{2} \, \sin^2\theta. \tag{4.18}$$

Using (3.16), the gain of the Hertzian dipole is written as

$$G(\theta, \phi) = e_a \left[4\pi \, \frac{U(\theta, \phi)}{P_{rad}} \right] = e_a \, D(\theta, \phi). \tag{4.19}$$

The power density in terms of the antenna directivity and radiated power is written as

$$P_d = \frac{\eta}{2} \left[\frac{k \, I_o \, \Delta z}{4\pi r} \sin\theta \right]^2 = \underbrace{\eta \, \frac{\pi}{3} \left[\frac{I_o \, \Delta z}{\lambda} \right]^2}_{\text{Power radiated}} \underbrace{\left[\frac{3}{2} \, \sin^2\theta \right]}_{\text{Directivity}} \underbrace{\left[\frac{1}{4\pi r^2} \right]}_{\text{Area of a sphere}} \qquad \text{(W/m}^2\text{)}. \tag{4.20}$$

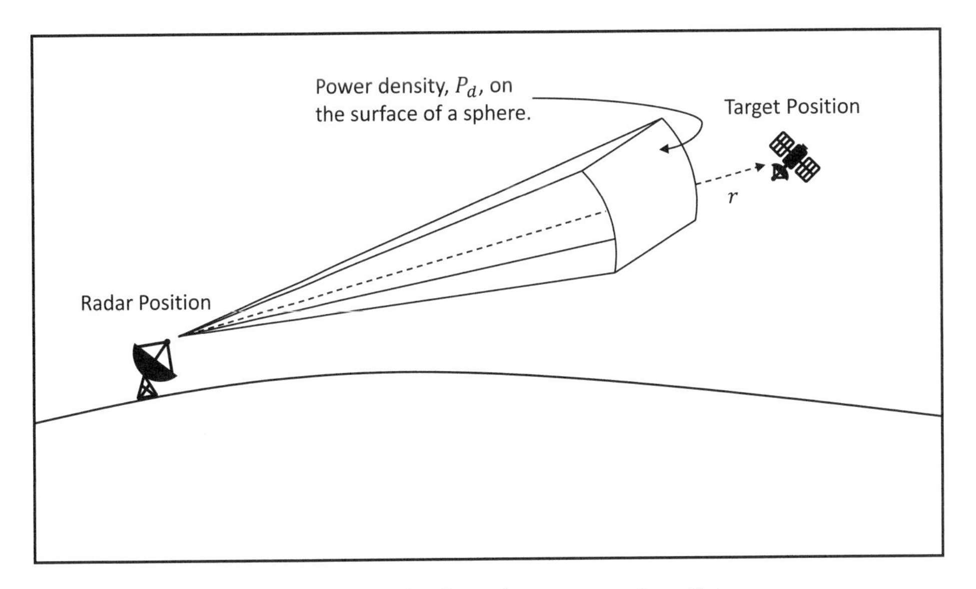

Figure 4.2 Power density at the target at a given distance.

Therefore, the power density in terms of antenna gain and input power is

$$P_d = \frac{P_{rad}\, D(\theta,\phi)}{4\pi r^2} = \frac{P_{in}\, G(\theta,\phi)}{4\pi r^2} \qquad (\text{W/m}^2). \tag{4.21}$$

The above equation is an important result as it gives the power density present at a target for a particular radar transmitter power and antenna gain. This result is used in deriving the full radar range equation.

4.2 BASIC RADAR RANGE EQUATION

From the previous section, the power density at the target at a given distance, as shown in Figure 4.2, is

$$P_d = \frac{P_{in}\, G(\theta,\phi)}{4\pi r^2} \qquad (\text{W/m}^2). \tag{4.22}$$

When the radiated power density from a radar system impinges on a target, the induced currents on the target reradiate or scatter electromagnetic energy in all directions, as shown in Figure 4.3. The amount of energy scattered is proportional to the target size, orientation, shape, material composition, and other factors. The effects from all of these factors are grouped together into a single target specific parameter know as *radar cross section* (RCS), denoted by σ. The radar cross section is defined as the ratio of the power

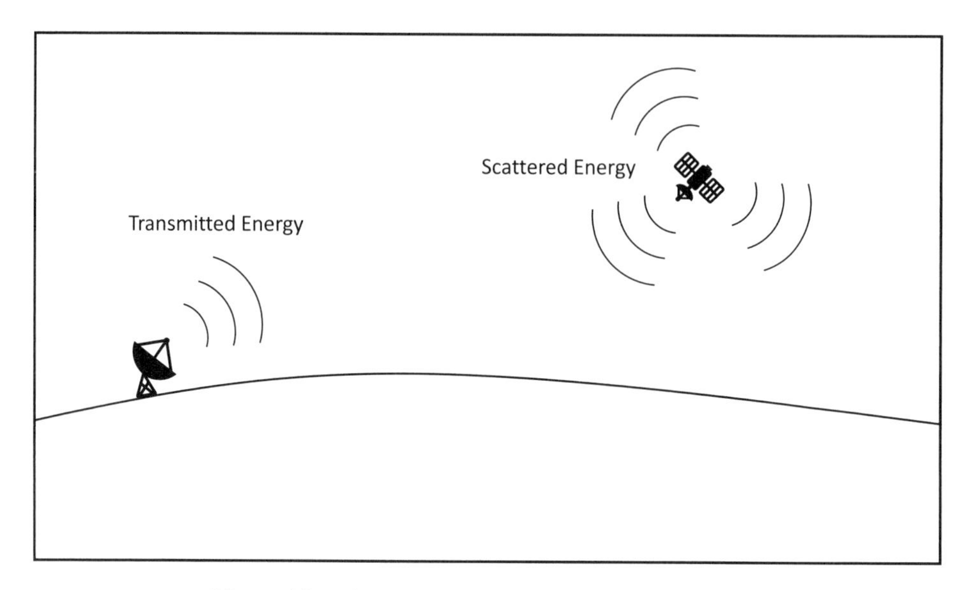

Figure 4.3 Electromagnetic energy scattered by a target.

reflected back to the radar to the power density incident on the target [1]. The radar cross section is written as

$$\sigma = \lim_{r\to\infty} 4\pi r^2 \frac{|\mathbf{E}^s|^2}{|\mathbf{E}^i|^2} = \frac{P_r}{P_d} \qquad (\text{m}^2), \tag{4.23}$$

where

$\mathbf{E}^s$ = scattered electric field intensity (V/m),
$\mathbf{E}^i$ = incident electric field intensity (V/m),
P_r = power reflected from the target (W),
P_d = power density incident upon the target (W/m^2).

Table 4.1 shows typical radar cross section values of various objects at X-band [2, 3]. The power scattered by the target is then found by multiplying the power density from (4.22) by the radar cross section. This gives

$$P_{target} = \frac{P_t\, G\, \sigma}{4\pi r^2} \qquad (\text{W}), \tag{4.24}$$

where P_t is the radar power input to the antenna. The power density at some distance from the target is found from (4.21) by replacing $P_{in}\, G(\theta,\phi)$ with P_{target}. The power density back at the antenna is then

$$P_{antenna} = \frac{P_t\, G\, \sigma}{(4\pi r^2)^2} \qquad (\text{W/m}^2). \tag{4.25}$$

Table 4.1
Typical RCS Values for Objects at X-Band

Object	*RCS* (m^2)	*RCS (dBsm)*
Container ship	$50,000$	47
Corner reflector	$20,000$	43
Automobile	100	20
Commercial jet	40	16
Cabin cruiser boat	10	10
Large fighter aircraft	6	7.78
Small fighter aircraft	2	3
Human	1	0
Surface-to-air missile	0.5	-3
Bird	0.01	-20
Insect	1×10^{-5}	-50
Advanced tactical fighter	1×10^{-6}	-60

The radar cross section is analogous to the antenna pattern. The antenna pattern is created by induced currents on the antenna structure from a transmitter through a feed and matching network. The radar cross section pattern is created by induced currents on the target structure as a result of the incident energy transmitted by the radar. The radiation mechanism for an antenna with associated currents is the same as the radiation from a target with associated currents. When solving these numerically, the source code is the same for the two cases; it is the source of the induced currents that is different. To find the total power received by the radar, multiply the power density at the antenna by the effective aperture of the antenna, which gives

$$P_{radar} = \frac{P_t\,G\,\sigma}{(4\pi r^2)^2}\,A_e \qquad \text{(W)}, \tag{4.26}$$

where A_e is the antenna effective aperture in m^2. The antenna effective aperture is related to the antenna gain by [2]

$$A_e = \frac{\lambda^2\,G}{4\pi} \qquad (\text{m}^2). \tag{4.27}$$

Substituting (4.27) into (4.26) results in

$$P_{radar} = \frac{P_t\,G^2\,\lambda^2\,\sigma}{(4\pi)^3\,r^4} \qquad \text{(W)}. \tag{4.28}$$

The power received by the radar is proportional to r^4. This means a doubling of the distance from the radar to the target results in a power loss of a factor of 16, which is approximately 12 dB. Similarly, to increase the detection range by a factor of 2 would require 16 times more power.

4.2.1 Maximum Detection Range

To determine the maximum detection range, replace the power at the radar with the minimum detectable signal power, P_{min}, which gives

$$P_{min} = \frac{P_t\, G^2\, \lambda^2\, \sigma}{(4\pi)^3\, r^4} \qquad \text{(W)}. \tag{4.29}$$

Solving (4.29) for the range gives

$$r_{max} = \left[\frac{P_t\, G^2\, \lambda^2\, \sigma}{(4\pi)^3\, P_{min}}\right]^{1/4} \qquad \text{(m)}. \tag{4.30}$$

4.2.2 Noise

To this point, the effects of noise have not been considered. Noise corrupts the signal received by the radar and decreases the maximum detection range. Noise is a random process and is typically specified by its *power spectral density* (PSD). The power spectral density is the power (variance) per frequency interval. It is a measure of a signal's power content per frequency. It is typically used to characterize broadband random signals such as noise. Its units are (watt/Hz $\rightarrow$ watt $\cdot$ second $\rightarrow$ joule). For thermal noise, the power spectral density for a resistive device is [4]

$$S = k\,T \qquad \text{(J)}, \tag{4.31}$$

where k is Boltzmann's constant, which has a value of $1.38064852 \times 10^{-23}$ (J/K), and T is the noise temperature of the device (K). The input noise power for a radar system with an operating bandwidth, B, is found by multiplying by the power spectral density as

$$N_{input} = S \times B = k\,T\,B \qquad \text{(W)}. \tag{4.32}$$

The fidelity of radar receivers is often given by the *noise figure* and *noise factor*. These are measures of how components in the receiver chain degrade the signal-to-noise ratio and will be covered in more detail in Chapter 5. The noise factor is the ratio of the input signal-to-noise ratio to the output signal-to-noise ratio. The noise figure is the noise factor expressed in decibels. The noise factor is given as

$$F = \frac{SNR_i}{SNR_o} = \frac{P_i/N_i}{P_o/N_o}, \tag{4.33}$$

where SNR_i and SNR_o are the input and output signal-to-noise ratios respectively. The input signal power, P_i, may be written in terms of the output signal-to-noise ratio as

$$P_i = k\,T_0\,B\,F\,SNR_o \qquad \text{(W)}, \tag{4.34}$$

where T_0 is the reference temperature. T_0 has been specified by the IEEE to be 290° K for defining noise factor. Setting the detection threshold to the minimum output signal-to-noise ratio required for target detection allows (4.34) to be written as

$$P_{min} = k\, T_0\, B\, F\, (SNR_o)_{min} \qquad \text{(W)}. \tag{4.35}$$

The expression for the maximum range given in (4.30) becomes

$$r_{max} = \left[\frac{P_t\, G^2\, \lambda^2\, \sigma}{(4\pi)^3\, k\, T_0\, B\, F\, (SNR_o)_{min}}\right]^{1/4} \qquad \text{(m)}. \tag{4.36}$$

The radar range equation is now written to give the output signal-to-noise ratio as

$$SNR_o = \frac{P_t\, G^2\, \lambda^2\, \sigma}{(4\pi)^3\, k\, T_0\, B\, F\, r^4}. \tag{4.37}$$

4.2.3 Losses

There are losses associated with radar systems that include transmission line loss, receiver line loss, filter loss, atmospheric loss, integration loss, signal processing loss, and several others. While some of these were studied in Chapter 2, the remaining ones will be treated in more detail in Chapter 5. For the purposes of the basic radar range equation, these effects are grouped together in a single loss term denoted by L. These losses reduce the overall output signal-to-noise ratio. Including loss, the minimum detectable signal is written as

$$P_{min} = k\, T_0\, B\, F\, L\, (SNR_o)_{min} \qquad \text{(W)}. \tag{4.38}$$

The maximum detection range then becomes

$$r_{max} = \left[\frac{P_t\, G^2\, \lambda^2\, \sigma}{(4\pi)^3\, k\, T_0\, B\, F\, L\, (SNR_o)_{min}}\right]^{1/4} \qquad \text{(m)}. \tag{4.39}$$

Finally, the expression for radar range equation is written as

$$SNR_o = \frac{P_t\, G^2\, \lambda^2\, \sigma}{(4\pi)^3\, k\, T_0\, B\, F\, L\, r^4}. \tag{4.40}$$

4.2.4 Radar Reference Range and Loop Gain

The radar reference range and radar loop gain are measures used to characterize a radar system's sensitivity and therefore its capability to detect and track targets. The radar reference range is the range at which the radar system achieves a required signal-to-noise ratio on a target with a specified radar cross section. For example, a radar system may produce a signal-to-noise ratio of 20 dB on a target with a radar cross section of

-10 dBsm at a range of 500 km. The radar loop gain, which is commonly denoted as C, combines these into a single value as

$$C = \frac{r_0^4 \, SNR_0}{\sigma_0} \qquad \text{(m}^2\text{)}, \tag{4.41}$$

where

r_0 = reference range (m),
SNR_0 = reference signal to noise ratio,
σ_0 = reference radar cross section (m^2).

Therefore, if the loop gain is known, the signal to noise ratio for a target at a particular range may be found from

$$SNR_T = C \times \frac{\sigma_T}{r_T^4} = SNR_0 \times \frac{\sigma_T}{\sigma_0} \times \frac{r_0^4}{r_T^4}, \tag{4.42}$$

where SNR_T is the signal-to-noise ratio for a target located at a range of r_T having a radar cross section of σ_T. Historically, various techniques have been used at different test ranges to determine the loop gain, and in some cases these methods have resulted in inconsistent estimates for loop gain [5].

4.3 SEARCH RADAR RANGE EQUATION

One of the first tasks a radar system performs is to search a specified volume of space for a target of interest. Therefore, a modified version of the radar range equation tailored to the search function of a radar system will be derived. Search volumes are typically given in terms of a solid angle, Ω, which has units of radians^2 or steradians. The *power aperture product* is commonly used to characterize a radar system's capability to perform search missions. The power aperture product is the product of the time average radiated power and the antenna effective aperture. If the azimuth and elevation extents, (Θ_a, Θ_e), of the search volume are relatively small, then Ω may be expressed as

$$\Omega = \Theta_a \times \Theta_e \qquad \text{(steradian)}. \tag{4.43}$$

The result of this expression is the search volume is rectangular in azimuth and elevation space. While the search volume could be any shape and may be represented in any coordinate system, a more common search volume is the surface of a sphere bounded in azimuth and elevation, as shown in Figure 4.4. The angular area of this type of search sector is given as

$$\Omega = 2\,\Theta_a \, \cos{(e_0)} \, \sin{(\Theta_e/2)} \qquad \text{(steradian)}, \tag{4.44}$$

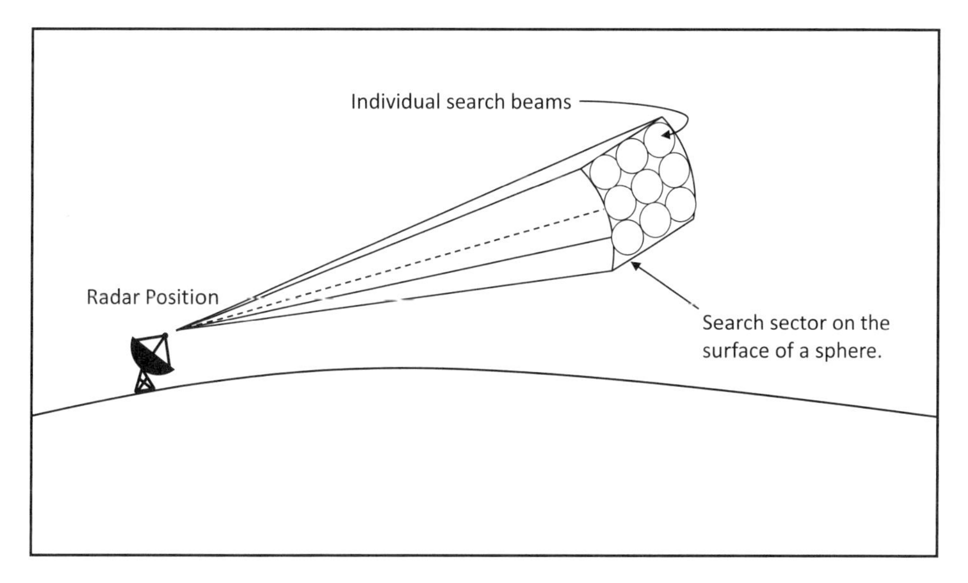

Figure 4.4 Spherical surface search volume.

where e_0 is the elevation angle corresponding to the center of the search sector. If the radar antenna has 3-dB beamwidths denoted by θ_a, θ_e, then the solid angle coverage for a single beam may be written as

$$\Omega_b = \theta_a \times \theta_e \qquad \text{(steradian)}. \tag{4.45}$$

Then the number of beams required to cover the solid angle is

$$N_b = \frac{\Omega}{\Omega_b}. \tag{4.46}$$

This is a simplistic approach to the number of beams as both the search sector and the beam coverage is rectangular in this case. A factor, P, known as the *packing factor* accounts for the fact the beams are not rectangular and that there is some overlap of the beams when filling the search sector. The required number of beams is then given by

$$N_b = P\,\frac{\Omega}{\Omega_b}. \tag{4.47}$$

Typical values for P depend on how the beams are placed and how much they overlap. For example, a packing factor of $4/3$ is typical for hexagonal beam packing with 3 dB overlap [6].

The relationship between the peak transmit power and average transmit power is shown in Figure 4.5 and is expressed mathematically as

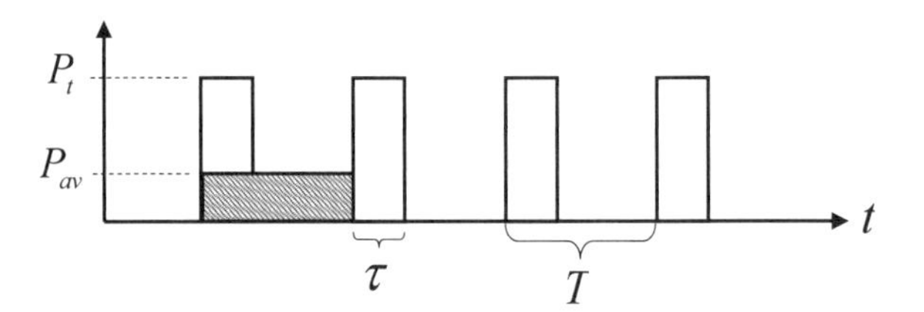

Figure 4.5 Duty cycle, peak, and average power.

$$P_t = P_{av}\,\frac{T}{\tau} \qquad \text{(W)}, \tag{4.48}$$

where T is the *pulse repetition interval* and τ is the pulse width. For a continuous wave pulse, the bandwidth is approximately the inverse of the pulse width,

$$B = \frac{1}{\tau} \qquad \text{(Hz)}. \tag{4.49}$$

Using the radar equation (4.40) along with (4.48) and (4.49) results in

$$SNR_o = \frac{P_{av}\,G^2\,\lambda^2\,\sigma\,T}{(4\pi)^3\,k\,T_s\,F\,L\,r^4}. \tag{4.50}$$

A requirement is for the radar system to cover the search volume in a specified amount of time, T_{scan}. Therefore, the radar must process N_b beams in T_{scan}. The time for each beam is then

$$T_{beam} = \frac{T_{scan}}{N_b} = \frac{T_{scan}\,\Theta_a\,\Theta_e}{\Omega} \qquad \text{(s)}. \tag{4.51}$$

Allowing one beam per pulse results in

$$SNR_o = \frac{P_{av}\,G^2\,\lambda^2\,\sigma\,T_{scan}}{(4\pi)^3\,k\,T_0\,F\,L\,r^4}\,\frac{\Theta_a\,\Theta_e}{\Omega}. \tag{4.52}$$

Using

$$G = \frac{4\,\pi\,A_e}{\lambda^2} = e\,\frac{4\,\pi}{\theta_a\,\theta_e}, \tag{4.53}$$

in (4.52) gives

$$SNR_o = \frac{P_{av}\,A_e\,\sigma}{(4\pi)^3\,k\,T_0\,F\,L\,r^4}\,\frac{T_{scan}}{\Omega}, \tag{4.54}$$

where $P_{av}\,A_e$ is the power aperture product. This is a useful form of the radar range equation as it provides a means of performing search radar design trade studies without the need for detailed parameters. However, it is stressed that this provides a first draft radar design. It gives sizing information about the radar system and should be used as a starting point for more detailed designs that include parameters not explicitly detailed in (4.54).

4.4 BISTATIC RADAR RANGE EQUATION

Another useful form of the radar range equation is the bistatic radar range equation. Figure 4.6 depicts a typical bistatic radar scenario in which the transmitter and receiver are not collocated as in the monostatic case. Recently, bistatic radar systems have received attention for their potential to detect stealth targets due to enhanced target forward scatter [7]. These systems have also found use in weather and atmospheric physics applications. An extension of the bistatic case is multistatic radar where there is typically a single transmitter and multiple receivers. Furthermore, multiple passive receivers allow radar receivers to detect and track targets without the need for transmission of a signal. These types of systems have found use in navigation, air traffic, and surveillance scenarios. Bistatic radar systems have properties that offer certain advantages for particular functions. First, the receiver is completely passive, which reduces its likelihood to be detected by the electronic measures, and it is safe from attack by antiradiation missiles or deliberate directional interference and jamming. Second, the receiver may be located in areas that may not be suitable to place transmitters, such as near humans or flammable materials. Next, there is no transmit-receive switch or duplexer, which are lossy, expensive, and heavy. Because bistatic systems do not suffer from the same range blindness as an equivalent monostatic system, less transmitter power and higher pulse repetition frequencies may be used. Finally, if the target angle can be measured at both sites as well as the bistatic range, data can be checked for self-consistency to aid in the removal of false alarms [7]. Referring to Figure 4.6 and recalling from (4.22), the power density at a target at a distance, r_t, from the transmitter is

$$P_d = \frac{P_t\, G_t(\theta, \phi)}{4\pi\, r_t^2} \qquad \text{(W/m}^2\text{)}, \tag{4.55}$$

where G_t is the gain of the transmitting antenna and r_t is the range from the transmitter to the target. As before, multiply by the radar cross section to get the power scattered by the target. The radar cross section is written explicitly as a function of (θ, ϕ), as the incident and scattering directions are different for the bistatic case. The power scattered by the target is written as

$$P_{target} = \frac{P_t\, G_t(\theta, \phi)\, \sigma(\theta, \phi)}{4\pi\, r_t^2} \qquad \text{(W)}. \tag{4.56}$$

Following the derivation in Section 4.2, the power density at the receiving antenna is written as

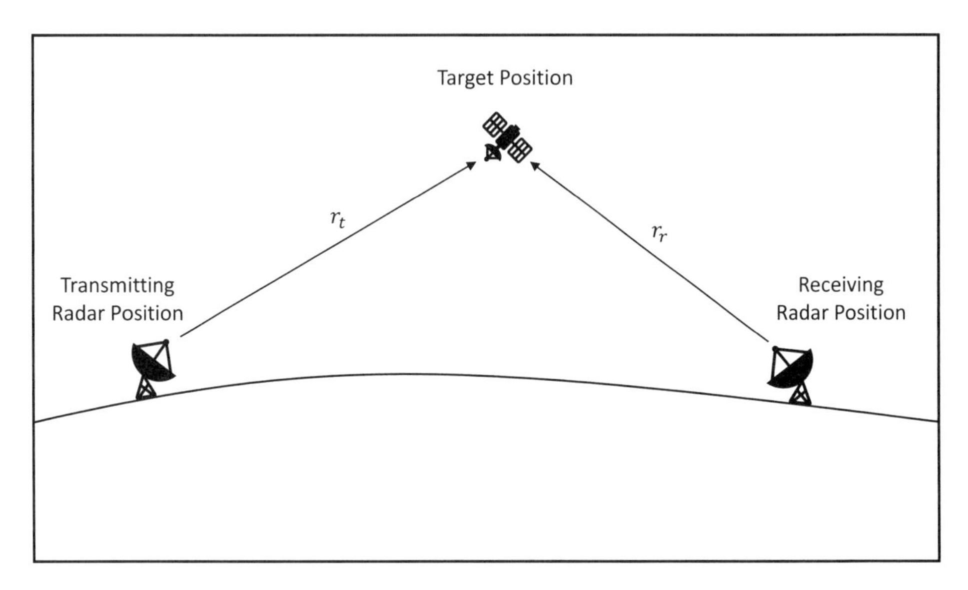

Figure 4.6 Typical bistatic radar configuration.

$$P_{antenna} = \frac{P_t\, G_t(\theta, \phi)\, \sigma(\theta, \phi)}{(4\pi)^2\, r_t^2\, r_r^2} \qquad \text{(W/m}^2\text{)}, \tag{4.57}$$

where r_r is the range from the target to the receiver. Multiplying the power at the antenna by the receiving antenna effective aperture gives the total power received by the radar, which is written as

$$P_{radar} = \frac{P_t\, G_t(\theta, \phi)\, \sigma(\theta, \phi)}{(4\pi)^2\, r_t^2\, r_r^2}\, A_e \qquad \text{(W)}. \tag{4.58}$$

Substituting the following expression for the antenna effective aperture,

$$A_e = \frac{\lambda^2\, G_r}{4\pi}, \tag{4.59}$$

gives the final expression for the basic bistatic radar range equation,

$$P_{radar} = \frac{P_t\, G_t(\theta, \phi)\, G_r(\theta, \phi)\, \sigma(\theta, \phi)\, \lambda^2}{(4\pi)^3\, r_t^2\, r_r^2} \qquad \text{(W)}. \tag{4.60}$$

As before, the basic radar range equation is modified to include the effects of noise and loss. Using (4.38) gives

$$SNR_o = \frac{P_t\, G_t(\theta, \phi)\, G_r(\theta, \phi)\, \sigma(\theta, \phi)\, \lambda^2}{(4\pi)^3\, k\, T_0\, B\, F L_t\, L_r\, r_t^2\, r_r^2}, \tag{4.61}$$

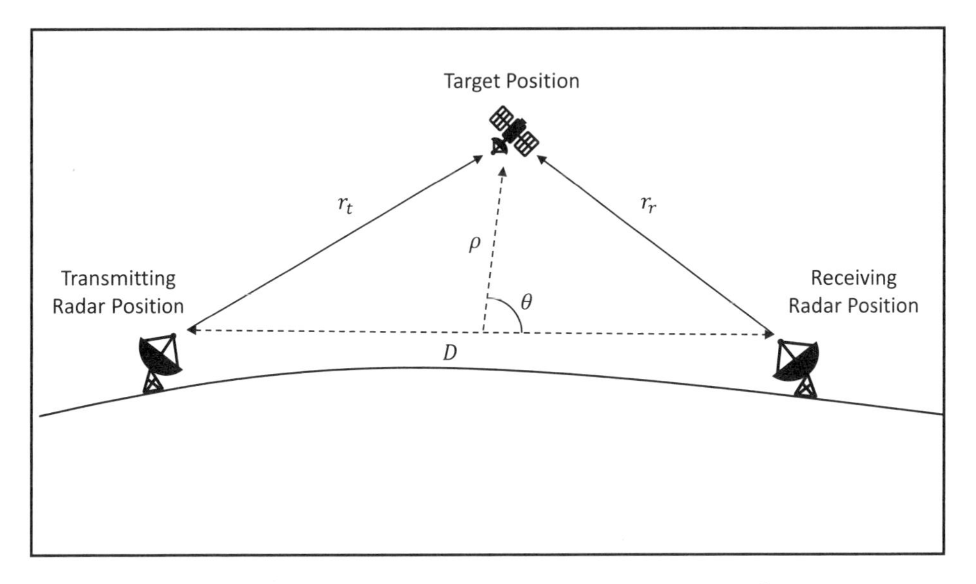

Figure 4.7 Typical bistatic radar configuration with polar coordinates.

where L_t is the loss in the transmitting radar and L_r is the loss in the receiving radar.

4.4.1 Maximum Detection Range

In the bistatic case, it is not possible to write a single maximum detection range as there are now two separate ranges. Instead, a maximum range product is used and is given by

$$(r_t\, r_r)_{max} = \left[\frac{P_t\, G_t(\theta,\phi)\, G_r(\theta,\phi)\, \sigma(\theta,\phi)\, \lambda^2}{(4\pi)^3\, k\, T_0\, B\, F\, L_t\, L_r\, (SNR_o)_{min}} \right]^{1/2} \quad (\mathrm{m}^2). \tag{4.62}$$

In two dimensions, the product in (4.62) belongs to a family of curves known as *Cassini ovals*. The Cassini ovals are a family of quartic curves, sometimes referred to as Cassini ellipses, described by a point such that the product of its distances from two fixed points a distance apart is a constant [8]. For bistatic systems, the system performance may be analyzed by plotting the Cassini ovals for various signal-to-noise ratios. With respect to Figure 4.7, r_t and r_r are converted to polar coordinates to aid in the plotting of the Cassini ovals. Using the geometrical relationships for two dimensions gives

$$r_t^2 = \left(\rho^2 + (D/2)^2\right) + \rho\, D\, \cos\theta, \tag{4.63}$$

$$r_r^2 = \left(\rho^2 + (D/2)^2\right) - \rho\, D\, \cos\theta, \tag{4.64}$$

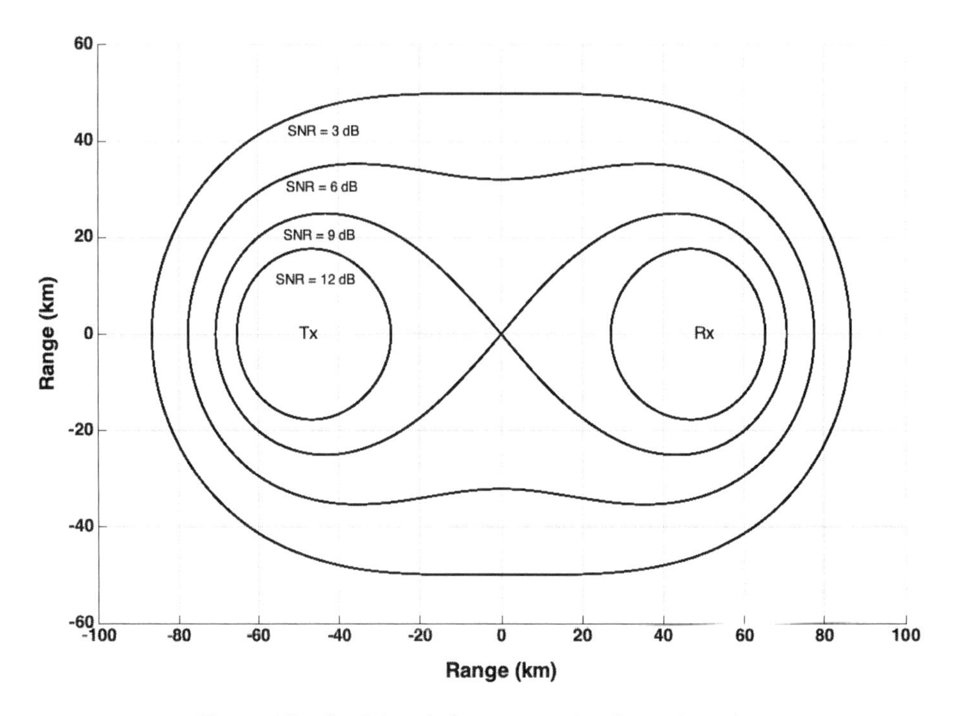

Figure 4.8 Cassini ovals for constant signal-to-noise ratios.

where

ρ = distance from the origin of the polar coordinates to the target (m),
D = distance between the transmitter and receiver (m),
θ = angle between the vector from the origin to the target and the vector from the origin to the receiver (rad).

Combining the expressions in (4.63) results in

$$(r_t\, r_r)^2 = \left(\rho^2 + (D/2)^2\right)^2 - \rho^2\, D^2 \cos^2\theta. \tag{4.65}$$

Using (4.65) in (4.62) allows the radar range equation to be written in polar coordinates, which is useful in plotting the Cassini ovals. Figure 4.8 shows an example of Cassini ovals for constant signal to noise ratio. As can be seen in this figure, there are three distinct regions:

1. Two separate ellipses enclosing the transmitter and receiver.
2. A single continuous ellipse enclosing both the transmitter and receiver.
3. A lemniscate with a cusp at the origin of the polar coordinate system.

The Cassini ovals are two-dimensional ellipses aiding in the understanding of the detection range for bistatic radar systems. In reality, the constant signal-to-noise ratio would be described by surfaces in three dimensions.

4.5 EXAMPLES

The examples in the following sections illustrate the concepts of this chapter with several Python examples. The examples for this chapter are in the directory *pyradar\Chapter04* and the matching MATLAB examples are in the directory *mlradar\Chapter04*. The reader should consult Chapter 1 for information on how to execute the Python code associated with this book.

4.5.1 Hertzian Dipole

For this example, calculate and display the electric and magnetic fields, power density, total power radiated, radiation intensity, and directivity for a Hertzian dipole of length 1×10^{-3} m with a current of 1 A operating at a frequency of 1 GHz. The field point is at a distance of 10 m from the dipole, and the region surrounding the dipole has a relative permittivity of 1.3.

Solution: The solution to the above example is given in the Python code *hertzian_dipole_example.py* and in the MATLAB code *hertzian_dipole_example.m*. Running the Python example code displays a GUI, allowing the user to enter the constitutive parameters for the region, operating frequency, current, length, and distance to the field point. The GUI allows the user to select which values to plot. The code then calculates and creates a plot of the selected values, as shown in Figure 4.9.

4.5.2 Basic Radar Range Equation

As illustrated in Section 4.2, a radar system radiates a certain amount of energy that impinges upon objects in the environment, including targets of interest. The following subsections give examples to illustrate how this energy interacts with these objects.

4.5.2.1 Power Density

For this example, consider a target at some distance from the radar. Calculate and display the power density generated by a radar system with a peak transmit power of 50 kW and an antenna gain of 20 dB. For this example, allow the target range to vary from 1 to 5 km.

Solution: The solution to the above example is given in the Python code *power_density_example.py* and in the MATLAB code *power_density_example.m*. Running the Python

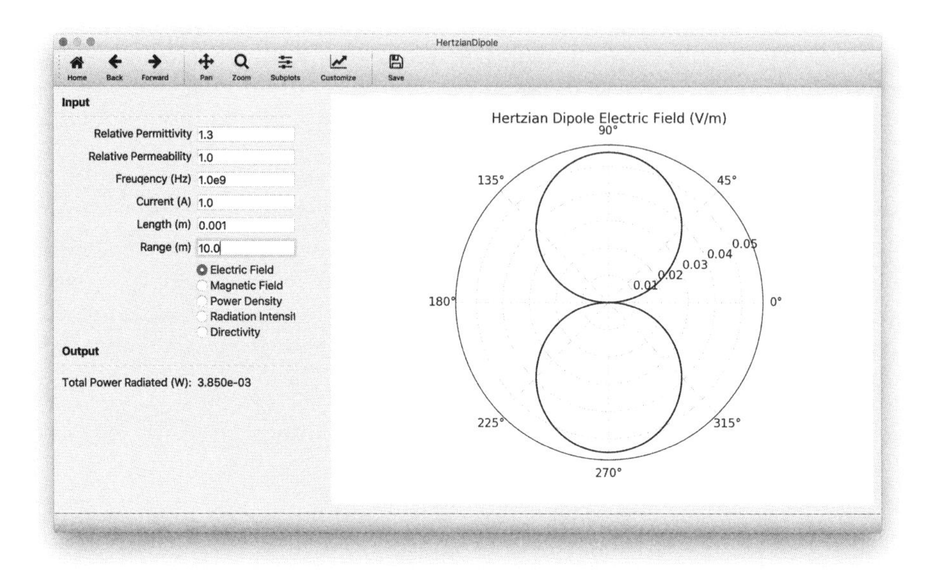

Figure 4.9 The electric field of a Hertzian dipole calculated by *hertzian_dipole_example.py*.

example code displays a GUI, allowing the user to enter the target minimum and maximum range, the peak transmitted power, and the antenna gain. The code then calculates and displays the power density as a function of target range, as shown in Figure 4.10.

4.5.2.2 Power at the Radar

Now that the energy incident upon a target has been calculated, find the energy delivered back to the radar. Calculate and display the power at the radar for a system with a peak transmit power of 50 kW, an antenna gain of 20 dB, and an operating frequency of 1 GHz. The target has a radar cross section of 10 dBsm, and the target range varies from 1 to 5 km.

Solution: The solution to the above example is given in the Python code *power_at_radar_example.py* and in the MATLAB code *power_at_radar_example.m*. Running the Python example code displays a GUI, allowing the user to enter the target minimum and maximum range, peak transmitted power, antenna gain, operating frequency, and target radar cross section. The code then calculates and displays the power back at the radar as a function of target range, as shown in Figure 4.11.

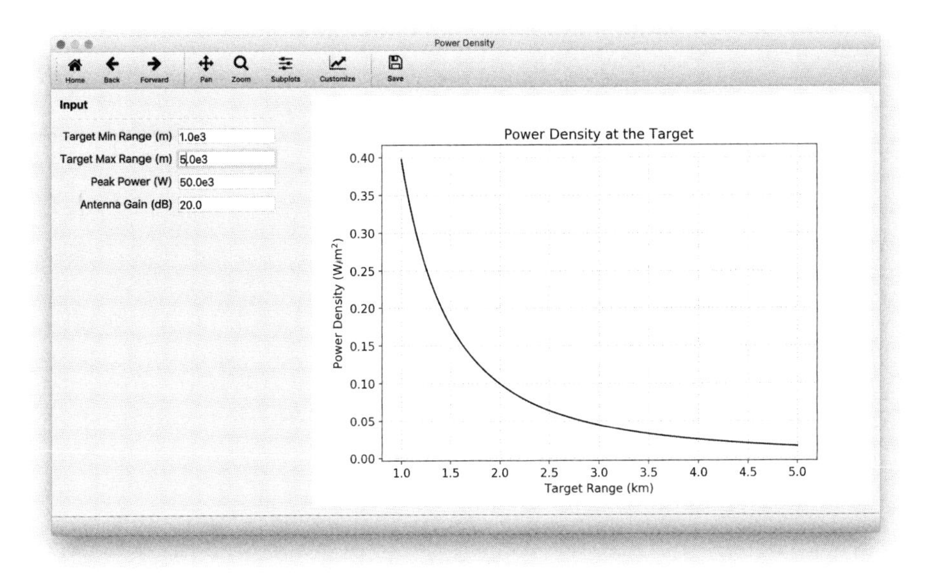

Figure 4.10 The incident power density calculated by *power_density_example.py*.

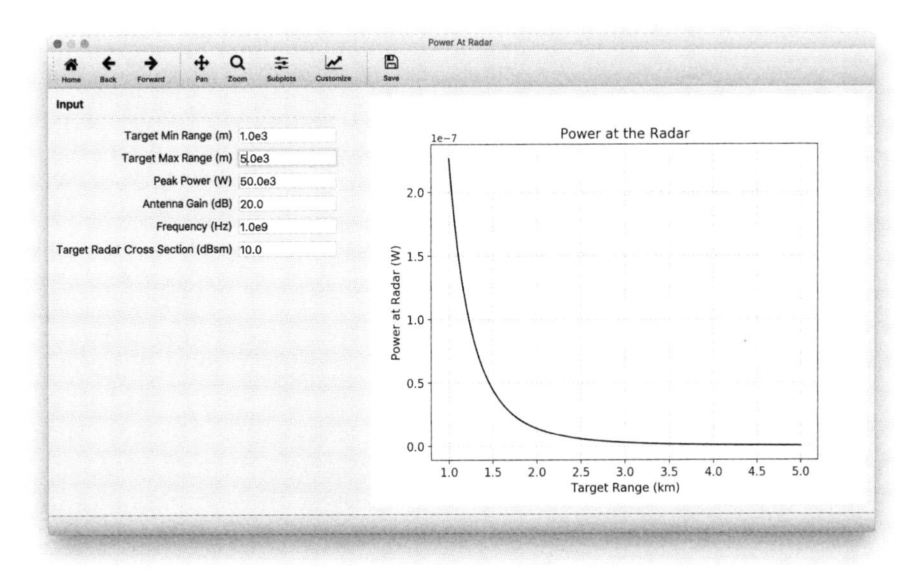

Figure 4.11 The power returned to the radar calculated by *power_at_radar_example.py*.

4.5.2.3 Minimum Detectable Signal

In the previous examples, the effects of noise and losses on the radar range were not considered. Adding these two effects, calculate and display the minimum detectable signal for a radar system having a system temperature of 290° K, a receiver bandwidth of 10 MHz, a noise figure of 6 dB, losses of 4 dB, and requires a signal-to-noise ratio of 20 dB for detection.

Solution: The solution to the above example is given in the Python code *minimum_detectable_signal_example.py* and in the MATLAB code *minimum_detectable_signal_example.m*. Running the Python example code displays a GUI, allowing the user to enter the radar parameters given above. The code then calculates and displays the minimum detectable signal, as shown in Figure 4.12.

4.5.2.4 Maximum Detection Range

For this example, calculate and display the maximum detection range as a function of signal-to-noise ratio for a radar system with a system temperature of 290° K, receiver bandwidth of 10 MHz, noise figure of 6 dB, losses of 4 dB, peak transmitter power of 50 kW, antenna gain of 20 dB, operating frequency of 1 GHz, and the target has a radar

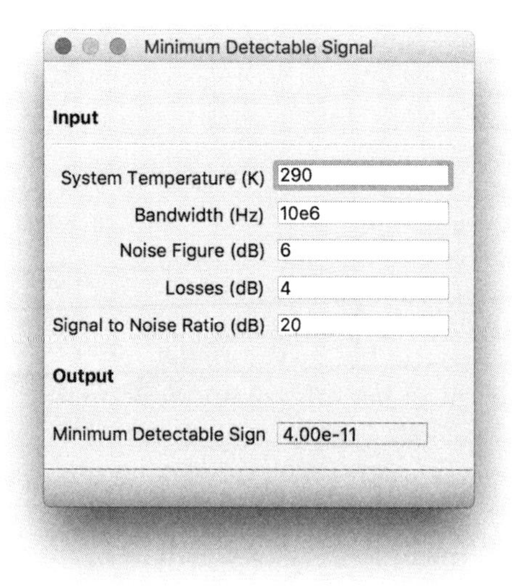

Figure 4.12 The minimum detectable signal calculated by *minimum_detectable_signal_example.py*.

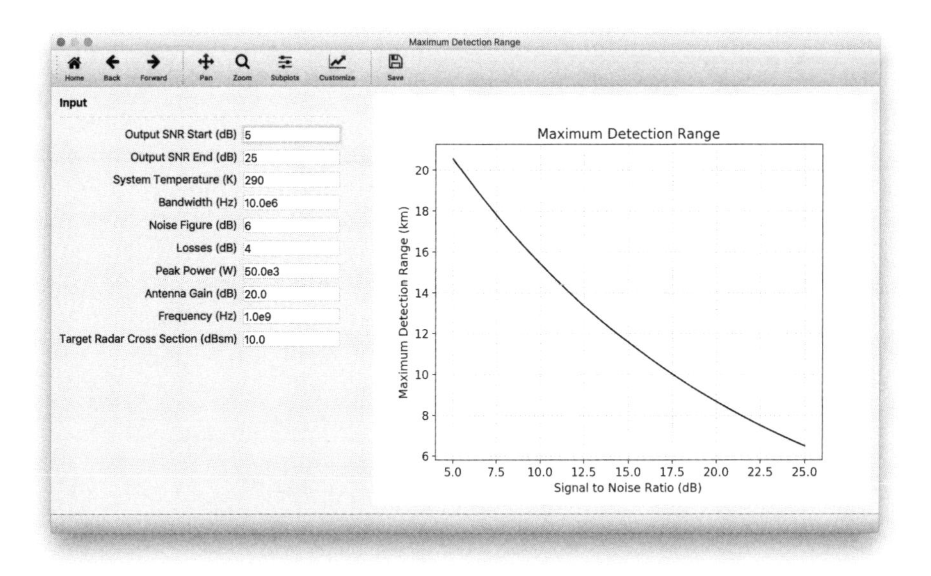

Figure 4.13 The maximum detection range calculated by *maximum_detection_range_example.py*.

cross section of 10 dBsm.

Solution: The solution to the above example is given in the Python code *maximum_detection_range_example.py* and in the MATLAB code *maximum_detection_range_example.m*. Running the Python example code displays a GUI, allowing the user to enter the radar and target parameters. The code then calculates and displays the maximum detection range as a function of signal-to-noise ratio, as shown in Figure 4.13.

4.5.2.5 Output Signal-to-Noise Ratio

In this example, calculate and display the output signal-to-noise ratio as a function of target range for a radar system with a system temperature of 290° K, receiver bandwidth of 10 MHz, noise figure of 6 dB, losses of 4 dB, peak transmitter power of 50 kW, antenna gain of 20 dB, operating frequency of 1 GHz, and the target has a radar cross section of 10 dBsm. Allow the target range to vary from 1 to 100 km.

Solution: The solution to the above example is given in the Python code *output_signal_to_noise_example.py* and in the MATLAB code *output_signal_to_noise_example.m*. Running the Python example code displays a GUI, allowing the user to enter the radar and target

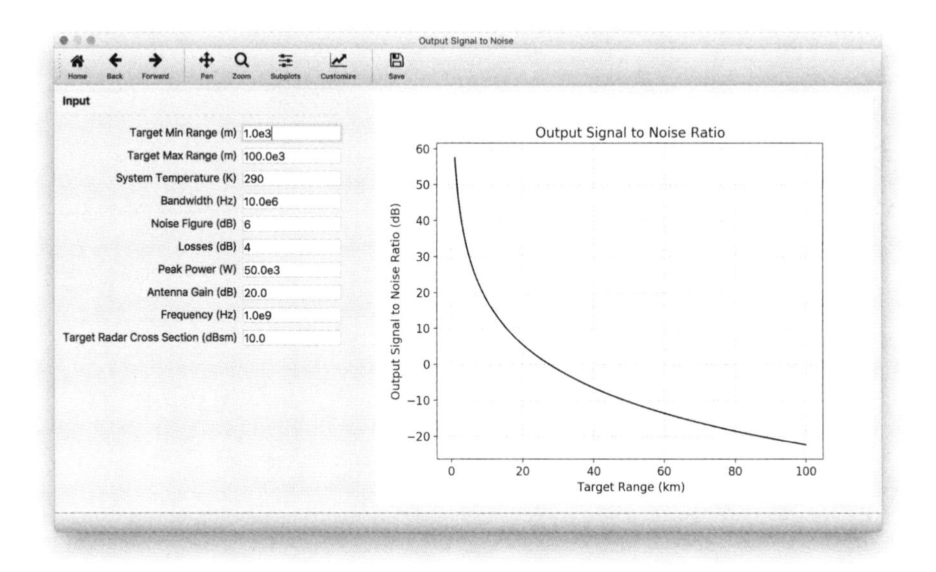

Figure 4.14 The output signal-to-noise ratio calculated by *output_signal_to_noise_example.py*.

parameters. The code then calculates and displays the output signal-to-noise ratio as a function of target range, as shown in Figure 4.14.

4.5.2.6 Loop Gain

This example studies radar sensitivity with the radar loop gain. Calculate and display the radar loop gain for a system that can achieve a signal-to-noise ratio of 20 dB on a target with a radar cross section of 10 dBsm at a range of 100 km.

Solution: The solution to the above example is given in the Python code *loop_gain_example.py* and in the MATLAB code *loop_gain_example.m*. Running the Python example code displays a GUI, allowing the user to enter the radar system's reference range, reference signal-to-noise ratio, and the reference target radar cross section. The code then calculates and displays the resulting loop gain, as shown in Figure 4.15.

4.5.3 Search Radar Range Equation

The subsections below illustrate a special form of the radar range equation, which was developed for search-specific missions and is described in Section 4.5.3. Python examples for calculating the power aperture and the output signal-to-noise ratio are given.

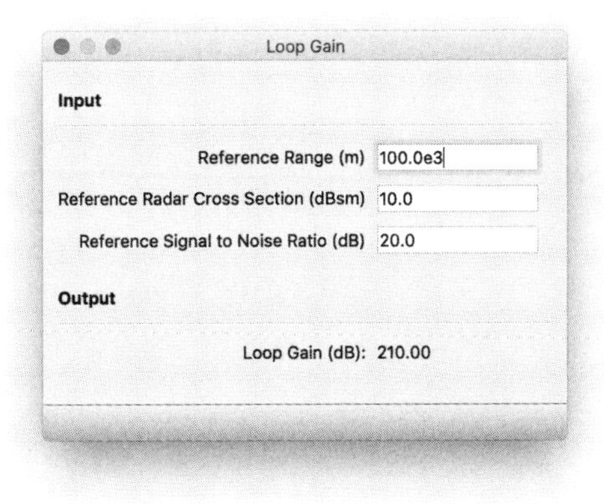

Figure 4.15 The radar loop gain calculated by *loop_gain_example.py*.

4.5.3.1 Power Aperture

For this example, calculate and display the required power aperture to perform a search mission as a function of target range. The radar system has a system temperature of 290° K, noise figure of 6 dB, losses of 4 dB. The system is required to search a volume of 4 steradian in a time of 2 seconds for a target of 10 dBsm and achieves a signal-to-noise ratio of 20 dB.

Solution: The solution to the above example is given in the Python code *power_aperture_example.py* and in the MATLAB code *power_aperture_example.m*. Running the Python example code displays a GUI, allowing the user to enter the target minimum and maximum range, system temperature, search volume, noise figure, losses, signal-to-noise, scan time, and radar cross section. The code then calculates and displays the required power aperture as a function of target range, as shown in Figure 4.16.

4.5.3.2 Output Signal-to-Noise Ratio

Similar to the basic radar equation, this example shows the output signal-to-noise ratio for the search form of the radar range equation. Calculate and display the output signal-to-noise ratio as a function of range for a radar system with a system noise temperature of 290° K, a search volume of 4 steradian, a noise figure of 6 dB, losses of 4 dB, a power aperture of 50 kW m^2, and a scan time of 2 seconds. The target has a radar cross section of 10 dBsm, and the target range varies from 1 to 100 km.

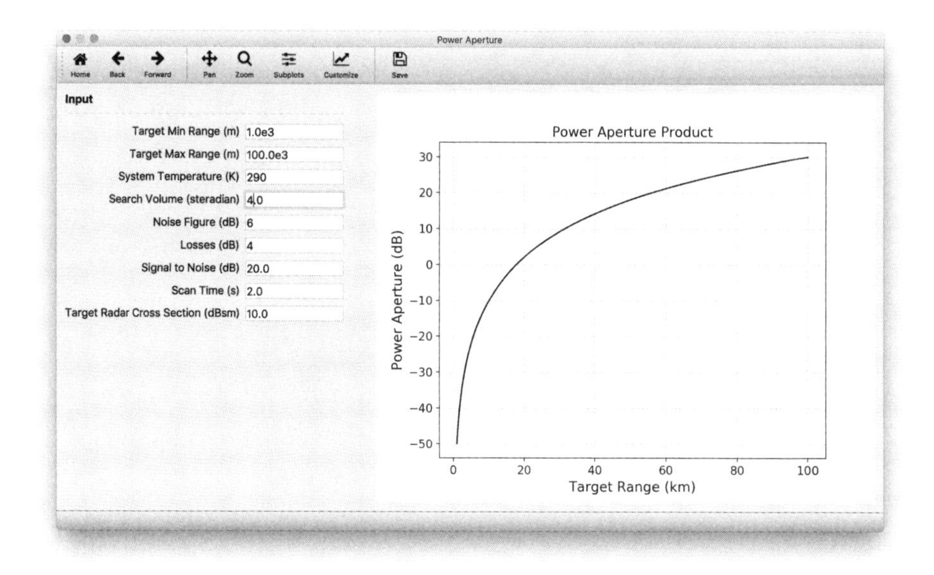

Figure 4.16 The power aperture product calculated by *power_aperture_example.py*.

Solution: The solution to the above example is given in the Python code *output_signal_to_noise_search_example.py* and in the MATLAB code *output_signal_to_noise_search_example.m*. Running the Python example code displays a GUI, allowing the user to enter the target and radar parameters as well as the search parameters. The code then calculates and displays the output signal-to-noise ratio as a function of target range, as shown in Figure 4.17.

4.5.4 Bistatic Radar Range Equation

The subsections below illustrate the bistatic radar range equation that was developed in Section 4.5.4. Python examples are given for the output signal-to-noise ratio, the maximum detection range, and the Cassini ovals.

4.5.4.1 Power at the Radar

Similar to the monostatic case, this example deals with the power back at the radar. However, this is now the power at the receiving radar, which is not collocated with the transmitting radar. Calculate and display the power at the receiving radar as a function of range product. The transmitting system has a peak power of 50 kW, antenna gain of 20 dB, and an operating frequency of 1 GHz. The receiving system has an antenna gain of

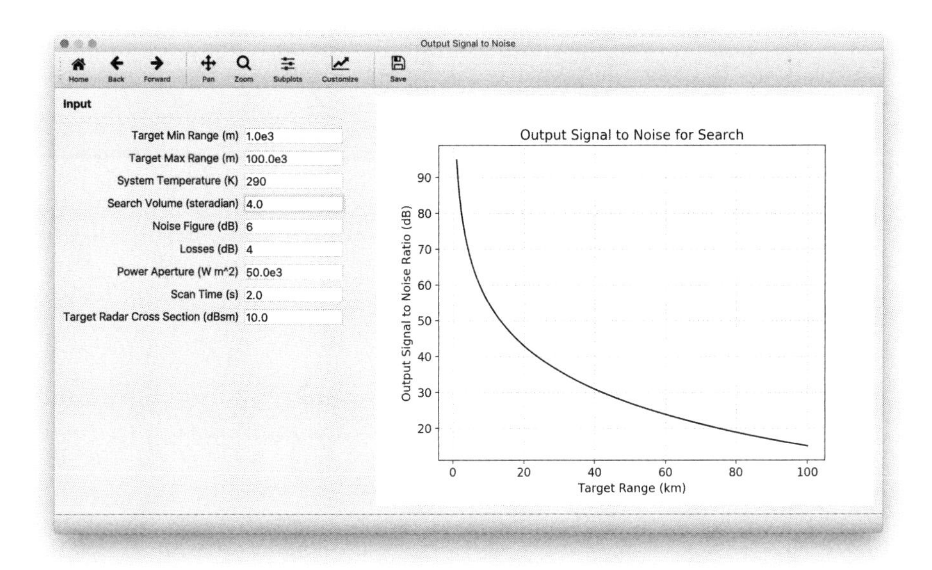

Figure 4.17 The output signal-to-noise ratio for the search equation calculated by *output_signal_to_noise_search_example.py*.

10 dB, and the target has a bistatic radar cross section of 10 dBsm.

Solution: The solution to the above example is given in the Python code *power_at_radar_bistatic_example.py* and in the MATLAB code *power_at_radar_bistatic_example.m*. Running the Python example code displays a GUI, allowing the user to enter the transmitting and receiving system parameters and the target parameters. The code then calculates and displays the power at the receiving radar as a function of range product, as shown in Figure 4.18.

4.5.4.2 Output Signal-to-Noise Ratio

For this example, calculate and display the output signal-to-noise ratio for a bistatic radar scenario in which the transmitting radar has a peak transmit power of 50 kW, antenna gain of 20 dB, losses of 4 dB, and operates at a frequency of 1 GHz. The receiving radar has a system temperature of 290° K, bandwidth of 10 MHz, losses of 2 dB, and an antenna gain of 10 dB. The target has a bistatic radar cross section of 10 dBsm. Plot the output signal-to-noise ratio as a function of range product.

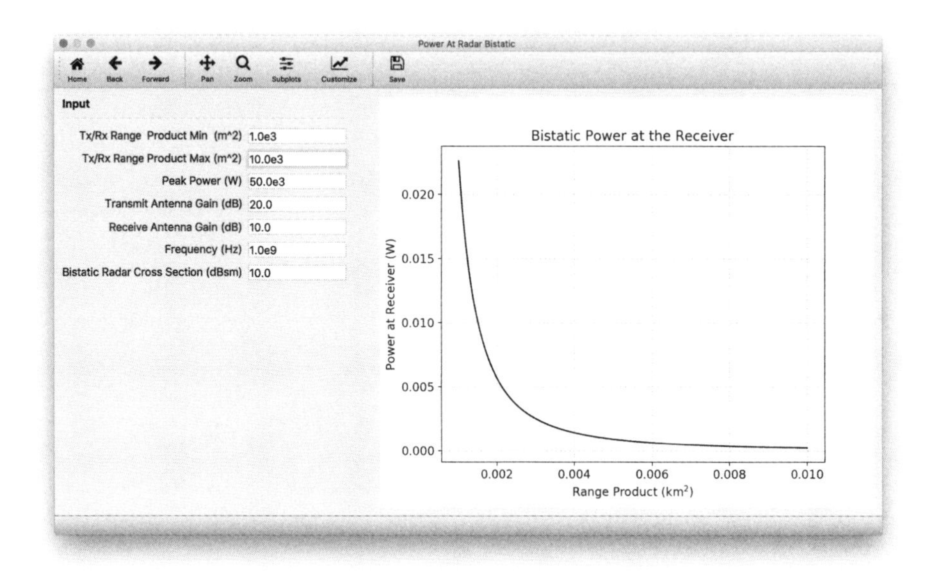

Figure 4.18 The power at the receiver for bistatic radar configurations calculated by *power_at_radar_bistatic_example.py*.

Solution: The solution to the above example is given in the Python code *output_signal_to_noise_bistatic_example.py* and in the MATLAB code *output_signal_to_noise_bistatic_example.m*. Running the Python example code displays a GUI, allowing the user to enter the system parameters for the transmitting and receiving radars as well as the target parameters. The code then calculates and displays the output signal-to-noise ratio as a function of range product, as shown in Figure 4.19.

4.5.4.3 Cassini Ovals

Cassini ovals provide a mechanism for studying the maximum detection range of a bistatic radar system. The Cassini ovals provide a better understanding than the range product alone. For this example, calculate and display the Cassini ovals for the three distinct regions outlined in Section 4.4.1.

Solution: The solution to the above example is given in the Python code *ovals_of_cassini_example.py* and in the MATLAB code *ovals_of_cassini_example.m*. Running the Python example code displays a GUI, allowing the user to enter the parameters for the transmitting and receiving systems and the target parameters. The code then calculates and displays the ovals of Cassini for varying signal-to-noise ratios covering the three regions of Section 4.4.1, as shown in Figure 4.20.

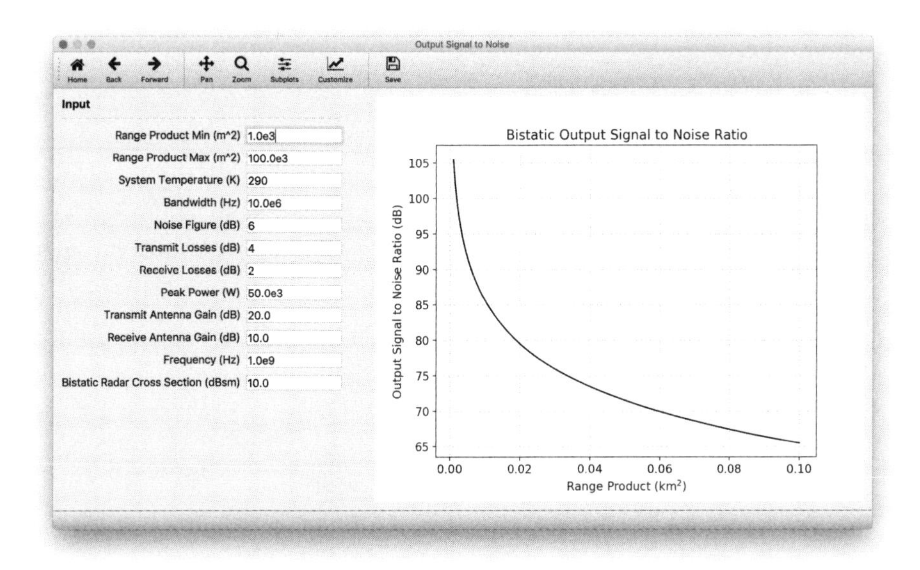

Figure 4.19 The output signal-to-noise ratio for bistatic radar configurations calculated by *output_signal_to_noise_bistatic_example.py*.

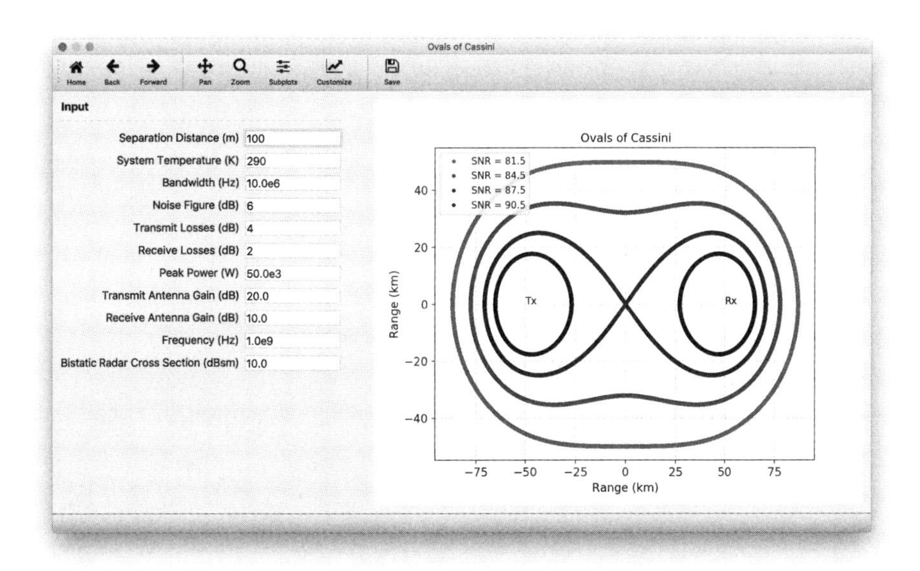

Figure 4.20 Ovals of Cassini for bistatic radar configurations calculated by *ovals_of_cassini_example.py*.

PROBLEMS

4.1 Explain the significance of the radar cross-section values in Table 4.1 in relationship to radar performance and antenna selection.

4.2 Consider a radar system employing a Hertzian dipole of length 0.1 m operating at a frequency of 100 MHz. Calculate the power density for a target at a range of 3 km and elevation of 15° above the $x - y$ plane of the antenna. Assume the current on the dipole is 1 A.

4.3 For the scenario in Problem 4.2, find the power scattered as the target is varied. Use radar cross-section values for a commercial jet, small fighter aircraft, and surface-to-air missile given in Table 4.1.

4.4 For each of the targets given in Problem 4.3, calculate the power received at the radar.

4.5 Assume the minimum power required for target detection is 1 μW and then determine the maximum detection range for each of the targets given in Problem 4.3.

4.6 Calculate the minimum signal power required for target detection when the system temperature is 290° K, bandwidth is 10 MHz, noise factor is 0.5, and the signal-to-noise ratio required for detection is 15 dB.

4.7 For a radar system with an operating temperature of 300° K and operating bandwidth of 10 MHz, find the input noise power.

4.8 Describe the noise factor and how it affects the detection range and received signal-to-noise ratio.

4.9 Find the maximum detection range for a target with a radar cross section of 10 dBsm. The radar system has a bandwidth of 20 MHz, noise figure of 6 dB, losses of 4 dB, antenna gain of 20 dB, transmits a peak power of 50 kW, and operates at a frequency of 1 GHz. Assume the system temperature is 380° K.

4.10 For the scenario given in Problem 4.9, calculate the output signal-to-noise ratio when the target is at a range of 10 km.

4.11 Suppose a radar system is capable of producing an output signal-to-noise ratio of 20 dB on a 10 dBsm target at a range of 750 km. Determine the output signal-to-noise ratio achieved for a -10 dBsm target at a range of 500 km.

4.12 A search radar is required to search a volume given by an azimuth extent of $\pi/4$ radians, and elevation extent of $\pi/6$ radians. Find the total number of beams required to perform the search. Assume the radar has 3-dB beamwidths of 0.1° and 0.2° in azimuth and elevation and uses hexagonal beam packing with 3 dB overlap.

4.13 Calculate the power aperture product for a search radar with an operating temperature of 290° K, noise figure of 3 dB, and losses of 3 dB. The radar is required to search a volume of 6 steradians in 2.0 s and must produce a signal-to-noise ratio of 12 dB for a 5 dBsm target.

4.14 For the radar parameters given in Problem 4.13, find the output signal-to noise ratio for a target with a radar cross section of 12 dBsm located at a range of 18 km.

4.15 Determine the output signal-to-noise ratio for a bistatic radar system given the transmitting radar has a peak power of 100 kW, antenna gain of 30 dB, and losses of 6 dB. The receiving radar has an operating bandwidth of 10 MHz, noise figure of 9 dB, losses of 2 dB, and antenna gain of 20 dB. The operating frequency is 10 GHz, the target has a radar cross section of -3 dBsm, and is located at a range product of $100,000\ \mathrm{m}^2$.

4.16 Using the radar and target parameters given in Problem 4.15, calculate the maximum detection range product when the minimum output signal-to-noise ratio required for detection is 10 dB.

4.17 Describe the ovals of Cassini and their significance in the design of bistatic radar systems. Also explain the conditions necessary for two separate ovals enclosing the transmitter and receiver.

References

[1] C. A. Balanis. *Advanced Engineering Electromagnetics,* 2nd ed. John Wiley and Sons, New York, 2012.

[2] C. A. Balanis. *Antenna Theory Analysis and Design,* 3rd ed. John Wiley and Sons, New York, 2005.

[3] D. Pozar. *Microwave Engineering,* 4th ed. John Wiley and Sons, New York, 2012.

[4] A. Doerry. Noise and noise figure for radar receivers. Technical Report SAND2016-9649, Sandia National Laboratories, Albuquerque, NM, October 2016.

[5] Joint Range Instrumentation Accuracy Improvement Group. Radar loop gain measurements. Technical Report 754-98, U.S. Army White Sands Missile Range, New Mexico 88002-5110, 1998.

[6] M. C. Budge and S. R. German. *Basic Radar Analysis.* Artech House, Norwood, MA, 2015.

[7] C. L. Teo. Bistatic radar system analysis and software development. Technical report, Naval Postgraduate School, Monterey, CA 93943, 2003.

[8] Eric W. Weisstein. *Cassini Ovals.* `http://mathworld.wolfram.com/CassiniOvals.html`.

Chapter 5

Radar Receivers

The function of the receiver is to take the signals received by the antenna, amplify, filter, downconvert, and digitize these signals, and deliver them to the radar signal and data processors. This chapter discusses how the receiver performs these functions, starting with common receiver configurations and components that make up the receive chain. Next, some of the fundamental parameters of receivers, including noise, dynamic range, and bandwidth, are presented. The filtering of unwanted signals, demodulation of the radar signal, and analog-to-digital conversion are covered as these steps prepare the return signal for further digital processing. Finally, a brief review of digital receivers is presented. The chapter concludes with several Python examples to emphasize the key functions of radar receivers.

5.1 CONFIGURATIONS

With modern antenna and receiver hardware, the reference boundaries of the radar receiver can become ambiguous. For example, conventional radar systems make use of a duplexer to provide the input to the receiver from the antenna. However, active array antennas include low noise amplifiers prior to forming receive beams. While these amplifiers are part of the antenna system, they are sometimes included as part of the receiver for analysis. For this book, the components illustrated in Figure 5.1 are considered to be the radar receiver. While there are many types of receivers, including crystal video, homodyne, and superregenerative, most radar systems employ a superheterodyne receiver, as shown in Figure 5.1. The superheterodyne receiver was developed by Edwin Armstrong [1–3] in 1918 during World War I. It was originally designed as a radio receiver that converts the received signal to an intermediate frequency (IF), which is useful for many reasons discussed in the following sections. The superheterodyne receiver adjusts the local oscillator to follow the tuning of the transmitter without disturbing the filtering

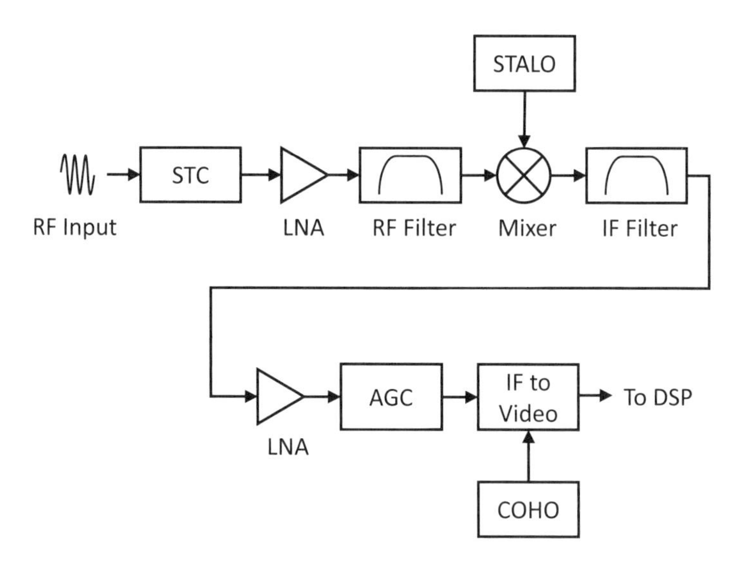

Figure 5.1 Typical superheterodyne radar receiver.

at the intermediate frequency. Since the intermediate frequency filters are all set to a fixed frequency, this makes their design, fabrication, and tuning simpler. This also allows the signals to occupy a wider percentage bandwidth, and to have very selective filters. Another advantage is transistors operating at lower frequencies generally have higher gain. To arrive at the final desired intermediate frequency, there may be more than one stage of conversion. Often double or triple conversion is necessary to avoid image frequency and spurious frequency problems occurring in the mixing process.

Referring to Figure 5.1, the receiver chain begins with a *sensitivity time control* (STC) attenuator. While other types of gain control may be used, their purpose is to increase the dynamic range over that of a fixed gain system. Radar systems receive signals of widely varying amplitude due to differences in radar cross section, atmospheric conditions, clutter, and range. The effect of range on the strength of the return signal is often much greater than other causes. This effect may be mitigated with the STC attenuator, which adjusts the attenuation to cause the received signal strength to be independent of range. The next element in the receiver chain is a *low-noise amplifier* (LNA), which provides gain with low noise in order to minimize signal degradation by subsequent components. Following the LNA is a bandpass filter that rejects out-of-band signals, including the image frequency. The mixer then uses a highly stable local oscillator (STALO) to downconvert the received signals to the intermediate frequency. After downconversion to the intermediate frequency, a bandpass filter is used to reject unwanted signals arising from the mixing process and to set the receiver processing bandwidth. An additional LNA is used to increase the signal level along with *automatic gain control* (AGC) attenuators to set the signal levels for analog-to-digital conversion. Finally, the IF to video conversion takes place, which produces data suitable for digital signal processing. This conversion makes use of a highly stable coherent oscillator (COHO) locked to the IF frequency.

5.2 NOISE

As illustrated in Chapter 4, noise is one of the fundamental limiting factors in the radar range equation. The received signal is often corrupted by noise generated internal to the receiver. The major source of noise in receivers is thermal noise that can obscure weak signals [4]. The noise generated in the receiver is random, and statistical techniques are used to characterize its effect. The noise level at the input to the receiver is primarily determined by the antenna noise temperature and its associated loss. The noise level is usually specified by a power spectral density as discussed in Section 4.2.2, or alternatively as an average power level over a specified bandwidth. The system noise is then the combined antenna noise and receiver noise:

$$T_s = T_a + T_r = T_a + L_r\, T_0\, (F - 1) \qquad \text{(K)} \tag{5.1}$$

where

T_s = total system noise temperature (K),
T_a = antenna noise temperature (K),
T_r = receiver noise temperature (K),
L_r = receiver loss,
T_0 = receiver operating temperature (K),
F = receiver noise factor.

The noise due to the receiver is typically small compared to the noise input to the receiver. Therefore, receiver noise only has a small effect on the total system noise temperature. This is important when calculating parameters such as signal-to-noise ratio as the SNR referenced to the receiver noise can be very different from the SNR referenced to the total system noise [5]. From Section 4.2.2, the noise figure is the ratio of the input signal-to-noise ratio to the output signal-to-noise ratio. For cascaded networks, the noise figure is the contribution from each stage in the receive chain as

$$F_{total} = F_1 + \sum_{i=2}^{N} \left(\frac{F_i - 1}{\prod_{j=1}^{i-1} G_j} \right). \tag{5.2}$$

For example, Figure 5.2 shows a five-stage network with the components and their parameters given in Table 5.1. Expanding (5.2), the total noise factor is then

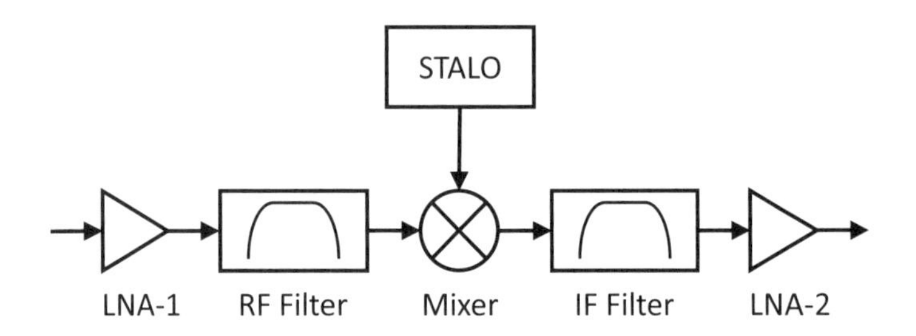

Figure 5.2 Five-stage network for noise figure analysis of a cascaded network.

Table 5.1
Receiver Components for Cascaded Noise Figure Analysis

Component	*Gain (dB)*	*Noise Figure (dB)*
LNA-1	20	3
RF filter	−0.5	0.5
Mixer	−6	6
IF filter	−1	1
LNA-2	30	5

$$F_{total} = F_1 + \frac{F_2 - 1}{G_1} + \frac{F_3 - 1}{G_1\,G_2} + \frac{F_4 - 1}{G_1\,G_2\,G_3} + \frac{F_5 - 1}{G_1\,G_2\,G_3\,G_4}$$

$$= 1.995 + \frac{0.122}{100} + \frac{2.981}{89.13} + \frac{0.259}{22.39} + \frac{2.162}{17.78} = 2.16, \tag{5.3}$$

or expressed in decibels, the noise figure is

$$F_{total} = 10\log_{10}(2.16) = 3.35 \qquad \text{(dB)}. \tag{5.4}$$

The analysis of radar systems if often concerned with ratios such as signal-to-noise ratio, signal-to-clutter ratio, signal-to-clutter plus noise, and others. However, knowledge of absolute voltages and currents is important when dealing with radar receivers. For example, knowledge of absolute voltages at the input of the analog-to-digital converter (ADC) is crucial to prevent saturating the components. Likewise, the level of the noise floor is also very important as this sets the absolute level for weak signals competing with the noise [6]. The absolute noise level at each stage in the receiver is found from

$$N_i = kT_0B_iF \qquad \text{(W)}, \tag{5.5}$$

where the following are at the output of each stage of the receiver

k	=	Boltzmann's constant, 1.380649×10^{-23} (J/K),
T_0	=	reference temperature, typically 290° (K),
F	=	noise factor,
B_i	=	bandwidth (Hz).

5.3 DYNAMIC RANGE

The dynamic range of a receiver is the range of signal strength over which the receiver is able to perform its functions without distortion [5]. It is the difference between the maximum and minimum usable signals through the receiver. The strength of the return signal to the radar can vary widely, and therefore a receiver with a large dynamic range is desirable. Referring to (4.28), the variation in the received signal strength is caused by differences in radar cross section of objects in the scene, the distance to the objects, and the antenna gain in the direction of the objects. The dynamic range is limited by components in the receive chain such as mixers and amplifiers as well as analog-to-digital converters. The minimum usable signal strength is limited by the noise floor, while the upper signal strength is limited by saturation of analog components. The instantaneous dynamic range is the difference between the maximum and minimum signals the receiver can discern at the same time (i.e., the greatest signal-to-noise ratio possible). The limiting component for the instantaneous dynamic range is often the ADC. The total dynamic range is achieved through programmed gain variations and may be much larger than the instantaneous dynamic range, as illustrated in Figure 5.3. Modern radar systems make use of linear receiver channels followed by digital signal processing. The output of a linear amplifier will increase proportionally to an increase in the amplifier input, as shown in Figure 5.4. The limit on the usable signal level is the *1-dB compression point*, illustrated in Figure 5.4. It is the input signal level at which the receiver gain is 1 dB less than the ideal linear signal gain would be. This type of compression may occur in amplifiers and mixers as well as other components.

The dynamic range of the receiver is also limited by spurious signals. These are unwanted signals not deliberately created and occupy noninteger multiples of the input frequency. The *spurious free dynamic range* is the difference between the maximum signal strength and the strength of the largest spurious signal created in the receiver and is shown in Figure 5.5. These spurious signals may appear as false targets or mask targets with weak return signals, and therefore it is desirable to suppress these while maintaining a large dynamic range. These spurious signals may be created in any nonlinear processes in the receiver, including intermodulation in the mixer, spurious

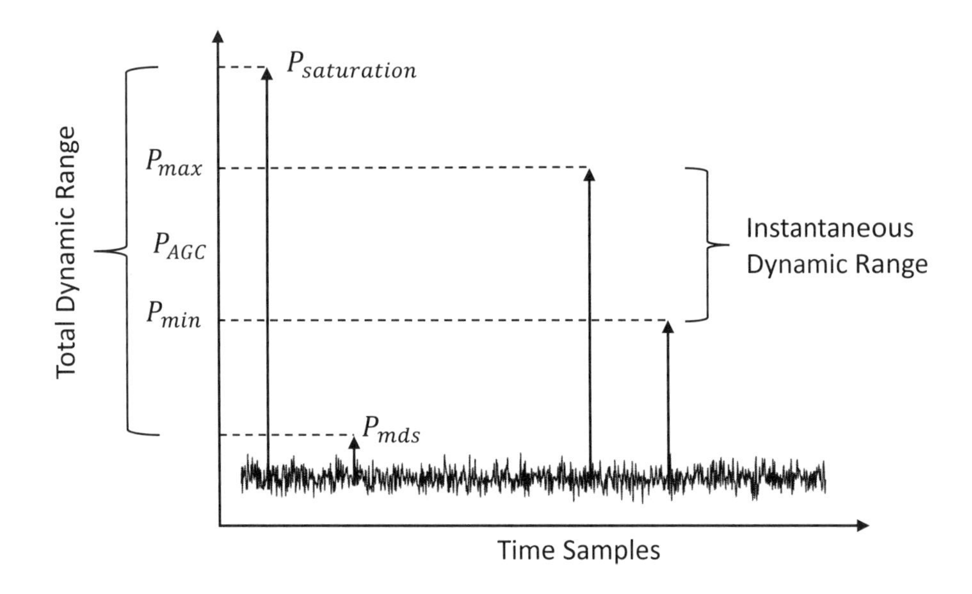

Figure 5.3 Receiver instantaneous and total dynamic range.

content of the local oscillators, the ADC, and other components. Nonlinear processes in the receiver result in *intermodulation distortion*, which generates signals at frequencies that are combinations of the fundamental frequencies of the input signals. Second- and third-order intermodulation becomes dominant at the upper end of the receiver output. Receiver performance is typically specified by the second- and third-order intercept points. The intercept point is the level at which the power of the intermodulation distortion equals that of the two fundamental signals, as shown in Figure 5.6. Given two input frequencies, f_1 and f_2, second-order intermodulation distortion creates signals at frequencies of 0, $f_1 - f_2$, $f_1 + f_2$, $2f_1$ and $2f_2$. Third-order intermodulation distortion creates signals at frequencies of $2f_1 - f_2$, $2f_2 - f_1$, $2f_1 + f_2$, $2f_2 + f_1$, $3f_1$ and $3f_2$. Only the frequencies $2f_1 - f_2$ and $2f_2 - f_1$ are in band for narrowband input signals. Therefore, third-order intermodulation distortion is the most dominant and is the primary concern in receivers. The power associated with third-order intermodulation distortion is given by [5]

$$P_{2f_1-f_2} = 2P_{f_1} + P_{f_2} - 2IP_3 \qquad \text{(dBW)}, \tag{5.6}$$

$$P_{2f_2-f_1} = 2P_{f_2} + P_{f_1} - 2IP_3 \qquad \text{(dBW)}, \tag{5.7}$$

where,

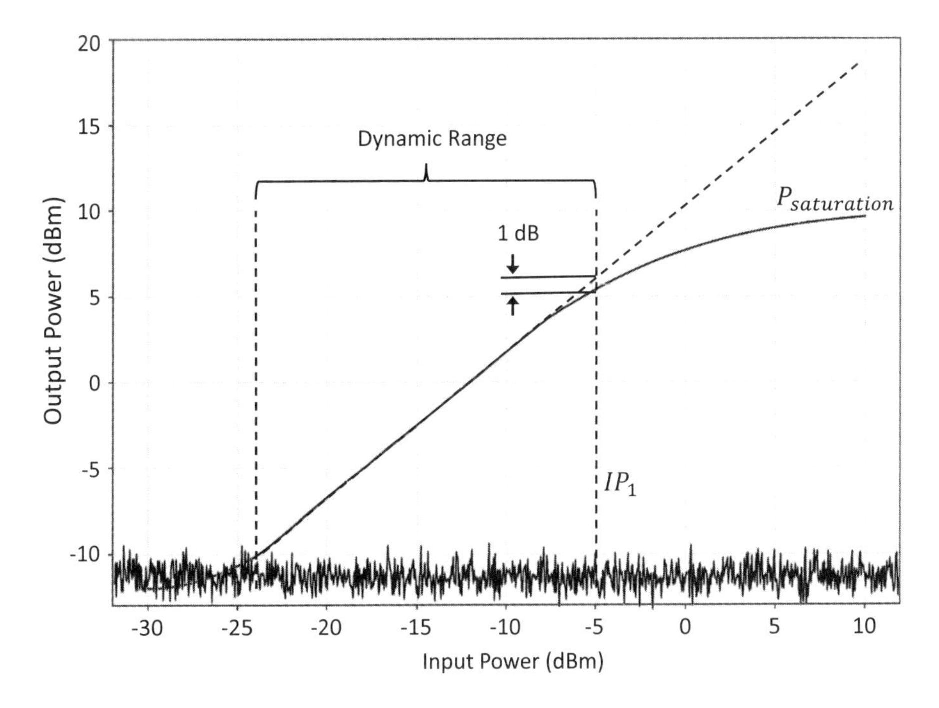

Figure 5.4 Linear receiver distortion vs input signal.

$P_{2f_1-f_2}$ = power of first intermodulation distortion frequency (dBW),
$P_{2f_2-f_1}$ = power of second intermodulation distortion frequency (dBW),
P_{f_1} = power of first input frequency (dBW),
P_{f_2} = power of second input frequency (dBW),
IP_3 = third-order intercept point (dBW).

5.4 BANDWIDTH

For radar receivers, a distinction must be made between instantaneous and operating bandwidth. The instantaneous bandwidth is the range of frequencies over which the receiver can process two or more signals simultaneously, whereas the operating bandwidth is the entire range of frequencies the radar receiver may tune, as illustrated in Figure 5.7. For radar systems, the term instantaneous bandwidth may refer to various parameters such as the RF processing bandwidth, waveform bandwidth, or IF processing bandwidth. When dealing with the receiver, the instantaneous bandwidth refers to the resulting bandwidth primarily set by the IF filtering stages. However, the RF portion cannot be ignored

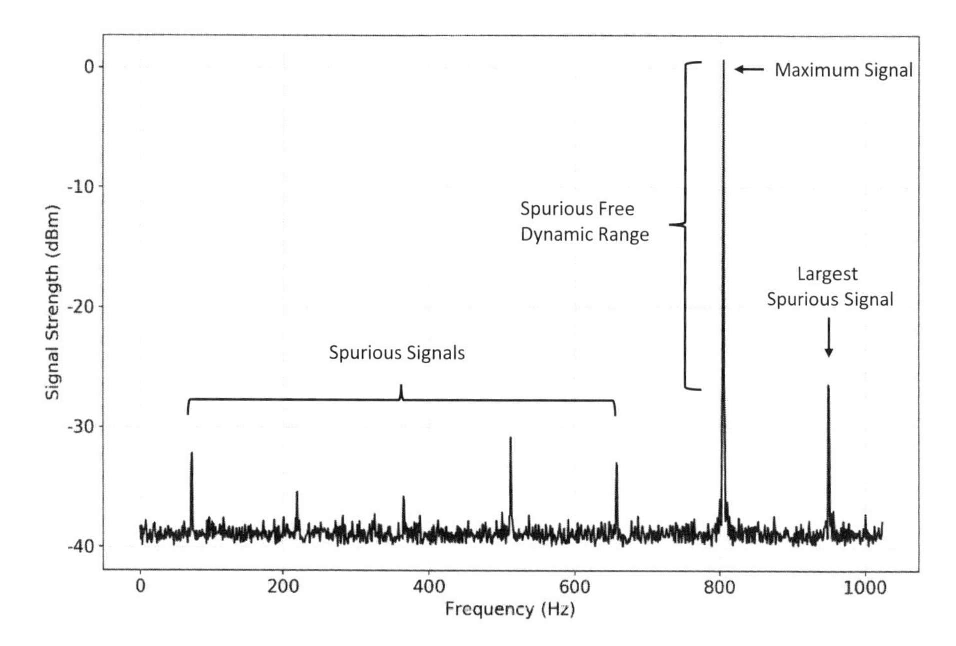

Figure 5.5 Spurious free dynamic range.

as wide bandwidth may have a negative impact on dynamic range. As discussed in Section 5.3, there may be out-of-band signals converted to the IF frequency by the spurious response of a mixer as well as wideband noise that raises the noise floor and thus lowers the signal to noise ratio.

5.5 GAIN CONTROL

As discussed in Section 5.3, the strength of the radar return signal can vary widely. This variation often exceeds the dynamic range of fixed gain systems. One method for overcoming this limitation is through the use of gain control. The first type of gain control considered is STC. For targets of constant radar cross section, their return signal strength decreases with range as $1/r^4$, where r is the range from the radar to the target. Decreasing the attenuation as a function of $1/r^4$ would maintain a return signal of constant amplitude versus range. Surface clutter tends to decrease as $1/r^3$ [7]. Therefore, the STC attenuator shown in Figure 5.1 is set to reduce the attenuation as $1/r^{3.5}$ to provide a good compromise between the targets and clutter. STC may be used with low and medium pulse repetition frequency (PRF) waveforms, such as those used for search and track, but not with high PRF waveforms as distant targets competing with nearby clutter would be attenuated [6]. With AGC, the attenuation is based on the strength of

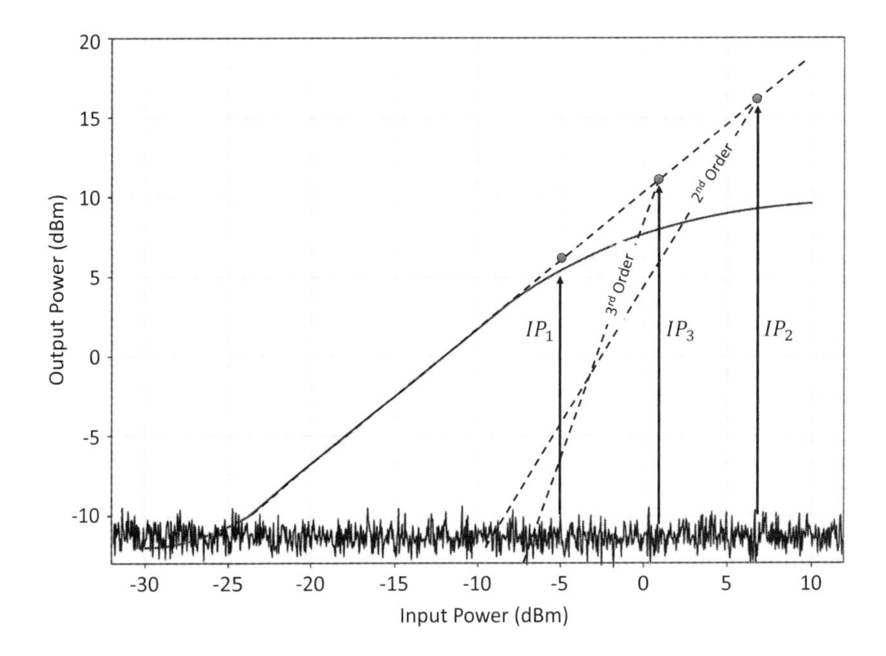

Figure 5.6 Receiver intermodulation distortion with intercept points.

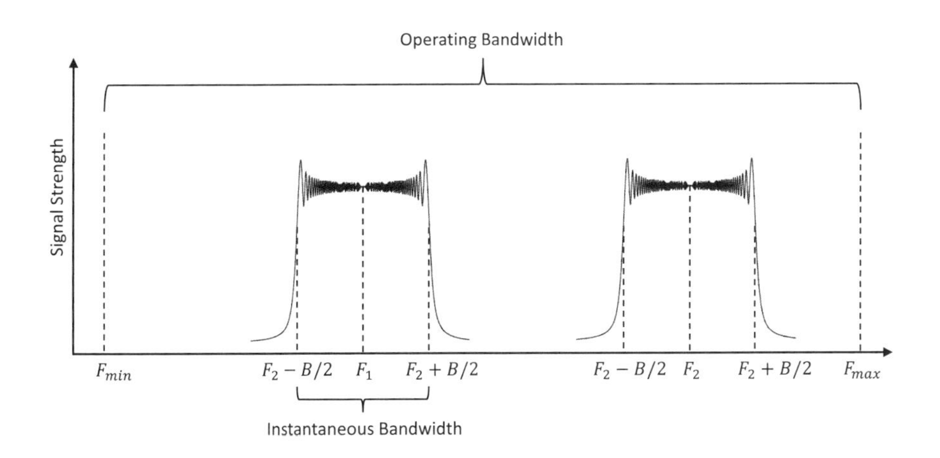

Figure 5.7 Receiver instantaneous and operating bandwidth.

the return signals. While the overall gain of the system may be set manually by the operator, this is not ideal for individual targets as this would require frequent adjustment by the operator. AGC is typically achieved by adjusting the gain through the use of an IF attenuator. When tracking targets, AGC may be used to continuously adjust the attenuation to provide a nearly constant output. If the AGC is used to separately adjust the attenuation for each target separated in range, then the system maintains a constant angle measurement sensitivity. Gain control may also be used for clutter map gain control, which is useful when operating in high clutter environments. As an alternative to STC, clutter map gain control uses a digital map to record the mean amplitude of the clutter in each cell, and then adjust the attenuation as necessary to ensure the clutter does not saturate the receiver. Another use for gain control is *gain normalization*. It is necessary to have accurate gain control in the receiver for measuring radar cross section, monopulse angle accuracy, maximizing dynamic range, and controlling the noise level. However, this can be challenging as the receiver response varies with changes in the environment, aging of components, and other factors. To overcome these variations in the receiver, a known test signal is injected and measured at the output of the receiver. Then calibration coefficients are calculated and stored. The attenuation is then set as a function of frequency, temperature, humidity, etc. Finally, gain control may be used for automatic control of the noise level. It is often desired to maintain a certain noise level input to the ADC. If the noise level is too small compared to the quantization of the ADC this may cause a reduction in the receiver sensitivity. This is particularly important in systems employing STC, as the noise level at close range may fall below the quantization level. Often, a noise source with an attenuator is used to inject noise into the system to overcome the noise reduction caused by the STC attenuator. To ensure a constant noise level at the ADC input, the injected noise is synchronized with the STC attenuator.

5.6 FILTERING

The main purpose of filtering in the receiver is to eliminate unwanted interference and is performed at different stages throughout the receive chain. Filtering is performed on the radio frequency, intermediate frequency, and baseband signals. As illustrated in Section 5.3, spurious signals may be generated in the mixing process and translated to the intermediate frequency. One method for reducing these unwanted spurious signals is to filter prior to the mixing stage. Filtering may also be performed in the digital signal processing function and is a key part of generating in-phase and quadrature data, as will be shown in Section 5.7.

Beginning at the RF level, the main purpose of filtering is to reject the image signal that is created during the first downconversion. The image frequency is shown in Figure 5.8. The RF carrier frequency is $1,000$ MHz, and the desired IF frequency is 30 MHz. The frequency of the local oscillator is 970 MHz. The image frequency is located at

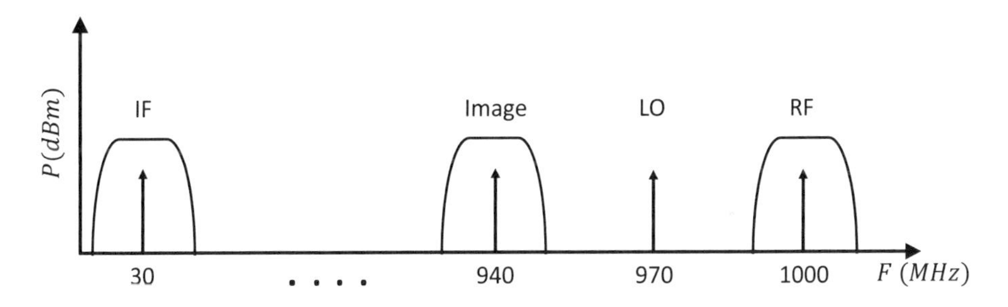

Figure 5.8 Image frequency and RF filtering.

$$f_{image} = f_{RF} - 2f_{IF} = 940 \qquad \text{(MHz)}, \tag{5.8}$$

where

f_{RF} = RF carrier frequency (MHz),
f_{IF} = desired IF frequency (MHz),
f_{image} = image frequency (MHz).

Many systems employ two- or three-step downconversion to help eliminate the image signals. This way, the image suppression problem is avoided, and a wide tuning band free of spurious signals is also achieved. It is also important to filter at the RF level to reject signals that would cause intermodulation distortion, as shown in Section 5.3. Tunable or switched RF filtering may be required if the receiver bandwidth is a substantial percentage of the RF bandwidth, as this would result in the image signal shifting in the receiver band. Another method for eliminating the image signal is to first upconvert the RF signal. This allows for the use of a single RF filter covering the entire operating bandwidth.

In receivers employing baseband conversion, the IF filtering sets the receiver bandwidth. This is followed by video filtering at an increased bandwidth to prevent an imbalance between the in-phase and quadrature channels. On the other hand, in IF sampling receivers, the final receiver bandwidth is set by digital filtering. This also prevents aliasing in the decimation of the in-phase and quadrature data. Digital filtering is advantageous as it can be customized to the desired passband and stopband characteristics. While many digital filters are linear phase filters, they may also be used to compensate for the amplitude and phase response of the RF and IF analog filters in the receiver. Figure 5.9 shows a comparison of common filter types. As can be seen, the filter responses are quite different. The type of filter chosen depends greatly on the application. For example, Butterworth filters have quite flat responses in the passband. This type of filter should be used for cases where minimal distortion of the signal is required, such as filtering a signal prior to analog-to-digital conversion. Chebyshev filters, on the other hand, should be chosen when the frequency content of the signal is more important than passband flatness. An example of this would be trying to separate signals closely spaced in frequency. Elliptic filters are

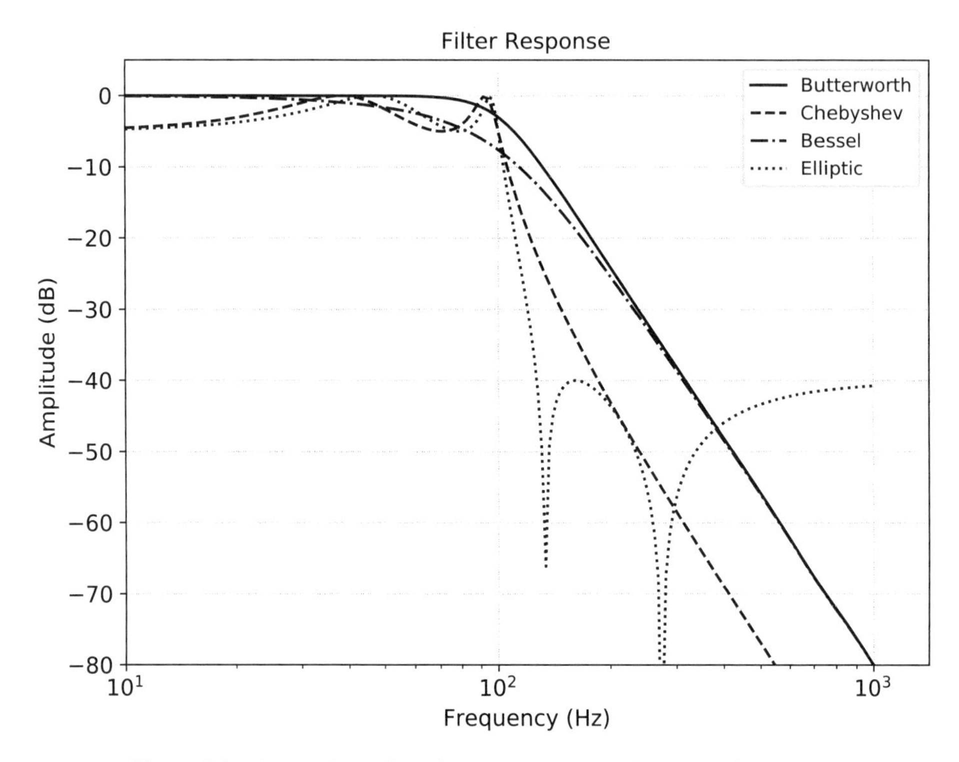

Figure 5.9 Comparison of the frequency response of common filter types.

more difficult to design but do have the advantage of providing the fastest roll-off for a given number of poles.

The filter responses above maybe calculated in Python using the SciPy signal processing module ***scipy.signal*** [8]. For example, the filter response for an analog, low-pass, fourth-order, Butterworth filter with a critical frequency of 100 Hz is given in Listing 5.1. In the listing, ***signal.freqs(b,a)*** computes the frequency response of an analog filter given the numerator, ***b***, and denominator, ***a***. The frequency response may then be displayed using Matplotlib's [9] plotting functions, as given in Listing 5.2. The Chebyshev, Bessel, and Elliptic filter responses shown in Figure 5.9 may be calculated in Python as illustrated in Listing 5.3.

Listing 5.1 Butterworth Low-Pass Filter Response

```
from scipy import signal
b, a = signal.butter(4, 100, 'low', analog=True)
w, h = signal.freqs(b, a)
```

Listing 5.2 Butterworth Filter Response Plot

```
import matplotlib.pyplot as plt
from scipy import log10
plt.semilogx(w, 20 * log10(abs(h)))
plt.title('Butterworth Filter Frequency Response')
plt.xlabel('Frequency (Hz)')
plt.ylabel('Amplitude (dB)')
plt.show()
```

Listing 5.3 Chebyshev, Bessel, and Elliptic Filter Response Calculations

```
from scipy import signal
"""
Chebyshev, 4th order, low pass, analog filter with maximum
passband ripple = 3 dB and critical frequency = 100 Hz.
"""
b, a = signal.cheby1(4, 3, 100, 'low', analog=True)

"""
Bessel, 4th order, low pass, analog filter with
critical frequency = 100 Hz.
"""
b, a = signal.bessel(4, 100, 'low', analog=True, norm='phase')

"""
Elliptic, 4th order, low pass, analog filter with maximum
passband ripple = 3 dB, stop band attenuation = 40 dB
and critical frequency = 100 Hz.
"""
b, a = signal.ellip(4, 3, 40, 100, 'low', analog=True)
```

5.7 DEMODULATION

Prior to the radar return signals being sent to the signal processor, the radar receiver must downconvert the signals to either baseband or an IF frequency and then perform either coherent or noncoherent detection. This section examines both coherent and noncoherent receivers and gives examples of both.

5.7.1 Noncoherent Detection

Noncoherent detection does not preserve the phase information contained in the return signal. Noncoherent detection results in the amplitude-only envelope of the IF signal. Noncoherent receivers are often found in simplistic radar systems as they do not require complicated hardware and software. Figure 5.10 shows the simplest of noncoherent detectors known as a peak detector or envelope detector. It consists of a diode and a low-pass filter. The diode serves to detect the amplitude modulation of the IF signal, and the low-pass filter removes the IF frequency component, as shown in Figure 5.10. To operate properly, the diode current must remain in the square law region. The ideal diode current is written as

$$I_D = I_s\left(e^{V_D/V_T} - 1\right) \qquad \text{(A)}, \tag{5.9}$$

where

I_s = reverse bias saturation current (A),
V_D = diode voltage (V),
V_T = thermal voltage (V),
I_D = diode current (A).

The thermal voltage is given by

$$V_T = \frac{kT}{q} \qquad \text{(V)}, \tag{5.10}$$

where

k = Boltzmann's constant (1.380649×10^{-23}) (J/K),
T = diode temperature (K),
q = electron charge ($1.60217662 \times 10^{-19}$) (C),
V_T = thermal voltage (V).

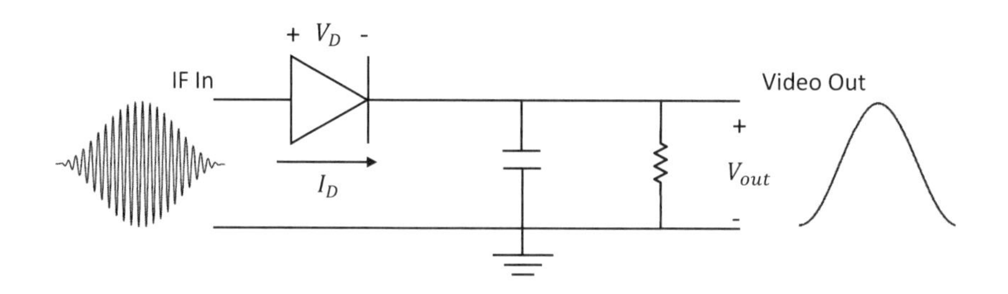

Figure 5.10 Simplistic diode envelope detector.

The diode current in (5.9) may be expanded as

$$I_D = I_s \left[\frac{V_D}{V_T} + \frac{1}{2}\left(\frac{V_D}{V_T}\right)^2 + \frac{1}{6}\left(\frac{V_D}{V_T}\right)^3 + \cdots \right] \quad \text{(A)}. \tag{5.11}$$

The idea is to keep the diode operating in the square law region so the second term on the right-hand side of (5.11) is dominant and the output current is proportional to the input power. The advantage of this type of detector is that it requires few components, making it simple, easy to implement, reliable, and low cost. However, there are also some disadvantages. The diode detector introduces distortion as it is nonlinear. Also, this type of detector suffers from poor sensitivity and selective fading.

5.7.2 Coherent Detection

Radar signal processing functions such as pulse compression, Doppler processing, monopulse comparison, moving target indication, synthetic aperture radar imaging, and space-time adaptive processing all require both amplitude and phase information. To perform coherent detection, the phase of the transmitted signal is locked to the phase of the local oscillator, as shown in Figure 5.1, and then the detector converts the signals from the IF frequency to a complex representation at baseband. Coherent detectors are sometimes referred to as IQ detectors, quadrature detectors, or synchronous detectors. Figure 5.11 illustrates a block diagram of a coherent detector. Mathematically, the band-limited IF signal may be written as

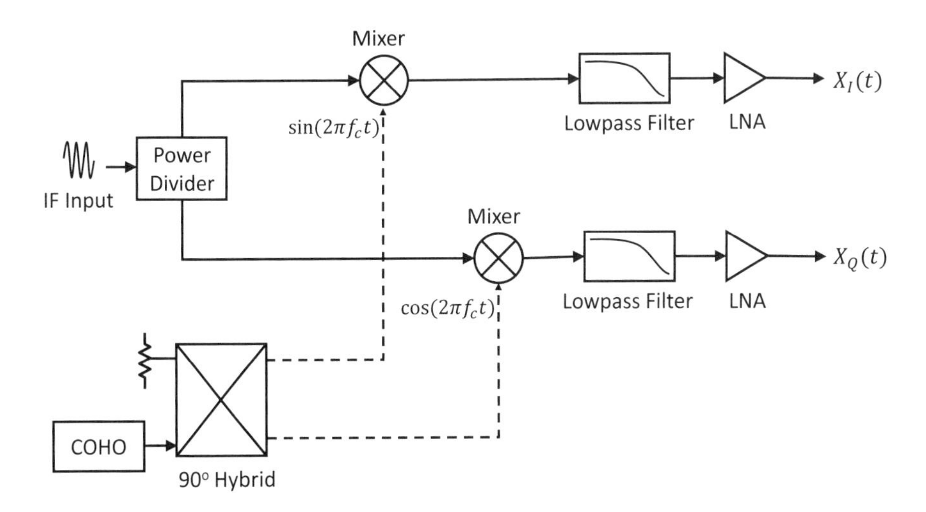

Figure 5.11 Block diagram of a coherent detector.

$$x(t) = a(t)\cos(2\pi f_0 t + \phi(t)) \qquad \text{(V)}, \tag{5.12}$$

where

$a(t)$ = amplitude modulation (envelope) (V),
$\phi(t)$ = phase modulation (rad),
f_0 = carrier frequency (Hz),
$x(t)$ = band limited IF signal (V).

Using the identity $\cos(a+b) = \cos(a)\cos(b) - \sin(a)\sin(b)$, the IF signal is then written as

$$\begin{aligned} x(t) &= a(t)\cos(2\pi f_0 t)\cos(\phi(t)) - a(t)\sin(2\pi f_0 t)\sin(\phi(t)) \\ &= X_I(t)\cos(2\pi f_0 t) - X_Q \sin(2\pi f_0 t), \end{aligned} \tag{5.13}$$

where

$X_I(t)$ = $a(t)\cos(\phi(t))$, in-phase component (I),
$X_Q(t)$ = $a(t)\sin(\phi(t))$, quadrature component (Q).

The baseband in-phase and quadrature signals are then sampled using a pair of ADCs, providing a representation of the IF signal that preserves the phase information relative to the transmitted signal. The amplitude, $a(t)$, and phase, $\phi(t)$, of the signal may be obtained from the in-phase and quadrature components by

$$a(t) = \sqrt{X_I(t)^2 + X_Q(t)^2}, \tag{5.14}$$

$$\phi(t) = \tan^{-1}\left(\frac{X_Q(t)}{X_I(t)}\right). \tag{5.15}$$

Modern radar systems are moving away from baseband coherent detectors and are moving toward sampling the IF signal directly and then employing digital signal processing to arrive at the baseband signal representation. While baseband coherent detectors are still common, their use is becoming limited to systems employing wide bandwidth signals, as the combination of dynamic range and bandwidth required to perform sampling at the IF frequency is not feasible for current hardware.

5.8 ANALOG-TO-DIGITAL CONVERSION

Modern radar systems make use of a large number of digital signal processing techniques, including those mentioned in Section 5.7. This has led to analog-to-digital converters with higher sampling rates and improved dynamic range. Also, many systems now directly sample the signals at IF, thereby reducing the number of analog components in the system [10]. Modern radar systems operate in various modes employing different bandwidths and having different requirements on dynamic range. Previously, it was common to have different ADCs in different stages of the system. State-of-the-art ADCs are also capable of operating in different modes with programmable sampling times, varying number of channels, interleaving of data, and others. The analog-to-digital conversion process transforms continuous time analog signals into discrete time digital signals by sampling in the time domain and then quantizing the signal levels, as shown in Figure 5.12.

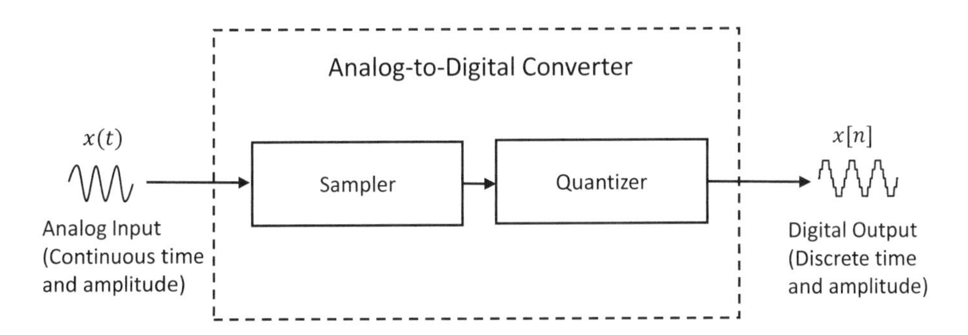

Figure 5.12 Block diagram of an analog-to-digital converter.

5.8.1 Sampling

The requirements on the sampling rate of the ADC is primarily determined by the IF bandwidth. For baseband sampling, illustrated in Figure 5.13, if the IF signal is band-limited with a bandwidth of B, then $X_I(t)$ and $X_Q(t)$ are band-limited with a bandwidth of $B/2$. Therefore, ADC sampling of these two signals at a frequency of $f_s \geq B$ meets the Nyquist criteria [11–13]. For IF sampling, shown in Figure 5.14, a sampling frequency of twice the IF bandwidth is required to avoid aliasing. However, a higher sampling rate is common as this reduces the requirements on alias rejection filtering. This also reduces the effects of quantization noise [5]. When directly sampling the IF signal, it is often intentionally aliased by under sampling with respect to the IF frequency, which places the signal in one of the Nyquist zones [14], as shown in Figure 5.14. Random jitter in the clocks used by the ADC for sampling introduces noise on the ADC output, thereby reducing the signal-to-noise ratio. When sampling at the IF level, clock jitter also

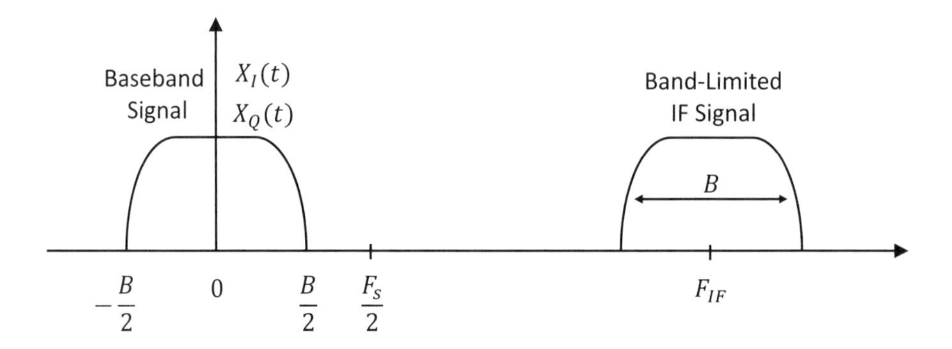

Figure 5.13 The resulting spectra for receiver baseband sampling.

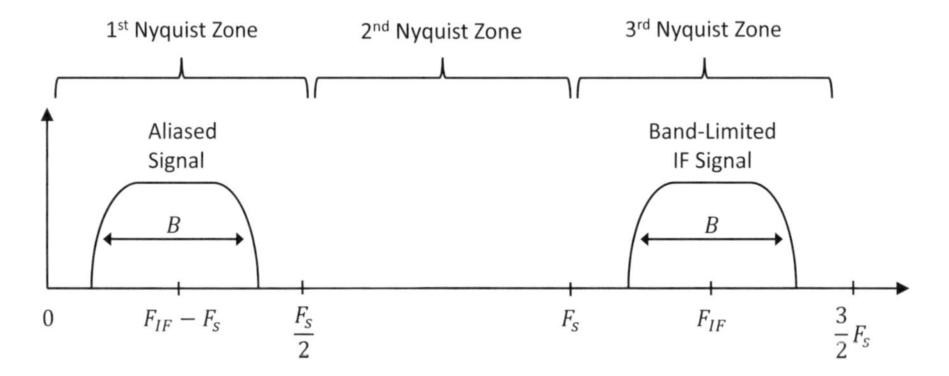

Figure 5.14 The resulting spectra for receiver IF sampling with Nyquist zones.

introduces phase error in the ADC output, which becomes predominant at high sampling rates [15].

5.8.2 Quantization

The full-scale voltage of an ADC is the difference between the maximum and minimum input voltages, as

$$V_{FS} = V_{max} - V_{min} \qquad \text{(V)}. \tag{5.16}$$

The resolution of the ADC is determined by the number of output bits per sample. For N bits, there are 2^N discrete levels. The step size between levels, and therefore the size of the least significant bit (LSB), is

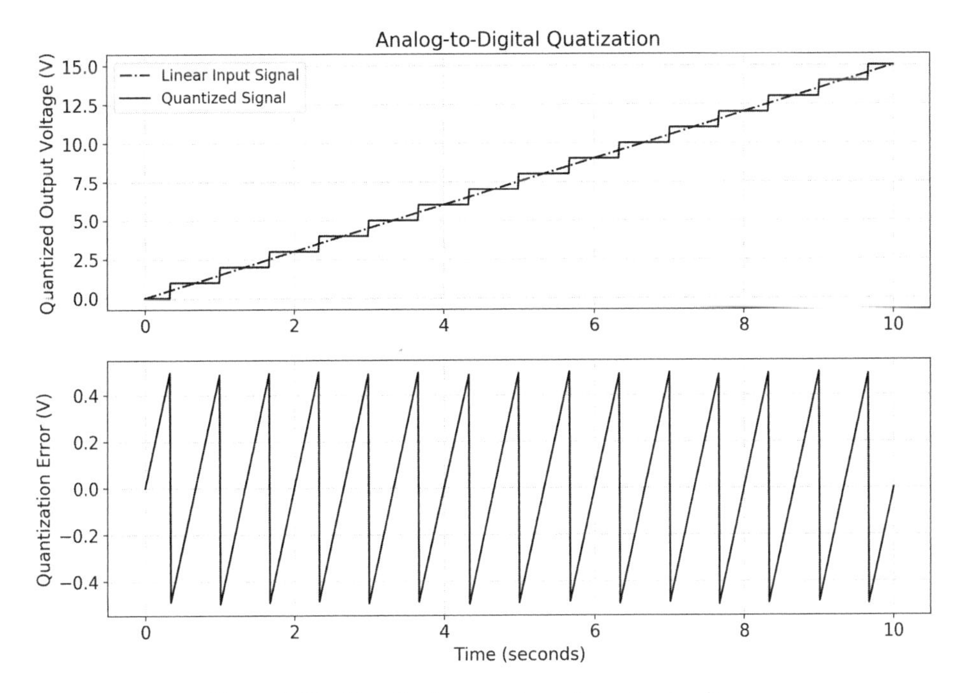

Figure 5.15 Quantization error for a linear input voltage.

$$Q = LSB = \frac{V_{FS}}{2^N - 1} \qquad \text{(V)}. \tag{5.17}$$

Since the output of the ADC is quantized, there is a difference between the input analog signal level and the output level of the ADC. This difference is known as *quantization error* and is illustrated in Figure 5.15. The signal-to-noise ratio of an ADC is the ratio of the root mean square (RMS) values of the signal amplitude to the RMS value of the noise power. For an ideal ADC, the only error present is the quantization error. For a sinusoidal waveform, an approximation to the RMS value of the quantization error is given as [16, 17]

$$E = \frac{Q}{\sqrt{12}} \qquad \text{(V)}. \tag{5.18}$$

The RMS voltage for a sinusoidal waveform is

$$V_{RMS} = \frac{V_{FS}}{2\sqrt{2}} = \frac{Q\,(2^N - 1)}{2\sqrt{2}} \qquad \text{(V)}. \tag{5.19}$$

Using (5.18) and (5.19), the SNR is found to be

$$SNR = \frac{V_{RMS}^2}{E^2} = \left((2^N - 1)\sqrt{3/2}\right)^2. \tag{5.20}$$

For $N > 5$, the SNR is approximated in logarithmic form as

$$SNR \approx 6.02N + 1.76 \qquad \text{(dB)}. \tag{5.21}$$

The expression in (5.21) is commonly found in literature [5, 6] and is for an ideal ADC. Practical ADCs have additional errors, including jitter and thermal noise, resulting in an SNR less than that of an ideal ADC given in (5.21). The errors in ADCs are dependent on the operating frequency, sampling rate, and input signal strength. Therefore, it is important to review the manufacturer's product data sheet to ensure the ADC will meet the requirements over the full range of expected operating conditions. Table 5.2 gives a sampling of various ADCs currently available. As can be seen in this table, the SNR is dependent on the sampling rate as well as the number of bits. Table 5.2 also gives the spurious free dynamic range. Recall from Section 5.3 that spurious signals are generated during the downconversion process and may appear as false targets or mask targets with weak return signals. Therefore, a high SFDR is critical for the detection of weak targets and overall radar sensitivity. While not always practical, some of the spurious signals may be avoided by using a large oversampling factor. This technique pushes the frequency of the spurious signals outside the receiver bandwidth. Also, the effect of spurious signals on the overall radar performance depends heavily on the operating mode. For example, a system employing *linear frequency modulation* (LFM), commonly called *chirp*, waveforms with large time-bandwidth products are less affected by spurious signals as these signals are rejected in the pulse compression process. On the other hand, pulse Doppler radars are very sensitive to spurious signals as they contain components appearing at various frequencies [5]. The term *effective number of bits* (ENOB) is often used to describe the dynamic range of an ADC and is based on the equation for the SNR of an ideal ADC, which is given in (5.21). Since the SNR of practical ADCs never reach this ideal value due to noise and distortion, the SNR may be related to that of an ideal ADC with fewer bits. ENOB specifies the number of bits of an ideal ADC achieving the same SNR as the ADC under consideration. Rewriting (5.21) gives the effective number of bits as

$$N_{eff} = \frac{SNR - 1.76}{6.02}. \tag{5.22}$$

5.9 DIGITAL RECEIVERS

With increasing sample rates available in analog-to-digital converters, direct sampling of the signals at IF is becoming common in modern radar receivers, with the ultimate goal

Table 5.2
Brief Survey of Analog-to-Digital Converters

Device	*Channels*	*Bits*	*Sampling Rate (Samples/Second)*	*SNR (dB)*	*SFDR (dBc)*
Texas Instruments - AD9224R	2	16	3M	88.5	95
Texas Instruments - ADS554J20	2	12	1G	65.6	85
Texas Instruments - ADS1675	1	24	4M	107	120
Analog Devices - AD9697	1	14	1.3G	65.6	80
Analog Devices - AD9213	1	12	10G	56.2	85
Linear Technology - LTC2208	1	16	130M	78	83
Maxim Integrated - MAX19777	2	12	3M	72.5	83
Maxim Integrated - MAX11284	2	24	4K	118	120

of directly sampling at the radar RF frequency. In digital receivers, a single ADC is used to sample the IF signal, and downconversion to baseband in-phase and quadrature signals is performed in digital signal processing. One of the main advantages of direct digital sampling is the virtual elimination of the amplitude and phase imbalance between the in-phase and quadrature channels. Other advantages include reduction of DC offset errors, improved linearity, flexibility of bandwidth, improved filtering, and reduction in size, weight, and power [5]. There are two main methods for using digital signal processing to generate baseband in-phase and quadrature signals from the sampled IF data. The first is through a process known as direct digital downconversion, and the second method is through the use of the Hilbert transform [18].

5.9.1 Direct Digital Downconversion

The process of direct digital downconversion is illustrated in Figure 5.16. The first step is to directly sample the band-limited IF signal discussed in Section 5.8.1. Next, the sampled signal is frequency shifted to zero frequency by complex multiplication with $e^{-j\omega_0 n}$, as shown in Figure 5.16. This step is typically accomplished through the use of a numerically controlled oscillator [5]. After this translation to baseband, the signals are passed through two low-pass digital filters to remove the unwanted image frequency. These low-pass filters are often implemented as *finite impulse response* (FIR) filters [19, 20]. SciPy provides various functions to aid in the design of FIR filters, including ***scipy.signal.firls()***, ***scipy.signal.firwin()***, and ***scipy.signal.firwin2()*** [8]. At this point in the receiver chain there is a complex baseband representation that may be used in the signal processor. However, to reduce the amount of data and computational burden, the signals are decimated by a factor that ensures the signals are still suitable for signal processing.

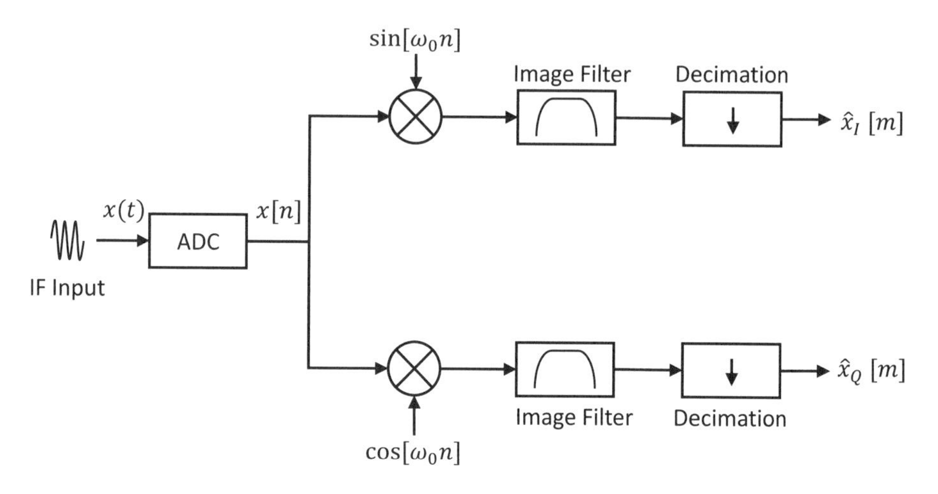

Figure 5.16 Direct digital downconversion.

5.9.2 Hilbert Transform

The second method for producing baseband I and Q signals employs the Hilbert transform [19, 21, 22], and is shown in Figure 5.17. The Hilbert transform has a phase shift property of $\pi/2$ to the digital signal. The Hilbert transform is implemented as two filters $h[n]$ and $g[n]$, as shown in Figure 5.17, with their frequency responses given as [5]

$$\left|H(\omega)\right| \approx \left|G(\omega)\right| \approx 1, \tag{5.23}$$

$$\frac{H(\omega)}{G(\omega)} = \begin{cases} -j & \text{for} \quad |\omega - \omega_0| \leq B \\ +j & \text{for} \quad |\omega + \omega_0| \leq B. \end{cases} \tag{5.24}$$

Referring to (5.23) and (5.24), this method of digital filtering produces outputs that form the complex I and Q representation of the signal centered at ω_0 and rejects the signal at $-\omega_0$. The final step in this type of receiver is to shift the spectrum of the signal located at ω_0 to zero frequency by decimation. For example, a signal spectrum centered at $\omega_0 = 2\pi n/D$, is shifted to zero frequency by decimating by a factor of D, as shown in Figure 5.18. The Hilbert transform of a periodic sequence $x[n]$ may be calculated using the SciPy implementation ***scipy.fftpack.hilbert(x)***. Also, the Hilbert transform of an analytic signal $x(t)$ may be computed with ***scipy.signal.hilbert(x)***. Note the two transforms above differ in sign. The transform of the periodic sequence does not have a factor -1 often found in the definition of the Hilbert transform, whereas the transform of the analytic signal does contain this factor [8].

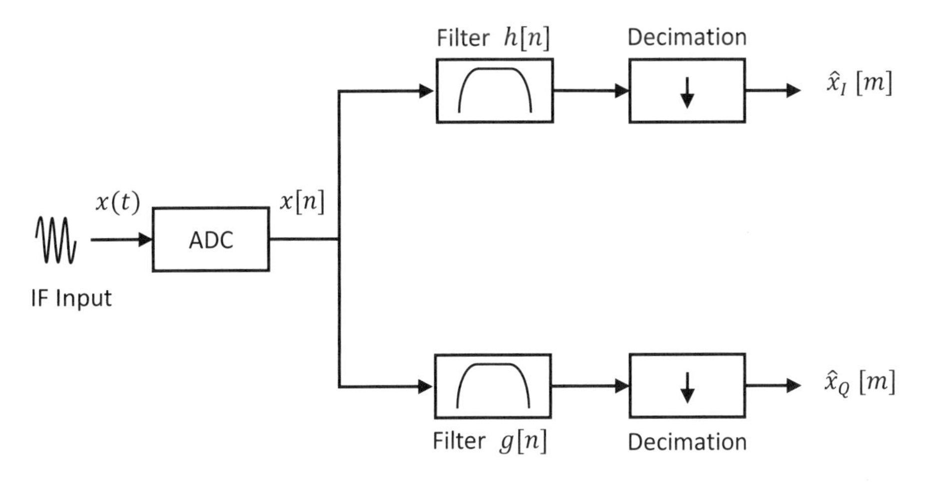

Figure 5.17 Generation of IQ data with the Hilbert transform.

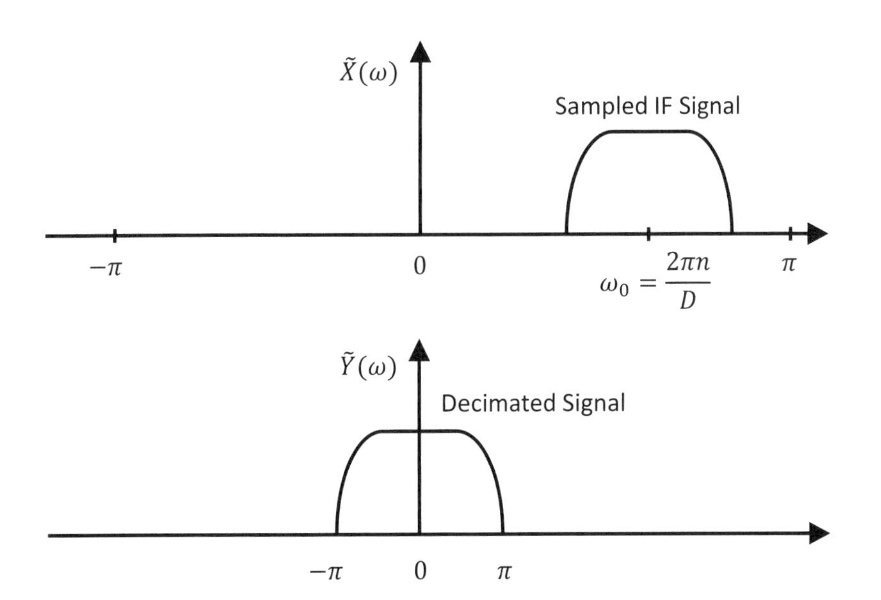

Figure 5.18 Signal spectra associated with the Hilbert transform receiver.

5.10 EXAMPLES

The sections below illustrate the concepts of this chapter with several Python examples. The examples for this chapter are in the directory *pyradar\Chapter05* and the matching MATLAB examples are in the directory *mlradar\Chapter05*. The reader should consult Chapter 1 for information on how to execute the Python code associated with this book.

5.10.1 Sensitivity Time Control

For this example, calculate and display the normalized attenuation for a transmitted waveform with a pulsewidth of $1 \mu s$ and a pulse repetition frequency of 30 KHz.

Solution: The solution to the above example is given in the Python code *sensitivity_time_control_example.py* and in the MATLAB code *sensitivity_time_control_ example.m*. Running the Python example code displays a GUI allowing the user to enter the pulsewidth and pulse repetition frequency. The code then calculates and creates a plot of the normalized attenuation versus range, as shown in Figure 5.19. This figure illustrates the attenuation is at maximum value during the transmitting period to help prevent receiver saturation.

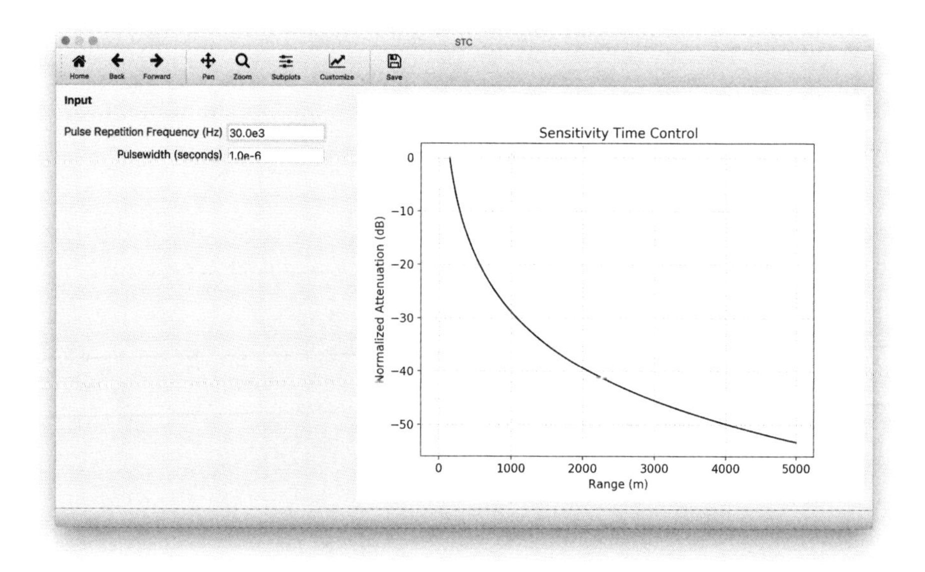

Figure 5.19 The normalized sensitivity time control attenuation calculated by *sensitivity_time_control_example.py*.

5.10.2 Noise Figure

Referring to Section 5.2, calculate the total noise figure for the five-stage network with component values given in Table 5.1.

Solution: The solution to the above example is given in the Python code *noise_figure_example.py* and in the MATLAB code *noise_figure_example.m*. Running the Python example code displays a GUI allowing the user to enter the gain and noise figure values for each component in the network. The code then calculates and displays the total noise figure for the network, as shown in Figure 5.20. As expected, the Python results match those given in Section 5.2.

5.10.3 Receiver Filtering

Referring to Section 5.6, calculate the filter response of a fourth-order, low-pass, Chebyshev filter with a critical frequency of 100 Hz, and a maximum passband ripple of 3 dB.

Solution: The solution to the above example is given in the Python code *low_pass_filter_example.py* and in the MATLAB code *low_pass_filter_example.m*. Running the Python example code displays a GUI allowing the user to select a filter type and enter the filter order, critical frequency, maximum passband ripple, and minimum required attenuation.

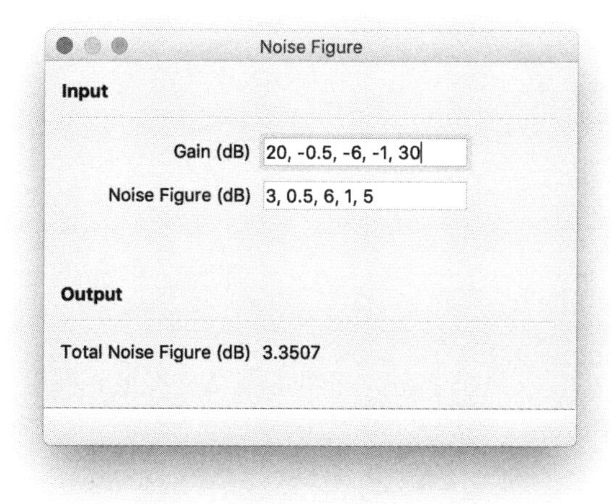

Figure 5.20 The total noise figure for a cascaded network calculated by *noise_figure_example.py*.

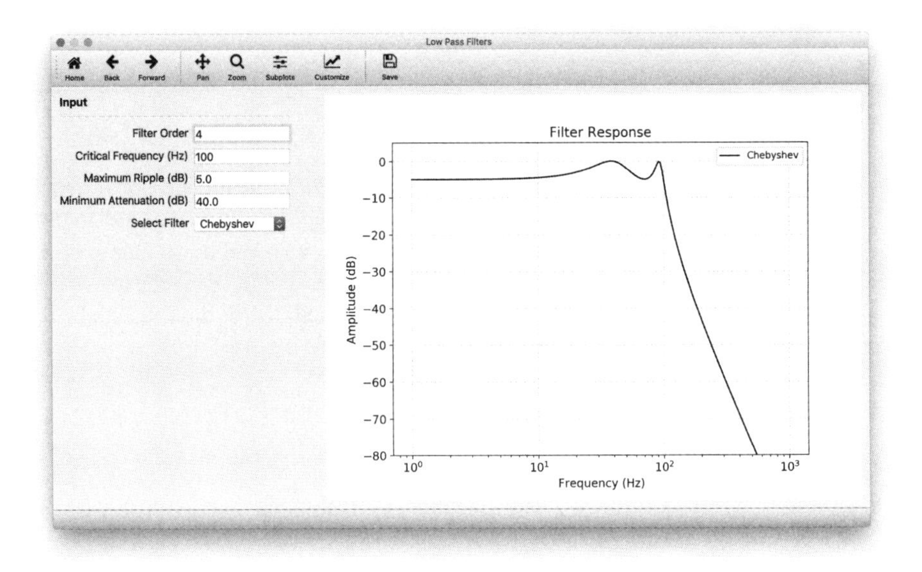

Figure 5.21 The frequency response of a fourth-order, low-pass, Chebyshev filter calculated by *low_pass_filter_example.py*.

The code then calculates and displays the frequency response of the selected filter, as shown in Figure 5.21.

5.10.4 Noncoherent Detection

For the simplistic envelope detector given in Figure 5.10, calculate and display the IF input signal and video output signal. The input signal is an amplitude modulated linear frequency modulated (chirp) waveform with a starting frequency of 20 Hz and an ending frequency of 80 Hz. The amplitude modulation has a relative amplitude of 0.1 and a modulation frequency of 4 Hz.

Solution: The solution to the above example is given in the Python code *envelope_detector_example.py* and in the MATLAB code *envelope_detector_example.m*. Running the Python example code displays a GUI allowing the user to enter the parameters for the amplitude modulated chirp waveform. The code then calculates and displays the original IF signal along with the output video signal, as shown in Figure 5.22.

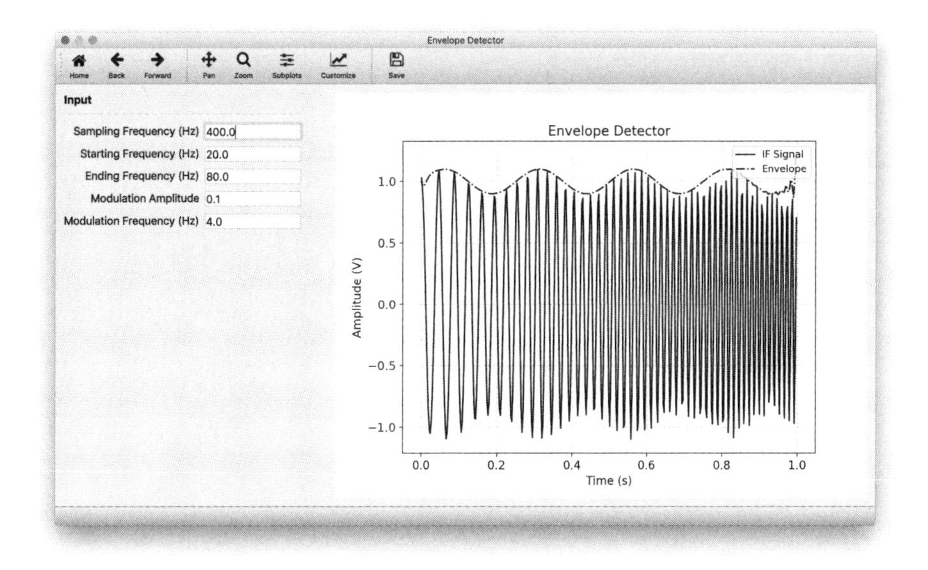

Figure 5.22 The input IF and output video signal of the simple diode envelope detector calculated by *envelope_detector_example.py*.

5.10.5 Coherent Detection

For the coherent detector given in Figure 5.11, calculate and display the resulting baseband in-phase and quadrature signals for an amplitude modulated linear frequency modulated (chirp) waveform with a starting frequency of 20 Hz and an ending frequency of 80 Hz. The amplitude modulation has a relative amplitude of 0.1 and a modulation frequency of 4 Hz.

Solution: The solution to the above example is given in the Python code *coherent_detector_example.py* and in the MATLAB code *coherent_detector_example.m*. Running the Python example code displays a GUI allowing the user to enter the parameters for the amplitude modulated chirp waveform. The code then calculates and displays the baseband in-phase and quadrature signals, as shown in Figure 5.23.

5.10.6 Analog-to-Digital Conversion

In this example, calculate and display the digital signal resulting from the analog-to-digital conversion of an amplitude modulated linear frequency modulated (chirp) waveform with a starting frequency of 1 Hz and an ending frequency of 4 Hz. The amplitude modulation has a relative amplitude of 0.1 and a modulation frequency of 4 Hz. Use a

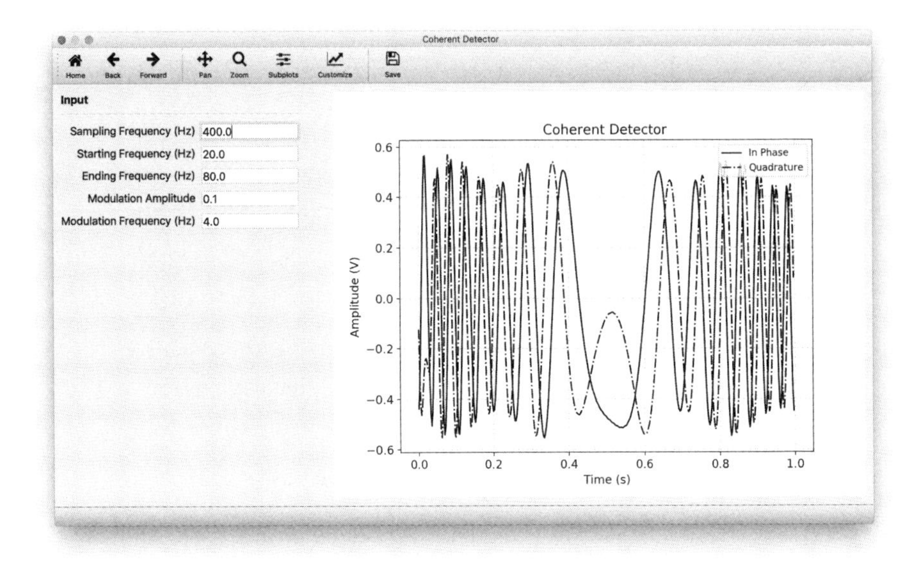

Figure 5.23 The baseband in-phase and quadrature signals from a coherent detector calculated by *coherent_detector_example.py*.

sampling rate of 100 Hz and 3 bits of resolution.

Solution: The solution to the above example is given in the Python code *adc_example.py* and in the MATLAB code *adc_example.m*. Running the Python example code displays a GUI allowing the user to enter the parameters for the ADC and for the amplitude modulated chirp waveform. The code then calculates and displays the original analog signal, the resulting sampled and quantized signal, and the associated quantization error output by the ADC, as shown in Figure 5.24.

5.10.7 Analog-to-Digital Resolution

Referring to (5.21) and (5.22), calculate the signal-to-noise ratio for an ideal analog-to-digital converter with 12 bits of resolution. Also, calculate the effective number of bits for an analog-to-digital converter with a measured signal-to-noise ratio of 63 dB.

Solution: The solution to the above example is given in the Python code *adc_resolution_example.py* and in the MATLAB code *adc_resolution_example.m*. Running the Python example code displays a GUI allowing the user to enter the number of bits for the ideal ADC and the measured SNR for a practical ADC. The code then calculates and displays the ideal SNR, and the resulting effective number of bits, as shown in Figure 5.25.

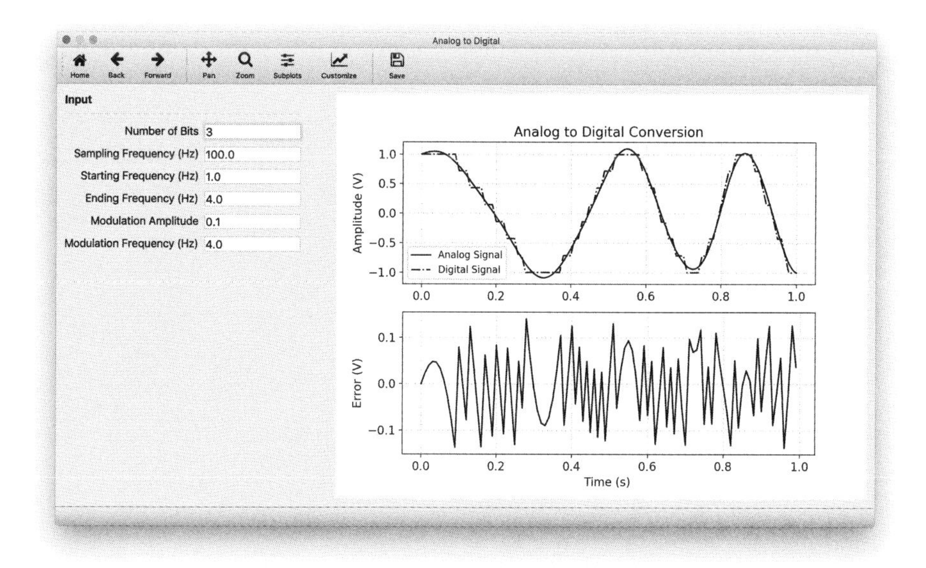

Figure 5.24 The sampled and quantized ADC output, and associated error calculated by *adc_example.py*.

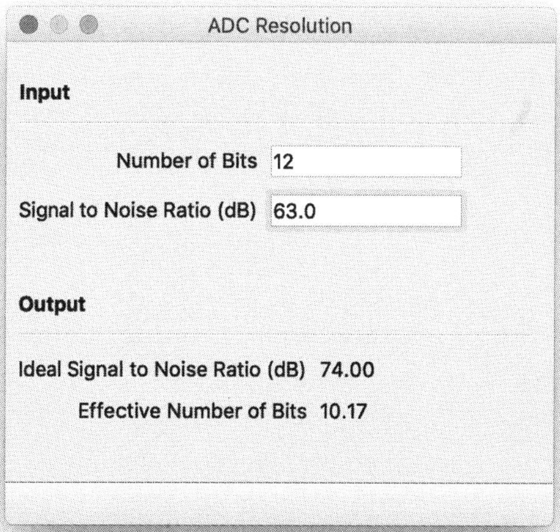

ADC Resolution

Input

Number of Bits 12

Signal to Noise Ratio (dB) 63.0

Output

Ideal Signal to Noise Ratio (dB) 74.00

Effective Number of Bits 10.17

Figure 5.25 The ideal SNR and effective number of bits calculated by *adc_resolution_example.py*.

PROBLEMS

5.1 Describe the major components of the superheterodyne receiver and explain its significance in the development of modern radar systems.

5.2 For a radar system transmitting a waveform with a pulsewidth of $2\,\mu$s at a pulse repetition frequency of 20 kHz, find the STC normalized attenuation at a range of 3.5 km.

5.3 Consider a cascaded network composed of four components; a low noise amplifier with a gain of 10 dB and noise figure of 2 dB, an RF filter with a gain of -1 dB and noise figure of 0.5 dB, a mixer with a gain of -3 dB and noise figure of 4 dB, and an IF filter with a gain of -0.5 dB and noise figure of 1 dB. Calculate the total noise figure for this cascaded network.

5.4 Given two input frequencies, $f_1 = 100$ MHz and $f_2 = 125$ MHz, find the second- and third-order intermodulation distortion frequencies.

5.5 Referring to Problem 5.4, the power associated f_1 is 1W, the power associated with f_2 is 10W and the third-order intercept point is 2W. Find the power associated with third-order intermodulation distortion.

5.6 Suppose the RF carrier frequency is 2 GHz and the local oscillator has a frequency of 1.9 GHz. Calculate the image frequency created during the first downconversion step.

5.7 Describe the difference in the filter responses shown in Figure 5.9 and explain the advantages and disadvantages of each.

5.8 Explain the difference between coherent and noncoherent detection and give examples of when it is advantageous to use each one.

5.9 Find the size of the least significant bit for the Texas Instruments - AD9224R analog-to-digital converter when the input signal is a sine wave with a full-scale voltage of 1.4V.

5.10 For a sine wave input signal, calculate the ideal signal-to-noise ratio for the Analog Devices - AD9697 analog-to-digital converter. Assuming the achieved signal to noise ratio is 3 dB less than the ideal value, calculate the effective number of bits.

5.11 Describe the process of direct digital downconversion and explain when this technique is desirable.

References

[1] E. H. Armstrong. Some recent developments in the audion receiver. *Proceedings of the Institute of Radio Engineers*, 3(3):215–238, September 1915.

[2] E. H. Armstrong, June 1920. United States Patent No. 1,342,885.

[3] E. H. Armstrong. The super-heterodyne - its origin, development, and some recent improvements. *Proceedings of the Institute of Radio Engineers*, 12(5):539–552, October 1924.

[4] D. Pozar. *Microwave Engineering,* 4th ed. John Wiley and Sons, New York, 2012.

[5] M. Skolnik. *Radar Handbook,* 3rd ed. McGraw-Hill Education, 2008.

[6] M. C. Budge and S. R. German. *Basic Radar Analysis.* Artech House, Norwood, MA, 2015.

[7] M. A. Richards, J. A. Scheer and W. A. Holm. *Principles of Modern Radar - Basic Principles.* SciTech Publishing, Raleigh, NC, 2010.

[8] E. Jones, et al. SciPy: Open source scientific tools for Python, 2001–. `http://www.scipy.org/`.

[9] J. D. Hunter. Matplotlib: A 2D graphics environment. *Computing in Science & Engineering*, 9(3):90–95, 2007.

[10] P. E. Pace. *Advanced Techniques for Digital Receivers.* Artech House, Norwood, MA, 2000.

[11] H. Nyquist. Certain factors affecting telegraph speed. *Transactions of the American Institute of Electrical Engineers*, XLIII:412–422, Jan 1924.

[12] H. Nyquist. Certain topics in telegraph transmission theory. *Transactions of the American Institute of Electrical Engineers*, 47(2):617–644, April 1928.

[13] C. E. Shannon. A mathematical theory of communication. *The Bell System Technical Journal*, 27(3):379–423, July 1948.

[14] P. Poshala. Why oversample when undersampling can do the job? 2013. Texas Instruments Application Report: SLAA594A.

[15] R. Zeijl, R. van Veldhoven and P. Nuijten. Sigma-Delta ADC clock jitter in digitally implemented receiver architectures. *Proceedings of the 9th European Conference on Wireless Technology*, September 2006.

[16] W. Kester. *The Data Conversion Handbook.* Newnes, New York, 2005.

[17] W. Kester. *Analog-Digital Conversion.* Analog Devices, 2004.

[18] S. Hahn. *Hilbert Transforms in Signal Processing.* Artech House, Norwood, MA, 1996.

[19] R. Lyons. *Understanding Digital Signal Processing,* 2nd ed. Prentice Hall, Upper Saddle River, NJ, 2004.

[20] S. Smith. *The Scientist and Engineer's Guide to Digital Signal Processing.* California Technical Publishing, San Diego, CA, 1997.

[21] R. N. Bracewell. *The Fourier Transform and Its Applications,* 3rd ed. McGraw-Hill, New York, 1999.

[22] H. Urkowitz. *Signal Theory and Random Processes.* Artech House, Dedham, MA, 1983.

Chapter 6

Target Detection

The radar detection process is a key part of any radar system as it distinguishes targets in the presence of noise. Also, the detection measurements resulting from this process are used for further processing such as tracking, synthetic aperture imaging, and clutter mitigation. In Chapter 4, the calculation of the maximum target detection range, given in (4.36), was based on the minimum signal-to-noise ratio required for detection. This chapter further investigates the selection and effect of the required signal-to-noise ratio on radar detection performance. The chapter begins with single pulse detection for both coherent and noncoherent receivers, where much of the early work in this area was performed by Rice [1]. Next, the improvement in radar detection performance resulting from the integration of multiple received pulses is studied. This area was largely influenced by Marcum [2], who extended the work of Rice for pulse integration. The detection of targets with fluctuating radar cross section is then analyzed. These concepts were originally developed by Swerling [3] and later extended for various target types [4]. The treatment of radar detection is extended to include constant false alarm rate processing. The chapter concludes with several Python examples to strengthen the ideas behind radar detection theory and processing. For readability, $\exp(x)$ is used for e^x notation in this chapter.

6.1 OPTIMAL DETECTION

As illustrated in Section 5.7.1, the noncoherent diode detector is one of the simplest radar receivers and is used as the starting point for discussions on target detection. The input signal to the receiver is the backscattered signal, $x(t)$, from the target, and the noise signal, $n(t)$. For now, it is assumed the noise is Gaussian, with zero mean and has a power (variance) of σ^2. The probability density function for a Gaussian distribution is given by [5]

$$p(x) = \frac{1}{\sigma\sqrt{2\pi}} \exp\left(-\frac{(x-\mu)^2}{2\sigma^2}\right), \tag{6.1}$$

where μ is the mean and σ is the standard deviation. This probability density function may be calculated using the SciPy implementation ***scipy.stats.norm.pdf(x, loc, scale)***, where the keyword *loc* is the mean and keyword *scale* is the standard deviation [6]. The noise is also assumed to be spatially incoherent and uncorrelated with the radar return signal. For each input signal, $x(t)$, the envelope detector produces an output, $y(t)$, which is passed through a detection process with one of two possible outcomes. These two hypotheses are denoted as H_0 and H_1, and their associated detection results and probability densities are given in Table 6.1, where N_0 is the noise level and $a(t)$ is the envelope of the radar return signal.

The design and implementation of detection algorithms is mainly focused on manipulating these two probability density functions to achieve radar detection performance that is in some sense optimal. Most radar designs make use of the *Neyman-Pearson criterion* [7], which specifies the probability of false alarm, P_{fa}, that is acceptable, and then maximizes the probability of detection, P_d, with respect to the signal-to-noise ratio. This leads to a detection threshold procedure, as shown in Figure 6.1. This threshold may be constant or dynamic and each sample in the radar data is compared to this threshold. If the sample is above the threshold level, then it is considered to be a target detection. On the other hand, if the sample is below the threshold, it is considered to be noise only. However, there are situations where these assumptions are incorrect. For example, the target return signal may be weak and below the detection threshold, and therefore a missed detection with probability $1 - P_d$. Also, a large spike in the noise may be above the detection threshold and therefore a false alarm with probability P_{fa}. This leads to four possible outcomes shown in Table 6.2. The *false alarm rate* (FAR) is defined as the number of false alarms per unit time, and is expressed as

$$FAR = P_{fa}\, T. \tag{6.2}$$

Thousands of pulses per second may pass through the detection process, resulting in a large number of opportunities for false alarms. The acceptable FAR depends on the particular system's parameters, emplacement, and mission, as false alarms require radar,

Table 6.1

Target Detection Hypotheses

Hypothesis	*Detection Result*	*Probability Density*
H_0 (Noise Only)	$y(t) = N_0$	$p(x\|H_0)$
H_1 (Target + Noise)	$y(t) = a(t) + N_0$	$p(x\|H_1)$

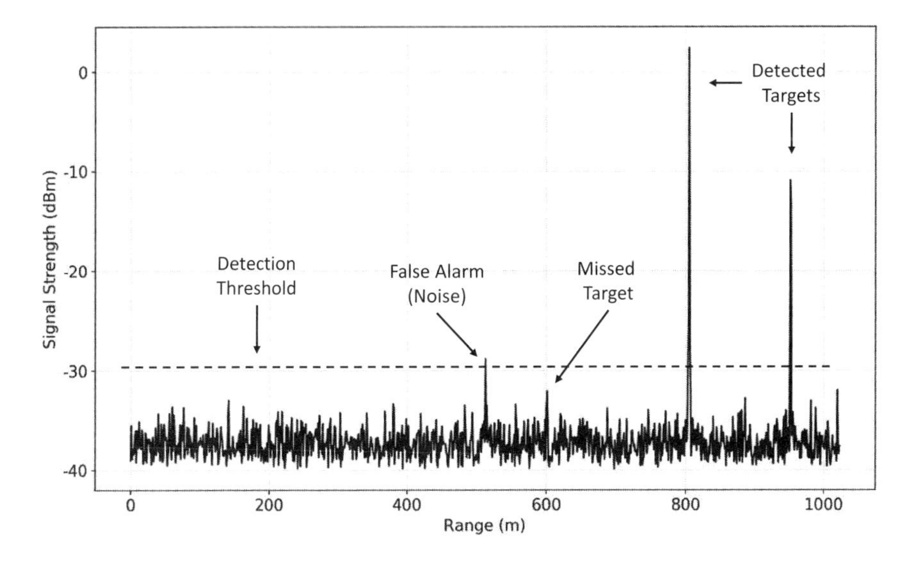

Figure 6.1 Basic target detection threshold.

Table 6.2
Basic Target Detection Hypotheses

Truth	*Detection Result*	*Outcome*
H_0	H_0	None
H_0	H_1	False alarm
H_1	H_0	Miss
H_1	H_1	Detection

and possibly operator resources, to make the determination that no target is present. P_{fa} values in the range of 10^{-4} to 10^{-8} are quite common.

6.1.1 Neyman-Pearson Lemma

The Neyman-Pearson lemma was introduced in 1933 by Jerzy Neyman and Egon Pearson [7–9] and results in the most powerful test for detection hypotheses. The previous section showed the output of the detection process, Y, according to one of two distributions $p(y|H_0)$ and $p(y|H_1)$. Based on the observation of Y, a testing procedure, T, is desired to be optimal in choosing between the hypotheses. Let R_T be a subset of the range of Y where the test chooses H_1. The probability of false alarm is then

$$P_{fa} = \int\limits_{R_T} p(y|H_0)\, dy, \tag{6.3}$$

and the probability of detection is

$$P_d = \int\limits_{R_T} p(y|H_1)\, dy. \tag{6.4}$$

Now consider the likelihood ratio test of the form

$$\frac{p(y \mid H_1)}{p(y \mid H_0)} \underset{H_0}{\overset{H_1}{\gtrless}} \lambda. \tag{6.5}$$

The subset of the range of Y where the test decides H_1 is denoted as

$$R(\lambda) = \{y : p(y|H_1) > \lambda\, p(y|H_0)\}, \tag{6.6}$$

which allows the probability of false alarm to be written as

$$P_{fa} = \int\limits_{R(\lambda)} p(y|H_0)\, dy. \tag{6.7}$$

It is apparent the set $R(\lambda)$ will vary with λ, and therefore the threshold can be set to achieve the desired probability of false alarm. For the likelihood ratio test given in (6.5), with the threshold, λ, chosen such that the probability of false alarm in (6.7) is at a significance level of α, another test does not exist with $P_{fa}(R_T) \leq \alpha$ and $P_d(R_T) > P_d(R(\lambda))$. In other words, the likelihood ratio test of (6.5) is the most powerful test with a probability of false alarm less than or equal to the significance level α. Proof of the Neyman-Pearson lemma may be found in various texts [7–9].

6.1.2 Noncoherent Detection

Figure 6.2 shows the probability density function for Gaussian noise given in (6.1) along with the detection threshold. The probability of false alarm is then the shaded area under the curve. To calculate the probability of false alarm for the Gaussian noise and given detection threshold, start with (6.7) and make use of (6.1) to give

$$P_{fa} = \int\limits_{R(\lambda)} p(y|H_0)\, dy = \int\limits_{V_t}^{\infty} \frac{1}{\sigma\sqrt{2\pi}} \exp\left(-\frac{y^2}{2\sigma^2}\right) dy. \tag{6.8}$$

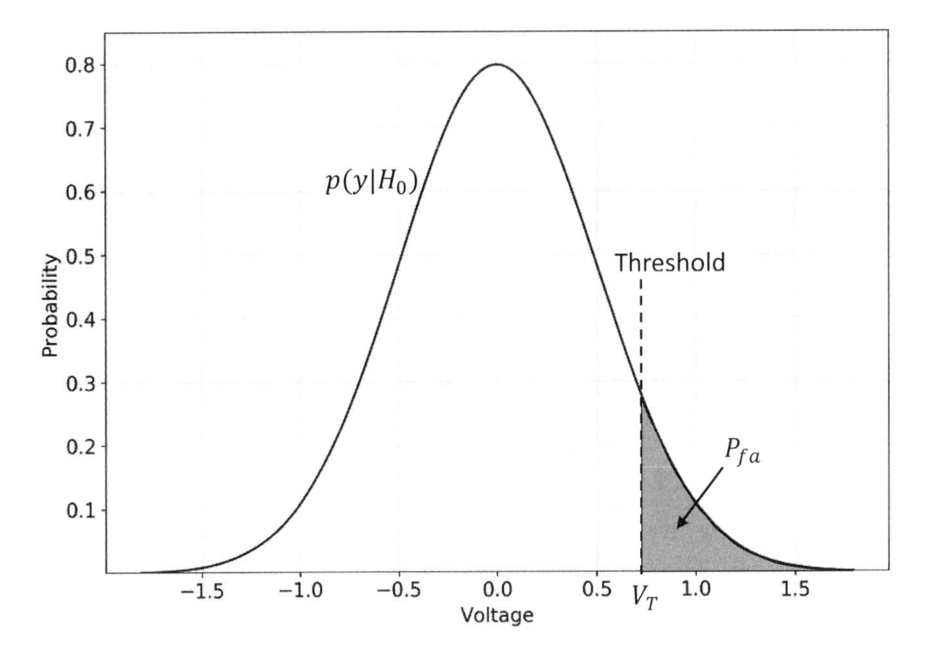

Figure 6.2 Gaussian probability density function with detection threshold and probability of false alarm.

The integral in (6.8) above may be found in various look-up tables and online calculators [10, 11], and is given here as

$$P_{fa} = \frac{1}{2}\left[1 - erf\left(\frac{V_T}{\sqrt{2}\sigma}\right)\right], \tag{6.9}$$

where erf is the *error function* [10]. The error function may be calculated using the SciPy implementation given in *scipy.special.erf()* [6]. A similar derivation for the probability of detection is followed. Assume the amplitude of the signal from the envelope detector is A, which allows the probability of detection to be written as

$$P_d = \int\limits_{R(\lambda)} p(y|H1)\, dy = \int\limits_{V_T}^{\infty} \frac{1}{\sigma\sqrt{2\pi}} \exp\left(\frac{-(y-A)^2}{2\sigma^2}\right) dy. \tag{6.10}$$

Performing the integration leads to

$$P_d = \frac{1}{2}\left[1 - erf\left(\frac{V_T - A}{\sqrt{2}\sigma}\right)\right]. \tag{6.11}$$

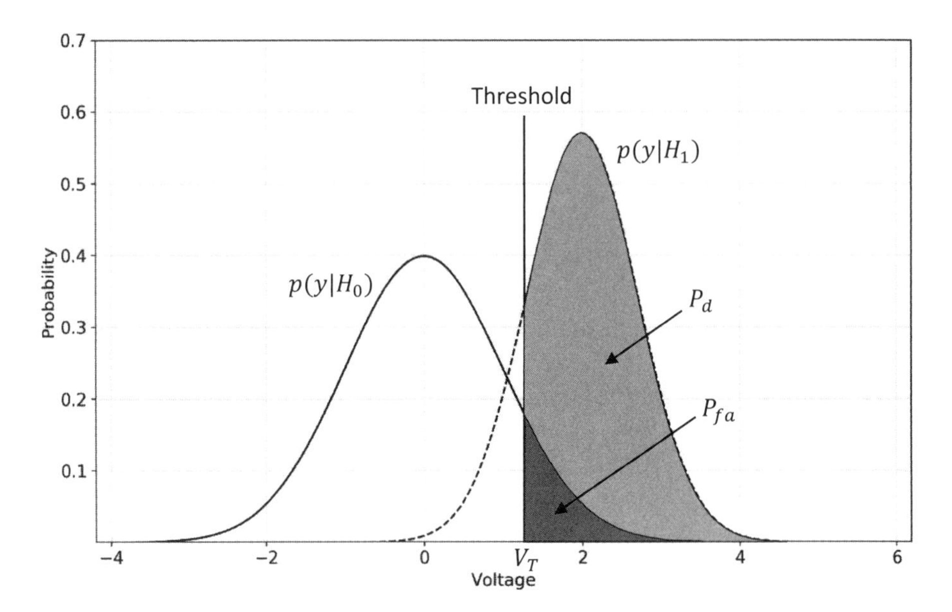

Figure 6.3 Gaussian probability density functions with detection threshold, probability of false alarm, and probability of detection.

Figure 6.3 shows both the probability of false alarm and the probability of detection for the case of a simple envelope detector with Gaussian distributions. The goal is for the probability of false alarm to be as small as possible and the probability of detection to be as large as possible. As can be seen in this figure, raising the detection threshold would lower probability of false alarm, but also lower the probability of detection for a given signal-to-noise level. Figure 6.4 illustrates an increase in the signal-to-noise ratio, which results in a higher probability of detection for a given detection threshold or probability of false alarm.

6.1.3 Coherent Detection

The previous section examined the probability of false alarm and the probability of detection for the envelope detector with Gaussian noise. For coherent detectors, illustrated in Figure 5.11, the noise is modeled at the output of the in-phase and quadrature channels as Gaussian, with zero mean and noise power of $\sigma^2/2$. The magnitude of the in-phase and quadrature channels is

$$z = \sqrt{X_I^2 + X_Q^2}, \tag{6.12}$$

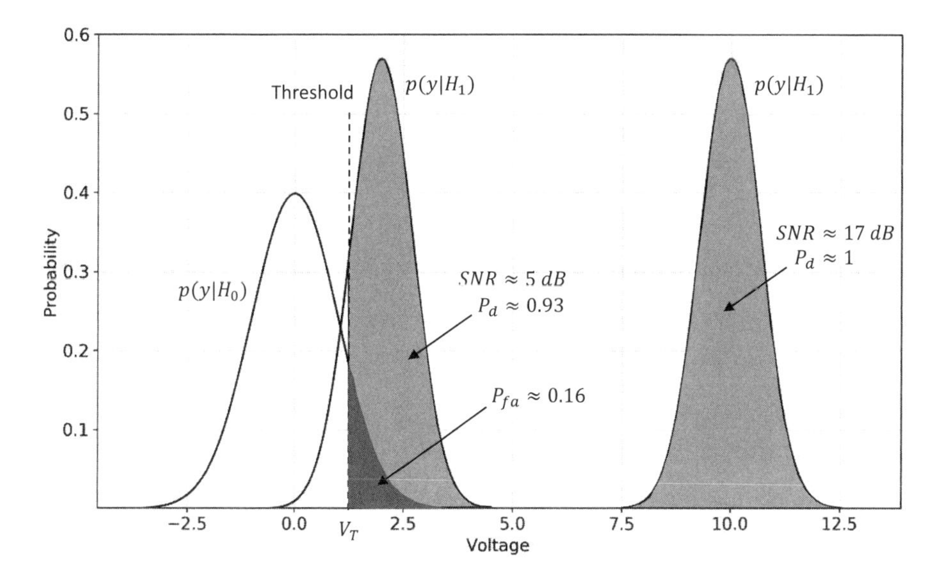

Figure 6.4 Probability of false alarm and probability of detection for various signal-to-noise ratios.

then the resulting probability density function follows a Rayleigh distribution given by [9, 12, 13]

$$p(z) = \frac{z}{\sigma^2} \exp\left(\frac{-z^2}{2\sigma^2}\right). \tag{6.13}$$

Figure 6.5 shows a comparison between the Gaussian and Rayleigh probability density functions for a variance of $\sigma^2 = 1$. Following the procedure in Section 6.1.2, the probability of false alarm for the coherent detector is written as

$$P_{fa} = \int_{V_T}^{\infty} \frac{z}{\sigma^2} \exp\left(-\frac{z^2}{2\sigma^2}\right) dz = \exp\left(-\frac{V_T^2}{2\sigma^2}\right). \tag{6.14}$$

Solving for the detection threshold gives

$$V_T = \sqrt{-2\sigma^2 \ln\left(P_{fa}\right)}. \tag{6.15}$$

Assuming a sinusoidal return signal with an amplitude of A, the probability density function of the signal plus noise is Rician [1] and the probability of detection is found from (6.4) as [13, 14]

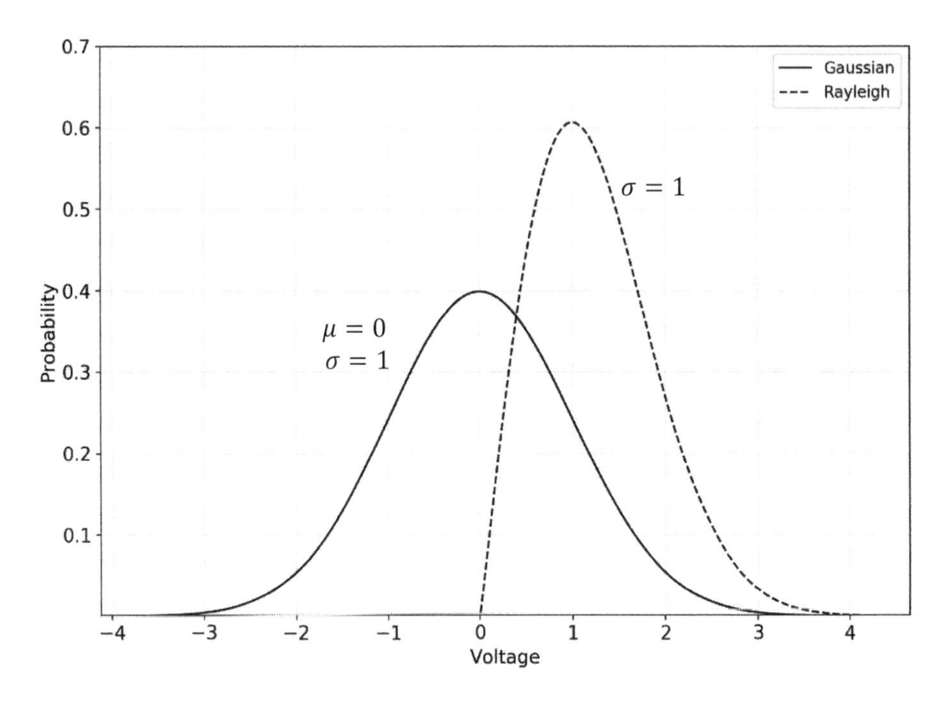

Figure 6.5 Probability density functions for Gaussian and Rayleigh distributions with $\sigma^2 = 1$.

$$P_d = \int_{V_T}^{\infty} \frac{z}{\sigma^2} I_0\left(\frac{zA}{\sigma^2}\right) \exp\left(-\frac{z^2 + A^2}{2\sigma^2}\right) dz = Q\left[\sqrt{\frac{A^2}{\sigma^2}}, \sqrt{-2\ln\left(P_{fa}\right)}\right], \tag{6.16}$$

where Q is *Marcum's Q function* [15], and I_0 is the modified Bessel function [10]. For a sinusoidal waveform, $SNR = \dfrac{A^2}{2\sigma^2}$, which results in

$$P_d = Q\left[\sqrt{2SNR}, \sqrt{-2\ln\left(P_{fa}\right)}\right]. \tag{6.17}$$

Marcum's Q function comes about frequently in radar detection problems, and while there is no known closed form expression for this integral, there are several numerical algorithms for its calculation [13, 16, 17]. Also, Cantrell and Ojha [18] give a comparison of several numerical algorithms used to calculate Marcum's Q function. Figure 6.6 shows the probability of detection versus signal-to-noise ratio for varying probability of false alarm.

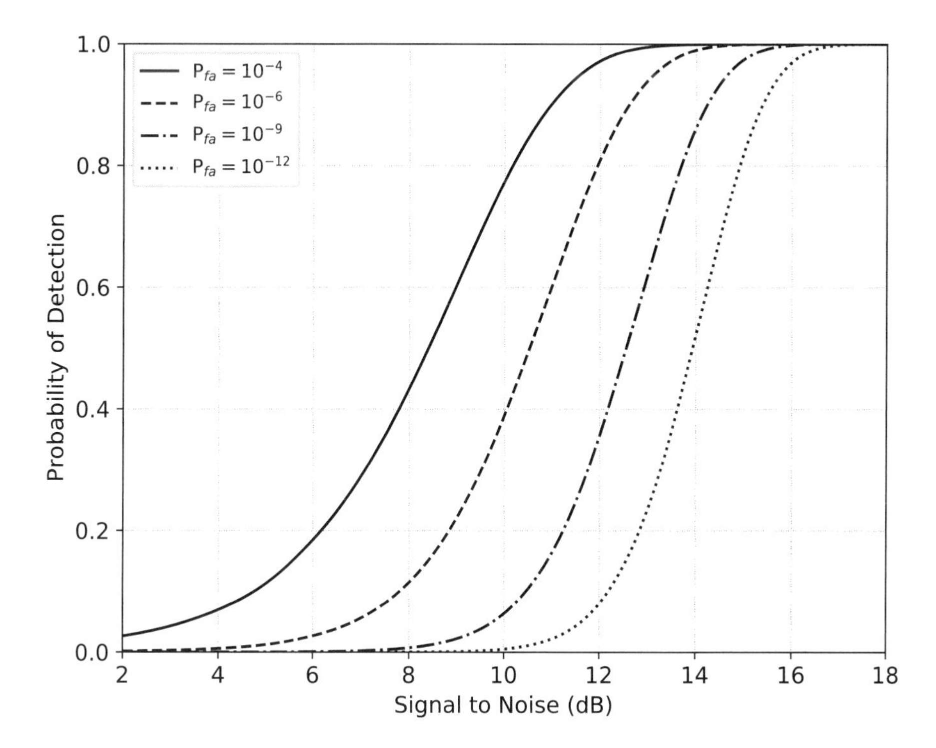

Figure 6.6 Probability of detection vs signal-to-noise ratio for varying probability of false alarm.

As expected, the probability of detection increases with signal-to-noise ratio for a fixed detection threshold, which is set by the probability of false alarm. Lowering the detection threshold lowers the signal-to-noise ratio required to produce the same probability of detection. However, this results in an increased probability of false alarm. The general procedure is to set the detection threshold based on (6.15) for a desired probability of false alarm and then compute the probability of detection versus signal-to-noise ratio for the associated probability of false alarm.

6.2 PULSE INTEGRATION

The previous sections were concerned with the detection of a measurement, y, from a single radar return or pulse. One method for improving detection performance is to integrate multiple pulses. A target may reflect several pulses during a single scan. Integrating the return signals increases the signal-to-noise ratio, and therefore improves detection probability. The number of return signals that may be integrated depends on several factors, including pulse repetition frequency, antenna scan rate, target motion, and relative velocity. Pulse integration may be performed coherently or noncoherently. While there are semicoherent integration schemes, these will not be discussed here. The reader is referred to two good resources on the subject [14, 17].

6.2.1 Coherent Integration

The first type of integration considered is coherent integration, which preserves the phase relationship between the received pulses. The integration is performed on the complex signals before being passed to the detector, as shown in Figure 6.7.

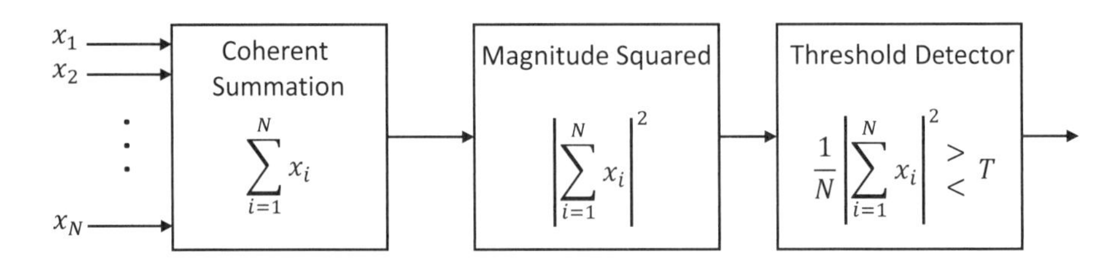

Figure 6.7 Coherent pulse integration.

Coherent integration would ideally result in an increase in the signal-to-noise ratio by a factor of N, which is written as

$$SNR_{ci} = N \cdot SNR_0, \tag{6.18}$$

where

N	=	number of pulses integrated coherently,
SNR_0	=	single pulse signal-to-noise ratio,
SNR_{ci}	=	signal-to-noise ratio after coherent integration.

However, integration loss always occurs in practical systems and this ideal value is not achieved. Integration loss may be due to component instability, environmental changes, target and platform motion, and fluctuations in target scattering [19]. Coherent integration is generally not employed over long time frames, especially for targets with quickly changing radar cross section or in the case of correlated clutter. Ideally, coherent integration may be achieved through direct summation of the signals. However, a more practical approach is through the use of the Fourier transform to determine the unknown initial phase [14].

6.2.2 Noncoherent Integration

The next type of integration is noncoherent integration shown in Figure 6.8. As shown in Section 5.7.1, noncoherent processing does not make use of phase information. Pulse integration is performed after the signal amplitude of each pulse has been found (i.e., after the signal has passed through the envelope detector).

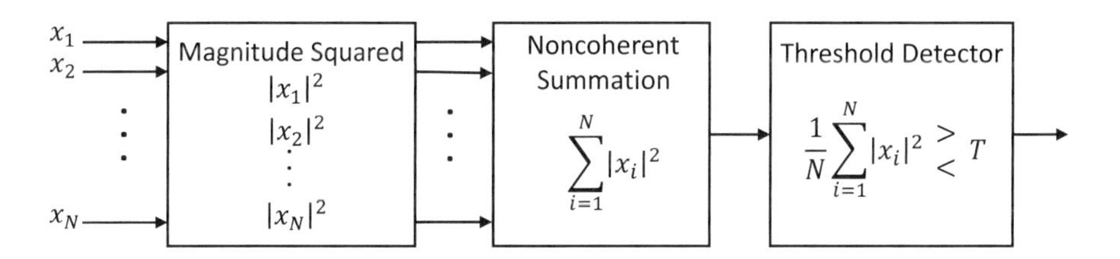

Figure 6.8 Noncoherent pulse integration.

Noncoherent integration is less efficient than coherent integration and much work has been performed in this area to characterize the degradation in performance. While closed-form expressions generally do not exist, there are some empirical approximations [20, 21]. One approach is to write the degradation as a loss factor compared to coherent integration as

$$SNR_{nci} = \frac{SNR_{ci}}{L_{nci}}, \tag{6.19}$$

where SNR_{nci} is the signal-to-noise ratio resulting from noncoherent integration and L_{nci} is the loss in integration as compared to coherent integration. One approximation for this loss factor is given as [21]

$$L_{nci} = \frac{1 + SNR_0}{SNR_0}, \tag{6.20}$$

where SNR_0 is the required single pulse signal-to-noise ratio for noncoherent detection. An expression for finding the single pulse signal-to-noise ratio given the number of pulses and signal-to-noise required to produce a specific probability of detection and probability of false alarm is written as [21]

$$SNR_0 = \frac{SNR_{nci}}{2N} + \sqrt{\frac{SNR_{nci}^2}{4N^2} + \frac{SNR_{nci}}{N}}. \tag{6.21}$$

Another approach is to express the signal-to-noise ratio for noncoherent integration as a gain over the single pulse signal-to-noise ratio. Sometimes in literature this is referred to as the noncoherent integration improvement factor [13]. In this case, the signal-to-noise ratio is

$$SNR_{nci} = G_{nci} + SNR_0 \qquad \text{(dB)}, \tag{6.22}$$

where G_{nci} is the noncoherent integration gain, and SNR_0 is the single pulse signal-to-noise ratio. An approximation for G_{nci} which has been shown to be accurate within 0.8 dB [21] is

$$G_{nci} = 6.79\,(1 + 0.235\,P_d)\left[1 + \frac{\log_{10}(1/P_{fa})}{46.6}\right]\log_{10}(N) \times \left[1 - 0.14\log_{10}(N) + 0.01831\log_{10}^2(N)\right] \qquad \text{(dB)}. \tag{6.23}$$

Figure 6.9 shows a comparison of the single pulse signal-to-noise ratio required for coherent integration and noncoherent integration. The figure includes both approximations for noncoherent integration. As illustrated in the figure, coherent integration outperforms noncoherent integration for increasing N. However, as the number of pulses becomes exceedingly large, coherent integration will start to degrade.

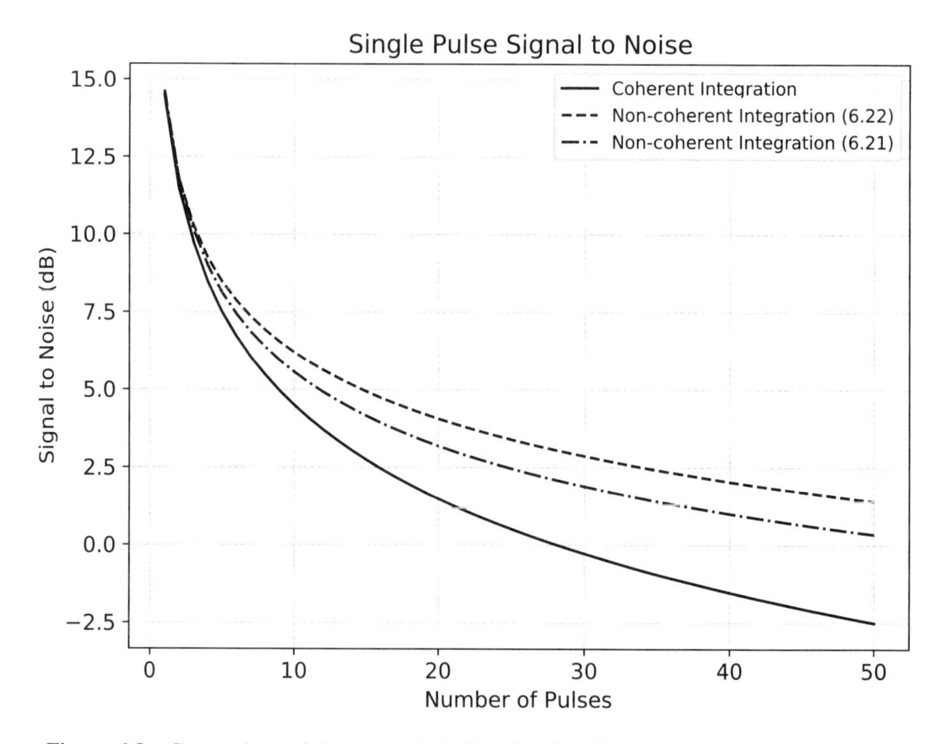

Figure 6.9 Comparison of the required single pulse signal-to-noise ratio for coherent and noncoherent pulse integration ($P_{fa} = 10^{-9}$, $P_d = 0.99$).

6.2.3 Binary Integration

Binary integration is another form of noncoherent integration, often referred to as M of N detection, and is shown in Figure 6.10. In this form of integration, each of the N return signals is passed separately through the threshold detector. There must be M individual detection declarations in order for a target to be declared as present. Binary integration is somewhat simpler to implement than coherent and noncoherent integration.

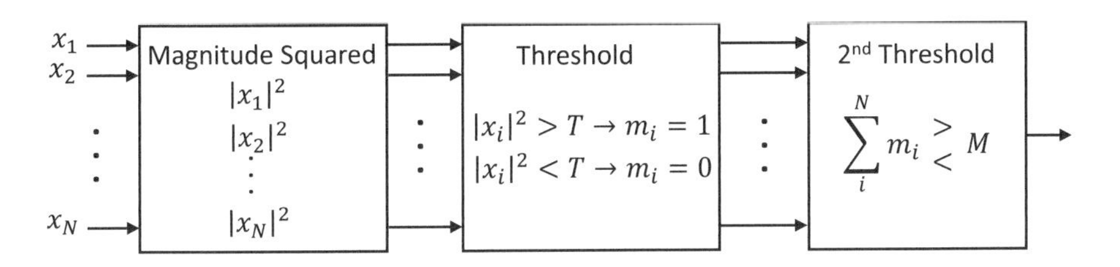

Figure 6.10 Noncoherent binary (M of N) integration.

The probability of detection for binary integration is written as [5, 22, 23]

$$P_{MN} = \sum_{k=M}^{N} C_{kN} \cdot p^k \cdot (1-p)^{N-k}, \tag{6.24}$$

where p is the single pulse probability of detection, and C_{kN} is the binomial coefficient given by

$$C_{kN} = \frac{N!}{k!\,(N-k)!}. \tag{6.25}$$

Similarly, the probability of false alarm is

$$F_{MN} = \sum_{k=M}^{N} C_{kN} \cdot f^k \cdot (1-f)^{N-k}, \tag{6.26}$$

where f is the single pulse probability of false alarm. For binary integration, some optimum value of M exists that minimizes the required signal-to-noise ratio for given probabilities of detection and false alarm, and a given N [22, 23]. Richards [19] provides a nice approximation for M for various target fluctuations, which is given here as

$$M_{opt} = 10^{\beta}\, N^{\alpha}, \tag{6.27}$$

where the values for α and β are given in Table 6.3.

Table 6.3
Optimum M for Binary Integration

Target Type	α	β
Swerling 0	0.8	−0.02
Swerling I	0.8	−0.02
Swerling II	0.91	−0.38
Swerling III	0.8	−0.02
Swerling IV	0.873	−0.27

6.2.4 Cumulative Integration

Cumulative integration, commonly called 1 of N detection, is another noncoherent integration technique. It is similar to binary integration except it only requires one detection declaration for every N received signals. The probability of detection for cumulative integration is written as [13, 23]

$$P_c = 1 - (1 - p)^N, \tag{6.28}$$

where p is the single pulse probability of detection, and N is the total number of pulses considered. This may also be found using (6.24) with $M = 1$. Figure 6.11 gives a comparison between binary and cumulative integration. As shown in this figure, cumulative integration requires the lowest signal-to-noise ratio for detection, while increasing M also increases the required signal-to-noise ratio. Figure 6.12 gives a comparison between coherent, noncoherent, binary, and cumulative integration. As expected, coherent integration provides the best performance.

6.3 FLUCTUATING TARGET DETECTION

The previous sections were concerned with the detection of targets with constant radar cross section. Statistical noise models were developed that allowed the probability of detection and probability of false alarm to be determined. As will be shown in Chapter 7, complex targets have various scattering mechanisms that can interfere destructively or constructively. Therefore, the energy backscattered from a target may vary from pulse to pulse or from scan to scan. This effect may be caused by small changes in the viewing aspect angle or frequency changes in the transmitted waveform. To more accurately predict radar detector performance, fluctuating target models are needed. A statistical model of the radar cross section, including both a probability density function and a decorrelation measure, are used along with the statistical noise model to determine the probability of detection, probability of false alarm and the signal-to-noise ratio. *Swerling*

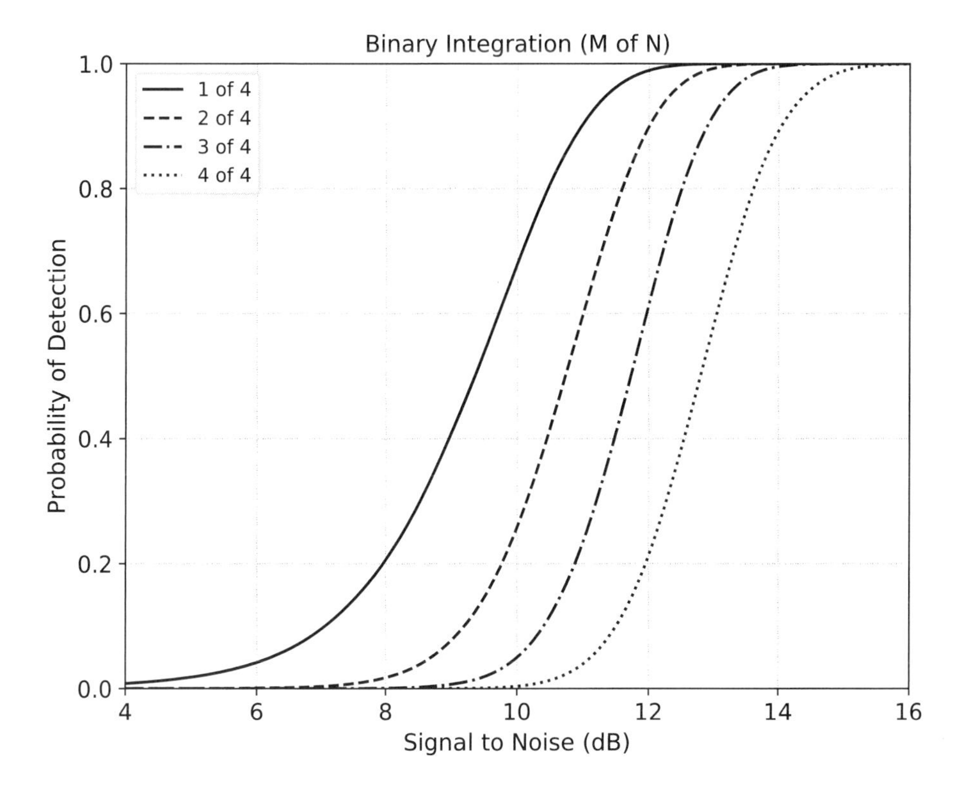

Figure 6.11 Comparison of probability of detection for binary and cumulative pulse integration ($P_{fa} = 10^{-6}$).

models [3] are an often-used set of statistical target models that are special cases of the chi-squared target models with specific degrees of freedom [5, 12]. Table 6.4 gives the four Swerling target models. Often the terms *Swerling 0* or *Swerling V* will be used to represent the constant radar cross-section model. The statistics associated with Swerling I and Swerling II target models are consistent with targets composed of several small scatterers of similar amplitude, while Swerling III and Swerling IV models are representative of targets comprised of a single dominant scatterer along with many small scatterers [3]. The probability density function for Swerling I and Swerling II target models is given as

$$p(\sigma) = \frac{1}{\sigma_{avg}} \exp\left(-\frac{\sigma}{\sigma_{avg}}\right), \tag{6.29}$$

where σ_{avg} is the average radar cross section of the target. The probability density function for Swerling III and Swerling IV target models is

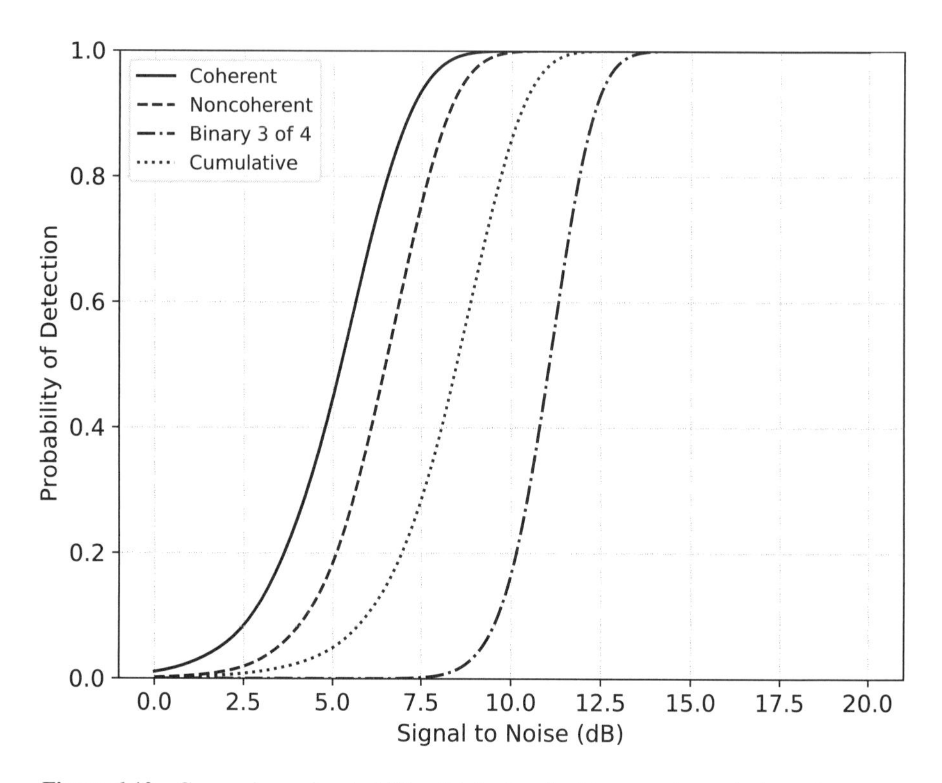

Figure 6.12 Comparison of probability of detection for binary and cumulative pulse integration ($P_{fa} = 10^{-6}$).

Table 6.4
Swerling Target Models

Scattering	*RCS Model*	Decorrelation *Scan to Scan*	*Pulse to Pulse*
Small similar scatterers	chi-squared (2 DOF)	Swerling I	Swerling II
One dominant scatterer plus small scatterers	chi-squared (4 DOF)	Swerling III	Swerling IV

$$p(\sigma) = \frac{4\sigma}{\sigma_{avg}^2} \exp\left(-\frac{2\sigma}{\sigma_{avg}}\right). \tag{6.30}$$

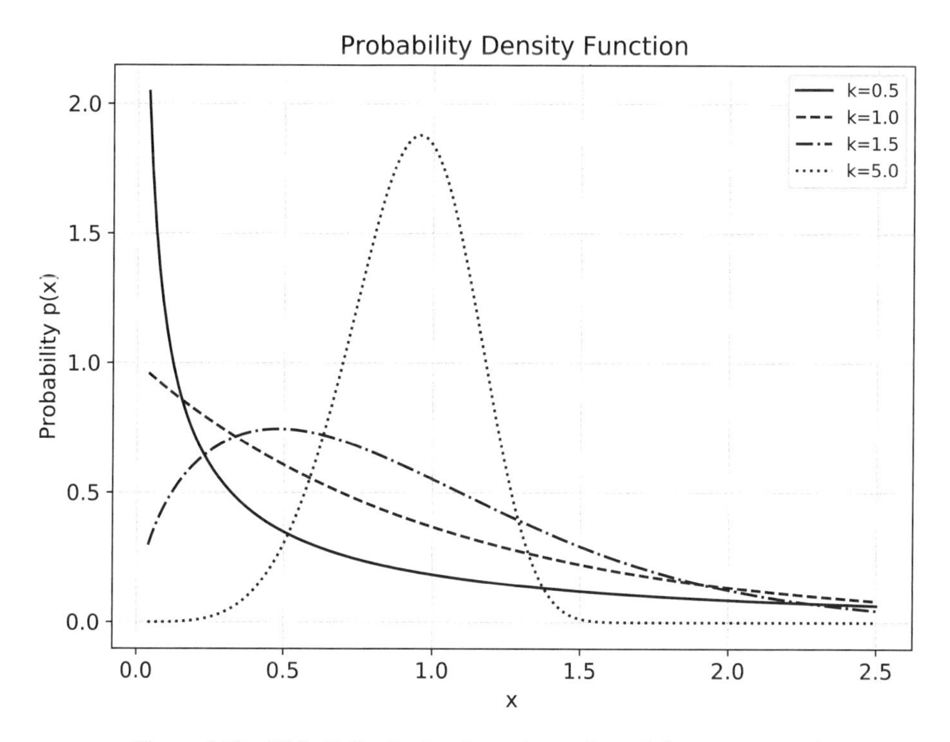

Figure 6.13 Weibull distribution for various values of shape parameter k.

Noncoherent integration can be applied to all five Swerling models. However, the target radar cross section decorrelates from pulse to pulse for Swerling II and Swerling IV. This means phase coherency is lost and coherent integration cannot be employed for those models. Other types of probability density functions may be used to develop statistical target models. These often include lognormal and Weibull distributions, which have long tails, as illustrated in Figure 6.13. For high-resolution radar systems, these types of distributions often give a better match to empirical data than the Swerling models [24]. The Weibull distribution may be calculated using the SciPy implementation in ***scipy.stats.weibull_min()*** and ***scipy.stats.weibull_max()*** [6].

For calculation of detection statistics when coherent integration in employed, the signal-to-noise ratio is replaced with the ideal value of $N \cdot SNR$. For example, the probability of detection for Swerling 0 targets given in (6.17) becomes

$$P_d = Q\left[\sqrt{2N \cdot SNR}, \sqrt{-2\ln\left(P_{fa}\right)}\right]. \tag{6.31}$$

Similarly, the probability of detection for Swerling I is given by

$$P_d = \exp\left(\frac{\ln(P_{fa})}{N \cdot SNR + 1}\right), \tag{6.32}$$

and for Swerling III

$$P_d = \left(1 - \frac{2N \cdot SNR \ln(P_{fa})}{(2 + N \cdot SNR)^2}\right) \exp\left(\frac{2\ln(P_{fa})}{2 + N \cdot SNR}\right). \tag{6.33}$$

6.3.1 Swerling 0

In Section 6.1.3 the probability of detection and probability of false alarm was derived for coherent detection in the presence of Gaussian noise. Those results are now extended for noncoherently integrating a number of pulses, N, from targets with constant radar cross section. For $N > 1$, the resulting probability density function $p(y|H_0)$ is the Erlang density [12]. This is given as

$$p(y|H_0) = \frac{y^{(N-1)}\, e^{-y}}{(N-1)!}. \tag{6.34}$$

Note that y now represents a sample of the squared magnitude. The expression in (6.34) reduces to the exponential probability density function. The probability of false alarm is then obtained by integrating (6.34) from the threshold to infinity as

$$P_{fa} = \int_{SNR_T}^{\infty} \frac{y^{(N-1)}\, e^{-y}}{(N-1)!}\, dy = 1 - \Gamma(N, SNR_T), \tag{6.35}$$

where Γ is the incomplete gamma function [10], and SNR_T is the signal-to-noise ratio of the detection threshold. Therefore, specifying a probability of false alarm sets the detection threshold as

$$SNR_T = \Gamma^{-1}(N, 1 - P_{fa}), \tag{6.36}$$

where Γ^{-1} is the inverse of the incomplete gamma function. The incomplete gamma function may be found using the SciPy function ***scipy.special.gammainc()***. Also, the inverse of the incomplete gamma function is found with ***scipy.special.gammaincinv()*** [6]. For example, the calculation of the signal-to-noise ratio, given the probability of false alarm and number of pulses, is shown in Listing 6.1. Conversely, the calculation of the probability of false alarm, given the signal-to-noise ratio threshold and number of pulses, is provided in Listing 6.2.

Listing 6.1 Calculation of SNR Threshold

```
from scipy import special
number_of_pulses = 10
Pfa = 1.0e-6
snr_threshold = special.gammaincinv(number_of_pulses, 1.0-Pfa)
print(snr_threshold)
...
32.71
```

Listing 6.2 Calculation of Probability of False Alarm

```
from scipy import special
number_of_pulses = 10
snr_threshold = 32.71
Pfa = 1.0 - special.gammainc(number_of_pulses, snr_threshold)
print(Pfa)
...
1.00e-6
```

The probability of detection for this Swerling case is found from the probability density function for $p(y|H_1)$ as [3, 19]

$$p(y|H_1) = \left(\frac{y}{N \cdot SNR}\right)^{(N-1)/2} \exp\left(-y - N \cdot SNR\right) \times I_{N-1}\left(2\sqrt{N \cdot SNR \cdot y}\right), \tag{6.37}$$

where I_{N-1} is the modified Bessel function of the first kind of order $N - 1$. The modified Bessel function may be found using the SciPy function ***scipy.special.iv()*** [6]. The probability of detection for this Swerling case is then found by integrating (6.37) from the threshold to infinity, which results in

$$P_d = Q\left[\sqrt{2N \cdot SNR}, \sqrt{2 \cdot SNR_T}\right] + \exp\left(-SNR_T - N \cdot SNR\right) \times \sum_{n=2}^{N} \left(\frac{SNR_T}{N \cdot SNR}\right)^{(n-1)/2} I_{n-1}\left(2\sqrt{N \cdot SNR \cdot SNR_T}\right). \tag{6.38}$$

Figure 6.14 shows a comparison of the probability of detection versus signal-to-noise ratio for a specified probability of false alarm and various number of pulses. As can

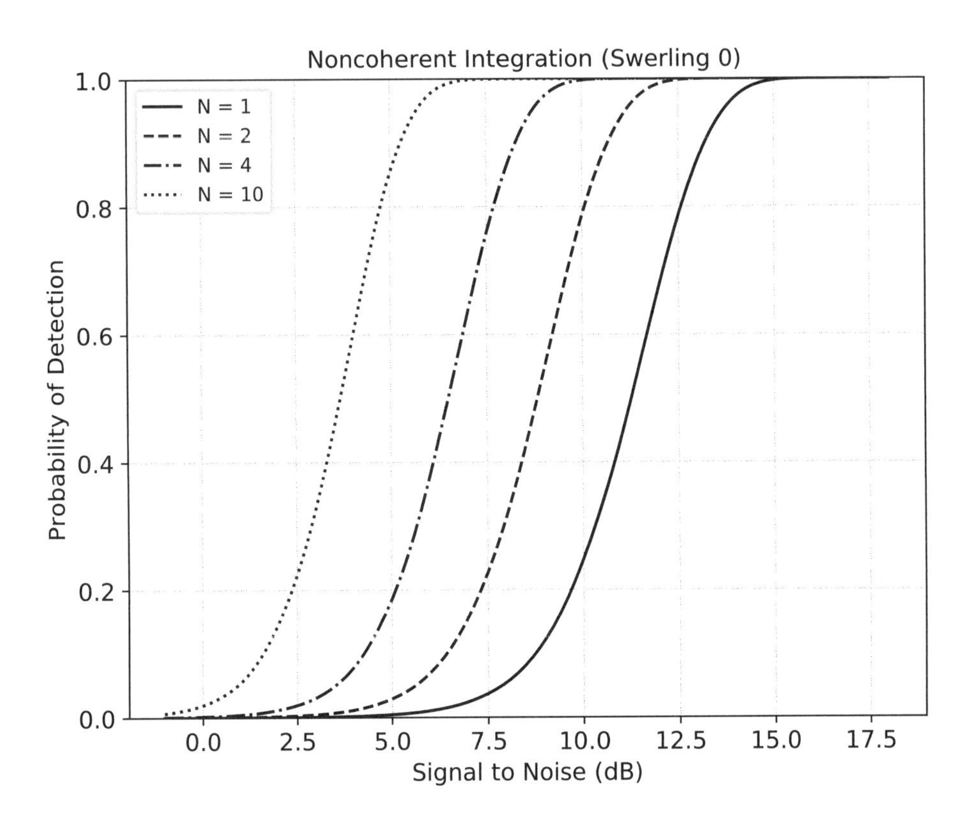

Figure 6.14 Probability of detection versus signal-to-noise ratio for Swerling 0 type targets ($P_{fa} = 10^{-6}$).

be seen, noncoherently integrating pulses reduces the required signal-to-noise ratio to achieve the same probability of detection.

6.3.2 Swerling I

The detailed derivation of the probability density function for a measurement, y, under H_1 for Swerling I type targets is beyond the scope of this book. The reader is referred to [19, 25] for these details, and the result is given here as

$$p(y|H_1) = \frac{1}{N \cdot SNR_{avg}} \left(1 + \frac{1}{N \cdot SNR_{avg}}\right)^{(N-2)} \exp\left(\frac{-y}{1 + N \cdot SNR_{avg}}\right)$$
$$\times\Gamma\left[\frac{y}{(1 + 1/(N \cdot SNR_{avg}))\sqrt{N-1}}, N-2\right], \tag{6.39}$$

where SNR_{avg} is the average signal-to-noise ratio over the N pulses. As before, integrate the probability density function in (6.39) from the voltage threshold to infinity to obtain the probability of detection. Performing the integration yields the following approximate expression [19, 25, 26]

$$P_d = 1 - \Gamma(N-1, SNR_T) + \left(1 + \frac{1}{N \cdot SNR_{avg}}\right)^{(N-1)} \Gamma\left(N-1, \frac{SNR_T}{1 + 1/(N \cdot SNR_{avg})}\right) \exp\left(\frac{-SNR_T}{1 + N \cdot SNR_{avg}}\right). \quad (6.40)$$

Since the noise model is the same in this case as for the Swerling 0 case, the probability of false alarm is given in (6.35). Figure 6.15 illustrates the probability of detection versus signal-to-noise ratio for Swerling I type targets. As in the Swerling 0 case, noncoherently integrating pulses reduces the requirement on the single pulse signal-to-noise ratio.

6.3.3 Swerling II

The probability density function for a measurement, y, under H_1 for Swerling II type targets is written as [19, 25]

$$p(y|H_1) = \frac{y^{(N-1)} \exp\left(\frac{y}{(1 + SNR_{avg})}\right)}{(1 + SNR_{avg})^N \, (N-1)!}. \quad (6.41)$$

As before, performing the integration of the probability density function in (6.41) results in the probability of detection for Swerling II type targets, given here as [19, 25, 26]

$$P_d = 1 - \Gamma\left[N, \frac{SNR_T}{1 + SNR_{avg}}\right]. \quad (6.42)$$

Again, the noise model is the same in this case as for the Swerling 0 case, and the probability of false alarm is given in (6.35). Figure 6.16 shows the probability of detection versus signal-to-noise ratio for Swerling II type targets. As in the previous cases, noncoherently integrating pulses reduces the requirement on the single pulse signal-to-noise ratio.

6.3.4 Swerling III

The derivation of the probability of detection for Swerling III type targets follows the same procedure as the Swerling I type targets except the probability density function for the target radar cross section follows a chi-squared probability with four degrees of

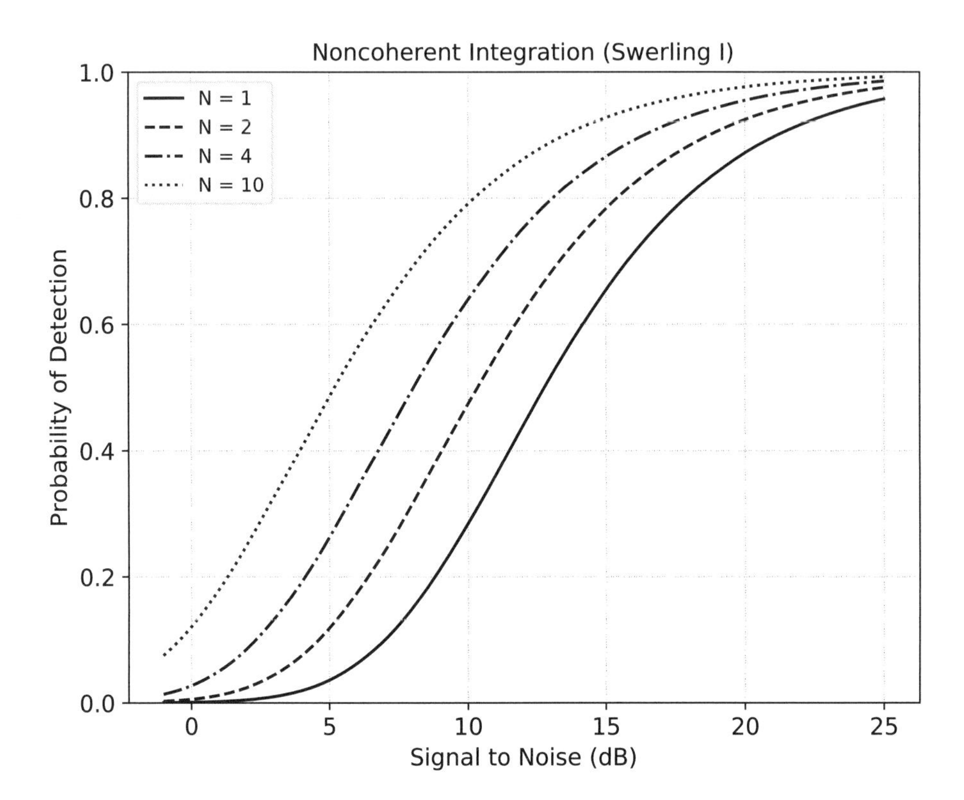

Figure 6.15 Probability of detection versus signal-to-noise ratio for Swerling I type targets ($P_{fa} = 10^{-6}$).

freedom rather than two. The details of this derivation may be found in [25, 27], and the result is given here as

$$P_d \approx \left(1 + \frac{2}{N \cdot SNR_{avg}}\right)^{(N-2)} \left[1 + \frac{SNR_T}{1 + (N \cdot SNR_{avg})/2} - \frac{2(N-2)}{(N \cdot SNR_{avg})}\right]$$

$$\times \exp\left(\frac{-SNR_T}{1 + (N \cdot SNR_{avg})/2}\right). \tag{6.43}$$

The expression in (6.43) is exact for $N = 1, 2$ and approximate for $P_{fa} \ll 1$ and $(N \cdot SNR_{avg})/2 > 1$. These conditions on the approximation are highly likely to be valid in practical situations. As in the previous cases, the noise model is the same in this case as for the Swerling 0 case, and the probability of false alarm is given in (6.35).

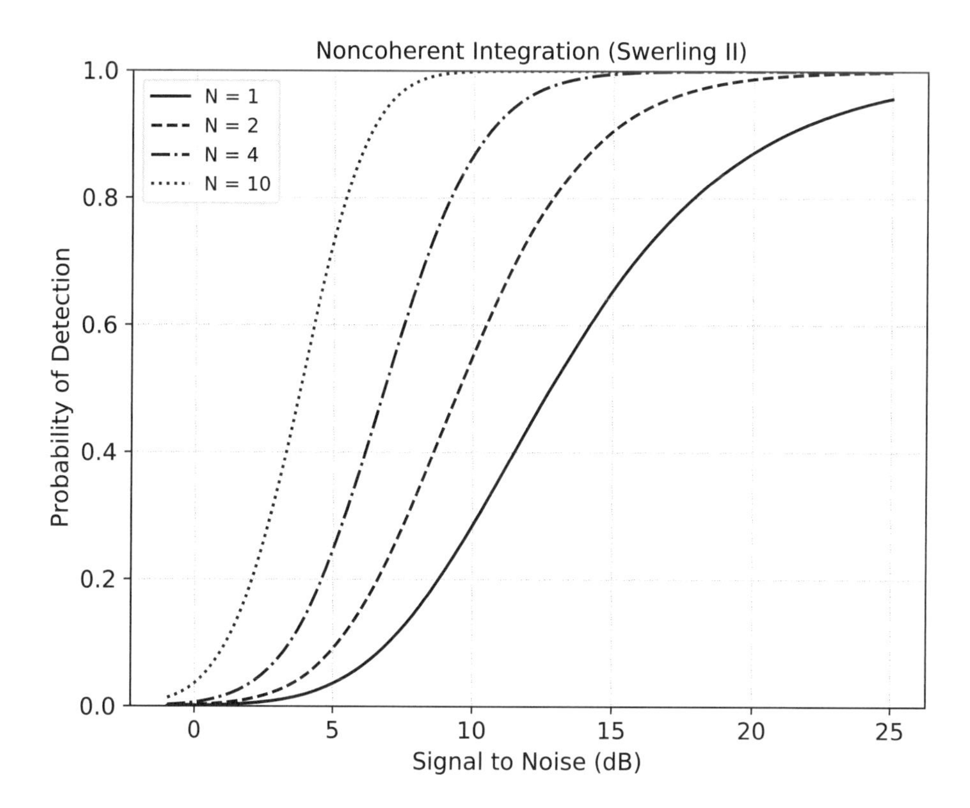

Figure 6.16 Probability of detection versus signal-to-noise ratio for Swerling II type targets ($P_{fa} = 10^{-6}$).

Figure 6.17 shows the probability of detection versus signal-to-noise ratio for Swerling III type targets. As in the previous cases, noncoherently integrating pulses reduces the requirement on the single pulse signal-to-noise ratio.

6.3.5 Swerling IV

The derivation of the probability of detection for Swerling IV type targets follows the same procedure as the Swerling II type targets except the probability density function for the radar cross section follows a chi-squared probability with four degrees of freedom rather than two. The details of this derivation may be found in [19, 25], and the results are given here as

$$P_d = 1 - \left(\frac{SNR_{avg}}{SNR_{avg}+2}\right)^N \sum_{k=0}^{N} \frac{N!}{k!(N-k)!} \left(\frac{SNR_{avg}}{2}\right)^{-k}$$

$$\times \Gamma\left[2N-k, \frac{2SNR_T}{SNR_{avg}+2}\right]. \tag{6.44}$$

Figure 6.18 shows the probability of detection versus signal-to-noise ratio for Swerling IV type targets. As in the previous cases, noncoherently integrating pulses reduces the requirement on the single pulse signal-to-noise ratio. Figure 6.19 provides a comparison between the five Swerling target types for 10 pulses. Swerling 0 has the lowest required signal-to-noise ratio compared to the other target models. This is expected since the target

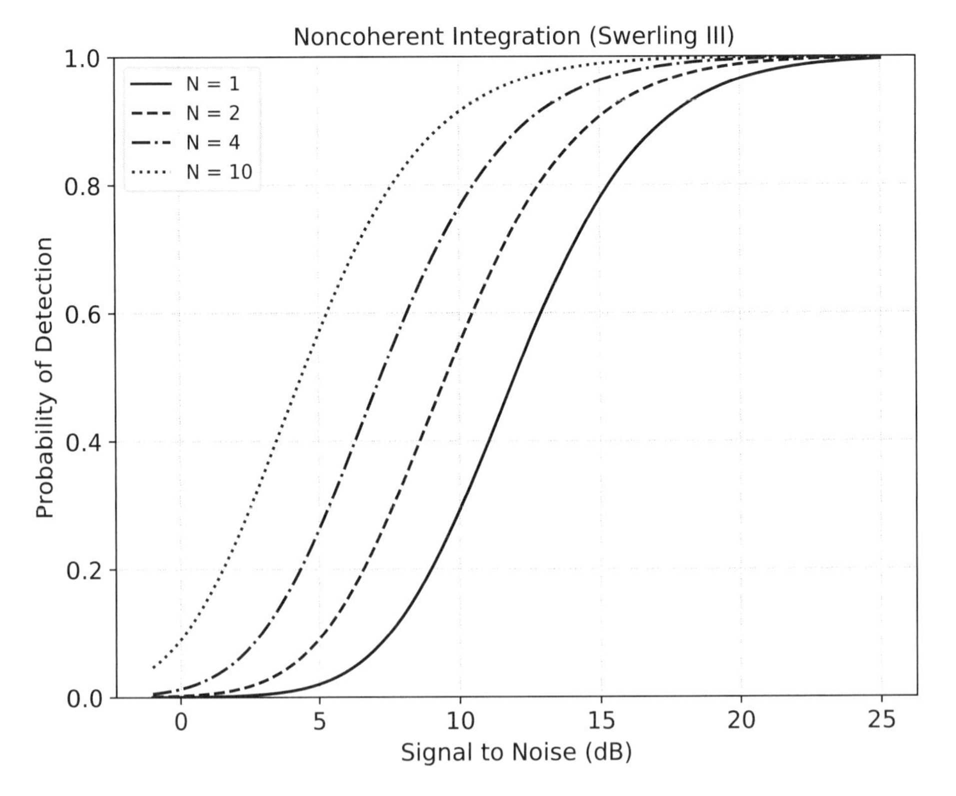

Figure 6.17 Probability of detection versus signal-to-noise ratio for Swerling III type targets ($P_{fa} = 10^{-6}$).

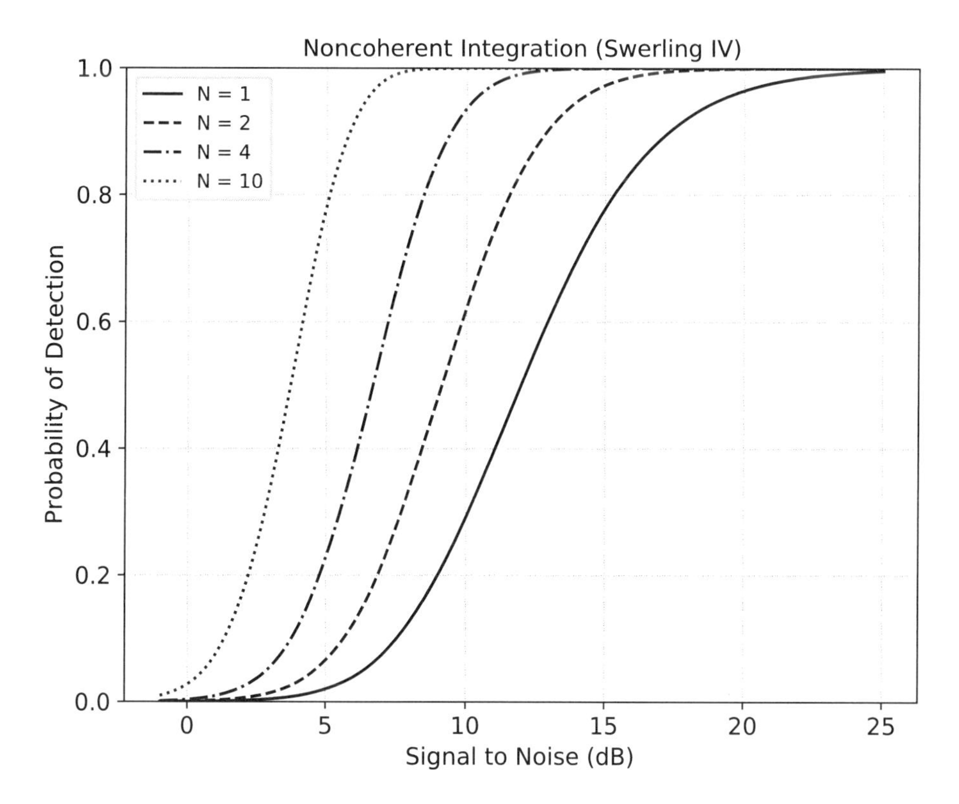

Figure 6.18 Probability of detection versus signal-to-noise ratio for Swerling IV type targets ($P_{fa} = 10^{-6}$).

is constant for each pulse. Swerling I and III require the highest signal-to-noise ratio as these models fluctuate from scan to scan.

6.3.6 Shnidman's Equation

As shown in the previous sections, the resulting analytical forms for the signal-to-noise ratio, probability of detection and probability of false alarm can become quite complicated. Shnidman [28] developed a set of equations, based on empirical data, for calculating the required signal-to-noise ratio for a specified probability of detection and probability of false alarm. These equations are applicable to all of the Swerling target models for either a single pulse or the noncoherent integration of N pulses. Shnidman's equation proceeds as follows. First, the parameter K is chosen based on the Swerling target type, then the parameter α is chosen based on the number of pulses as

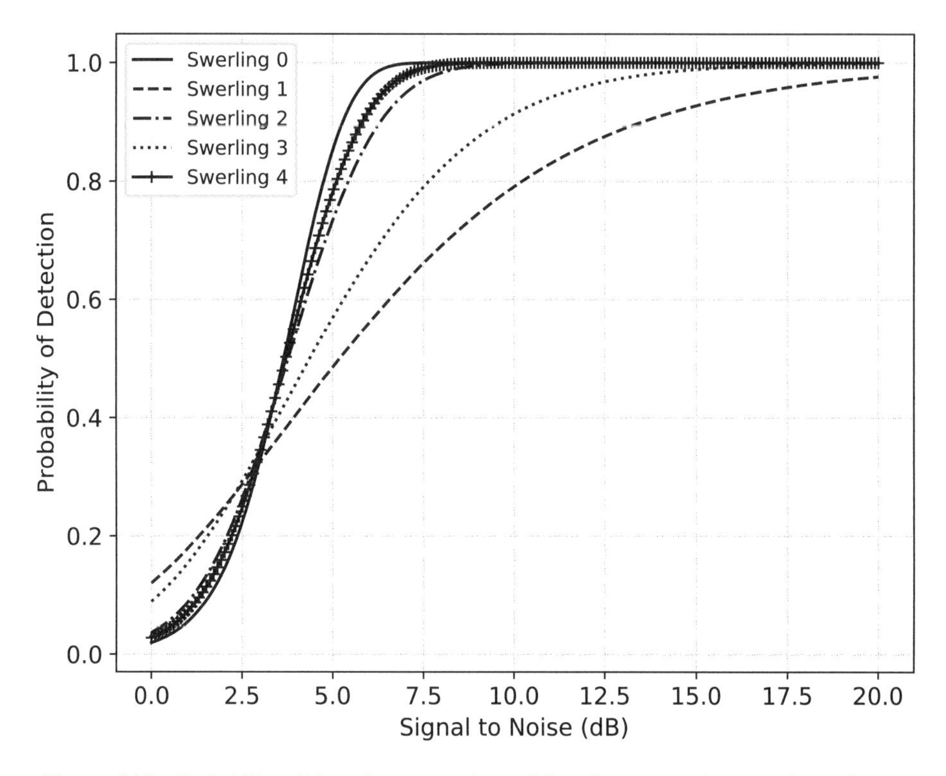

Figure 6.19 Probability of detection comparison of Swerling targets for noncoherently integrating 10 pulses ($P_{fa} = 10^{-6}$).

$$K = \begin{cases} \infty & \text{Swerling 0} \\ 1 & \text{Swerling I} \\ N & \text{Swerling II} \\ 2 & \text{Swerling III} \\ 2N & \text{Swerling IV} \end{cases}, \qquad \alpha = \begin{cases} 0 & N < 40 \\ \frac{1}{4} & N \geq 40 \end{cases}. \tag{6.45}$$

Then the two parameters η and X are computed from

$$\eta = \sqrt{-0.8\ln(4P_{fa}(1 - P_{fa}))} + \text{sign}(P_d - 0.5)\sqrt{-0.8\ln(4P_d(1 - P_d))} \tag{6.46}$$

and

$$X = \eta\left[\eta + 2\sqrt{\frac{N}{2} + \left(\alpha - \frac{1}{4}\right)}\right]. \tag{6.47}$$

The following constants are now computed

$$C_1 = \left(\left[(17.7006P_d - 18.4496)P_d + 14.5339\right]P_d - 3.525\right)/K, \tag{6.48}$$

$$C_2 = \frac{1}{K}\Bigg(\exp\left(27.31P_D - 25.14\right) + \left(P_d - 0.8\right) \times \left[0.7\ln\left(\frac{10^{-5}}{P_{fa}}\right) + \frac{(2N - 20)}{80}\right]\Bigg), \tag{6.49}$$

$$C_{dB} = \begin{cases} C_1 & \text{for } 0.1 \leq P_d \leq 0.872, \\ C_1 + C_2 & \text{for } 0.872 \leq P_d \leq 0.99, \end{cases} \tag{6.50}$$

$$C = 10^{(C_{dB}/10)}. \tag{6.51}$$

Finally, the signal-to-noise ratio is

$$SNR = 10\log_{10}\left(\frac{C \cdot X}{N}\right) \qquad \text{(dB)}. \tag{6.52}$$

Shnidman's equation has been shown to be accurate to within 0.5 dB within the following bounds

$$0.1 \leq P_d \leq 0.99, \tag{6.53}$$

$$10^{-9} \leq P_{fa} \leq 10^{-3}, \tag{6.54}$$

$$1 \leq N \leq 10. \tag{6.55}$$

Figure 6.20 shows the error in the signal-to-noise calculation when using Shnidman's equation for all five Swerling target types. As can be seen, Shnidman's equation provides accurate results for first cut type of radar system calculations.

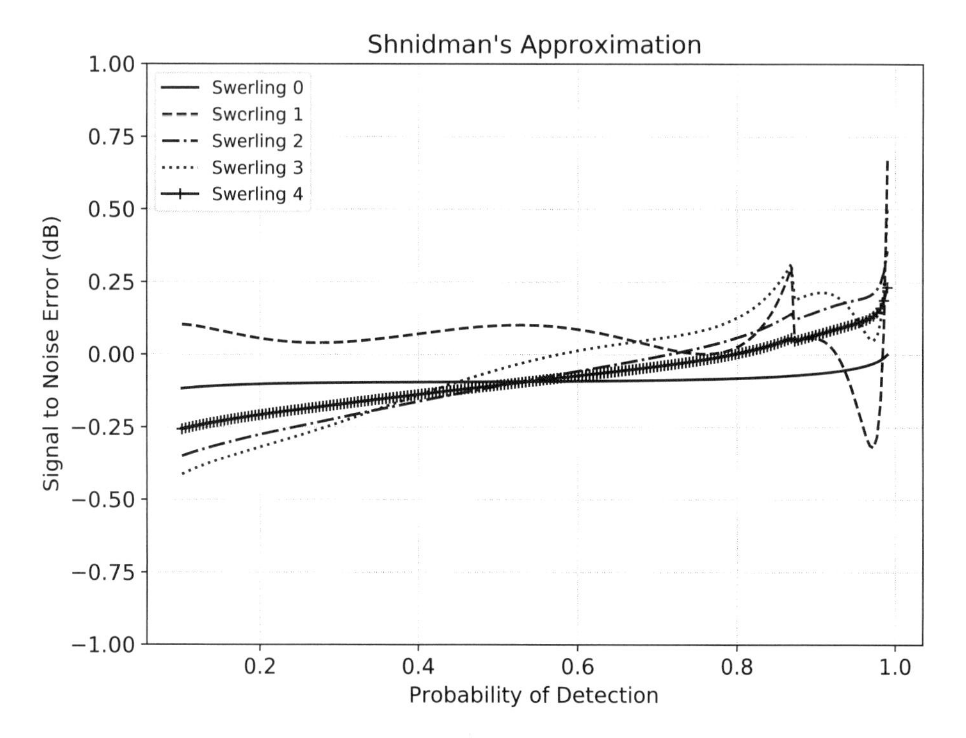

Figure 6.20 Error in the approximate signal-to-noise ratio using Shnidman's equation ($P_{fa} = 10^{-6}$, $N = 5$).

6.4 CONSTANT FALSE ALARM RATE

The previous sections were concerned with the calculation of detection statistics based on a chosen probability density function for the noise. A voltage threshold based on the probability of false alarm for a single pulse was calculated, as given in (6.15). Also calculated was the signal-to-noise ratio threshold using noncoherent integration of a number of pulses, given in (6.36). These types of fixed value thresholds satisfy the condition on the probability of false alarm as long as the noise power is constant. However, in practical systems, factors such as clutter, interfering sources, and nearby targets cause the noise level to change both spatially and temporally. Therefore, a continuously changing threshold is needed where the threshold is raised or lowered to maintain a constant probability of false alarm. This is typically referred to as *constant false alarm rate* (CFAR) processing. CFAR lowers the number of false alarms the radar system must process but may also prevent the detection of some targets. A fundamental circuit for CFAR processing was first described in 1968 by Finn and Johnson [29]. While the term CFAR processing refers to a common form of adaptive algorithms used to detect target returns against a background of noise, clutter, and interference, there are several approaches to accomplish this. The following sections present some of the more common CFAR techniques.

6.4.1 Cell Averaging CFAR

Cell averaging CFAR (CA-CFAR) is one of the simplest CFAR processing schemes. The estimated noise level around the cell under test (CUT) is found by taking a block of range and/or Doppler cells around the CUT and computing the average power level, as shown in Figure 6.21. The average noise power level for CFAR processing on one-dimensional measurements may be expressed as

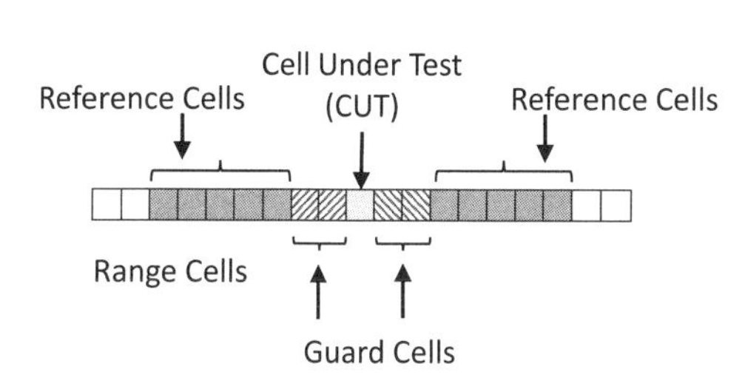

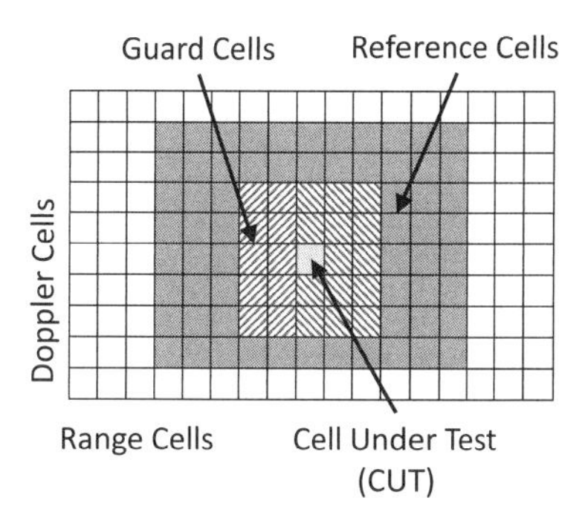

Figure 6.21 Range Doppler cell averaging CFAR.

$$Z = \frac{1}{N}\sum_{n=1}^{N} X_n, \tag{6.56}$$

where

N = number of range cells in CFAR average,
X_n = signal power in the nth cell,
Z = local estimate of the noise power.

The cells immediately surrounding the CUT are omitted from the calculation to avoid corruption of the noise estimate with power from the CUT. These cells are commonly referred to as *guard cells*. The local estimate of the noise power is generally increased to account for limited sample sizes. Consider the conventional cell averaging CFAR processing block diagram shown in Figure 6.22. A detection is declared in the CUT when

$$X \geq CZ, \tag{6.57}$$

where C is the CFAR bias. This bias is added to the local noise estimate due to limited sample sizes. The CFAR threshold is generally not computed for each cell in the range and/or Doppler window. Typically, a minimum detection threshold is first set. For each cell with an amplitude above this minimum detection threshold, a CFAR threshold is then calculated and applied. Cells with amplitudes above the CFAR threshold are considered detections, as shown in Figure 6.23. For illustration purposes, the CFAR threshold is calculated and plotted for every cell in the range data. This can be thought of as allowing the CFAR computation window, which consists of the cell under test, guard cells, and reference cells, to slide through the range data, as shown in Figure 6.24. Figure 6.25 shows the CFAR threshold calculated for every range cell in the range window. As can be seen, the CFAR threshold is not constant but varies with the input signal. Also, the CFAR threshold is larger on either side of the cells containing target returns. This helps to reduce the detections from range sidelobes or spillover from the cell under test. One disadvantage of CA-CFAR is that sharp clutter or interference boundaries tend to increase the number of false alarms.

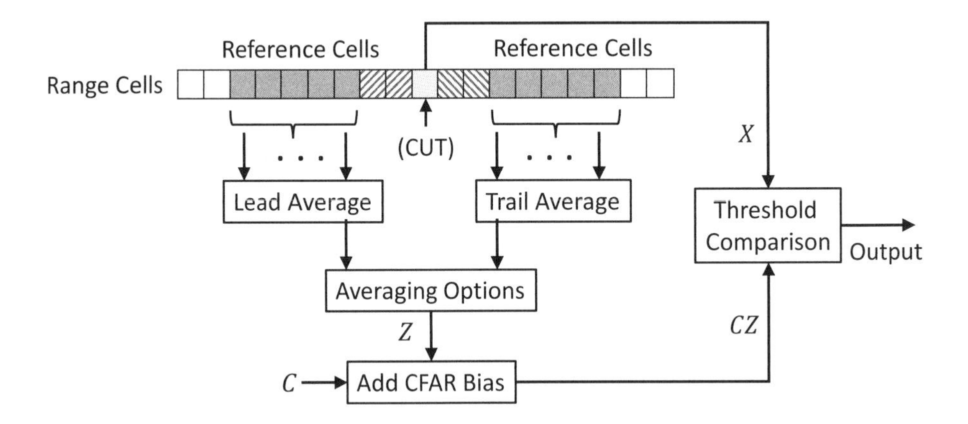

Figure 6.22 Conventional cell averaging CFAR processing.

6.4.2 Cell Averaging Greatest of CFAR

Conventional cell averaging CFAR estimates the local noise power by averaging over all the reference cells. In the case of cell averaging greatest of CFAR (CAGO-CFAR), the local noise power is estimated by first computing the average of the lead and trail reference cells separately, as shown in Figure 6.26. Then the CFAR threshold value is

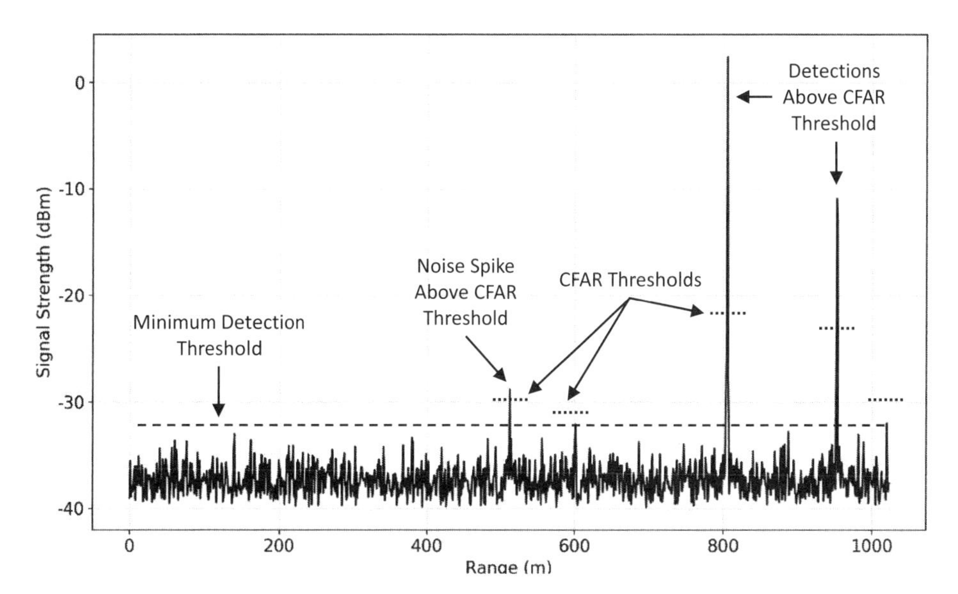

Figure 6.23 CFAR processing based on minimum detection threshold.

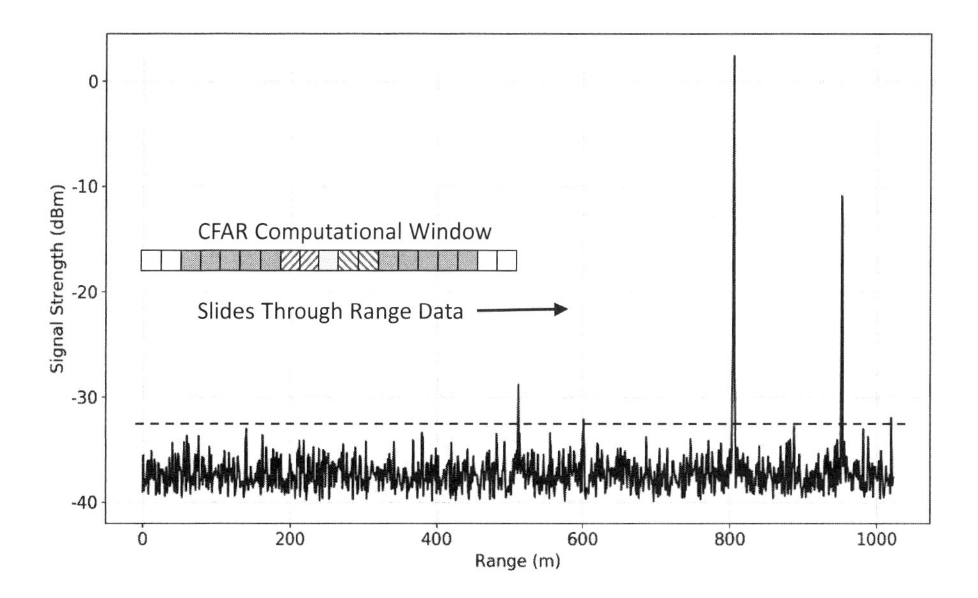

Figure 6.24 Cell averaging sliding window.

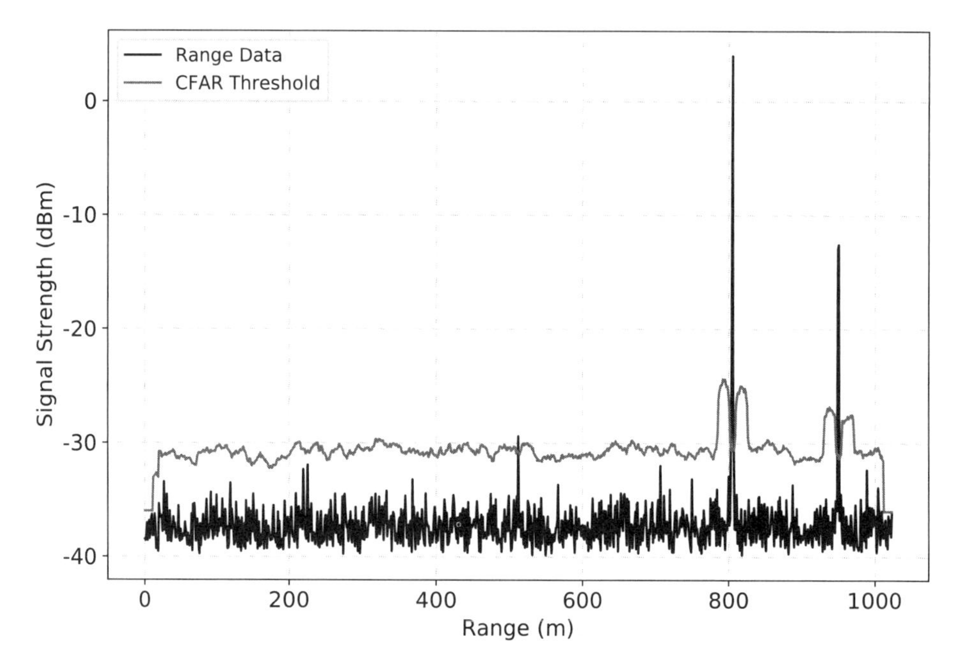

Figure 6.25 Cell averaging CFAR threshold computed over all range cells.

taken to be the maximum of these two separate averages. The estimated value of the noise power is then written as

$$Z = \frac{2}{N} \max\left(\left[\sum_{n=1}^{N/2} X_n\right], \left[\sum_{n=N/2+1}^{N} X_n\right]\right). \tag{6.58}$$

Again, for illustration, the CAGO-CFAR threshold for every range cell is calculated and shown in Figure 6.27. The main advantage of CAGO-CFAR, as compared to CA-CFAR, is improved performance for sharp clutter boundaries in nonhomogeneous environments. However, the CAGO-CFAR threshold can still mask two adjacent targets.

6.4.3 Censored Greatest of CFAR

One method to help alleviate the masking problem and improve detection performance is to compute the estimate for the local noise in the same fashion as the CAGO-CFAR case but remove the largest M samples before computing each average, as illustrated in Figure 6.28.

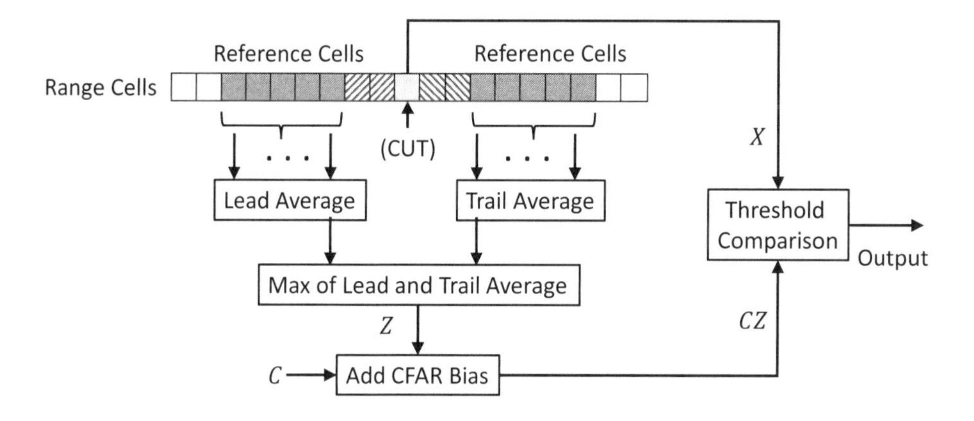

Figure 6.26 Cell averaging greatest of CFAR.

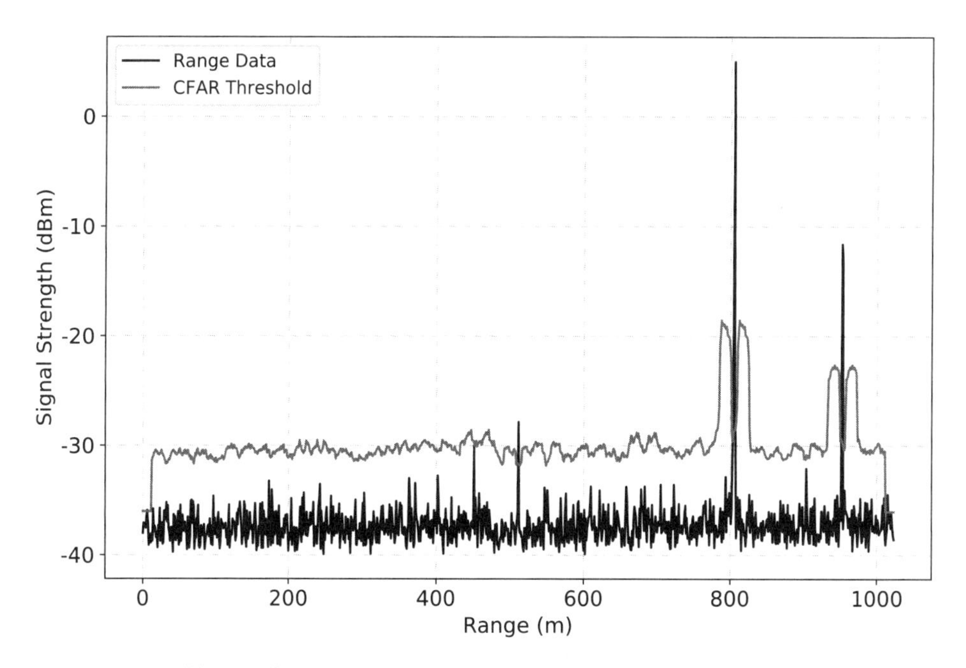

Figure 6.27 Cell averaging greatest of CFAR (CAGO-CFAR).

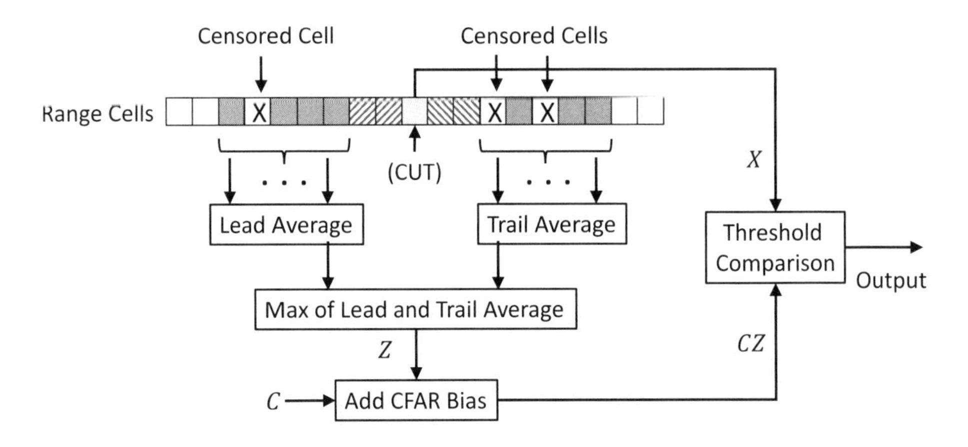

Figure 6.28 Censored cell averaging greatest of CFAR.

6.4.4 Cell Averaging Smallest of CFAR

Cell averaging small of CFAR (CASO-CFAR) is very similar to CAGO-CFAR except the minimum of the two averages is used rather than the maximum. This is expressed as

$$Z = \frac{2}{N} \min \left(\left[\sum_{n=1}^{N/2} X_n \right], \left[\sum_{n-N/2+1}^{N} X_n \right] \right). \tag{6.59}$$

The CASO-CFAR threshold for each range cell is shown in Figure 6.29. The main difference between CASO-CFAR and CAGO-CFAR is a reduced chance of masking from two adjacent targets.

6.4.5 Ordered Statistic CFAR

The CFAR methods of the previous sections do not set a threshold that deals with homogeneous and nonhomogeneous interference equally well. Therefore, the method of ordered statistic CFAR (OS-CFAR) was developed [30]. In this method, the measurements from the reference cells are rank ordered by decreasing magnitude. As in the censored CFAR case presented in Section 6.4.3, the largest M samples are removed prior to further processing. For OS-CFAR, the arithmetic mean of (6.56) is replaced by a single rank, k, of the ordered statistic

$$X_1 \leq \cdots \leq X_k \leq \cdots \leq X_N. \tag{6.60}$$

OS-CFAR can also be applied to the lead and trail reference gates by replacing the mean calculations in (6.58) and (6.59) with the ordered statistic k. The OS-CFAR threshold

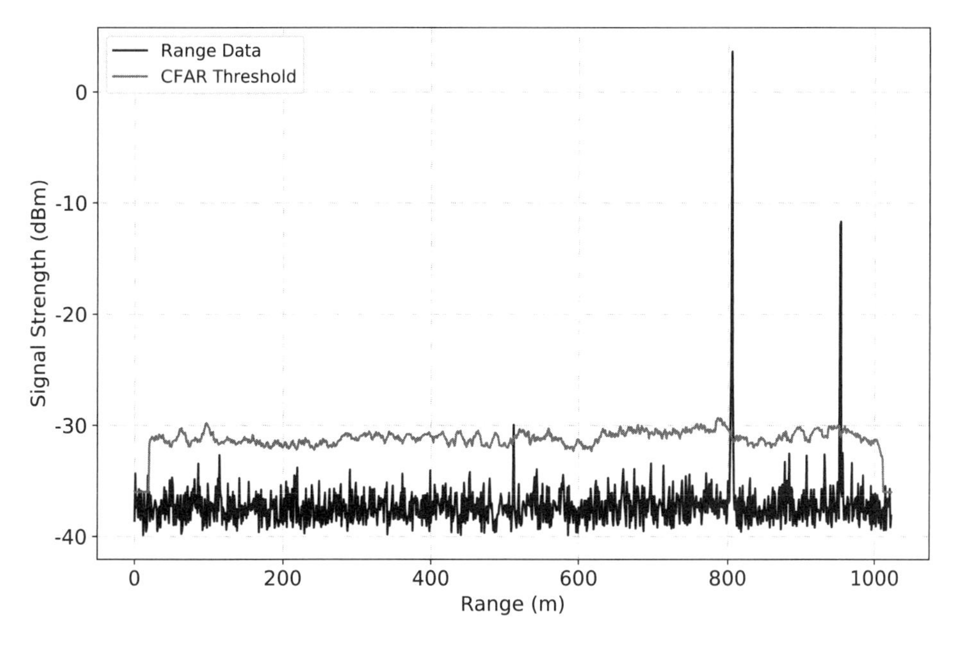

Figure 6.29 Cell averaging smallest of CFAR (CASO-CFAR).

for each range cell is shown in Figure 6.30. There are many advantages to OS-CFAR as it does not assume homogeneous clutter in the reference windows. It almost completely eliminates the masking problem associated with CA-CFAR. Unlike the CA-CFAR, which is based on the square law detector, OS-CFAR can be applied to both linear and square law detectors simultaneously without adaptation. Also, it is not necessary to have test and guard cells that are omitted from the reference window, and the window length itself is much less important. However, OS-CFAR does have high processing requirements mostly due to the sorting algorithm. Also, this processing must be completed in real time in the radar signal processing as the threshold must be calculated before target detection.

6.4.6 Cell Averaging Statistic Hofele CFAR

The cell averaging statistic Hofele (CASH-CFAR) is a statistical method based on a series of summing elements associated with each set of L storage cells. This is followed by a maximum-minimum process that selects a value, S_r, representative of the background interference, from the set of sums, $S_1, S_2, ...S_N$, as shown in Figure 6.31 [31]. The CASH-CFAR method has many advantages over other CFAR approaches. It is a much better fit to the clutter scenario based on the quality factor given by [31]

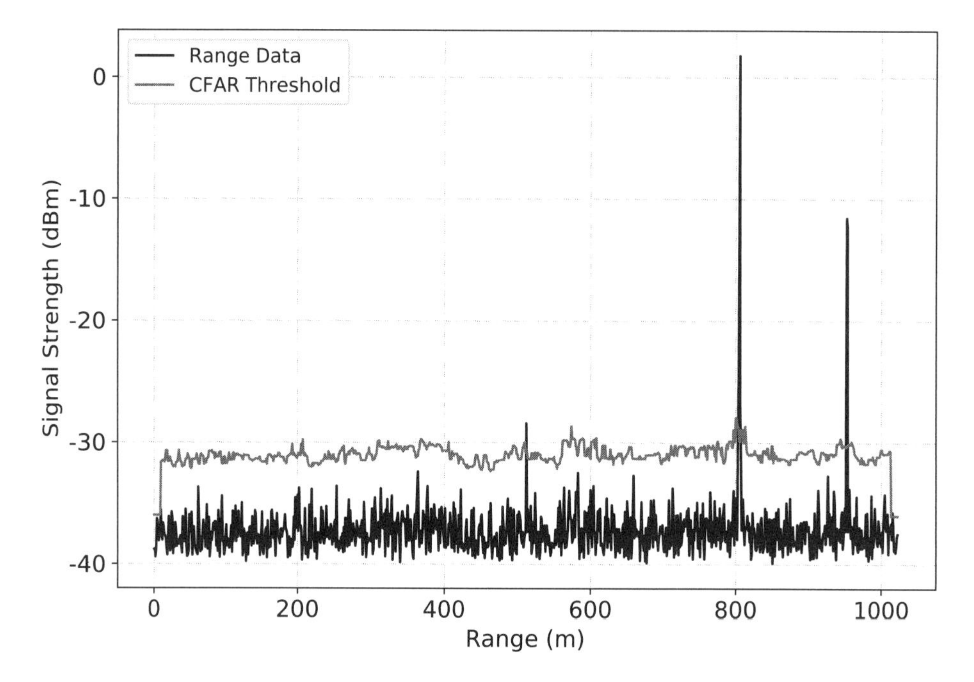

Figure 6.30 Ordered statistic CFAR (OS-CFAR).

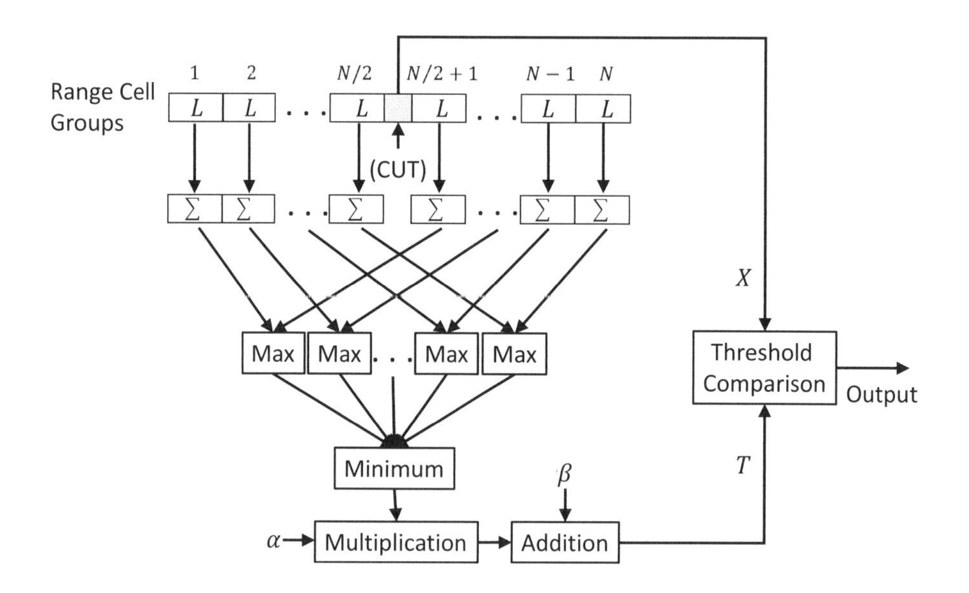

Figure 6.31 Cell averaging statistic Hoefele CFAR (CASH-CFAR).

$$q = \sum_{i=1}^{N} \Big[s(i) - c(i) \Big], \tag{6.61}$$

where $s(i)$ is the background interference for the scenario and $c(i)$ is the CFAR threshold. Also, the CASH-CFAR handles quick changes in the background interference without long delays. The range sidelobes resulting from pulse compression techniques are covered by the threshold as there is no alternate covering and aggregation of objects. Finally, the processing required by the CASH-CFAR is significantly less than that of rank selection type algorithms, including OS-CFAR and CAOS-CFAR.

6.5 EXAMPLES

The sections below illustrate the concepts of this chapter with several Python examples. The examples for this chapter are in the directory *pyradar\Chapter06* and the matching MATLAB examples are in the directory *mlradar\Chapter06*. The reader should consult Chapter 1 for information on how to execute the Python code associated with this book.

6.5.1 Probability Distributions

For this example, calculate and display the probability distribution function for a chi-squared random variable with a location of 2, scale of 1, and shape factor equal to 10.

Solution: The solution to the above example is given in the Python code *probability_distributions_example.py* and in the MATLAB code *probability_distributions_example.m*. Running the Python example code displays a GUI allowing the user to select from Gaussian, Rayleigh, Rice, Weibull, and chi-squared distributions. The GUI also allows the user to enter the location, scale, and shape factor for each distribution. The code then calculates and displays the probability density function spanning the 0.001 to 0.999 confidence interval for the selected distribution. Figure 6.32 shows the probability distribution function for the chi-squared example. The user is encouraged to study the distribution of other random variables available in this code and SciPy [6].

6.5.2 Detection Probability with Gaussian Noise

Referring to Section 6.1.2, calculate and display the probability of detection for a noncoherent receiver with Gaussian distributed noise. The probability of false alarm is 10^{-6} and the signal-to-noise ratio spans -3 to 18 dB.

Solution: The solution to the above example is given in the Python code *gaussian_noise_pd_example.py* and in the MATLAB code *gaussian_noise_pd_example.m*. Running the Python example code displays a GUI allowing the user to enter the range of

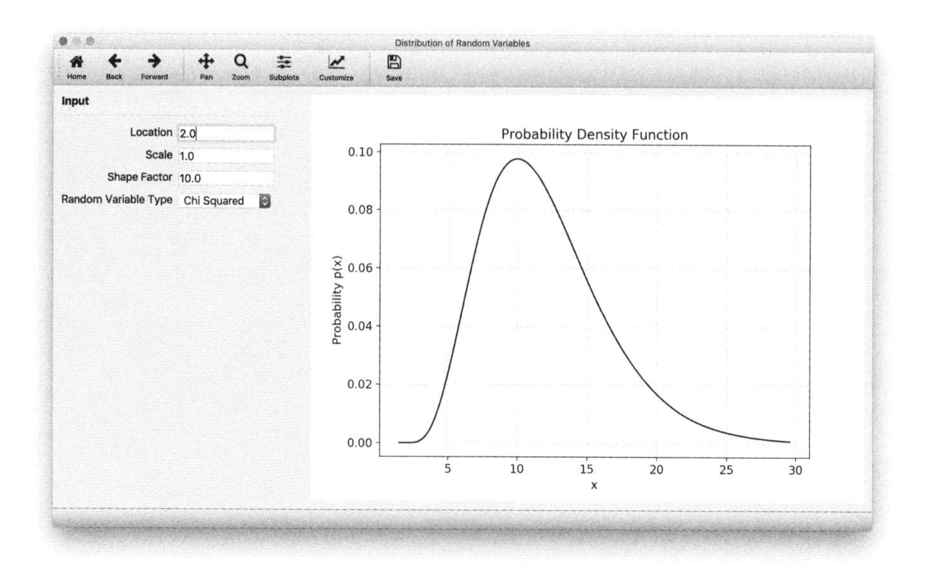

Figure 6.32 The probability density function for a chi-squared random variable with location = 2, scale = 1, and shape factor = 10 calculated by *probability_distributions_example.py*.

signal-to-noise values and the probability of false alarm. The code then calculates and displays the probability of detection over the signal-to-noise range and for the specified probability of false alarm, as shown in Figure 6.33.

6.5.3 Detection Probability with Rayleigh Noise

In the previous example of Section 6.5.2, the probability of detection for a noncoherent receiver was examined. This example deals with the probability of detection for a coherent receiver with Rayleigh distributed noise. Calculate and display the probability of detection when the probability of false alarm is 10^{-6} and the signal-to-noise ratio spans -3 to 18 dB.

Solution: The solution to the above example is given in the Python code *rayleigh_noise_pd_example.py* and in the MATLAB code *rayleigh_noise_pd_example.m*. Running the Python example code displays a GUI allowing the user to enter the range of signal-to-noise values and the probability of false alarm. The code then calculates and displays the probability of detection over the signal-to-noise range and for the specified probability of false alarm, as shown in Figure 6.34.

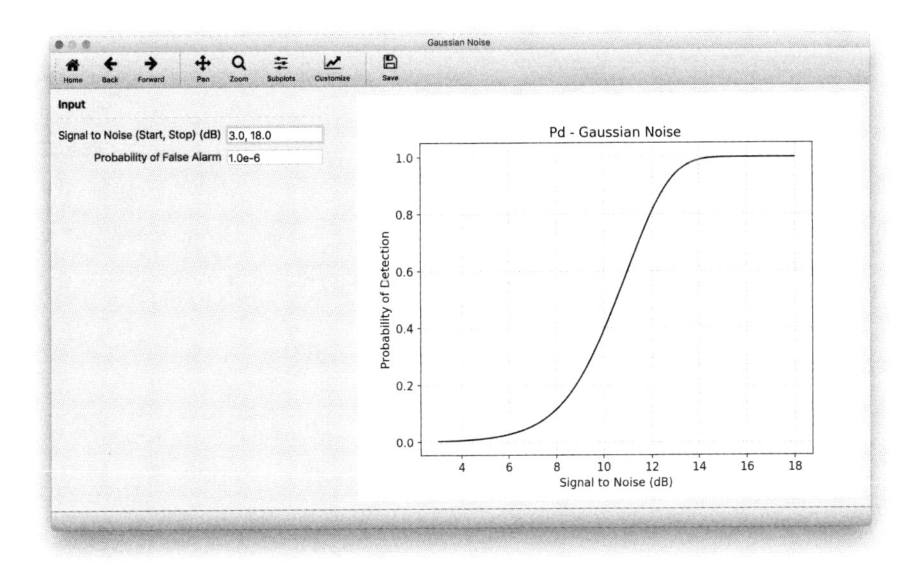

Figure 6.33 The probability of detection for a noncoherent receiver with Gaussian noise and $P_{fa} = 10^{-6}$ calculated by *gaussian_noise_pd_example.py*.

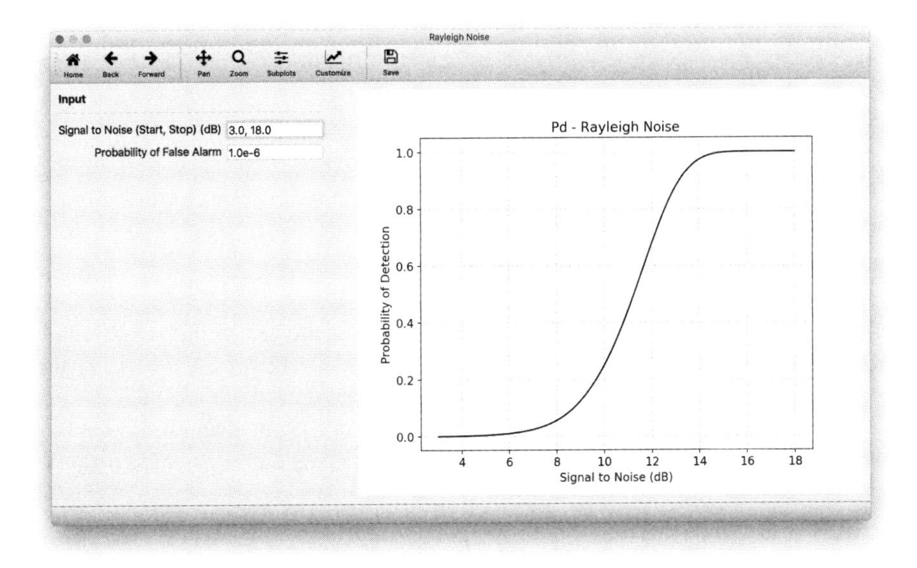

Figure 6.34 The probability of detection for a coherent receiver with Rayleigh noise and $P_{fa} = 10^{-6}$ calculated by *rayleigh_noise_pd_example.py*.

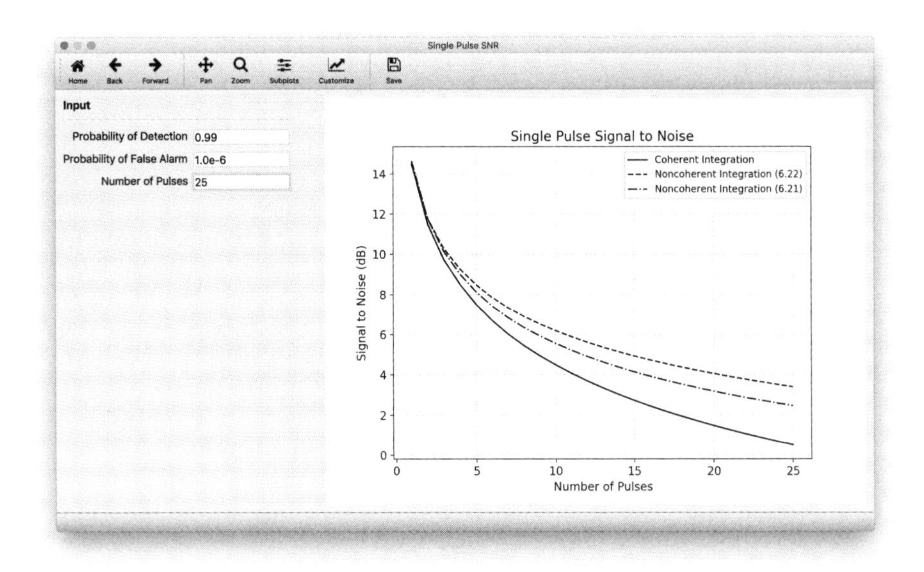

Figure 6.35 The single pulse signal-to-noise ratio required to produce $P_{fa} = 10^{-6}$ and $P_d = 0.99$ for both coherent and noncoherent integration calculated by *single_pulse_snr_example.py*.

6.5.4 Single Pulse signal-to-noise

For this example, calculate and display the required single pulse signal-to-noise ratio for both coherent and noncoherent integration of 25 pulses when the probability of false alarm is 10^{-6} and probability of detection is 0.99.

Solution: The solution to the above example is given in the Python code *single_pulse_snr _example.py* and in the MATLAB code *single_pulse_snr_example.m*. Running the Python example code displays a GUI allowing the user to enter the probability of detection, probability of false alarm, and the number of pulses to be integrated. The code then calculates and displays the single pulse signal-to-noise ratio required to meet the specified detection parameters, as shown in Figure 6.35. For noncoherent integration, the code calculates and display both approximations from (6.21) and (6.22).

6.5.5 Binary Integration

In this example, calculate and display the probability of detection for 3 of 5 binary integration with a probability of false alarm of 10^{-6}. The signal-to-noise ratio spans 0 to

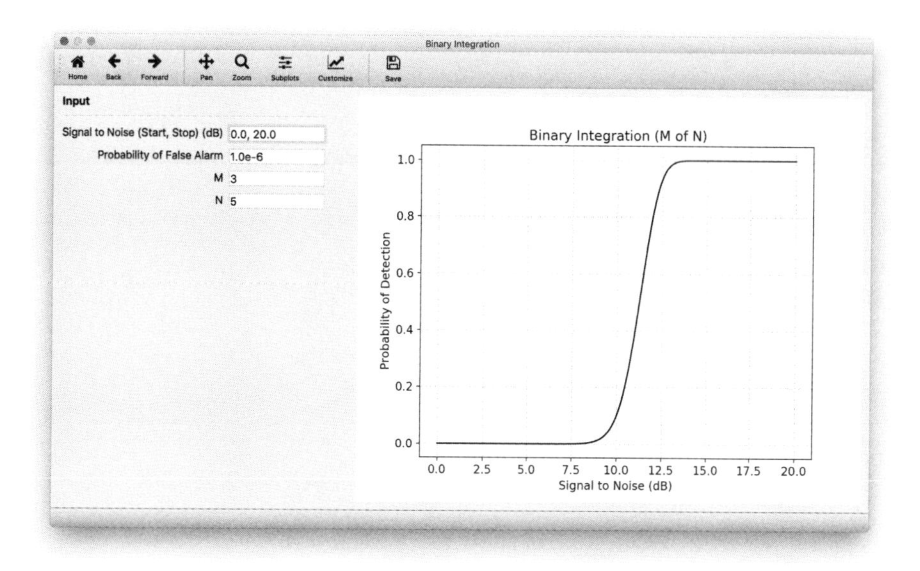

Figure 6.36 The probability of detection resulting from 3 of 5 binary integration with $P_{fa} = 10^{-6}$ calculated by *binary_integration_example.py*.

20 dB.

Solution: The solution to the above example is given in the Python code *binary_ integration_example.py* and in the MATLAB code *binary_integration_example.m*. Running the Python example code displays a GUI allowing the user to enter the probability of detection, range of signal-to-noise values, and the M of N values for binary integration. The code then calculates and displays the probability of detection for the specified parameters, as shown in Figure 6.36. The reader is encouraged to vary both M and N to see the effects of these parameters on the probability of detection.

6.5.6 Optimum Binary Integration

In the previous example of Section 6.5.5,the probability of detection for a given choice of M and N was studied. This example shows the optimum choice for the parameter M for a specified N and Swerling target type. Calculate and display the optimum choice for the parameter M for a Swerling I type target. Allow the number of pulses, N, to vary from 1 to 10.

Solution: The solution to the above example is given in the Python code *optimum_binary _example.py* and in the MATLAB code *optimum_binary_example.m*. Running the Python

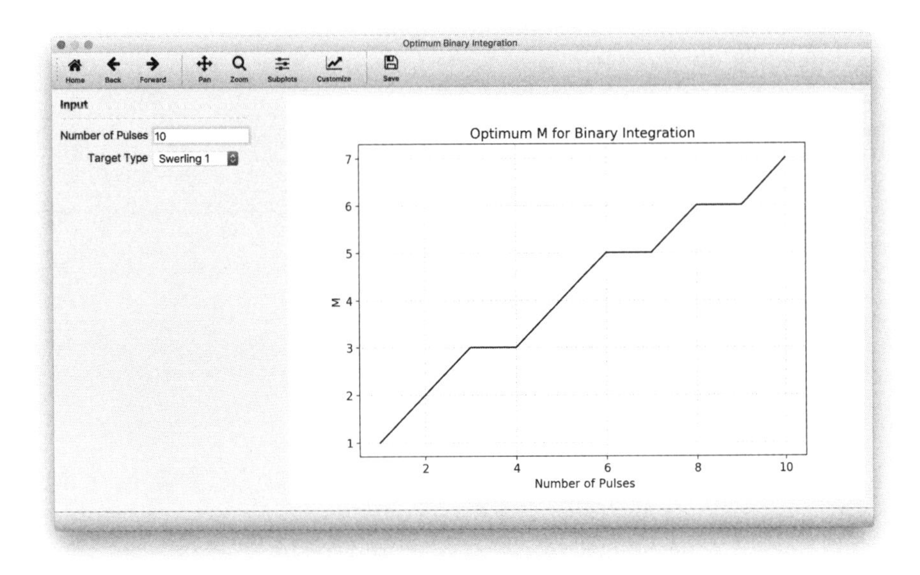

Figure 6.37 The optimum choice for the parameter M for a given number of pulses and Swerling target type calculated by *optimum_binary_example.py*.

example code displays a GUI allowing the user to enter the number of pulses and Swerling target type. The code then calculates and displays the optimum choice for M, as shown in Figure 6.37. The reader is encouraged to use this tool in conjunction with *binary_integration_example.py* from the previous example to further study the probability of detection for binary integration.

6.5.7 Coherent Pulse Integration

For this example, calculate and display the probability of detection for coherently integrating 10 pulses for a Swerling I type target with a probability of false alarm of 10^{-9}. Allow the signal-to-noise ratio to vary from -4 to 20 dB.

Solution: The solution to the above example is given in the Python code *coherent_integration_example.py* and in the MATLAB code *coherent_integration_example.m*. Running the Python example code displays a GUI allowing the user to enter the range of signal-to-noise values, the number of pulses, probability of false alarm, and Swerling target type. As discussed earlier, coherent pulse integration may only be applied to Swerling 0, Swerling I or Swerling III type targets. The code then calculates and displays the probability of detection over the signal-to-noise range, as shown in Figure 6.38.

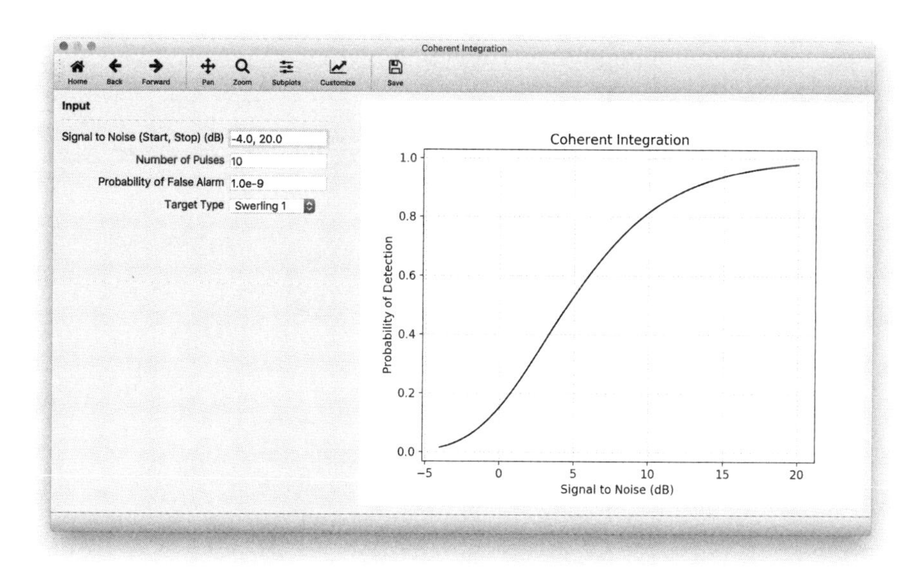

Figure 6.38 The probability of detection for coherent integration of 10 pulses, $P_{fa} = 10^{-9}$, and Swerling I type target calculated by *coherent_integration_example.py*.

6.5.8 Noncoherent Pulse Integration

In the previous example in Section 6.5.7, the probability of detection resulting from coherent pulse integration was examined. This example studies the probability of detection resulting from noncoherent pulse integration. Calculate and display the probability of detection for noncoherently integrating 10 pulses for a Swerling II type target with a probability of false alarm of 10^{-9}. Allow the signal-to-noise ratio to vary from 0 to 12 dB.

Solution: The solution to the above example is given in the Python code *non_coherent_integration_example.py* and in the MATLAB code *non_coherent_integration_example.m*. Running the Python example code displays a GUI allowing the user to enter the range of signal-to-noise values, the number of pulses, probability of false alarm, and Swerling target type. In the previous example only Swerling 0, Swerling I, and Swerling III type targets were allowed. For the noncoherent case, all five Swerling target types are permissible. The code then calculates and displays the probability of detection over the signal-to-noise range, as shown in Figure 6.39.

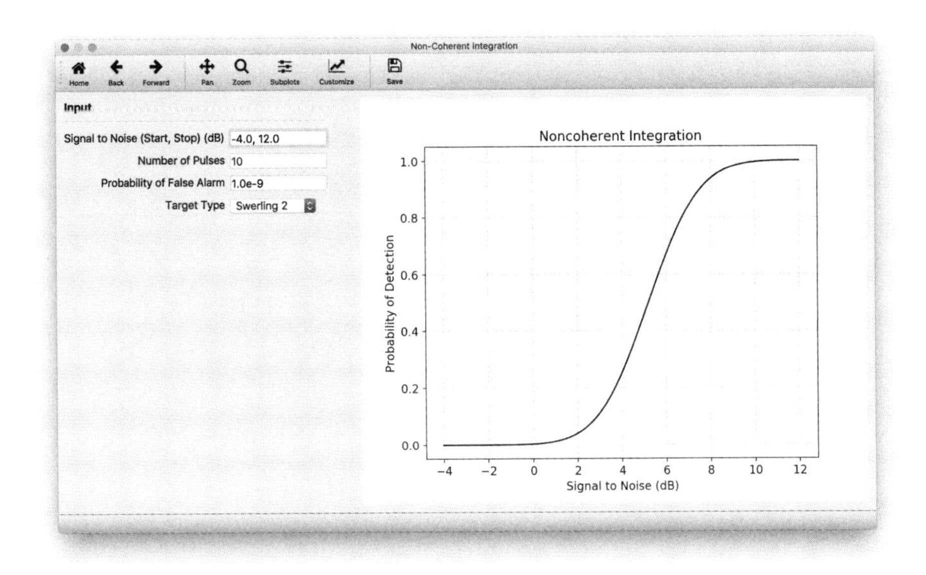

Figure 6.39 The probability of detection for noncoherent integration of 10 pulses, $P_{fa} = 10^{-9}$, and Swerling II type target calculated by *non_coherent_integration_example.py*.

6.5.9 Shnidman's Approximation

For this example, calculate and display the error for Shnidman's approximation for the required single pulse signal-to-noise ratio, as given in Section 6.3.6. Display the error for a Swerling III target type, probability of false alarm of 10^{-6}, 10 pulses, and allow the probability of detection to vary from 0.1 to 0.99.

Solution: The solution to the above example is given in the Python code *shnidman_example.py* and in the MATLAB code *shnidman_example.m*. Running the Python example code displays a GUI allowing the user to enter the range of probability of detection, number of pulses, probability of false alarm, and the Swerling target type. The code then calculates and displays the error in the required single pulse signal-to-noise ratio, as shown in Figure 6.40.

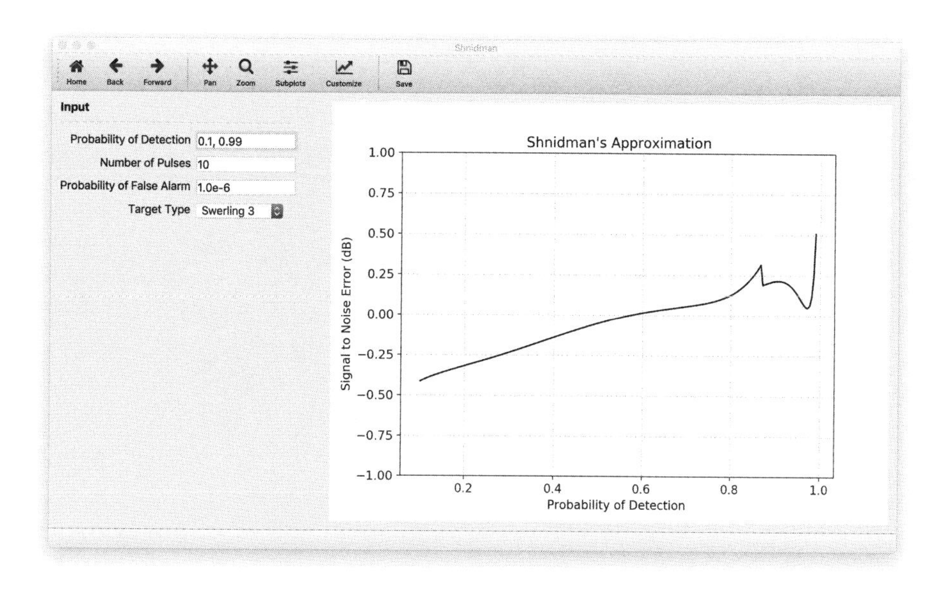

Figure 6.40 The error in the required single pulse signal-to-noise ratio using Shnidman's equation for a Swerling III type target with $P_{fa} = 10^{-9}$ calculated by *shnidman_example.py*.

6.5.10 Constant False Alarm Rate

For this example, calculate and display the CFAR threshold for CA-CFAR, CAGO-CFAR, CASO-CFAR, and OS-CFAR as described in Section 6.4. In each case, use 8 guard cells, 15 reference cells, and a bias of 4 dB.

Solution: The solution to the above example is given in the Python code *cfar_example.py* and in the MATLAB code *cfar_example.m*. Running the Python example code displays a GUI allowing the user to enter the number of guard cells, number of reference cells, bias, and the CFAR type. The code then calculates and displays the CFAR threshold along with the reference signal. The four CFAR thresholds for the example are shown in Figures 6.41 - 6.44.

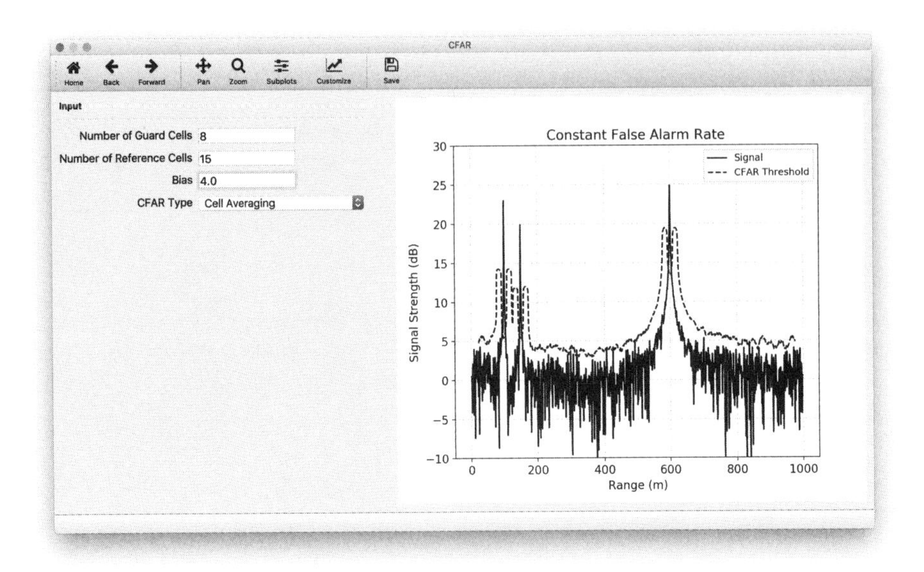

Figure 6.41 The cell averaging CFAR threshold (CA-CFAR) calculated by *cfar_example.py*.

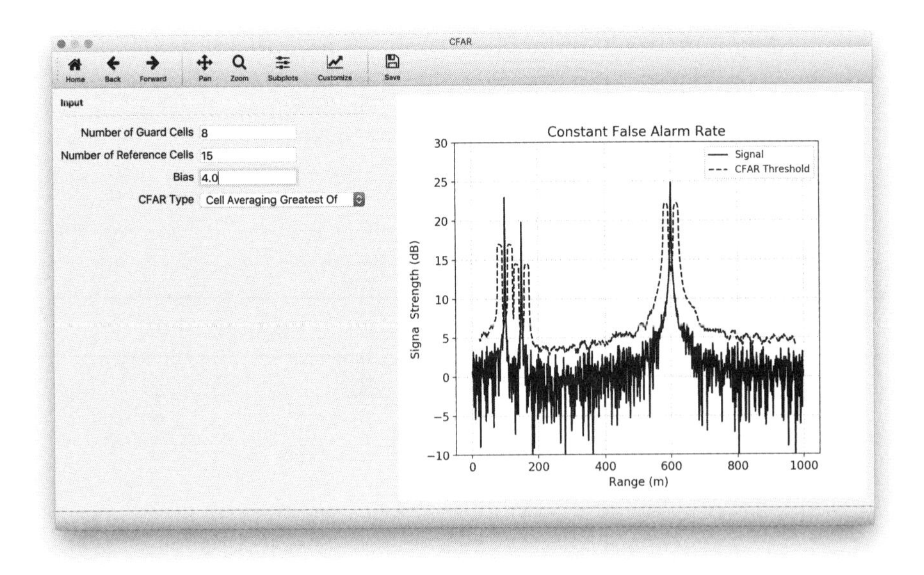

Figure 6.42 The cell averaging greatest of CFAR threshold (CAGO-CFAR) calculated by *cfar_example.py*.

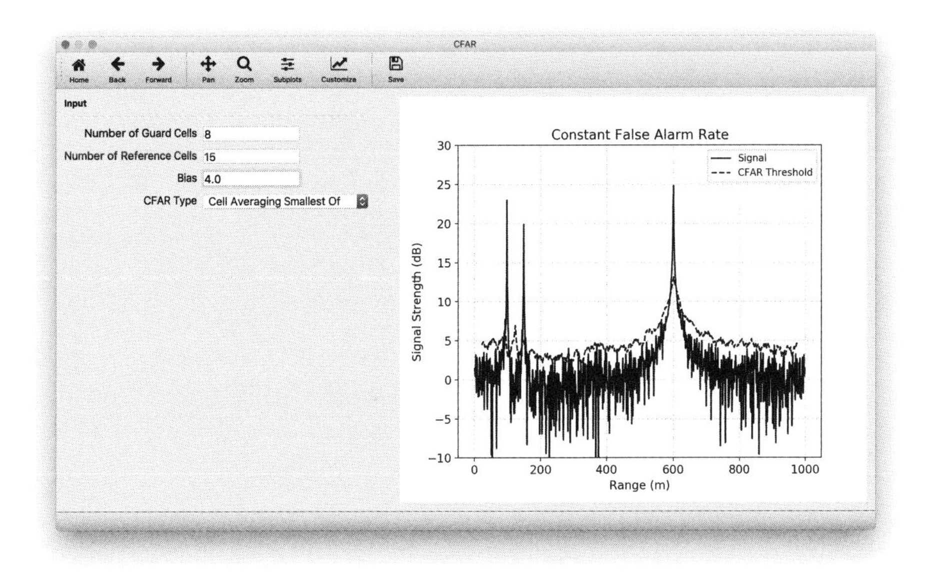

Figure 6.43 The cell averaging smallest of CFAR threshold (CASO-CFAR) calculated by *cfar_example.py*.

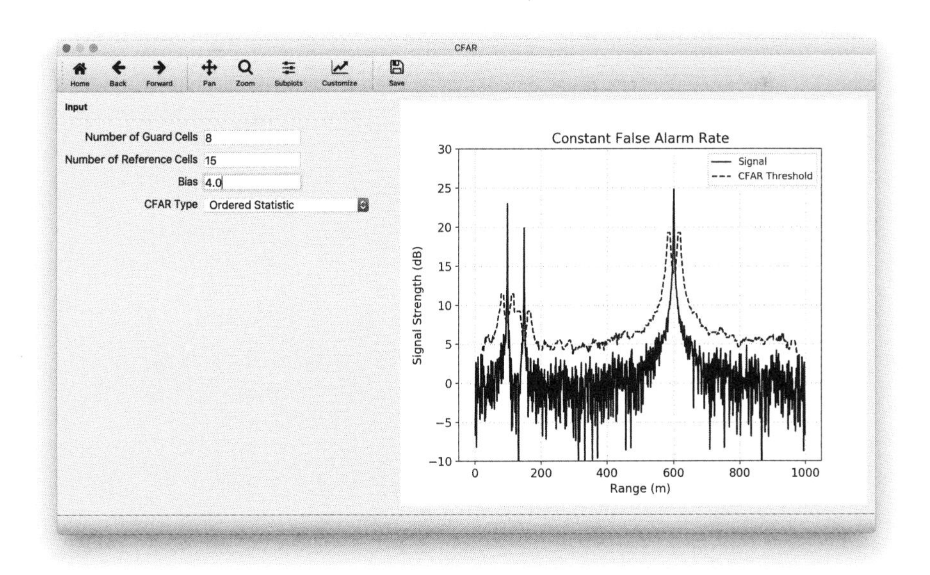

Figure 6.44 The ordered statistic CFAR threshold (OS-CFAR) calculated by *cfar_example.py*.

PROBLEMS

6.1 Calculate the false alarm rate given a probability of false alarm of 10^{-6} and a pulse repetition frequency of 10 kHz.

6.2 Explain why the Neyman-Pearson lemma results in the most powerful test for detection hypotheses. Why is this significant to the radar detection process?

6.3 For noncoherent detection, calculate the probability of false alarm given a voltage threshold of 0.7V. Assume the noise is represented by a zero mean Gaussian distribution with a variance of 2 V^2.

6.4 Suppose a signal has an amplitude of 1.3V; find the probability of detection when the voltage threshold is 1.1V and the noise is represented by a zero mean Gaussian distribution with a variance of 1 V^2.

6.5 For coherent detection, determine the voltage threshold required for a probability of false alarm of 10^{-9}. Assume the noise is represented by a zero mean Gaussian distribution with a variance of 4 V^2.

6.6 Given the noise and voltage threshold in Problem 6.5, calculate the probability of detection when the signal-to-noise ratio is 10 dB.

6.7 Calculate the probability of detection for both Gaussian noise and Rayleigh noise and compare the results. The probability of false alarm is 10^{-4} and the signal-to-noise ratio is 10 dB.

6.8 Describe the process of pulse integration for coherent and noncoherent detection and give the advantages and disadvantages of each.

6.9 Compare the signal-to-noise ratio for coherent and noncoherent pulse integration of 20 pulses.

6.10 For binary integration (M of N), find the optimum value of M that minimizes the required signal-to-noise ratio for 10 pulses and a Swerling II type target.

6.11 What is the difference between binary pulse integration and cumulative pulse integration? Describe a situation where cumulative pulse integration would be advantageous.

6.12 Describe the difference between the five Swerling fluctuating target models and give examples of practical radar targets each model represents.

6.13 Calculate the probability of detection for Swerling III type target model using both coherent and noncoherent integration of 32 pulses. The probability of false alarm is

10^{-9} and the signal-to-noise ratio is 18 dB. Compare the two results and explain the differences.

6.14 Determine the required single pulse signal-to-noise ratio to achieve a probability of false alarm of 10^{-6} with a probability of detection of 0.99 when noncoherently integrating 20 pulses for a Swerling II type target model. Compare this result with the Shnidman approximation.

6.15 Describe the purpose of guard cells in constant false alarm rate processing and explain how the number of guard cells affects CFAR performance.

6.16 Explain the major differences between cell averaging CFAR and censored greatest of CFAR.

6.17 Describe the conditions, and give an example scenario, where it is advantageous to use CASH-CFAR over OS-CFAR.

References

[1] S. O. Rice. Mathematical analysis of random noise. Technical Report 24, Bell System Technical Journal, 1945.

[2] J. I. Marcum. A statistical theory of target detection by pulsed radar. Technical report, RAND Corporation, RM-754, Santa Monica, CA, 1947.

[3] P. Swerling. Probability of detection for fluctuating targets. *IRE Transaction on Information Theory*, pages 269–308, April 1960.

[4] P. Swerling. Radar probability of detection for some additional fluctuating target cases. *IEEE Transactions on Aerospace and Electronic Systems*, pages 698–708, April 1997.

[5] S. Ross. *A First Course in Probability,* 9th ed. Pearson Education, Inc., Boston, MA, 2006.

[6] E. Jones, et al. SciPy: Open source scientific tools for Python, 2001–. `http://www.scipy.org/`.

[7] R. Nowak. Neyman-Pearson detectors. ECE 830 Statistical Signal Processing, University of Wisconsin - Madison.

[8] P. B. Hoel, S. C. Port and C. J. Stone. *Testing Hypotheses.* Houghton Mifflin, New York, 1971.

[9] A. D. Whalen. *Detection of Signals in Noise.* Academic Press, 1971.

[10] M. Abramowitz and I. A. Stegun. *Handbook of Mathematical Functions: with Formulas, Graphs, and Mathematical Tables.* U.S. National Bureau of Standards, Applied Mathematics Series, Washington, DC, 1964.

[11] Wolfram Alpha. *Online Integral Calculator.* `https://www.wolframalpha.com/calculators/integral-calculator/`.

[12] A. Papoulis and S. U. Pillai. *Probability, Random Variables, and Stochastic Processes,* 4th ed. McGraw-Hill, New York, 2002.

[13] B. R. Mahafza. *Radar Systems Analysis and Design Using MATLAB,* 3rd ed. CRC Press, Boca Raton, FL, 2013.

[14] C. Yang, et al. Comparative study of coherent, non-coherent, and semi-coherent integration schemes for GNSS receivers. Technical report, Air Force Research Lab, Wright-Patterson Air Force Base, OH, April 2007.

[15] J. I. Marcum. Table of Q functions. Technical report, U.S. Air Force RAND Research Memorandum M-339, Santa Monica, CA, January 1950.

[16] S. Parl. A new method of calculating the generalized Q function. *IEEE Transactions on Information Theory*, 32:121–124, January 1980.

[17] S. M. Kay. *Fundamentals of Radar Signal Processing*. McGraw-Hill, New York, 2005.

[18] P. E. Cantrell and A. K. Ojha. Comparison of generalized Q-function algorithms. *IEEE Transactions on Information Theory*, 33(4):591–596, July 1987.

[19] M. A. Richards. Coherent integration loss due to white Gaussian phase noise. *IEEE Signal Processing Letters*, 10(7), July 2003.

[20] D. K. Barton. *Modern Radar System Analysis.* Artech House, Norwood, MA, 1988.

[21] G. R. Curry. *Radar System Performance Modeling*. Artech House, Norwood, MA, 2001.

[22] M. Schwartz. A coincidence procedure for signal detection. *IRE Transactions on Information Theory*, pages 135–139, December 1956.

[23] T. J. Frey. An approximation for the optimum binary integration threshold for Swerling II targets. *IEEE Transactions on Aerospace and Electronic Systems*, 32(3):1181–1185, July 1996.

[24] M. A. Richards, J. A. Scheer and W. A. Holm. *Principles of Modern Radar - Basic Principles.* SciTech Publishing, Raleigh, NC, 2010.

[25] D. P. Meyer and H. A. Mayer. *Radar Target Detection.* Academic Press, New York, 1973.

[26] M. C. Budge and S. R. German. *Basic Radar Analysis.* Artech House, Norwood, MA, 2015.

[27] J. V. DiFranco and W. L. Rubin. *Radar Detection.* Artech House, Dedham, MA, 1980.

[28] D. A. Shnidman. Determination of required SNR values. *IEEE Transactions on Aerospace and Electronic Systems*, 38(3):1059–1064, July 2002.

[29] H. M. Finn and R. S. Johnson. Adaptive detection mode with threshold control as a function of spatially sampled clutter level estimates. *RCA Review*, 29(3):414–464, September 1968.

[30] H. Rohling. Ordered statistic CFAR technique - an overview. *2011 12th International Radar Symposium (IRS)*, pages 631–638, 2011.

[31] F. H. Hofele. An innovative CFAR algorithm. *2001 CIE International Conference on Radar Proceedings*, pages 329–333, October 2001.

Chapter 7

Radar Cross Section

Chapters 4 and 6 showed how the radar cross section plays an important role in received signal power, probability of detection, and probability of false alarm. The radar cross section also influences tracking accuracy, classification, and discrimination. Therefore, the study of radar cross section is fundamental to the understanding of radar system analysis and design. The chapter begins with the definition of radar cross section and some of its dependencies. Next, the scattering matrix is introduced along with some of the dominant scattering mechanisms for complex targets. Various methods for predicting the radar cross section from various objects are presented. Finally, techniques for reducing the radar cross section of a given target are discussed, and a brief overview of electronic countermeasures is given. The chapter concludes with several Python examples to further illustrate the concepts of radar cross section theory, prediction, and measurement.

7.1 DEFINITION

When electromagnetic energy transmitted by a radar system impinges on a target, the induced currents on the target reradiate or scatter energy in all directions, as shown in Figure 4.3. From Chapter 4, the radar cross section is defined by (4.23) and repeated here as

$$\sigma = \lim_{r \to \infty} 4\pi r^2 \frac{|\mathbf{E}^s|^2}{|\mathbf{E}^i|^2} = \frac{P_r}{P_d} \qquad (\mathrm{m}^2). \tag{7.1}$$

where

$\mathbf{E}^s$ = scattered electric field intensity (V/m),
$\mathbf{E}^i$ = incident electric field intensity (V/m),
P_r = power reflected from the target (W),
P_d = power density incident on the target (W/m^2).

The incident electric field intensity, $\mathbf{E}^i$, is determined by the waveform transmitted by the radar. The scattered electric field intensity, $\mathbf{E}^s$, must be found in order to calculate the radar cross section. From Section 2.4.2, the scattered electric field intensity may be found by first calculating the scalar and vector potentials of (2.49) and (2.50) and then solving for the scattered fields by (2.32) and (2.35). The analytic solution of these equations quickly becomes intractable for complex targets of interest to radar systems. However, the radar cross section for simple canonical shapes may be calculated, and this topic is covered in Section 7.4. This leads to other methods for predicting the radar cross section of complex targets such as numerical methods, scale model measurements, and other techniques in Section 7.4.

The incident and scattered electric field intensities in (7.1) are, in general, functions of range, angle, frequency, and polarization. Each of these parameters is dependent on the radar system being used, the geometry of the scenario, as well as the target itself. Also, these parameters affect the resulting radar cross section differently in different situations. For example, the radar cross section of a particular target as a function of angle tends to vary more rapidly at higher frequencies.

7.1.1 Angle Variation

The radar transmitter and receiver may be collocated by using the same antenna for transmit and receive or they may be separated, and each have their own antenna. These two cases are referred to as monostatic and bistatic, as discussed in Section 4.5.4 and shown in Figure 7.1. The angles formed by the direction of the transmitted waveform and the target orientation are referred to as the incident angles (θ_i, ϕ_i). Similarly, the angles formed by the direction of the scattered energy and the target orientation are referred to as the observation angles (θ_o, ϕ_o). For objects other than an ideal point target or a perfect sphere, the monostatic scattered energy is a function of incident angle. For the bistatic case, the scattered energy is a function of both the incident and observation angles. Figure 7.2 illustrates the radar cross section as a function of angle for the bistatic scattering from a flat rectangular plate. As can be seen, the radar cross section is highly dependent on both the angle of incidence and observation.

7.1.2 Frequency Variation

The radar cross section variation of a fixed object as a function of frequency may generally be broken into three regions. These regions are referred to as *Rayleigh*, *resonance*, and *optical*. While the term Rayleigh region or Rayleigh scattering is often used in the field of radar cross section, there is much ambiguity surrounding its meaning. The Rayleigh region is generally considered to be the region where the frequency of the incident energy is low, and the therefore the wavelength is long. Here, the term "long" is in reference to the largest dimension of the object. Born and Wolf [1] define a Rayleigh

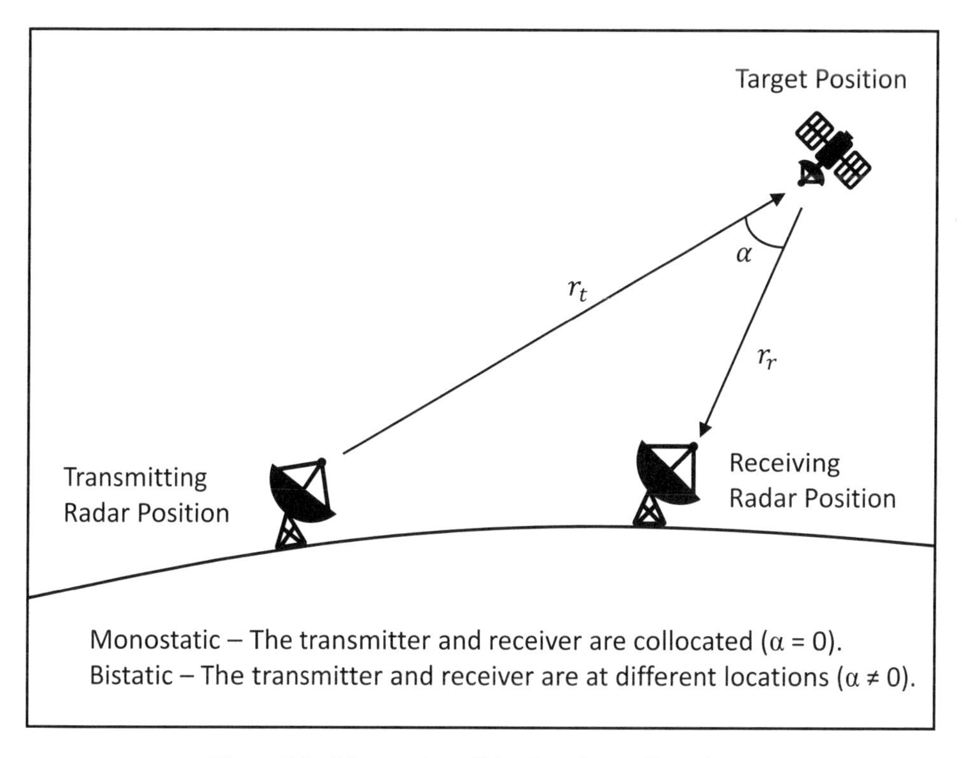

Figure 7.1 Monostatic and bistatic radar configurations.

scatterer as one that does not change the frequency of the incident energy and where the far-field scattering is linearly polarized and varies with the inverse of the square of the wavelength. Twersky [2] provides a broader definition of Rayleigh scattering which is not just limited to electromagnetic phenomena. Kleinman [3] gives a rigorous mathematical definition of the Rayleigh region as the range of wavelengths for which the quantity of interest may be expanded in a convergent series in positive integral powers of the wavenumber. This type of expansion is known as a Rayleigh series, and in all practical radar scattering problems such expansions exist and have finite radii of convergence. However, there is still no precise definition and care must be taken when using approximate solutions in these regions to ensure accurate results. Similar to the Rayleigh region, the resonance region is also somewhat ambiguous. It is the region where the frequency of the incident energy is medium. The radar cross section in this region tends to vary rapidly with changes in the frequency of the incident energy. In the optical region, the frequency of the incident energy is considered high, and the wavelength is small compared to the

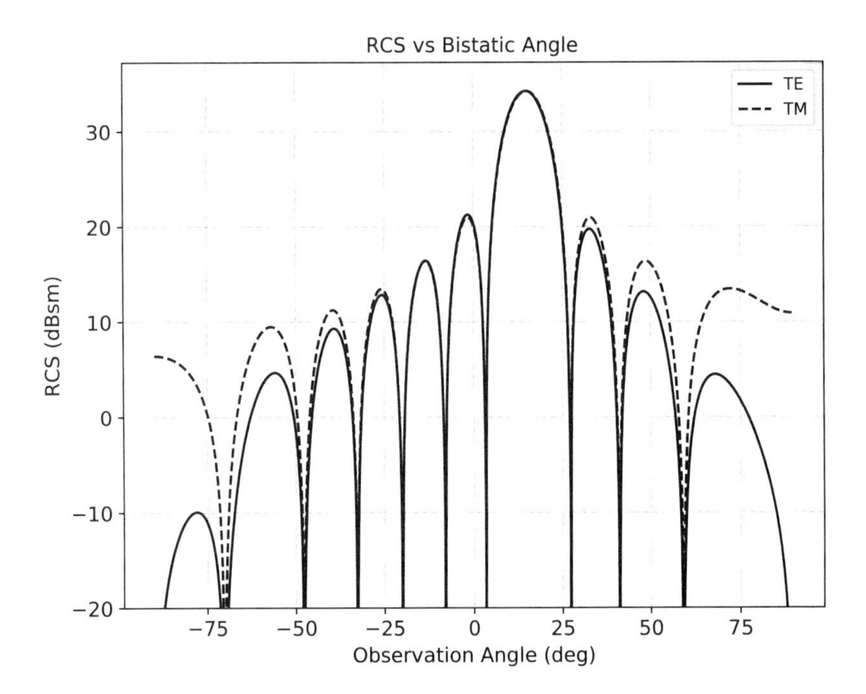

Figure 7.2 Bistatic radar cross section for a $3\lambda \times 5\lambda$ rectangular plate with an incident angle of $\theta_i = \ 15^{o}$.

dimension of the object. In the optical region, the radar cross section becomes less dependent on the frequency of the incident energy. Figure 7.3 illustrates these regions for a perfectly conducting sphere.

7.1.3 Polarization Variation

For a given aspect angle and frequency, objects reflect incident energy in a particular manner depending on the polarization. For many objects, the polarization of the scattered energy will be different from the polarization of the incident energy. This depolarization of the electromagnetic field is often measured by radar systems and is referred to as cross polarization. For example, Beckman [4] presented the depolarization of electromagnetic waves backscattered from the lunar surface. From Section 2.6, the polarization of the reflected and transmitted electromagnetic energy from a material interface depends the material, the aspect angle and the incident polarization. For example, a right-hand circularly polarized incident wave is left-hand circularly polarized upon reflection by

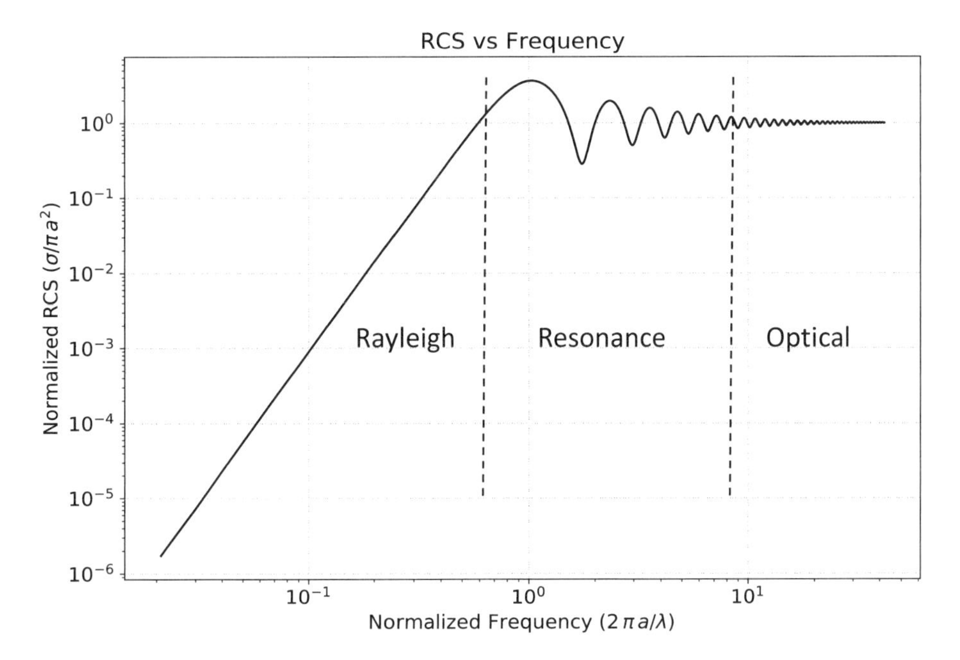

Figure 7.3 Scattering regions for a perfectly conducting sphere.

a planar perfect conductor. Furthermore, a complete definition of the radar cross section of a target would require a radar capable of transmitting and receiving both orthogonal polarizations. This will be examined further in the following sections.

7.2 SCATTERING MATRIX

As discussed in the previous section, the radar cross section is dependent on polarization and requires the full orthogonal polarization information for a complete definition. Begin with the vector scattered electric field as this provides more information about the nature of the scattering, such as relative phase, than the scalar radar cross section. The incident and scattered electric field intensities may be decomposed into spherical components as

$$\mathbf{E}^i = E^i_\theta \, \hat{\boldsymbol{\theta}} + E^i_\phi \, \hat{\boldsymbol{\phi}}, \tag{7.2}$$

$$\mathbf{E}^s = E^s_\theta \, \hat{\boldsymbol{\theta}} + E^s_\phi \, \hat{\boldsymbol{\phi}}. \tag{7.3}$$

where

$\mathbf{E}^i$ = incident electric field intensity (V/m),
$\mathbf{E}^s$ = scattered electric field intensity (V/m),
$\hat{\boldsymbol{\theta}}$ = unit vector in the θ direction,
$\hat{\boldsymbol{\phi}}$ = unit vector in the ϕ direction.

These spherical components are illustrated in Figure 7.4. When a $\hat{\boldsymbol{\theta}}$-polarized electric field is incident upon a target, the backscattered electric field can have nonzero $\hat{\boldsymbol{\theta}}$- and $\hat{\boldsymbol{\phi}}$-polarized components. This is also true for a $\hat{\boldsymbol{\phi}}$-polarized incident electric field. Therefore, the $\hat{\boldsymbol{\theta}}$- and $\hat{\boldsymbol{\phi}}$-polarized components form a basis to define the electric field, and the scattering of the target is fully described by the scattering matrix. This is expressed as

$$\begin{bmatrix} E_\theta^s \\ E_\phi^s \end{bmatrix} = \begin{bmatrix} S_{\theta\theta} & S_{\theta\phi} \\ S_{\phi\theta} & S_{\phi\phi} \end{bmatrix} \begin{bmatrix} E_\theta^i \\ E_\phi^i \end{bmatrix}, \tag{7.4}$$

where

$S_{\theta\theta}$ = $\hat{\boldsymbol{\theta}}$-polarized incident wave and $\hat{\boldsymbol{\theta}}$-polarized scattered wave,
$S_{\theta\phi}$ = $\hat{\boldsymbol{\phi}}$-polarized incident wave and $\hat{\boldsymbol{\theta}}$-polarized scattered wave,
$S_{\phi\theta}$ = $\hat{\boldsymbol{\theta}}$-polarized incident wave and $\hat{\boldsymbol{\phi}}$-polarized scattered wave,
$S_{\phi\phi}$ = $\hat{\boldsymbol{\phi}}$-polarized incident wave and $\hat{\boldsymbol{\phi}}$-polarized scattered wave.

This matrix can be thought of as a transformation of the incident electric field to the scattered electric field. The scattering matrix gives a complete definition of the scattering

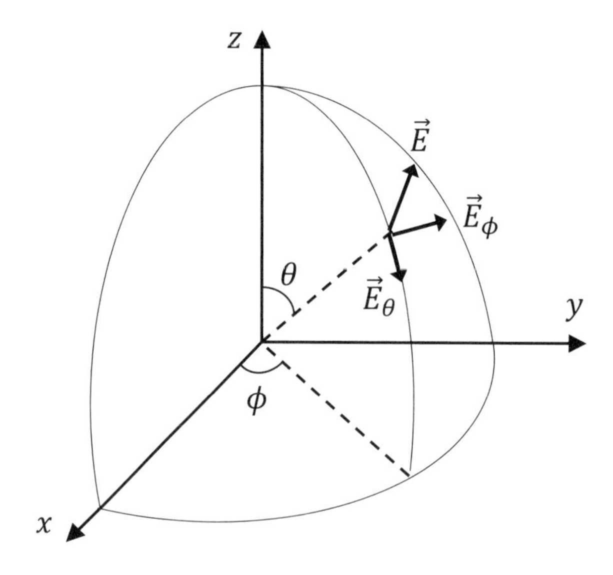

Figure 7.4 Electric field spherical components.

properties of an object. The components of the scattering matrix are complex and may be obtained from the magnitudes and phases measured by a polarimetric radar system. A single instance of the scattering matrix is valid for the incident and observation angles and frequency. Once the scattering matrix is determined, the scattered electric field intensity may be computed for any combination of viewing angles, frequency, and polarization using (E^i_θ, E^i_ϕ) as the basis. For monostatic radar systems, the scattering matrix is reciprocal for most targets. However, this is not true for bistatic systems. Using the definition of radar cross section in (7.1) results in

$$\begin{bmatrix} \sigma_{\theta\theta} & \sigma_{\theta\phi} \\ \sigma_{\phi\theta} & \sigma_{\phi\phi} \end{bmatrix} = 4\pi r^2 \begin{bmatrix} |S_{\theta\theta}|^2 & |S_{\theta\phi}|^2 \\ |S_{\phi\theta}|^2 & |S_{\phi\phi}|^2 \end{bmatrix}. \tag{7.5}$$

Given the scattering matrix for a particular target, Mikulski's theorem [5] may be used to determine the combination of transmit and receive polarization that maximizes the received signal power. Mikulski's theorem states the polarization is found by first forming $\mathbf{A} = \mathbf{S}^* \, \mathbf{S}$. The polarization is then determined by the eigenvector of $\mathbf{A}$ corresponding to the largest eigenvalue of $\mathbf{A}$.

A scattering matrix may be transformed from one coordinate system to another by a unitary transformation. Also, the scattering matrix may be transformed from one set of orthogonal polarizations to another set of orthogonal polarizations through a similar procedure. In many radar applications it is convenient to define the scattering matrix in terms of linear or circular polarizations, which is expressed as

$$\begin{bmatrix} E^s_H \\ E^s_V \end{bmatrix} = \begin{bmatrix} S_{HH} & S_{HV} \\ S_{VH} & S_{VV} \end{bmatrix} \begin{bmatrix} E^i_H \\ E^i_V \end{bmatrix}, \tag{7.6}$$

and

$$\begin{bmatrix} E^s_R \\ E^s_L \end{bmatrix} = \begin{bmatrix} S_{RR} & S_{RL} \\ S_{LR} & S_{LL} \end{bmatrix} \begin{bmatrix} E^i_R \\ E^i_L \end{bmatrix}. \tag{7.7}$$

The subscripts (V, H) represent vertical and horizontal polarization, and the subscripts (R, L) represent right-hand circular and left-hand circular polarization. The linear and circular field components are related to each other through

$$\begin{bmatrix} E_R \\ E_L \end{bmatrix} = \frac{1}{\sqrt{2}} \begin{bmatrix} 1 & -j \\ 1 & +j \end{bmatrix} \begin{bmatrix} E_H \\ E_V \end{bmatrix} = [T] \begin{bmatrix} E_H \\ E_V \end{bmatrix}, \tag{7.8}$$

and

$$\begin{bmatrix} E_H \\ E_V \end{bmatrix} = \frac{1}{\sqrt{2}} \begin{bmatrix} 1 & 1 \\ +j & -j \end{bmatrix} \begin{bmatrix} E_R \\ E_L \end{bmatrix} = [T]^{-1} \begin{bmatrix} E_R \\ E_L \end{bmatrix}. \tag{7.9}$$

Using (7.8) and (7.9), the circularly polarized scattering matrix is written in terms of the linear scattering matrix as

$$\begin{bmatrix} S_{RR} & S_{RL} \\ S_{LR} & S_{LL} \end{bmatrix} = [T] \begin{bmatrix} S_{HH} & S_{HV} \\ S_{VH} & S_{VV} \end{bmatrix} \begin{bmatrix} 1 & 0 \\ 0 & -1 \end{bmatrix} [T]^{-1}. \tag{7.10}$$

Similarly, the linearly polarized scattering matrix is written in terms of the circular scattering matrix as

$$\begin{bmatrix} S_{HH} & S_{HV} \\ S_{VH} & S_{HH} \end{bmatrix} = [T]^{-1} \begin{bmatrix} S_{RR} & S_{RL} \\ S_{LR} & S_{LL} \end{bmatrix} \begin{bmatrix} 1 & 0 \\ 0 & -1 \end{bmatrix} [T]. \tag{7.11}$$

When describing the propagation of electromagnetic energy in a three-dimensional space, one must account for the coordinate system changing the handedness upon reflection from a target. Therefore, the coordinate space has to be defined for the incident and scattered energy, which has led to two conventions. The first is the forward scattering alignment (FSA) in which the positive z axis is in the same direction as the direction of propagation for both the incident and the scattered wave. The second is the back scattering alignment (BSA), where the positive z axis points toward the target for both the incident and scattered waves. The BSA convention is more commonly used for monostatic radar systems. This difference in convention leads to a slightly different form in the scattering matrix. In the FSA convention, the scattering matrix is the Jones matrix, while in the BSA convention, the scattering matrix is the Sinclair matrix.

7.3 SCATTERING MECHANISMS

As discussed in previous sections, the energy backscattered from an object may be found from the currents induced on the object. The mechanism by which these currents are induced and reradiate energy is quite complicated. Referring to Figure 7.5, complex targets have several scattering mechanisms resulting in different signal strengths and the scattered energy may be thought of as a coherent sum of the contributions of the individual scattering mechanisms. A few of the more important scattering mechanisms for radar targets are

1. Specular reflection, sometimes referred to as regular reflection, is a mirror-like reflection of incoming energy. From Chapter 2, this scattering mechanism obeys Snell's law as each incident wave is reflected at the same angle on the opposite side of the surface normal. At the frequencies of interest to radar systems, returns from specular reflections tend to be very large.

2. Diffraction is generally the second greatest contributor to the radar cross section and occurs when the incident energy encounters discontinuities such as edges. Unlike specular reflection, the scattered energy from diffraction travels in many directions.

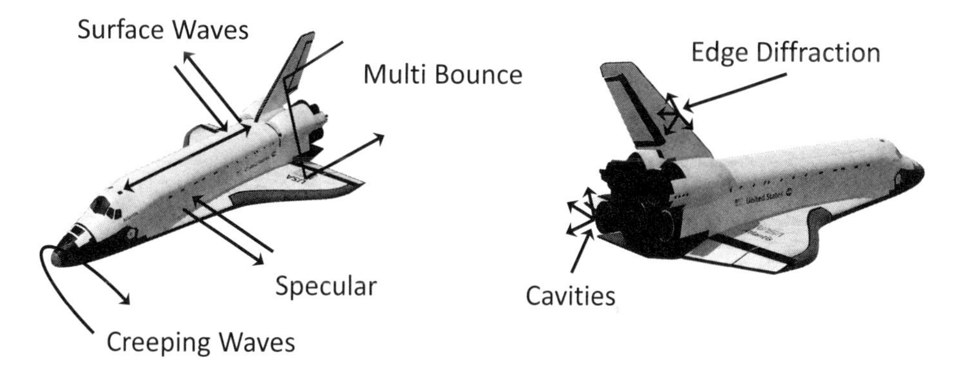

Figure 7.5 Radar target scattering mechanisms (NASA).

3. Creeping waves act as continuous diffraction around a smooth section of the object. Creeping waves and diffracted waves both allow energy to interact with portions of the target that may otherwise be shadowed from the incident energy.

4. Surface waves occur when the object acts like a transmission line and guides waves along surface. These waves will continue until a discontinuity is encountered and may set up as standing waves on the object. Surface waves can reduce the energy of the return signal as the conductivity of the object will attenuate the signal.

5. Cavities on an object act as resonators and produce extended returns. Much research has been conducted in prediction of cavity returns, and in the reduction of these returns [6–8].

6. Multibounce scattering occurs when the reflections are due to the interaction of the incident energy with multiple parts of the body. Multibounce scattering plays a large role in the polarization of the backscattered energy.

7.4 PREDICTION METHODS

The following sections study various radar cross-section prediction techniques. These will include analytic techniques, numerical methods, and measurement techniques. The advantages and disadvantages of each method will be examined, along with target types, geometries, and frequency ranges best suited for each method.

7.4.1 Analytical Techniques

While the analytical solution for complex targets of interest to radar systems remains intractable, there are approximate solutions for many objects that are in good agreement with measured results [9]. Also, superposition of simple objects may be used to form more complex composite objects. However, with advances in numerical techniques and computing power, superposition techniques are quickly being replaced. The following sections present the analytic solution to several objects. Table 7.1 gives the approximate radar cross section for simple shapes with axial incidence. These expressions are in the optical region ($D \gg \lambda$), where D is a characteristic dimension of the object. While these expressions are good for back-of-the-envelope type calculations and first cut analysis, more rigorous prediction techniques are needed for more detailed system analysis.

7.4.1.1 Two-Dimensional Strip

For two-dimensional geometries, where the object is infinitely long in one dimension, the scattering parameter is sometimes referred to as the *scattering width* or as the radar cross section per unit length. For this book, this is simply referred to this as the two-dimensional radar cross section denoted by σ_{2D}. The two-dimensional radar cross section is defined as

$$\sigma_{2D} = \lim_{r \to \infty} 2\pi r \frac{|\mathbf{E}^s|^2}{|\mathbf{E}^i|^2} = \lim_{r \to \infty} 2\pi r \frac{P_d^s}{P_d^i} \qquad \text{(m)}. \tag{7.12}$$

where

$\mathbf{E}^s$ = scattered electric field intensity (V/m),
$\mathbf{E}^i$ = incident electric field intensity (V/m),
P_d^s = scattered power density (W/m),
P_d^i = incident power density (W/m).

Note that the units for the two-dimensional radar cross section is meters. Consider the geometry of an infinitely long conducting strip of width w, illustrated in Figure 7.6. The incident fields for TM^z polarized waves are written as

$$\mathbf{E}^i = E_0 \, e^{jk(x\cos\phi_i + y\sin\phi_i)} \, \hat{\mathbf{z}} \qquad \text{(V/m)}, \tag{7.13}$$

$$\mathbf{H}^i = \frac{E_0}{\eta} e^{jk(x\cos\phi_i + y\sin\phi_i)} \left(-\sin\theta_i \, \hat{\mathbf{x}} + \cos\theta_i \, \hat{\mathbf{y}}\right) \qquad \text{(A/m)}. \tag{7.14}$$

The two-dimensional bistatic radar cross section for the TM^z polarization is given by [10]

$$\sigma_{2D} = \frac{2\pi w^2}{\lambda} \left[\sin\phi_i \left(\frac{\sin\left(0.5\, kw(\cos\phi_o + \cos\phi_i)\right)}{0.5\, kw(\cos\phi_o + \cos\phi_i} \right) \right]^2 \qquad \text{(m)}, \tag{7.15}$$

where

Table 7.1
RCS of Simple Shapes with Axial Incidence

Object	*RCS (m^2)*	*Geometry*
Dipole ($l = \lambda$)	$0.93\lambda^2$	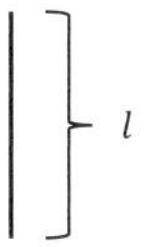 l
Sphere	πa^2	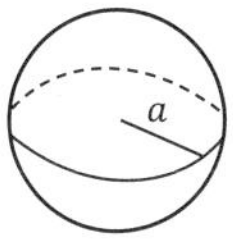 a
Circular disc	$4\pi \frac{(\pi a^2)^2}{\lambda^2}$	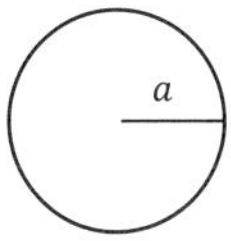 a
Rectangular plate	$4\pi \frac{(ab)^2}{\lambda^2}$	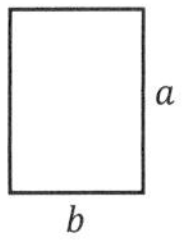 a, b
Circular reflector	$4\pi \frac{(\pi a^2)^2}{\lambda^2}$	 a
Corner reflector	$8\pi \frac{(ab)^2}{\lambda^2}$	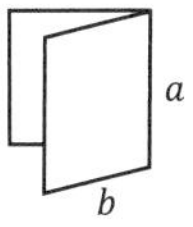 a, b

Triangular reflector	$4\pi \dfrac{a^4}{\lambda^3}$	a
Cylinder	$4\pi \dfrac{ab^2}{\lambda}$	b, a
Simple ogive	$\dfrac{\lambda^2}{16\pi} \tan^4(\theta)$	θ, a, b

$$
\begin{aligned}
k &= \text{wavenumber (rad/m)},\\
w &= \text{strip width (m)},\\
\phi_o &= \text{observation angle (rad)},\\
\phi_i &= \text{incident angle (rad)}.
\end{aligned}
$$

The incident fields for TE^z polarized waves are written as

$$\mathbf{H}^i = H_0 \, e^{jk(x \cos\phi_i + y \sin\phi_i)} \, \hat{\mathbf{z}} \qquad \text{(A/m)}, \tag{7.16}$$

$$\mathbf{E}^i = \eta H_0 e^{jk(x \cos\phi_i + y \sin\phi_i)} \left(\sin\theta_i \, \hat{\mathbf{x}} - \cos\theta_i \, \hat{\mathbf{y}}\right) \qquad \text{(V/m)}. \tag{7.17}$$

The two-dimensional bistatic radar cross section for the TE^z polarization is given by [10]

$$\sigma_{2D} = \frac{2\pi w^2}{\lambda} \left[\sin\phi_o \left(\frac{\sin\left(0.5\, kw(\cos\phi_o + \cos\phi_i)\right)}{0.5\, kw(\cos\phi_o + \cos\phi_i} \right) \right]^2 \qquad \text{(m)}. \tag{7.18}$$

The difference in the bistatic two-dimensional radar cross section for the TM^z and TE^z polarizations is the $\sin^2\phi_i$ term is replaced with $\sin^2\phi_o$. For the monostatic case, the expressions in (7.15) and (7.18) reduce to the same equation, given here by

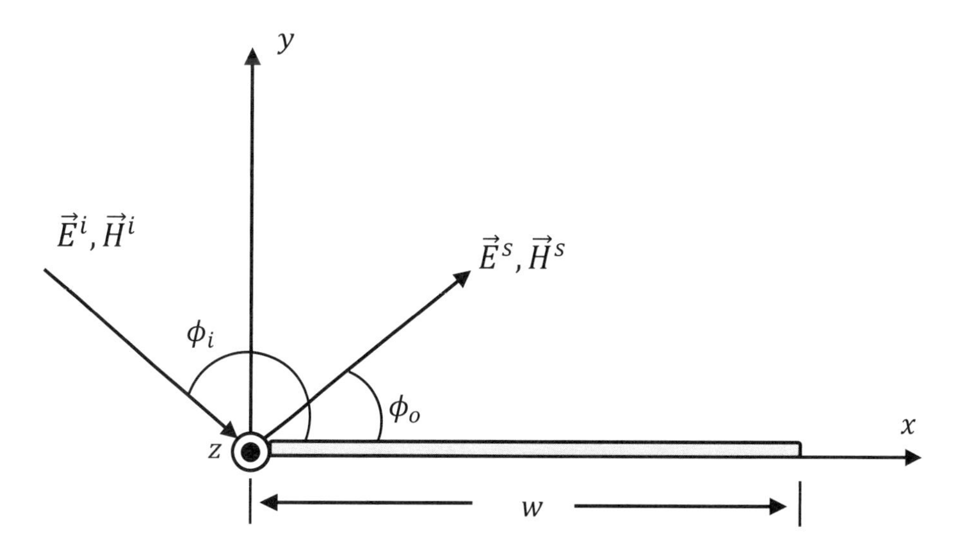

Figure 7.6 Two-dimensional strip geometry.

$$\sigma_{2D} = \frac{2\pi w^2}{\lambda}\left[\sin\phi_i\left(\frac{\sin\left(kw\cos\phi_i\right)}{kw\cos\phi_i}\right)\right]^2 \qquad \text{(m)}. \qquad (7.19)$$

As the width of the strip increases, the maximum value of the radar cross section also increases, as does the number of lobes in the radar cross section pattern.

7.4.1.2 Two-Dimensional Cylinder

Cylinders are commonly used to represent scatterers such as the fuselage of missiles, airplanes, and other objects. This makes circular cylinders a particularly important class of objects for radar cross-section investigation. The solution of the scattering from circular cylinders is given in terms of Bessel and Hankel functions, for which there are numerous tabulated values [11] as well as libraries such as SciPy [12] for their calculation. For example, the Hankel function of the second kind of real order, ν, and complex argument, z, may be found with ***scipy.special.hankel2(v,z)***. Beginning with the two-dimensional conducting circular cylinder shown in Figure 7.7, the incident electric field for TM^z polarized waves is written as

$$\mathbf{E}^i = E_0\, e^{-jkx}\,\hat{\mathbf{z}} \qquad \text{(V/m)}. \qquad (7.20)$$

Note that the wave is incident from $\phi = 270^\circ$. The two-dimensional bistatic radar cross section for the TM^z polarization is given as [10]

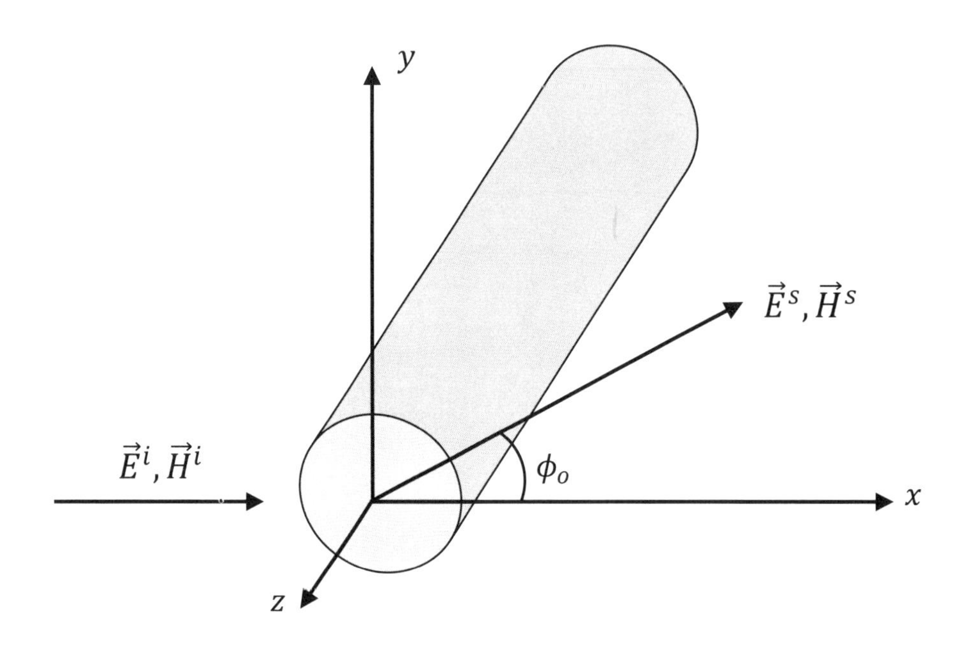

Figure 7.7 Two-dimensional circular cylinder geometry.

$$\sigma_{2D} = \frac{2\lambda}{\pi}\left|\sum_{n=0}^{\infty} \epsilon_n \frac{J_n(ka)}{H_n^{(2)}(ka)} \cos(n\,\phi_o)\right|^2 \qquad \text{(m)}, \tag{7.21}$$

where

ϵ_n = 1 for $n = 0$, 2 for $n \neq 0$,
k = wavenumber (rad/m),
a = cylinder radius (m),
J_n = Bessel function of order n,
$H_n^{(2)}$ = Hankel function of the second kind of order n,
ϕ_o = observation angle (rad).

The incident magnetic field for TE^z polarized waves is given as

$$\mathbf{H}^i = H_0\, e^{-jkx}\, \hat{\mathbf{z}} \qquad \text{(A/m)}. \tag{7.22}$$

The two-dimensional bistatic radar cross section for the TE^z polarization is given as [10]

$$\sigma_{2D} = \frac{2\lambda}{\pi}\left|\sum_{n=0}^{\infty} \epsilon_n \frac{J_n'(ka)}{H_n^{(2)\prime}(ka)} \cos(n\,\phi_o)\right|^2 \qquad \text{(m)}. \tag{7.23}$$

The difference in the two-dimensional radar cross section for the TM^z and TE^z polarizations is the first derivative of the Bessel and Hankel functions for the TE^z case.

For finite length cylinders with normal incidence, the three-dimensional radar cross section is related to the two-dimensional radar cross section by

$$\sigma_{3D} \approx \sigma_{2D}\frac{2L^2}{\lambda} \qquad (\text{m}^2), \tag{7.24}$$

where L is the cylinder length. The three-dimensional radar cross section for the TM^z and TE^z polarizations is then

$$\sigma_{TM} \approx \frac{4L^2}{\pi}\left|\sum_{n=0}^{\infty} \epsilon_n \frac{J_n(ka)}{H_n^{(2)}(ka)} \cos(n\,\phi_o)\right|^2 \qquad (\text{m}^2), \tag{7.25}$$

and

$$\sigma_{TE} \approx \frac{4L^2}{\pi}\left|\sum_{n=0}^{\infty} \epsilon_n \frac{J_n'(ka)}{H_n^{(2)\prime}(ka)} \cos(n\,\phi_o)\right|^2 \qquad (\text{m}^2). \tag{7.26}$$

7.4.1.3 Two-Dimensional Cylinder Oblique

Section 7.4.1.2 examined infinitely long cylinders with normal incidence. This section studies the scattering from infinite length cylinders with oblique incidence, as shown in Figure 7.8. The incident electric field for TM^z polarized waves is given by

$$\mathbf{E}^i = E_0\, e^{-jk(x\sin\theta_i - z\cos\theta_i)}\,(\cos\theta_i\,\hat{\mathbf{x}} + \sin\theta_i\,\hat{\mathbf{z}}) \qquad \text{(V/m)}. \tag{7.27}$$

Again, the wave is incident at $\phi = 270^\circ$. The two-dimensional bistatic radar cross section for the TM^z polarization is given by [10]

$$\sigma_{2D} = \frac{2\lambda}{\pi\sin\theta_i}\left|\sum_{n=0}^{\infty} \epsilon_n \frac{J_n(ka\sin\theta_i)}{H_n^{(2)}(ka\sin\theta_i)} \cos(n\,\phi_o)\right|^2 \qquad \text{(m)}, \tag{7.28}$$

where

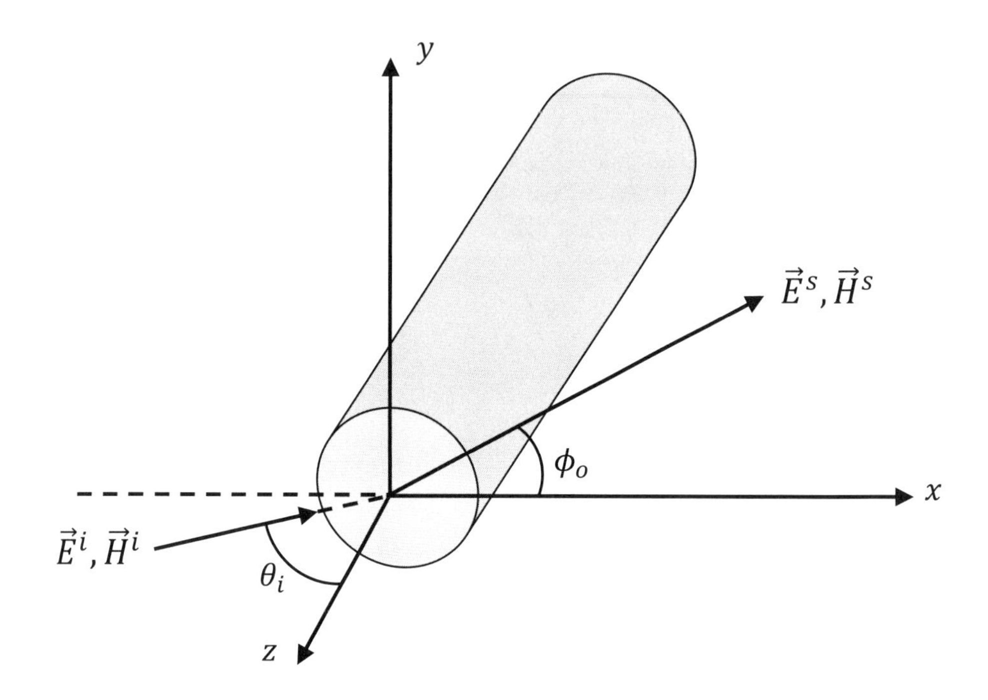

Figure 7.8 Two-dimensional circular cylinder oblique incidence.

ϵ_n = 1 for $n = 0$, 2 for $n \neq 0$,
k = wavenumber (rad/m),
a = cylinder radius (m),
J_n = Bessel function of order n,
$H_n^{(2)}$ = Hankel function of the second kind of order n,
θ_i = incident theta angle (rad),
ϕ_o = observation angle (rad).

The incident magnetic field for TE^z polarized waves is expressed as

$$\mathbf{H}^i = H_0 \, e^{-jk(x \sin\theta_i - z\cos\theta_i)} \, (\cos\theta_i \, \hat{\mathbf{x}} + \sin\theta_i \, \hat{\mathbf{z}}) \qquad \text{(A/m)}. \tag{7.29}$$

The two-dimensional bistatic radar cross section for the TE^z polarization is given by [10]

$$\sigma_{2D} = \frac{2\lambda}{\pi \sin\theta_i} \left| \sum_{n=0}^{\infty} \epsilon_n \frac{J_n'(ka\sin\theta_i)}{H_n^{(2)\prime}(ka\sin\theta_i)} \cos(n\,\phi_o) \right|^2 \qquad \text{(m)}. \tag{7.30}$$

As in the case of normal incidence, an approximation of the three-dimensional radar cross section from the two-dimensional radar cross section is

$$\sigma_{TM} \approx \frac{4L^2}{\pi}\frac{\sin^2\theta_o}{\sin\theta_i}\left|\sum_{n=0}^{\infty}\epsilon_n \frac{J_n(ka\sin\theta_i)}{H_n^{(2)}(ka\sin\theta_i)}\cos(n\,\phi_o)\right|^2$$

$$\times\left[\frac{\sin\left(0.5kL(\cos\theta_i+\cos\theta_o)\right)}{0.5kL(\cos\theta_i+\cos\theta_o)}\right]^2 \qquad (\text{m}^2), \qquad (7.31)$$

and

$$\sigma_{TE} \approx \frac{4L^2}{\pi}\sin\theta_i\left|\sum_{n=0}^{\infty}\epsilon_n \frac{J_n'(ka\sin\theta_i)}{H_n^{(2)\prime}(ka\sin\theta_i)}\cos(n\,\phi_o)\right|^2$$

$$\times\left[\frac{\sin\left(0.5kL(\cos\theta_i+\cos\theta_o)\right)}{0.5kL(\cos\theta_i+\cos\theta_o)}\right]^2 \qquad (\text{m}^2). \qquad (7.32)$$

7.4.1.4 Rectangular Plate

The previous sections were concerned with two-dimensional scatterers, where the object is infinitely long in one dimension. Now consider the scattering from a three-dimensional object; specifically, a rectangular conducting plate, as shown in Figure 7.9. The incident fields for TE^x polarized waves are written as

$$\mathbf{E}^i = \eta H_0\, e^{-jk(y\sin\theta_i - z\cos\theta_i)}\left(\cos\theta_i\,\hat{\mathbf{y}} + \sin\theta_i\,\hat{\mathbf{z}}\right) \qquad (\text{V/m}), \qquad (7.33)$$

$$\mathbf{H}^i = H_0\, e^{-jk(y\sin\theta_i - z\cos\theta_i)}\,\hat{\mathbf{x}} \qquad (\text{A/m}). \qquad (7.34)$$

Note that the incident wave is from $\phi_i = 270^o$. For TE^x polarization, the bistatic radar cross section can be written as [10]

$$\sigma = 4\pi\left(\frac{ab}{\lambda}\right)^2\left(\cos^2\theta_o\sin^2\phi_o + \cos^2\phi_o\right)\left[\frac{\sin(X)}{X}\right]^2\left[\frac{\sin(Y)}{Y}\right]^2 \quad (\text{m}^2), \qquad (7.35)$$

where

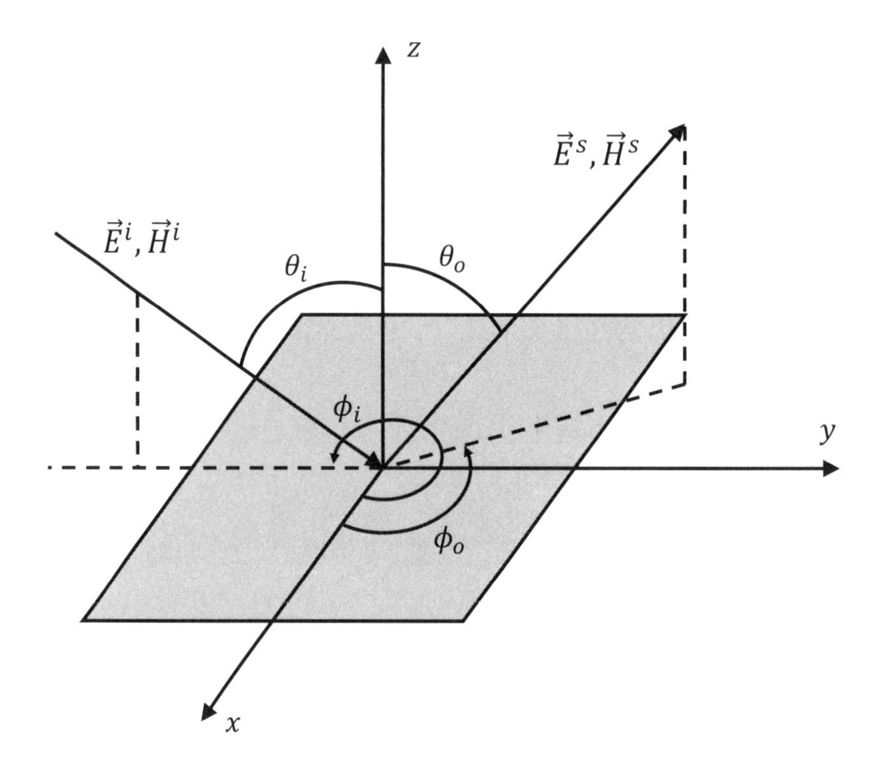

Figure 7.9 Rectangular conducting plate geometry.

$$
\begin{aligned}
X &= 0.5\,ka\sin\theta_o\cos\phi_o \text{ (rad)},\\
Y &= 0.5\,kb(\sin\theta_o\sin\phi_o - \sin\theta_i) \text{ (rad)},\\
a &= \text{plate } x \text{ dimension (m)},\\
b &= \text{plate } y \text{ dimension (m)},\\
\theta_o, \phi_o &= \text{observation angles (rad)},\\
\theta_i, \phi_i &= \text{incident angles (rad)}.
\end{aligned}
$$

The incident fields for TM^x polarized waves are written as

$$
\mathbf{E}^i = E_0\, e^{-jk(y\sin\theta_i - z\cos\theta_i)}\,\hat{\mathbf{x}} \qquad \text{(V/m)}, \tag{7.36}
$$

$$
\mathbf{H}^i = \frac{-E_0}{\eta}\, e^{-jk(y\sin\theta_i - z\cos\theta_i)}\,(\cos\theta_i\,\hat{\mathbf{y}} + \sin\theta_i\,\hat{\mathbf{z}}) \qquad \text{(A/m)}. \tag{7.37}
$$

For the TM^x polarization, the bistatic radar cross section can be written as [10]

$$\sigma = 4\pi \left(\frac{ab}{\lambda}\right)^2 \left[\cos^2\theta_i(\cos^2\theta_o\cos^2\phi_o + \sin^2\phi_o)\right] \times \left[\frac{\sin(X)}{X}\right]^2 \left[\frac{\sin(Y)}{Y}\right]^2 \quad (\text{m}^2), \tag{7.38}$$

For the monostatic case, the expressions in (7.35) and (7.38) reduce to the same equation, given here as

$$\sigma = 4\pi \left(\frac{ab}{\lambda}\right)^2 \cos^2\theta_i \left[\frac{\sin(kb\sin\theta_i)}{kb\sin\theta_i}\right]^2 \quad (\text{m}^2), \tag{7.39}$$

For the TM^x polarization, the maximum of scattered field always occurs at the specular angle. However, this is not true for the TE^x polarization, although it does approach the specular angle as the dimensions of the plate become large compared to the wavelength. As well, the monostatic radar cross section for the flat plate is the same for both TM^x and TE^x polarizations. This is also true for the two-dimensional plate. Measurements have shown that these two values are slightly different.

7.4.1.5 Stratified Sphere

Mie gives the exact solution for scattering from a sphere [13]. The solution is composed of vector wave functions defined in a spherical coordinate system. The terms of the Mie series are obtained from boundary value techniques. Therefore, the Mie formulation may be employed regardless of the composition of the sphere. This section examines the scattering from a radially stratified sphere, as shown in Figure 7.10. The incident field is a plane wave traveling in the negative z direction, as illustrated in Figure 7.10, and the electric and magnetic fields are given by

$$\mathbf{E}^i = E_0\, e^{-jkz}\,\hat{\mathbf{x}} \qquad (\text{V/m}), \tag{7.40}$$

$$\mathbf{H}^i = -\frac{E_0}{\eta}\, e^{-jkz}\,\hat{\mathbf{y}} \qquad (\text{A/m}). \tag{7.41}$$

To calculate the radar cross section of a sphere, use the Mie formulation along with far field approximations to give

$$S_1(\theta_o) = \sum_{n=1}^{\infty} (j)^{n+1} \left[A_n \frac{P_n^1(\cos\theta_o)}{\sin\theta_o} - jB_n \frac{d}{d\theta_o} P_n^1(\cos\theta_o)\right], \tag{7.42}$$

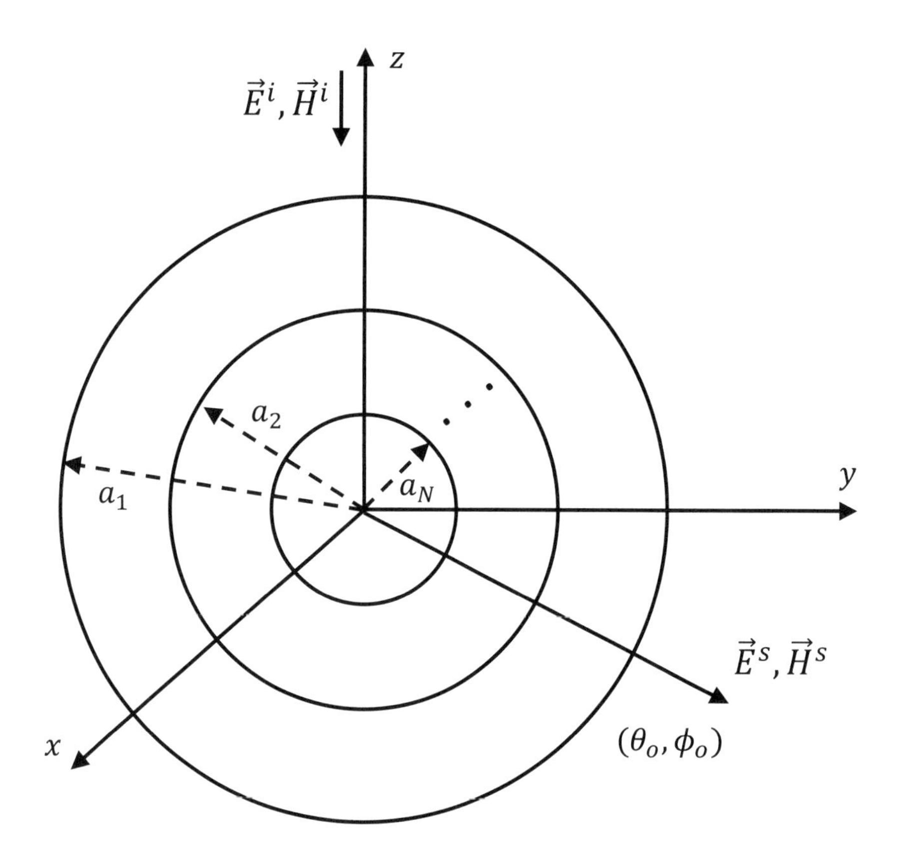

Figure 7.10 Stratified sphere with N concentric layers.

$$S_2(\theta_o) = \sum_{n=1}^{\infty} (j)^{n+1} \left[A_n \frac{d}{d\theta_o} \frac{P_n^1(\cos\theta_o)}{\sin\theta_o} - jB_n P_n^1(\cos\theta_o) \right], \tag{7.43}$$

where P_n^1 is the associated Legendre polynomial and may be calculated using the SciPy implementation ***scipy.special.lpmn(m, n, z)*** . $S_1(\theta_o)$ and $S_2(\theta_o)$ are the complex far-field scattered radiation values for the $\hat{\boldsymbol{\theta}}$ and $\hat{\phi}$ directions. The radar cross section for the $\hat{\boldsymbol{\theta}}$ and $\hat{\phi}$ polarization states is then found to be

$$\sigma_\theta = \frac{4\pi}{k_0^2} S_1(\theta_o)\cos^2(\phi_0) \qquad (\text{m}^2), \tag{7.44}$$

$$\sigma_\phi = \frac{4\pi}{k_0^2} S_2(\theta_o) \sin^2(\phi_0) \qquad (\text{m}^2). \tag{7.45}$$

For the N-layer concentric sphere, use the Mie coefficients of the following form [9]

$$A_n = -(j)^n \frac{2n+1}{n(n+1)} \frac{k_0 a_0 J_n(k_0 a_0) + j Z_n(k_0 a_0)(k_0 a_0 J_n'(k_0 a_0)}{k_0 a_0 H_n(k_0 a_0) + j Z_n(k_0 a_0)(k_0 a_0 H_n'(k_0 a_0)}, \tag{7.46}$$

$$B_n = (j)^n \frac{2n+1}{n(n+1)} \frac{k_0 a_0 J_n(k_0 a_0) + j Y_n(k_0 a_0)(k_0 a_0 J_n'(k_0 a_0)}{k_0 a_0 H_n(k_0 a_0) + j Y_n(k_0 a_0)(k_0 a_0 H_n'(k_0 a_0)}, \tag{7.47}$$

where

a_0	=	outermost radius of the sphere (m),
k_0	=	wavenumber (rad/m),
J_n	=	Bessel function of order n,
H_n	=	Hankel function of order n,
$Z_n(k_0 a_0)$	=	normalized modal surface impedance,
$Y_n(k_0 a_0)$	=	normalized modal surface admittance.

Ruck et al. showed that the modal surface impedance and admittance can be derived using an iterative technique similar to the method used for transmission lines [9]. To begin, the impedance at the interface between the core and the first layer, Z_n^N, is determined independently. Then, the impedance at the second interface, Z_n^{N-1}, is determined from Z_n^N. This process continues until the impedance at the outermost surface, Z_n^0, is found. Then $Z_n(k_0 a_0) = j(Z_n^0/\eta)$. Following the same process for the admittance, $Y_n(k_0 a_0)$ may also be calculated. The impedance and admittance are used in the Mie coefficients of (7.46) for the scattering radiation calculation in (7.42). Finally, the radar cross section is obtained from (7.44).

7.4.1.6 Cone

This section deals with the monostatic radar cross section for conducting, right circular cones based on the geometrical theory of diffraction. The reader is referred to Keller [14] for a detailed analysis of this problem, and Bechtel [15], who gives corrections to Keller's original work. The geometry of the right circular cone is given in Figure 7.11. For axial incidence, the radar cross section is independent of polarization, and is written as

$$\sigma = \frac{\lambda^2}{\pi} \frac{\left(\dfrac{ka \sin(\pi/n)}{n}\right)^2}{\left(\cos(\pi/n) - \cos(3\pi/n)\right)^2} \qquad (\text{m}^2), \tag{7.48}$$

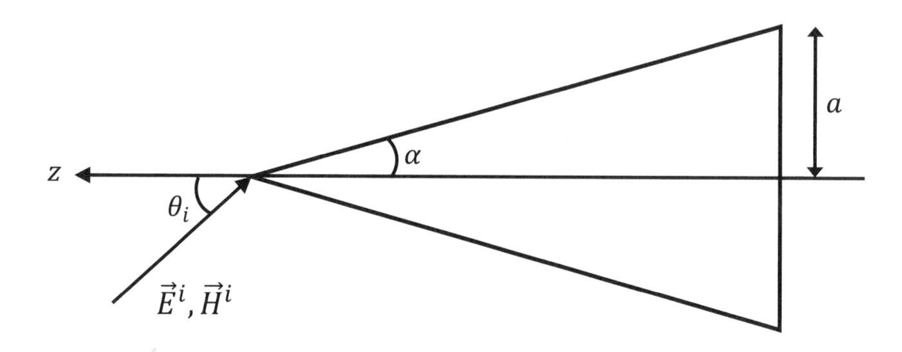

Figure 7.11 Geometry for right circular cones.

where

k	=	wavenumber (rad/m),
a	=	base radius (m),
λ	=	wavelength (m),
n	=	$3/2 + \alpha/\pi$ (rad),
α	=	half angle of the cone (rad).

Equation (7.48) is for first-order diffraction only, which is valid when $ka \gg 1$. Taking double diffraction into account, a more accurate expression for the axial radar cross section is

$$\sigma = \frac{\lambda^2}{\pi}\left(\frac{ka\sin(\pi/n)}{n}\right)^2 \left|\frac{1}{\left(\cos(\pi/n) - \cos(3\pi/n)\right)^2} + \frac{\sin(\pi/n)\exp\left(j(2ka - \pi/4)\right)}{n\sqrt{\pi ka}\left(\cos(\pi/n) - \cos(3\pi/2n)\right)^2}\right|^2 \quad (\mathrm{m}^2). \quad (7.49)$$

If the angle of the incident energy is normal to the generator of the cone, $\theta_i = 90^\circ - \alpha$, then the geometrical theory of diffraction equations are no longer valid. Instead, an expression based on the asymptotic expansion of the physical optics equation is used. This is written as

$$\sigma = \frac{8\lambda^2\pi}{9\sin^2\alpha\cos\alpha}\left(\frac{a}{\lambda}\right)^3 \quad (\mathrm{m}^2). \quad (7.50)$$

The other special case is when the cone is viewed from the base, $\theta_i = 180^\circ$. For this case, the physical optics expression for a circular disc is used; see Table 7.1, and repeated here as

$$\sigma = \frac{\lambda^2 (ka)^4}{4\pi} \qquad (\mathrm{m}^2). \tag{7.51}$$

For all other incident angles, the radar cross section depends on the polarization of the incident energy, and is given by the following equations [15]

$$\begin{aligned}\sigma = &\frac{\lambda^2 ka}{4\pi^2 \sin\theta_i}\left(\frac{\sin(\pi/n)}{n}\right)^2 \times \left| \exp\left[-j(2ka\sin\theta_i - \frac{\pi}{4})\right]\left[\left(\cos\frac{\pi}{n} - 1\right)^{-1}\right.\right.\\ &\left.\pm\left(\cos\frac{\pi}{n} - \cos\frac{3\pi - 2\theta_i}{n}\right)^{-1}\right] + \exp\left[j(2ka\sin\theta_i - \frac{\pi}{4})\right]\left[\left(\cos\frac{\pi}{n} - 1\right)^{-1}\right.\\ &\left.\left.\pm\left(\cos\frac{\pi}{n} - \cos\frac{3\pi + 2\theta_i}{n}\right)^{-1}\right]\right|^2, \quad 0 < \theta_i < \alpha;\end{aligned} \tag{7.52}$$

$$\begin{aligned}\sigma = \frac{\lambda^2 ka}{4\pi^2 \sin\theta_i}\left(\frac{\sin(\pi/n)}{n}\right)^2 &\left[\left(\cos\frac{\pi}{n} - 1\right)^{-1} \pm\right.\\ &\left.\left(\cos\frac{\pi}{n} - \cos\frac{3\pi - 2\theta_i}{n}\right)^{-1}\right]^2, \quad \alpha < \theta_i < \pi/2;\end{aligned} \tag{7.53}$$

$$\begin{aligned}\sigma = &\frac{\lambda^2 ka}{4\pi^2 \sin\theta_i}\left(\frac{\sin(\pi/n)}{n}\right)^2 \times \left| \exp\left[-j(2ka\sin\theta_i - \frac{\pi}{4})\right]\left[\left(\cos\frac{\pi}{n} - 1\right)^{-1}\right.\right.\\ &\left.\pm\left(\cos\frac{\pi}{n} - \cos\frac{3\pi - 2\theta_i}{n}\right)^{-1}\right] + \exp\left[j(2ka\sin\theta_i - \frac{\pi}{4})\right]\left[\left(\cos\frac{\pi}{n} - 1\right)^{-1}\right.\\ &\left.\left.\pm\left(\cos\frac{\pi}{n} - \cos\frac{2\theta_i - \pi}{n}\right)^{-1}\right]\right|^2, \quad \pi/2 < \theta_i < \pi.\end{aligned} \tag{7.54}$$

The positive sign is used for horizontal polarization and the negative sign is used for vertical polarization in (7.52), (7.53), and (7.54). For comparison of the above approximate equations with measured data, the reader is referred to [15].

For cones with rounded nose tips, as shown in Figure 7.12, the physical optics approximation with axial incidence is given by [9]

$$\sigma = \pi b^2 \left[1 - \frac{\sin\left(2kb(1 - \sin\alpha)\right)}{kb\cos^2\alpha} + \frac{1 + \cos^4\alpha}{4(kb)^2\cos^4\alpha}\right.$$

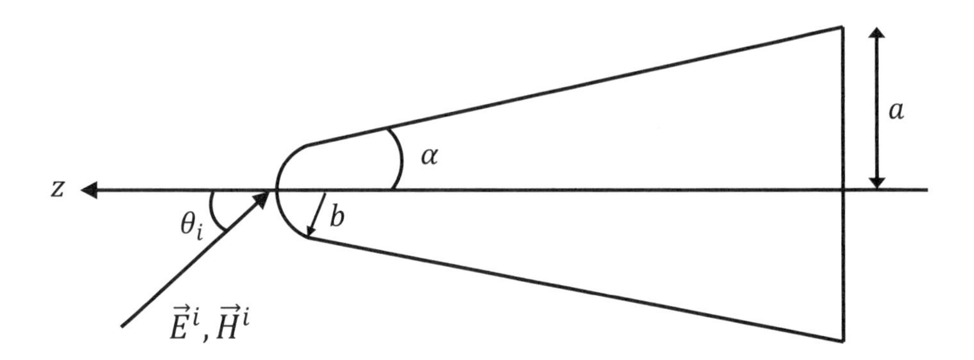

Figure 7.12 Geometry of a cone with a rounded nose tip.

$$- \frac{\cos\left(2kb(1 - \sin\alpha)\right)}{2(kb^2)\cos^2\alpha}\Bigg] \qquad (\text{m}^2), \qquad (7.55)$$

where b is the radius of the rounded nose tip. For incident angles other than axial, but less than the cone angle ($\theta_i \leq \alpha$),

$$\sigma = \pi b^2 \frac{1 + \theta_i^2}{4(kb)^2}\Big[A_1 + A_2 \cos\left(2k\cos\theta_i(1 - \sin\alpha)\right) + A_3 \sin\left(2k\cos\theta_i(1 - \sin\alpha)\right)\Big] \qquad (\text{m}^2), \qquad (7.56)$$

where

$$A_1 = 2 + 2\alpha^2 - 2\theta_i^2 + \alpha^4 - \alpha^2\theta_i^2 + 0.5\theta_i^4 + 2\alpha^4\theta_i^2 + 4(kb)^2 - 2(kb)^2\theta_i^2 - 8(kb)^3\theta_i^2 + (kb)^2\theta_i^4 + 6(kb)^2\alpha^2\theta_i^2 + 8(kb)^3\theta_i^4 + 13(kb)^4\theta_i^4, \qquad (7.57)$$

$$A_2 = -2 - 2\alpha^2 + 2\theta_i^2 + \alpha^2\theta_i^2 - 0.5\theta_i^4 - 6(kb)^2\theta_i^2 + 8(kb)^4\theta_i^3 + 3(kb)^2\theta_i^4, \qquad (7.58)$$

$$A_3 = -4\left(1 + \alpha^2 - 0.5\theta_i^2 + 3(kb\theta_i)^2\right)\left(kb - kb\theta_i^2 - (kb\theta_i)^2\right) - 4(kb\theta_i)^3. \qquad (7.59)$$

7.4.1.7 Frustum

Considering the frustum geometry shown in Figure 7.13, an approximation for the radar cross section due to linearly polarized incident energy is given by [16]

$$\sigma = \frac{b\,\lambda}{8\pi \sin\theta_i}\tan^2(\theta_i - \alpha) \qquad (\text{m}^2), \tag{7.60}$$

where

a = small end radius (m),
b = large end radius (m),
α = half angle of the cone ($\tan^{-1}(b-a)/L$) (rad),
L = length of the frustum (m).

When the incident angle is normal to the side of the frustum, $\theta_i = 90 + \alpha$, an approximation for the radar cross section is [16]

$$\sigma = \frac{8\pi\left(z_2^{1.5} - z_1^{1.5}\right)^2 \sin\alpha}{9\lambda\cos^4\alpha} \qquad (\text{m}^2). \tag{7.61}$$

When the incident angle is either 0° or 180°, the radar cross section is approximated by a flat circular plate, as given in Table 7.1. A comparison of the radar cross section calculated from (7.60) and (7.61) to the radar cross section obtained from physical optics is given in the examples in Section 7.6.

7.4.2 Numerical Techniques

As shown in the previous sections, analytic solutions are only available for a small set of canonical objects. While there are approximate solutions for many objects, which are in good agreement with measured results, these are still quite limited compared to the complex targets of interest to radar systems. In the past, the design and analysis of electromagnetic scattering from objects was largely experimental. However, with the computing power and programming languages currently available, computational electromagnetics has seen ever-increasing usage. There have been many advances in the area of computational electromagnetics, allowing the solution of larger and more complex problems [17, 18]. While measured results provide valuable data and insight into the scattering of complex objects, they are quite costly in terms of hardware, labor, and time. Numerical techniques allow for the prediction of the radar cross section of complex objects before they are constructed. This provides the opportunity to optimize certain parameters, and for the evaluation of "what if" scenarios, which would be prohibitive if done experimentally.

7.4.2.1 Exact Methods

Exact methods are numerical techniques without any implicit approximation in their treatment of the problem. Therefore, they solve Maxwell's equations in an exact sense and can be highly accurate. This is desirable since many radar systems make use of the radar cross section for target classification and discrimination. However, these methods are

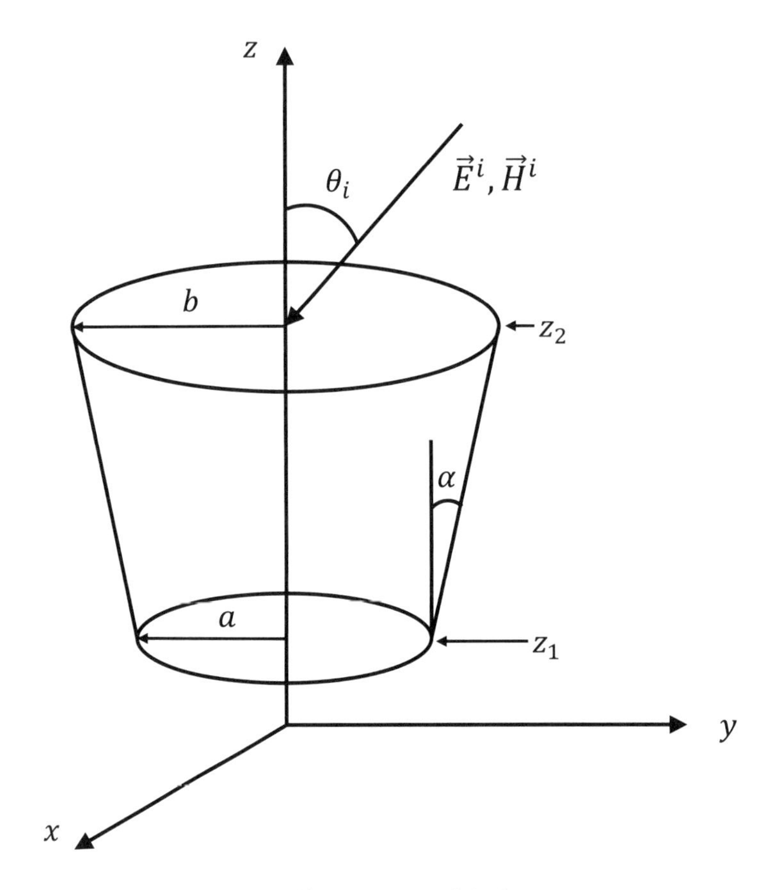

Figure 7.13 Geometry of the frustum.

often limited to objects of small electrical size due to computer memory and computation time constraints. Exact methods are often termed *low-frequency* techniques, as their practical application is limited to low-frequency regimes for targets of interest to radar systems, such as tanks, aircraft, and ships. Exact methods include the finite difference time domain (FDTD) method, finite element method (FEM), and method of moments (MOM). Each of these has their own advantages and disadvantages discussed in the following sections.

7.4.2.1.1 Finite Difference Time Domain

The FDTD method is one of the simplest of the exact methods [19–21]. It is straightforward in its concept and implementation. FDTD uses finite differences as approximations to both the spatial and temporal derivatives in Maxwell's equations. Penetrable dielectric and magnetic materials are handled in FDTD as well as complex geometries. Since FDTD is a time domain technique, it can produce wideband responses rather than single frequency responses of other techniques. Also, the FDTD method is highly parallelizable [22], which greatly improves run times.

Yee first proposed an FDTD algorithm in 1966 employing second-order central differences [23]. The algorithm is summarized as

1. Replace derivatives in Ampere's and Faraday's laws with finite differences.
2. Discretize space and time to offset the electric and magnetic fields in both domains.
3. Solve the resulting difference equations for the future fields in terms of past fields.
4. Update the magnetic fields one time step in the future.
5. Update the electric fields one time step in the future.
6. Repeat the two previous steps over the desired time duration.

For illustration, the FDTD update equations in one dimension are derived. Fields with variations only in the x direction, and an electric field which only has a z component are considered. Faraday's law (2.1) for this case may be written as

$$-\mu \frac{\partial \mathbf{H}}{\partial t} = \nabla \times \mathbf{E} = -\frac{\partial E_z}{\partial x}\,\hat{\mathbf{y}}. \tag{7.62}$$

Similarly, Ampere's law (2.2) for this case is written as

$$\epsilon \frac{\partial \mathbf{E}}{\partial t} = \nabla \times \mathbf{H} = \frac{\partial H_y}{\partial x}\,\hat{\mathbf{z}}. \tag{7.63}$$

Next, replace the derivatives in (7.62) and (7.63) with finite differences. Figure 7.14 illustrates the offset of the electric and magnetic fields in both space and time. This allows (7.62) to be written as

$$\mu\,\frac{H_y^{t+1/2}(n+1/2) - H_y^{t-1/2}(n+1/2)}{\Delta t} = \frac{E_z^t(n+1) - E_z^t(n)}{\Delta x}, \tag{7.64}$$

which leads to the update equation for the magnetic field, given by

$$H_y^{t+1/2}(n+1/2) = H_y^{t-1/2}(n+1/2) + \frac{\Delta t}{\mu \Delta x}\Big(E_z^t(n+1) - E_z^t(n)\Big). \tag{7.65}$$

Starting with (7.63) and following a similar procedure results in the update equation for the electric field as

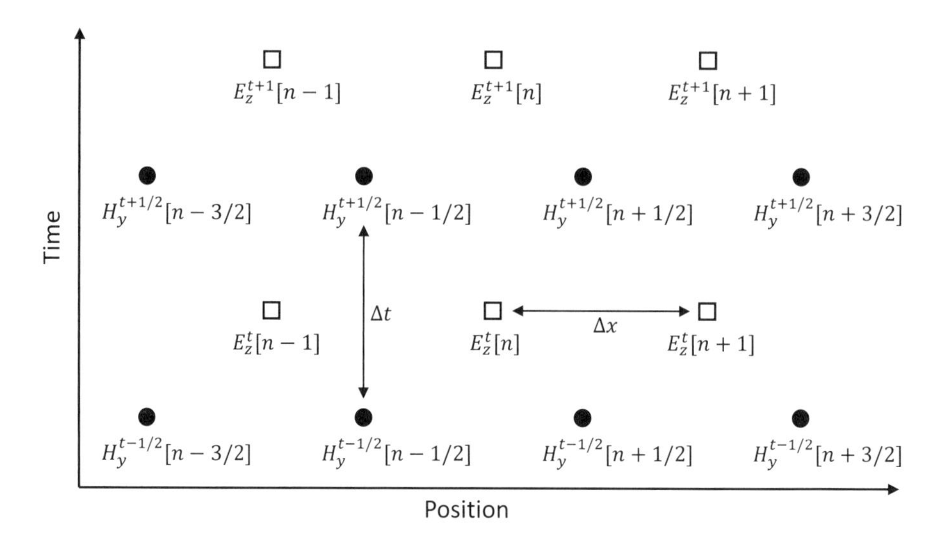

Figure 7.14 Electric and magnetic fields in time and space for the Yee algorithm.

$$E_z^{t+1}(n) = E_z^t(n) + \frac{\Delta t}{\epsilon \Delta x}\Big(H_y^{t+1/2}(n+1/2) - H_y^{t+1/2}(n-1/2)\Big). \tag{7.66}$$

In the update equations (7.65) and (7.66), there are two terms, Δx and Δt, which are the spatial and time discretization. The fundamental constraint on the spatial cell size is that it must be much less than the smallest wavelength. The constraint has to do with reducing the dispersion error in the grid. One figure often given in literature is a cell size of $\lambda/10$ for accurate results. However, for very accurate determination of radar cross section, a cell size of $\lambda/20$ or smaller may be necessary. Also, the cell size must be appropriate to accurately model the geometry being considered. For example, the modeling of a smooth surface with a rectangular cell may result in staircase errors and may require very small cells. Once the spatial step size has been determined, the maximum time step, Δt, needs to be found. While it is desirable for the time step to be large to reduce the required number of calculations, if the time step too large the algorithm is unstable, and results tend to infinity. Much work has been performed in the area of stability conditions for the FDTD method and the reader is referred to several good references on the subject [24–27]. For the purposes of this book, the fields are not allowed to propagate more than one spatial step, in one time step, which leads to the stability condition, $v_p\,\Delta t \leq \Delta x$, where v_p is the maximum velocity of propagation of light in any medium in the problem geometry.

The FDTD computational volume is finite in size, and when the fields reach the outer boundary they are reflected back into the computational domain. For scattering problems, such as the computation of the radar cross section, this causes errors in the results. One approach is to make the computational volume very large and stop the time stepping before the reflections cause significant error. However, this approach is not practical as the number of spatial cells required would be prohibitive. A more appealing approach is to terminate the FDTD volume with some type of reflectionless boundary. Much work has been performed in the area of absorbing type boundaries [28–30], with perfectly matched layers (PML) being considered state of the art. The PML was first proposed by Berenger in 1994 [31]. While there are many different formulations for PMLs, the basic idea is for the layers to act as lossy material and absorb energy traveling away from the interior of the FDTD volume. The PML is constructed such that it is anisotropic and there is no loss tangential to the interface but is always lossy in the direction normal to the interface. A highly attractive PML formulation is the convolutional-PML [28]. This formulation constructs the PML from an anisotropic, dispersive material that has a straightforward implementation.

7.4.2.1.2 Finite Element Method

FEM is a numerical technique for solving boundary value problems in engineering and mathematical physics. It was first proposed in the early 1940s by Hrennikoff [32] and Courant [33] and began being used for aircraft design in the 1950s. The method was developed and used extensively in the 1960s and 1970s, with NASA sponsoring the original version of NASTRAN for the aerospace industry [34]. Since then, the finite element method has seen much interest in areas such as structural analysis, heat transfer, fluid flow, mass transport, and electromagnetics [35–37]. The basic steps for solving boundary value problems with the finite element method are

1. Discretize the problem domain.
2. Choose the basis functions.
3. Formulate the set of system equations.
4. Solve the set of system equations.

Discretization of the problem domain involves subdividing Ω into some number of smaller domains, which are referred to as elements. Figure 7.15 shows traditional finite elements chosen for one-, two-, and three-dimensional problem domains. In two dimensions, the triangular element works well for modeling targets with arbitrary geometry. Similarly, in three dimensions the tetrahedral element is well suited for arbitrary geometries. There are several tools for creating discretized problem domains suitable for analysis with the finite element method, including Gmsh [38], Rhinoceros© [39], and PTC Creo© [40].

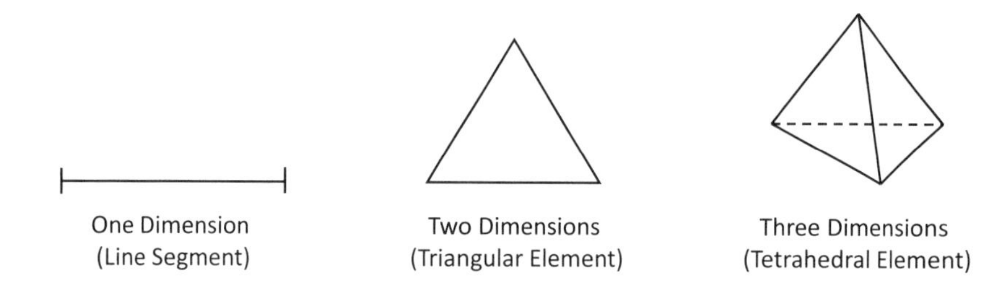

Figure 7.15 Traditional finite elements for one-, two-, and three-dimensional problem domains.

The choice of basis functions depends on a few factors, including the problem domain, complexity, relative error, and numerical dispersion. Basis functions are usually chosen to be either linear, quadratic, or higher-order polynomials. While the higher-order basis functions produce quite accurate solutions, the formulation becomes more complicated. Entire domain basis functions, including wavelet and wavelet-like functions, have also been examined by the Harrison et al. [41–46]. Once the basis functions have been chosen, the unknown solution in an element is written as

$$\tilde{\phi}^e = \sum_{i=1}^{M} N_i^e \, \phi_i^e, \tag{7.67}$$

where

N = number of nodes in the element,
ϕ_i^e = the value of ϕ at node i,
N_i^e = the basis function for node i.

The next step in the finite element process is the formulation of the system of equations. Boundary value problems may be expressed as

$$\mathcal{L}\phi = f, \tag{7.68}$$

where $\mathcal{L}$ is a differential operator, and f if the forcing function. This is subject to boundary condition Γ on problem domain Ω, as shown in Figure 7.16. The form of $\mathcal{L}$ and f for scattering and radiation type problems can be the scalar wave equation of (2.43) or the vector wave equation of (2.42). The boundary condition, Γ, for the scattering problems is typically an impedance or radiation condition. The two most widely used approximate methods for solving (7.68) are the Ritz variational and Galerkin methods [37].

The final step in the finite element procedure is solving the system of equations, which may take on one of two forms, given as

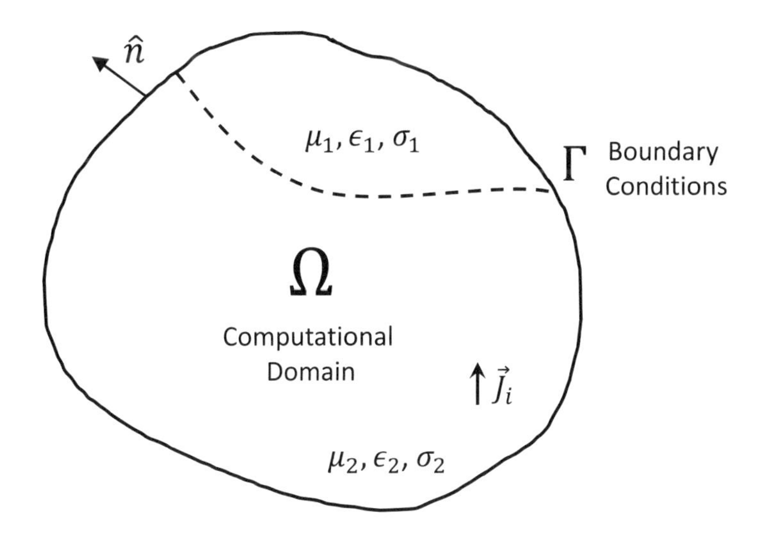

Figure 7.16 Finite element method computational domain.

$$\mathbf{A}\,\mathbf{x} = \mathbf{b}, \tag{7.69}$$

or

$$\mathbf{A}\,\mathbf{x} = \lambda\,\mathbf{B}\,\mathbf{x}. \tag{7.70}$$

The form given in (7.69) is deterministic and results from open radiation and scattering type problems. The second form, given in (7.70), is an eigenvalue system resulting from waveguide and resonant cavity type problems. There are generally two categories of methods used to solve linear equations of the form (7.69), direct and iterative. There are many direct methods, which are based on Gaussian elimination. Decomposition methods, such as LU decomposition, are best suited for finite element systems of equations. The direct solution by LU decomposition may be calculated with the SciPy implementation, ***scipy.linalg.lu_factor()***, which performs the LU decomposition, and ***scipy.linalg.lu_solve()***, which solves the matrix equation given the LU factorization [12]. Iterative methods for solving problems of the form given in (7.69) do not modify the system matrix, but rather begin with an initial approximation of the solution vector, **x**, and minimize the residual vector

$$\mathbf{r}_i = \mathbf{A}\,\tilde{\mathbf{x}}_i - \mathbf{b}, \tag{7.71}$$

where $\tilde{\mathbf{x}}_i$ is the approximate solution at each iteration. There are a few iterative solvers, including conjugate gradient, biconjugate gradient, and generalized minimum residual [37], suitable for use in the solution of finite element systems of equations. These iterative methods, along with a few others, are available in the SciPy module ***scipy.sparse.linalg***. The iterative solvers in SciPy also accept a preconditioner matrix as an input.

As an example of the direct and iterative methods, consider the code given in Listing 7.2. This demonstrates the solution to (7.69) with the direct LU decomposition method and the iterative conjugate gradient method. The two solutions differ slightly, which is a result of the convergence criteria based on the residual in (7.71) for the conjugate gradient method.

7.4.2.1.3 Method of Moments

MOM is a powerful numerical technique for solving boundary value problems in electromagnetics. It was first applied to electromagnetics problems in the mid-1960s [47–49], and Harrington unified the formulation of the method in 1968 [50]. In 1982, Rao, Wilton, and Glisson presented the electromagnetic scattering by surfaces of arbitrary shape using MOM [51]. Gibson's excellent text is a bridge between the theory and software implementation of the method and provides many implementation details [52]. The basic idea of MOM is to convert a set of integro-differential equations into a system of equations that can be solved numerically. Consider again the boundary value problem

$$\mathcal{L}\phi = f. \tag{7.72}$$

Expanding ϕ in terms of basis functions results in

$$\phi = \sum_{n=1}^{N} c_n \, v_n, \tag{7.73}$$

where v_n are the basis functions, and c_n are the unknown coefficients. Substituting (7.73) into (7.72) results in

$$\sum_{n=1}^{N} c_n \, \mathcal{L} \, v_n = f. \tag{7.74}$$

To find c_n, take the inner product of (7.74) with a set of testing functions w_n, which results in

$$\sum_{n=1}^{N} c_n < w_m, \mathcal{L} \, v_n >=< w_m, f > \qquad m = 1, 2, \cdots M. \tag{7.75}$$

This may be expressed in matrix form as

$$\mathbf{Ax} = \mathbf{b}, \tag{7.76}$$

where

$$\mathbf{A} =< w_m, \mathcal{L}\, v_n >, \tag{7.77}$$

and

$$b =< w_m, f > . \tag{7.78}$$

Listing 7.1 Example Direct and Iterative Methods

```
from scipy.sparse.linalg import cg
from scipy.linalg import lu_factor, lu_solve
from scipy import array

# System matrix
A = array([[1, 2], [3, 4]])

# RHS vector
b = array([10, -7])

# Perform factorization
lu, piv = lu_factor(A)

# Perform forward and backward substitution
x = lu_solve((lu, piv), b)

# Print the solution vector
print('Direct Solver Solution: ', x)

# Solution with conjugate gradient
x = cg(A, b)

# Print the solution vector
print('Iterative Solver Solution: ', x[0])
...
Direct Solver Solution:  [-27.   18.5]
Iterative Solver Solution:  [-27.0000928   18.50004516]
```

The matrix equation in (7.76) may be solved through direct or iterative solvers. However, there are powerful fast solvers for use with MOM discussed in the later sections. There are four basic steps to solving boundary value problems in electromagnetics with the method of moments:

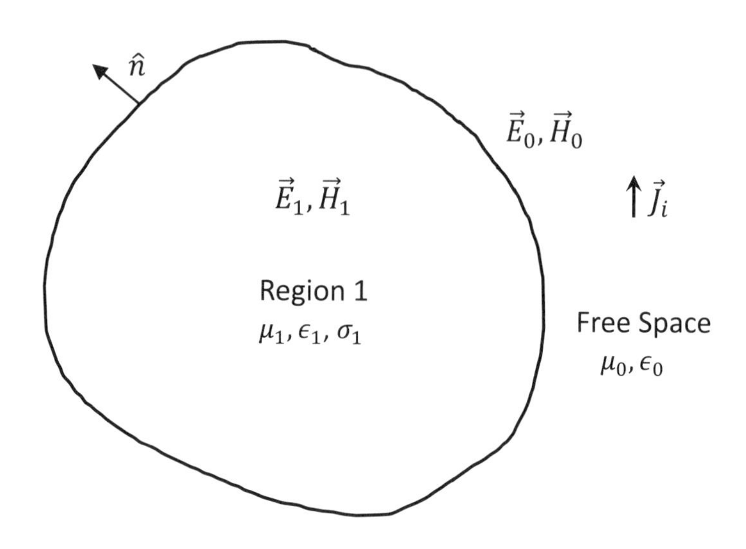

Figure 7.17 Arbitrarily shaped object in free space.

1. Formulate the problem in terms of integral equations.
2. Represent the solution in terms of basis functions.
3. Convert the integral equations to a matrix equation with the use of testing functions.
4. Solve the resulting matrix equations efficiently.

To formulate the integral equations, begin by considering the geometry of Figure 7.17, where there are electromagnetic fields produced by an electric current density, $\mathbf{J}_i$, in the presence of an arbitrarily shaped object situated in free space. In the region exterior to the object, the electric and magnetic fields obey the vector wave equation, given by

$$\nabla \times \nabla \times \mathbf{E} - k^2\mathbf{E} = -j\omega\mu\mathbf{J}, \tag{7.79}$$

$$\nabla \times \nabla \times \mathbf{H} - k^2\mathbf{H} = \nabla \times \mathbf{J}. \tag{7.80}$$

The three-dimensional Green's function [10] satisfies the Helmholtz equation,

$$\nabla^2 G(\mathbf{r}, \mathbf{r}') + k^2 G(\mathbf{r}, \mathbf{r}') = -\delta(\mathbf{r} - \mathbf{r}'). \tag{7.81}$$

The solution to (7.81) is given in many texts [10, 53, 54], and is repeated here as

$$G(\mathbf{r}, \mathbf{r}') = \frac{e^{-jk|\mathbf{r}-\mathbf{r}'|}}{4\pi||\mathbf{r} - \mathbf{r}'|}. \tag{7.82}$$

The incident fields are expressed as

$$\mathbf{E}^i = -j\omega\mu \iiint_V \left(\mathbf{J}\,G + \frac{1}{k^2}\nabla' \cdot \mathbf{J}\,G\right) dV', \tag{7.83}$$

$$\mathbf{H}^i = -\iiint_V \mathbf{J} \times G\,dV'. \tag{7.84}$$

To write the expressions more compactly, two operators are introduced:

$$\mathcal{L}(\mathbf{x}) = jk \oiint_S \left(\mathbf{x}\,G + \frac{1}{k^2}\nabla' \cdot \mathbf{x}\nabla G\right) dS', \tag{7.85}$$

$$\mathcal{K}(\mathbf{x}) = \oiint_S \mathbf{x} \times \nabla G\,dS'. \tag{7.86}$$

Recalling the electromagnetic boundary conditions from Section 2.3, the equivalent surface currents are

$$\mathbf{J}_s = \hat{\mathbf{n}} \times \mathbf{H}, \tag{7.87}$$

$$\mathbf{M}_s = \mathbf{E} \times \hat{\mathbf{n}}. \tag{7.88}$$

Letting $\mathbf{r}$ approach S and performing some mathematical manipulation results in

$$\frac{1}{2}\mathbf{M}_s - \hat{\mathbf{n}} \times \mathcal{L}(\mathbf{J}_s) + \hat{\mathbf{n}} \times \tilde{\mathcal{K}}(\mathbf{M}_s) = -\hat{\mathbf{n}} \times \mathbf{E}^i, \tag{7.89}$$

$$\frac{1}{2}\mathbf{J}_s + \hat{\mathbf{n}} \times \tilde{\mathcal{K}}(\mathbf{J}_s) + \mathbf{n} \times \mathcal{L}(\mathbf{M}_s) = \hat{\mathbf{n}} \times \mathbf{H}^i. \tag{7.90}$$

$\tilde{\mathcal{K}}$ is the same operator as (7.86), except the singular point ($\mathbf{r} = \mathbf{r}'$) has now been removed. The expression in (7.89) is known as the electric field integral equation (EFIE), and (7.90) is known as the magnetic field integral equation (MFIE).

When using the method of moments, a common discretization of the surface is through the use of triangular elements. A popular choice for expanding the surface current is the Rao-Wilton-Glisson (RWG) basis function [51], illustrated in Figure 7.18. The RWG basis function is defined over two triangular elements joined at a common edge, and is given here as

$$
\boldsymbol{\Lambda}_n = \begin{cases} \dfrac{\ell_n}{2A_n^+}\boldsymbol{\rho}_n^+ & \mathbf{r} \in T_n^+, \\ \\ \dfrac{\ell_n}{2A_n^-}\boldsymbol{\rho}_n^- & \mathbf{r} \in T_n^-, \\ \\ 0 & \text{otherwise} \end{cases} \tag{7.91}
$$

where

ℓ_n = length of the common edge between two triangular elements,
$T_n^{\pm}$ = two triangular elements with the common edge l_n,
$A_n^{\pm}$ = area of the two triangular elements,
$\rho_n^{\pm}$ = vectors associated with each element.

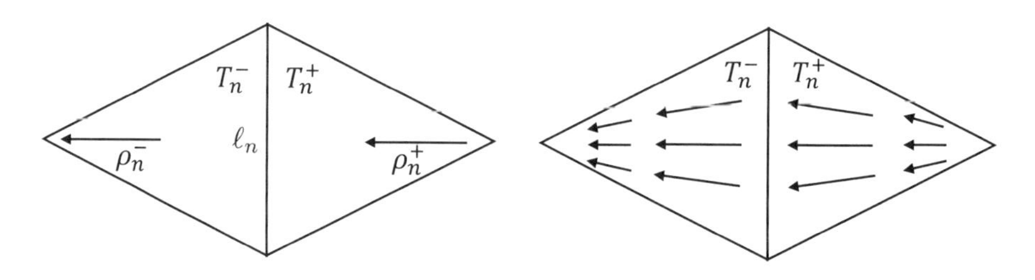

Figure 7.18 Rao-Wilton-Glisson basis function.

The RWG basis function guarantees the continuity of current flow over all edges, as the normal component of the basis function at edge l_n is a constant and the normal components of the basis function to the other edges is zero. As with the finite element method, MOM requires a well-constructed and well-connected mesh.

The final step in the MOM process is the solution of the system matrix equation of the form $\mathbf{A}\,\mathbf{x} = \mathbf{b}$. Unlike FEM, the MOM system matrix is fully populated. To solve the MOM matrix equation with a direct solver, such as LU decomposition, the operation count is $O(N^3)$ and the memory requirement is $O(N^2)$. To solve the MOM matrix equation with an iterative solver, both the operation count and memory requirements are $O(N^2)$. If the number of iterations required for convergence is relatively small, then the iterative methods are faster than the direct solvers. However, the iterative solution must be repeated for each right-hand side (forcing function), whereas the LU decomposition may be performed once, and the forward and backward substitutions are performed for each forcing function. Such computational complexity limits the size of problems feasible with either of the solution techniques. However, much work has been performed to reduce the complexity from $O(N^2)$ to $O(N^{1.5})$ and even $O(N \log N)$. To this end, the fast multipole method (FMM), first proposed by Greengard and Rohklin for particle simulations [55], was extended to solve problems acoustic wave scattering and eventually electromagnetic problems. The memory requirement and computational complexity are both $O(N^{1.5})$. Both of these can be reduced to $O(N \log N)$ by using the multilevel fast multipole method (MLFMM) [17]. Another method for reducing the time are memory requirement for MOM is the adaptive cross approximation (ACA) method, which was first proposed by Bebdndorf [56]. In MLFMM, the formulation, implementation, and sometimes performance depend on a priori knowledge of the integral equation kernel. ACA is purely algebraic in nature and is therefore independent of the integral equation kernel. This allows ACA to be modular and easily implemented into existing MOM code. The computational complexity of ACA has been shown to be $O(N \log N)$ for electrically small problems [37]. Lee et al. showed the memory requirement and computation complexity of ACA to be $O(N^{1.5} \log N)$ for moderately sized problems.

7.4.2.2 Approximate Methods

Objects of interest to radar systems are often very large compared to the wavelength of the incident energy. In order to solve the scattering from electrically large objects, many approximate techniques have been developed. These methods are largely based on edge diffraction and physical optics and provide accurate results for many classes of objects. Many of the approximate methods, when used in combination, can produce results accurate to within a few decibels of the true scattering result. These methods can be used for first cut design and analysis problems. Since these methods are used for electrically large objects, they are often referred to as *high-frequency* techniques. These methods include physical theory of diffraction (PTD), geometrical theory of diffraction (GTD), geometrical optics (GO), physical optics (PO), and shooting and bouncing rays (SBR).

7.4.2.2.1 Geometrical Optics

This section begins with the basic concepts of the well-known geometrical optics method for solving electromagnetic problems. The Luneburg-Kline [57] form of the solutions for the electric and magnetic fields are given by

$$\mathbf{E}(\mathbf{r}, \omega) \approx e^{-jk\phi(\mathbf{r})} \sum_{n=0}^{\infty} \frac{\mathbf{E}_n(\mathbf{r})}{(j\omega)^n}, \tag{7.92}$$

$$\mathbf{H}(\mathbf{r}, \omega) \approx e^{-jk\phi(\mathbf{r})} \sum_{n=0}^{\infty} \frac{\mathbf{H}_n(\mathbf{r})}{(j\omega)^n}. \tag{7.93}$$

These represent the expansion of the fields in inverse powers of ω. As the frequency approaches ∞, the only terms remaining are for $n = 0$, and the fields are written as

$$\mathbf{E}(\mathbf{r}, \omega) \approx \mathbf{E}_0 \, e^{-jk\phi(\mathbf{r})}, \tag{7.94}$$

$$\mathbf{H}(\mathbf{r}, \omega) \approx \mathbf{H}_0 \, e^{-jk\phi(\mathbf{r})}. \tag{7.95}$$

These are the geometrical optic rays and play an important role in solving high-frequency problems. If (7.92) and (7.93) are placed into Maxwell's equations, relationships between the coefficients of $\mathbf{E}_n$ and $\mathbf{H}_n$ may be obtained. Geometrical optic rays are defined as curves tangent to the direction of power flow. The fields $\mathbf{E}$ and $\mathbf{H}$ are required to be orthogonal to one another and to $\hat{\ell}$. Recall from Chapter 2 that this is the same as the field relationship for a plane wave. The geometrical optic ray does not need to be a plane wave, but at every point the orthogonality holds. For example, the geometrical optic ray fields can be diverging from the source, as shown in Figure 7.19. Important properties of geometrical optic rays are

1. All rays travel in straight lines.
2. In a homogeneous medium, the polarization of the fields along the ray path is unchanged.
3. Phase changes along the ray path are accounted for by $e^{-jk\phi(\ell)} = e^{-jk\phi(0)}\, e^{-jk\ell}$, where ℓ is the distance along the ray path.
4. Amplitude and phase of the fields along the ray path is given by

$$E = E_0 \, e^{-jk\phi(0)} \sqrt{\frac{\rho_1 \rho_2}{(\rho_1 + \ell)(\rho_2 + \ell)}} \, e^{-jk\ell}, \tag{7.96}$$

where ρ_1 and ρ_2 are the radii of curvature in the principal planes.

As an example, consider the calculation of the radar cross section of a sphere of radius a with geometrical optics. For calculating the monostatic radar cross section, only the ray at the specular point needs to be considered. The radii of curvature for the sphere are

$$\rho_1 = \rho_2 = \frac{a}{2} \qquad \text{(m)}. \tag{7.97}$$

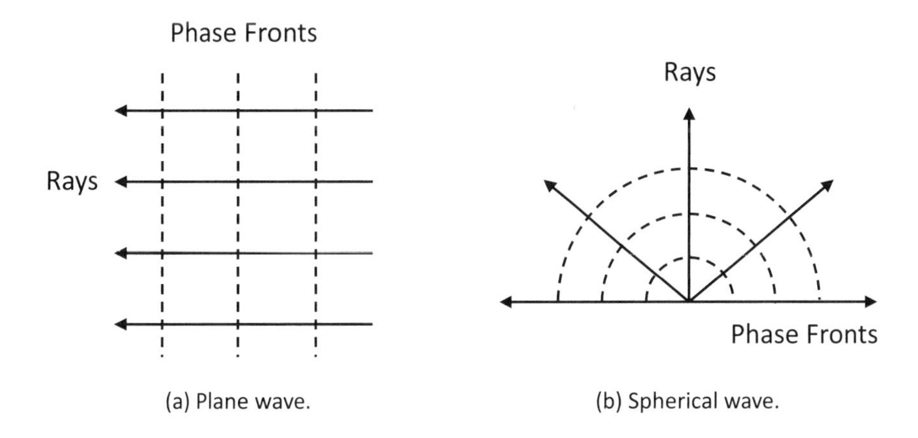

Figure 7.19 Geometrical optic ray fields for plane and spherical waves.

Using (7.96), the field in the backscattered direction is

$$E^s = -E_0 \left(\frac{a/2}{a/2 + \ell} \right) e^{-jk\ell}. \tag{7.98}$$

Using the definition of radar cross section in (7.1) allows the radar cross section of the sphere to be written as

$$\sigma = \lim_{\ell \to \infty} 4\pi\ell^2 \left(\frac{a/2}{a/2 + \ell} \right)^2 = \pi a^2 \qquad (\text{m}^2). \tag{7.99}$$

Comparing the result in (7.99) to Figure 7.3, the two are in good agreement as the frequency becomes large. This is to be expected since geometrical optics assumes the target is much larger than the wavelength.

7.4.2.2.2 Physical Optics

In the geometrical optic method, the surface currents introduced by the incident wave were never considered. In physical optics, it is assumed the field at the surface of the scattering body is the geometrical optic field. This implies the scattering occurs at each point as if there were an infinite tangent plane at that point. For a perfect electrical conductor, the physical optics surface current is

$$\mathbf{J} = \begin{cases} \hat{\mathbf{n}} \times \mathbf{H}_{total} & \text{illuminated,} \\ 0 & \text{shadowed.} \end{cases} \tag{7.100}$$

From image theory, the scattering body is replaced by equivalent currents in free space, and the tangential component of the magnetic field at a prefect electrical conductor are twice those from the source. This is written as

$$\mathbf{J} = 2\hat{\mathbf{n}} \times \mathbf{H}^i. \tag{7.101}$$

The scattered magnetic field is written as

$$\mathbf{H}^s = \nabla \times \iint_{S'} \mathbf{J}\, \psi \, dS', \tag{7.102}$$

where ψ is the free space Green's function given in (7.82). Mathematical manipulation results in

$$\mathbf{H}^s = e^{-jkr} \iint_{S'} \left(\mathbf{J} \times \hat{\mathbf{r}}\right) \frac{jk}{4\pi r}\, e^{jk\hat{\mathbf{r}}\cdot\mathbf{r}'}\, dS'. \tag{7.103}$$

Writing the radar cross section in terms of the magnetic field gives

$$\sigma = \lim_{r\to\infty} 4\pi r^2 \frac{|\mathbf{H}^s|^2}{|\mathbf{H}^i|^2} = \frac{4\pi}{\lambda^2}\left| \iint_{S'} \left(\hat{\mathbf{r}} \cdot \hat{\mathbf{n}}\right) e^{j2k\hat{\mathbf{r}}\cdot\mathbf{r}'}\, dS' \right|^2 \quad (\text{m}^2). \tag{7.104}$$

As an example, consider the calculation of the monostatic radar cross section of a sphere of radius a. If the incident wave is traveling in the negative z direction, then

$$\hat{\mathbf{z}} \cdot \hat{\mathbf{n}} = \frac{a - \ell'}{a}, \tag{7.105}$$

$$\hat{\mathbf{z}} \cdot \hat{\mathbf{r}} = a - \ell', \tag{7.106}$$

where ℓ' is the distance along z from a point on the sphere surface, as shown in Figure 7.20. Substituting (7.105) and (7.106) into (7.104) results in

$$\sigma = \frac{4\pi}{\lambda^2}\left| 2\pi\, e^{j2ka} \int_0^a e^{-j2k\ell'} (a - \ell')\, d\ell' \right|^2 \quad (\text{m}^2). \tag{7.107}$$

Performing the integration leads to

$$\sigma = \frac{4\pi}{\lambda^2}\left| \frac{a\lambda}{2j} \left[\left(1 + \frac{j}{2ka}\right) e^{j2ka} - \frac{j}{2ka} \right] \right|^2 \quad (\text{m}^2). \tag{7.108}$$

Allowing $ka \to \infty$ in (7.108) results in $\sigma \approx \pi a^2$. In the high-frequency limit, the radar cross section of the sphere as calculated by physical optics reduces to the radar cross section from geometrical optics.

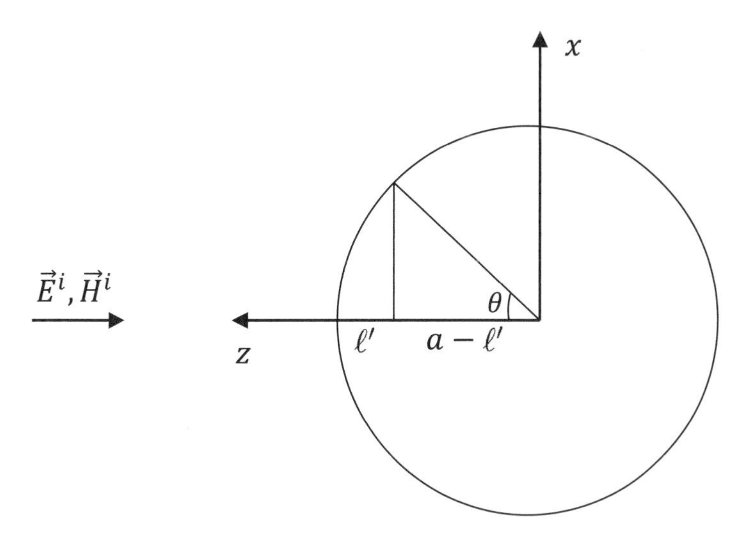

Figure 7.20 Sphere geometry for physical optics calculation.

7.4.2.2.3 Geometrical Theory of Diffraction

Physical optics and geometrical optics were concerned with backscattering. In the forward scattering direction, geometrical optics is incapable of predicting a nonzero field in the shadow region. Therefore, geometrical optics is extended with diffraction rays that do permit the calculation of scattering fields in the shadow region. Keller's geometrical theory of diffraction treats wedge diffraction and is an important extension of geometrical optics [14]. Consider the scattering by an infinitely thin infinitely long half plane illustrated in Figure 7.21. Using Huygen's principle, each point on the primary wavefront at $z = 0$ is the source of a secondary wave [1]. Therefore,

$$E(P) = \int_{x=a}^{\infty} dE, \tag{7.109}$$

where dE is the electric field at P, and is written as [57]

$$dE = \frac{C}{\sqrt{\ell + \delta(x)}} e^{-jk(\ell+\delta(x))}\, dx, \tag{7.110}$$

where C is a constant. Substituting (7.110) into (7.109), and solving for the constant C results in

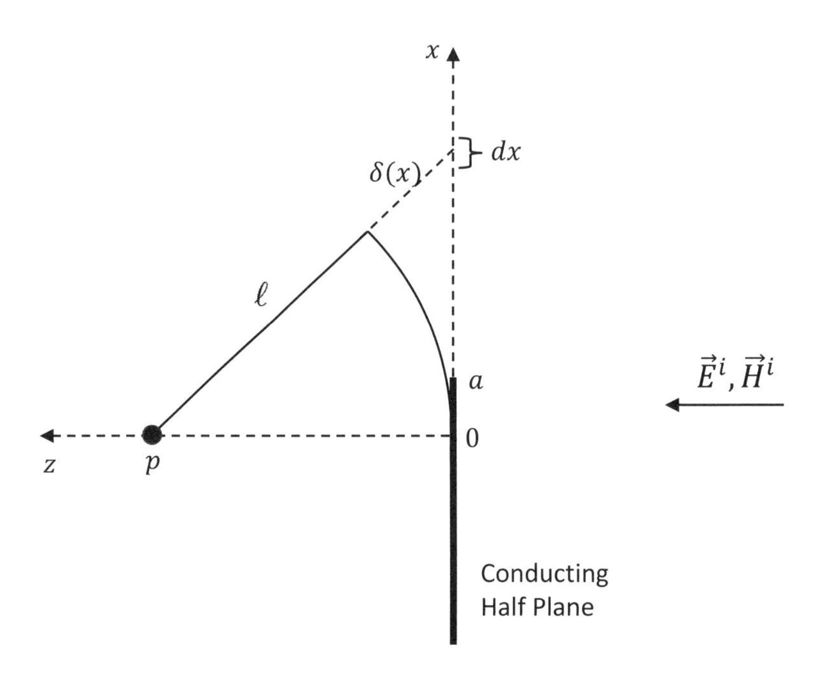

Figure 7.21 Conducting half plane for plane wave diffraction.

$$E(P) \approx \frac{E_0}{\sqrt{2}}\, e^{j(\pi/4)}\, e^{-jk\ell} \int_{\gamma a}^{\infty} e^{-j\pi/2u^2}\, du, \tag{7.111}$$

where $\gamma^2 = 2/\lambda\ell$. The diffracted field in the vicinity of the shadow boundary is shown in Figure 7.22.

7.4.2.2.4 Uniform Theory of Diffraction

While there are complex diffraction scattering mechanisms, including vertex diffraction, tip diffraction, and others, GTD may be divided into two basic problems: wedge diffraction and curved surface diffraction. The uniform theory of diffraction (UTD) accurately predicts the diffracted field in transition regions and near edges without the need to handle each type of incident field separately. UTD applies to all situations consistent with the postulates of GTD. Kouyoumjian et al. developed a generalized version of the scalar diffraction coefficients for parallel and perpendicular polarization [58], which are given here as

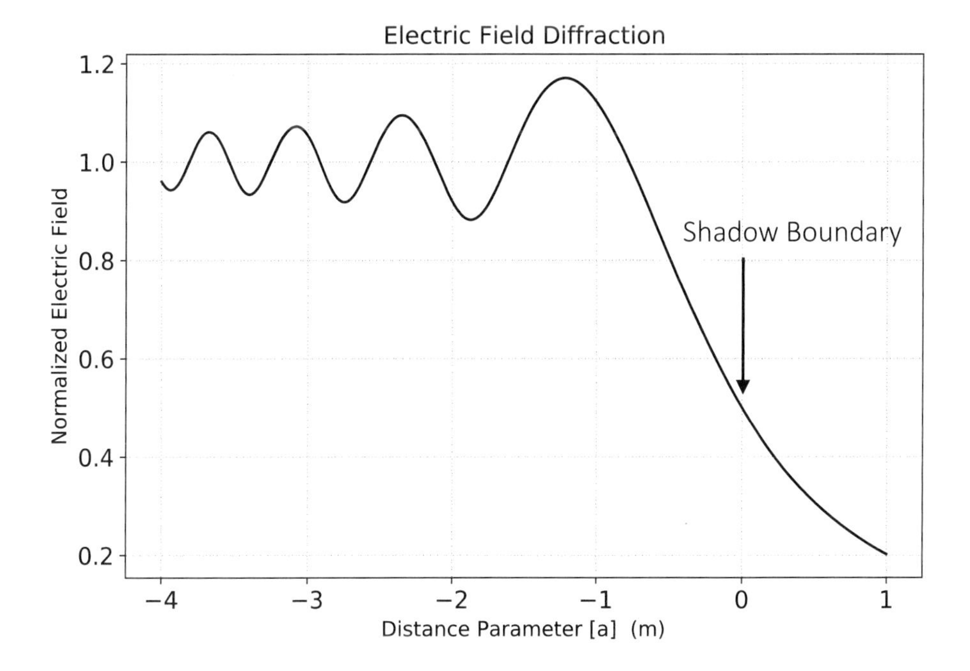

Figure 7.22 The diffracted electric field near the shadow boundary.

$$D_{\perp,\parallel} = \frac{-e^{-j\pi/4}}{2n\sqrt{2\pi k}\sin\gamma_0'}\Bigg[\cot\left(\frac{\pi + (\phi - \phi')}{2n}\right)F\big(kLa^{+}(\phi - \phi')\big)$$
$$+ \cot\left(\frac{\pi - (\phi - \phi')}{2n}\right)F\big(kLa^{-}(\phi - \phi')\big)$$
$$\pm \Bigg\{\cot\left(\frac{\pi + (\phi + \phi')}{2n}\right)F\big(kLa^{+}(\phi + \phi')\big)$$
$$+ \cot\left(\frac{\pi - (\phi + \phi')}{2n}\right)F\big(kLa^{-}(\phi + \phi')\big)\Bigg\}\Bigg], \tag{7.112}$$

where the function $F(x)$ is given by

$$F(x) = 2j|\sqrt{x}| \int_{|\sqrt{x}|}^{\infty} e^{-j\tau^2}\, d\tau, \tag{7.113}$$

and the distance parameter L is

$$L = \begin{cases} s \sin^2 \gamma_0' & \text{for plane waves,} \\ \dfrac{\rho' \rho}{\rho' + \rho} & \text{for cylindrical waves,} \\ \dfrac{s' s \sin^2 \gamma_0'}{s' + s} & \text{for spherical waves.} \end{cases} \tag{7.114}$$

The term in the argument of F is [57]

$$a^{\pm}(\phi \pm \phi') = 2\cos^2\left(\frac{2n\pi N^{\pm} - (\phi \pm \phi')}{2}\right), \tag{7.115}$$

where the integers, $N^{\pm}$, satisfy

$$2\pi n N^{+} - (\phi \pm \phi') = \pi, \qquad 2\pi n N^{-} - (\phi \pm \phi') = -\pi. \tag{7.116}$$

Finally, the diffracted fields are

$$\begin{bmatrix} E_{\parallel}^{d} \\ E_{\perp}^{d} \end{bmatrix} = \begin{bmatrix} -D_{\parallel} & 0 \\ 0 & -D_{\perp} \end{bmatrix} \begin{bmatrix} E_{\parallel}^{i} \\ E_{\perp}^{i} \end{bmatrix} A(s) e^{-jks}, \tag{7.117}$$

where $A(s)$ is the spatial attenuation term, given by

$$A(s) = \begin{cases} \dfrac{1}{\sqrt{s}} & \text{for plane and cylindrical incident waves,} \\ \sqrt{\dfrac{s'}{s(s' + s)}} & \text{for spherical incident waves.} \end{cases} \tag{7.118}$$

The second basic problem addressed by UTD is diffraction by curved surfaces, as illustrated in Figure 7.23. At the tangent point for the incident wave, P_1, the incident wave undergoes diffraction. However, some of the incident energy is trapped and a wave propagates on the surface of the scattering object. This wave is referred to as a *creeping wave* and may be described as continuous diffraction along the surface of the scattering object. In two dimensions, the creeping wave may be written as [57]

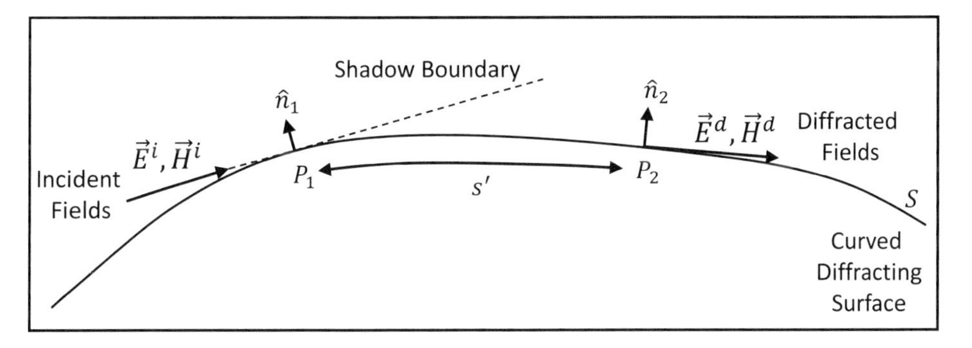

Figure 7.23 Diffraction by a curved surface.

$$E(s') = E^i(P_1)D(P_1)G(s')\exp\left[-\int \gamma(s')ds'\right], \tag{7.119}$$

where

$E(s')$ = field of the creeping wave along the path,
$D(P_1)$ = diffraction coefficient at P_1,
$G(s')$ = ray divergence factor along the path,
$\gamma(s')$ = propagation factor $\alpha(s') + j\beta(s')$ along the path,
s' = arc length along the surface of the scattering object.

As an example, consider the calculation of the two-dimensional radar cross section of a right circular cylinder, as shown in Figure 7.7. The incident field is written as

$$\begin{bmatrix} E^i_\parallel \\ E^i_\perp \end{bmatrix} = E_0\, e^{jk\rho}. \tag{7.120}$$

The two-dimensional GO field is first calculated as

$$\begin{bmatrix} E^r_\parallel \\ E^r_\perp \end{bmatrix} = \begin{bmatrix} E^i_\parallel \\ E^i_\perp \end{bmatrix} [R] \sqrt{\frac{a}{2\rho}}. \tag{7.121}$$

For $\ell \gg a$,

$$\sigma = \lim_{\rho\to\infty} 2\pi\rho \frac{|\mathbf{E}^s|^2}{|\mathbf{E}^i|^2} = \lim_{\rho\to\infty} 2\pi\rho \left[\frac{a}{2\rho}\right] = \pi a \qquad (\text{m}^2). \tag{7.122}$$

Next, the fields from the creeping waves are included. At the point P_1,

$$E(P_1) = E^i(P_1)D(P_1). \tag{7.123}$$

There is no transverse spreading of the waves on the cylinder surface, which leads to $G(s') = 1$. Therefore, the field at the point Q_2 is expressed as

$$
\begin{aligned}
E(P_2) &= E^i(P_1)D(P_1)e^{-jk\pi a}\exp\left[-\int_0^{\pi a} \alpha ds'\right] \\
&= E^i(P_1)D(P_1)e^{-jk\pi a}e^{-\alpha\pi a}. \qquad (7.124)
\end{aligned}
$$

Multiplying (7.124) by $D(P_2)$ results in the fields radiated by the creeping wave. By reciprocity, $D(P_1) = D(P_2)$, which results in

$$
\begin{bmatrix} E^c_{\parallel} \\ E^c_{\perp} \end{bmatrix} = \begin{bmatrix} E^i_{\parallel} \\ E^i_{\perp} \end{bmatrix} \begin{bmatrix} D_{\parallel} \\ D_{\perp} \end{bmatrix}^2 e^{-jk\pi a}e^{-\pi a\alpha}\sqrt{\frac{a}{2\rho}}. \qquad (7.125)
$$

Since there are shadow boundaries at the top and bottom of the cylinder, the fields from the creeping wave are doubled, and the total backscattered field becomes

$$
\begin{bmatrix} E^s_{\parallel} \\ E^s_{\perp} \end{bmatrix} = \begin{bmatrix} E^r_{\parallel} \\ E^r_{\perp} \end{bmatrix} + 2\begin{bmatrix} E^c_{\parallel} \\ E^c_{\perp} \end{bmatrix} \qquad (7.126)
$$

7.4.2.3 Hybrid Methods

Hybrid techniques typically use one method for solving a particular portion of the problem domain and a different method for the remaining portion. One example is the finite element boundary integral (FE-BI) method. This method has been widely used in the analysis of unbounded electromagnetic scattering and radiation from complicated structures and composite materials [59]. This method combines the capability of the finite element method in modeling highly complicated geometries and inhomogeneous materials with the efficiency of integral equations for open radiation and scattering. Recently, Jin et al. demonstrated an accurate and efficient GPU-accelerated multilevel fast multipole algorithm (MLFMA) FE-BI method for the three-dimensional analysis of complex objects in free space [60]. Figure 7.24 illustrates the typical domain computation for the FE-BI method.

7.4.3 Measurement Techniques

As shown in previous sections, analytic and numerical methods all suffer some limitations, some of which are only overcome by experimental techniques or measurement methods. Along with this is the need to verify new prediction techniques and theories. While radar cross-section measurement is very costly in terms of hardware, labor, and time, it provides valuable insight into scattering from complex targets not otherwise

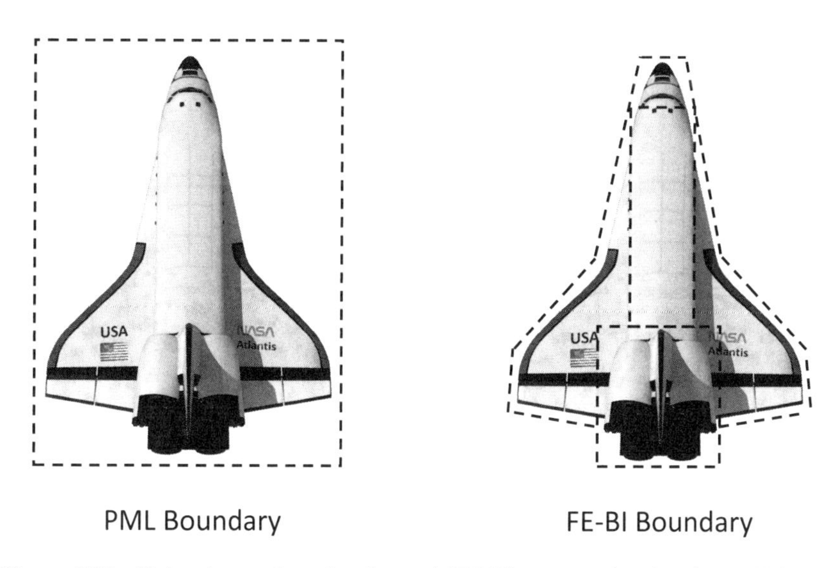

Figure 7.24 Finite element boundary integral (FE-BI) computational regions (NASA).

achievable. Measurement techniques for radar cross section may be performed with full-size targets or with scaled model targets. For scaled model targets, certain other parameters must also be scaled, as shown in Table 7.2, where s is the scale factor. While there are ranges designed to perform scaled model measurements [61], one of the drawbacks is the scaling of the parameters, particularly the frequency and the conductivity. However, with advances in materials and electronic equipment, more reliable scaled model measurements are becoming available. For radar cross-section measurements performed with full-size targets, an indoor or outdoor range may be used. Indoor ranges have the advantage of being a controlled space for both privacy and security as well as not being affected by the environment. However, they are often limited in size, which limits the types of targets that can be measured.

7.4.3.1 Outdoor Range

One problem of early outdoor ranges involved mounting a large target and isolating it from nearby clutter, especially ground reflections. This makes the accurate measurement of low RCS targets very difficult. Some designs include the use of fences to reduce the ground reflections, while others make use of the ground as a method of illuminating the target, as shown in Figure 7.25. The advantage of a ground reflection range is the elimination of the ground as clutter, and the addition of up to 12 dB in signal strength as there are four paths of the direct and reflected signals. However, there are also disadvantages associated with ground reflection ranges. The first is the need for a perfectly flat site for the range. Another disadvantage is that the height of the transmitting

Table 7.2
Scaled Model Radar Cross-Section Measurement Parameters

Parameter	*Full-Scale*	*Sub-Scale*
Dimension	D	$D_s = D/s$
Frequency	F	$F_s = s\,F$
Time	t	$t_s = t/s$
Permittivity	ϵ	$\epsilon_s = \epsilon$
Permeability	μ	$\mu_s = \mu$
Conductivity	C	$C_s = s\,C$
RCS	σ	$\sigma_s = \sigma/s^2$

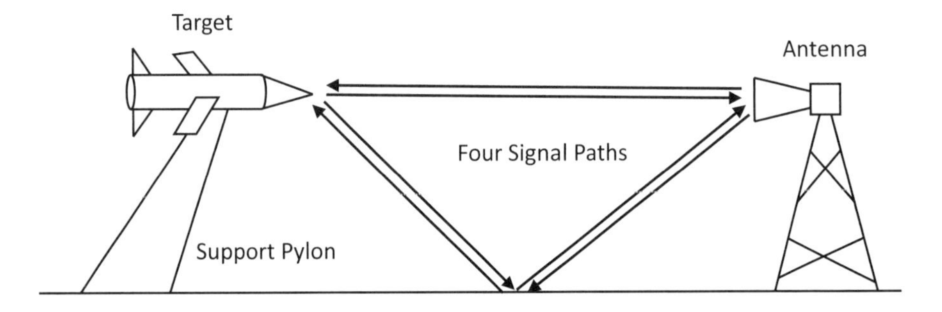

Figure 7.25 Outdoor ground reflection range.

antenna system is highly dependent on the operating frequency as the signals from the direct path and the reflected paths must arrive in phase at the target. Even with these difficulties, ground reflection ranges are still common. Another source of interference is the mounting structure for the target being measured. The structure must be capable of supporting and rotating targets weighing thousands of pounds while producing very little backscatter. Pylons have been developed that have very low backscatter. Typical backscatter values are -25 to -45 dBsm [62]. These pylons are often tapered and are ogive in shape. Many are also titled to further reduce the scattering in the direction of the measuring equipment.

7.4.3.2 Indoor Range

For indoor ranges, there are three typical designs; far-field chambers, tapered chambers, and compact ranges. Indoor far-field ranges, illustrated in Figure 7.26, are problematic due to size limitations as the transmitting antenna and the target under test must be sufficiently separated such that good plane wave conditions exist. Recalling from Chapter

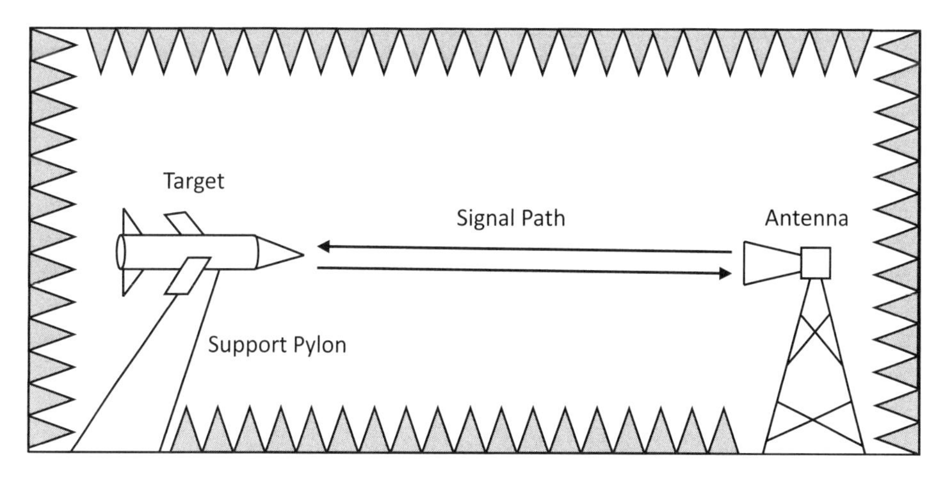

Figure 7.26 Indoor far-field range.

2 the far-field condition of $R \geq 2D^2/\lambda$ and assuming a target of dimension 3 meters and a frequency of 10 GHz requires a distance of $R = 600$ meters. This is prohibitively large for indoor far-field ranges. For tapered chambers, illustrated in Figure 7.27, the tapered region acts as a horn antenna. These types of indoor ranges are most effective at low frequencies. The typical design of a compact range is shown in Figure 7.28. It is designed such that a plane wave is generated by an offset reflector. These allow for much shorter distances between the transmitting antenna and the target under test. The main reflector is a large paraboloid surface fed by a horn antenna at the focus. The power at the receiver is [62]

$$P_r = P_t \frac{\sigma}{(4\pi)^3\lambda^2}\left(\frac{\lambda}{R_0}\right)^4 G_f^2 \qquad \text{(W)}, \tag{7.127}$$

where

P_r = power at receiver (W),
P_t = transmit power (W),
G_f = gain of the feed horn,
R_0 = range from feed horn to reflector (m).

The expression in (7.127) is similar to the radar range equation in (4.28), except the range term in the denominator is now the range from the feed horn to the reflector. Single-feed compact ranges have proven difficult to implement as the feed horn needs to have a very wide beamwidth to achieve uniform amplitude in the quite zone and possess a wide bandwidth, nonringing response. Therefore, some compact ranges operate with a smaller hyperboloid reflector to illuminate the large primary reflector. Sources of error in compact range measurements are shown in Figure 7.29. These errors are represented in time in

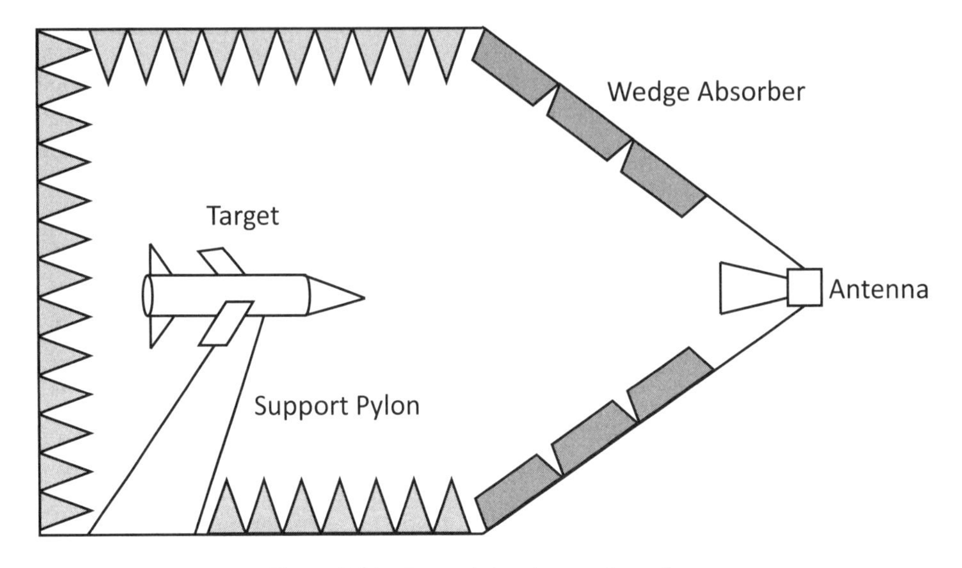

Figure 7.27 Tapered chamber configuration.

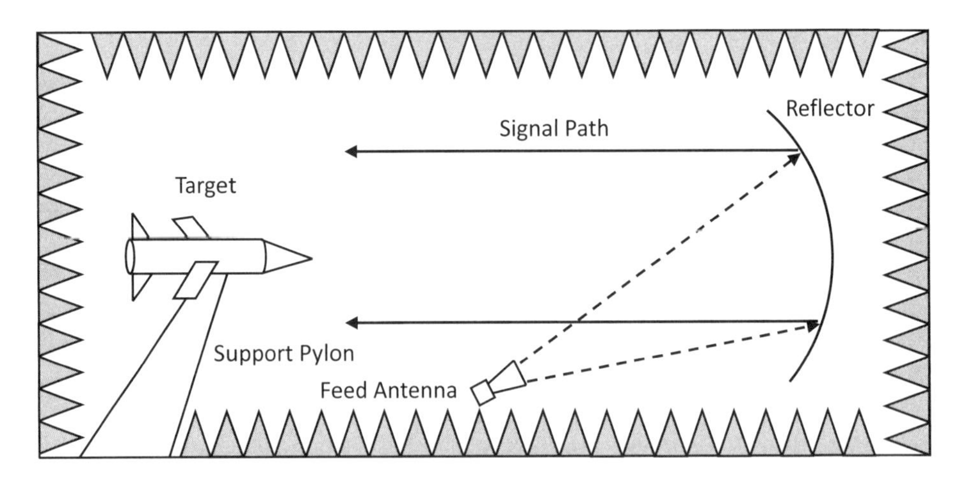

Figure 7.28 Compact range configuration.

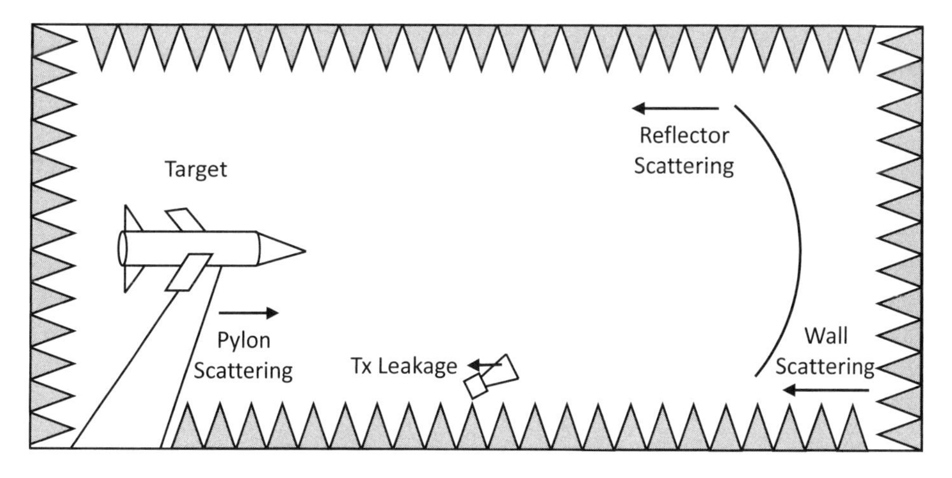

Figure 7.29 Sources of errors in compact ranges.

Figure 7.30. One method of alleviating some of these errors is through time domain gating with a pulsed transmit signal. This requires a large instantaneous bandwidth and the requirement for the pulsewidth is [62]

$$\tau \geq \tau_{rise} + 2\tau_L + \tau_{meas} + \tau_{fall} \qquad \text{(s)}, \tag{7.128}$$

where

τ	=	required pulsewidth (s),
τ_{rise}	=	rise time of the pulse (s),
τ_L	=	width corresponding to target length (s),
τ_{meas}	=	width of measurement window (s),
τ_{fall}	=	fall time of the pulse (s).

Also, the height of the target is chosen such that the reflected pulse does not illuminate the target during the period of time the direct pulse is illuminating the target.

$$height \geq \frac{1}{2}(2L_{target} + \tau_{meas}) \qquad \text{(m)}. \tag{7.129}$$

Another method for reducing the errors in compact ranges is through vector subtraction of the background. First, a calibration target of known scattering is used to calibrate the measurement equipment. The calibration target is then removed and the background signal with no target is measured. Next, the scattering of the desired target is measured. Finally, the background signal is coherently subtracted from the scattering measurements with the target present. This method does not account for interactions between the target and the chamber, but these are usually considered second-order effects.

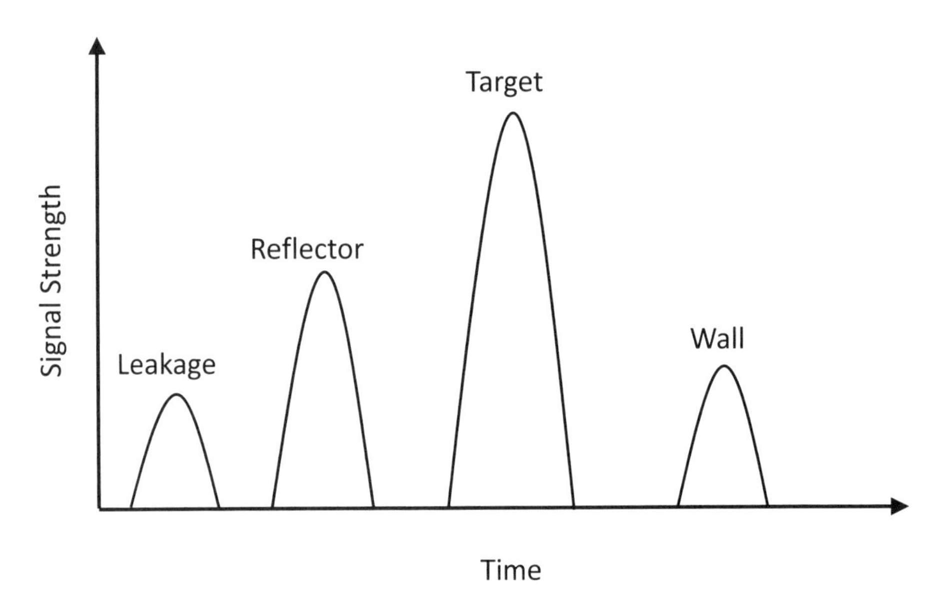

Figure 7.30 Time response of errors in compact ranges.

An important aspect of all indoor ranges is absorbing material placed in areas producing large unwanted scattering, such as walls, floors, doors, and mounting structures. The design of the absorber may be pyramidal, wedge, truncated pyramidal, convoluted, ferrite tile, and others [63]. Each has its own advantages and disadvantages. For example, wedge absorbers reduce scattering in areas where high incident angles occur, whereas convoluted absorbers are very effective in minimizing front face reflections at higher frequencies [63]. Typical values of reflectivity are from −20 to −50 dB, depending on the design, bandwidth, and application [63].

7.5 RADAR CROSS-SECTION REDUCTION

The desire to reduce the radar cross section of a target has been around since the early days of radar in World War II. As shown in Chapter 6, reducing the radar cross section, and therefore the signal-to-noise ratio in the radar receiver, can greatly reduce the probability of detection. Initial attempts to reduce the radar cross section were through the use of wood and other materials less reflective than metal. This quickly led to the idea of shaping as a method of reducing the radar cross section.

7.5.1 Shaping

A prime example of shaping is the F-117 stealth aircraft. The body and edges are angled to direct energy in directions other than the backscatter direction. There are special shape design ideas implemented on the seams, bay doors, and engine duct. This contrast in shape, compared to traditional aircraft design, led to considerable reduction in the radar cross section. Shaping is generally most effective at high frequencies where the energy is reflected in specular directions. Also, diffraction from edges and corners contribute to large radar cross section values. These effects can be reduced by rounding the edges, or sometimes applying a sawtooth design to the edge. Finally, care must be taken when designing low radar cross-section targets as they must still meet other requirements such aerodynamics, weight, structural, and thermal.

7.5.2 Radar Absorbing Material

Radar absorbing material (RAM) coatings have been in use since the 1950s [64]. The basic idea of RAM materials is to reduce the backscattered energy by converting the incident energy into heat. Since RAM materials have relatively high conductivity, a change in the polarization of the reflected wave can occur. RAM coatings should be lightweight, anticorrosive, temperature invariant, wideband, and effective in all directions. While these requirements are difficult to achieve simultaneously, advances have been made in chemical composition and production methods for these materials [63]. RAM is typically applied with robotic sprayers that can very accurately control the thickness of the material. Aerobotix, Inc. provides a range of integrated robotic systems capable of coating fully assembled aircraft [65]. Each system is equipped with precision fluid handling equipment to accurately apply performance coatings to all areas of the aircraft's outer mold line. The robotic systems can be outfitted with noncontact sensors to examine and verify coating thickness. The constitutive parameters of the material must also be tightly controlled to ensure maximum effectiveness.

7.5.3 Passive Cancellation

The main idea of passive cancellation is to place scatterers on the target to reflect the incident energy in such a way as to cancel the reflected energy from a different scatterer on the target. There have also been attempts to add parasitic elements and lumped impedance elements to passively dissipate the incident energy [64]. Passive techniques are limited to narrow frequency bands and small angular sectors. Furthermore, complex targets have many sources of scattering, making passive cancellation of very limited use and effectiveness.

7.5.4 Active Cancellation

In active cancellation techniques, an attempt is made to retransmit the incoming radar waveform to mask the return of the target itself. This requires the use of transmitting and receiving antennas, RF circuitry, and processing equipment. Active cancellation is similar to passive cancellation except that the use of an active circuit to generate the response increases the range of frequencies, angles, and dynamic range over which the technique is effective. For many targets, the structure is not the only source of scattering contributing to the radar cross section: electronics, antennas, and other sensors mounted on the target may have very large contributions to the radar cross section. For antenna systems, active cancellation seeks to adaptively weight the antenna elements to place a null in the transmit and receive antenna patterns in the direction of the radar.

7.5.5 Electronic Countermeasures

While the goal of the methods above is to lower the radar cross section of a target, there are much more advanced countermeasure techniques, including self-screening jammers, stand-off jammers, cross-eye jamming, range gate pull-off, and velocity gate pull-off. The methods in the sections above are also narrowband in nature, and more advanced techniques must be employed for wideband radar waveforms. Chapter 11 provides more details on jamming, and the reader is referred to several good sources on electronic countermeasures for further reading [16, 64, 66].

7.6 EXAMPLES

The sections below illustrate the concepts of this chapter with several Python examples. The examples for this chapter are in the directory *pyradar\Chapter07* and the matching MATLAB examples are in the directory *mlradar\Chapter07*. The reader should consult Chapter 1 for information on how to execute the Python code associated with this book.

7.6.1 Two-Dimensional Strip

As a first example, consider the two-dimensional strip of Figure 7.6. Calculate and display the two-dimensional radar cross section for a strip of width 3 meters at a frequency of 300 MHz with an incident angle of 60°.

Solution: The solution to the above example is given in the Python code *infinite_strip_example.py* and in the MATLAB code *infinite_strip_example.m*. Running the Python example code displays a GUI allowing the user to enter the width of the strip, the incident angle, and the operating frequency. The code then calculates and displays the two-dimensional bistatic radar cross section as a function of observation angle, as shown

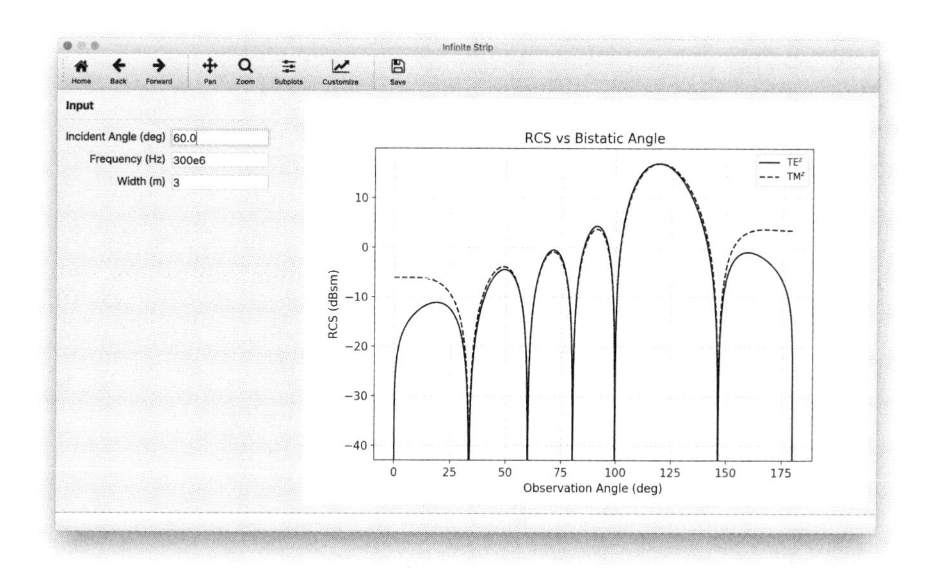

Figure 7.31 The two-dimensional bistatic radar cross section of an infinite strip calculated by *infinite_strip_example.py*.

in Figure 7.31. The user is encouraged to vary the parameters to see how this affects the radar cross section.

7.6.2 Two-Dimensional Cylinder

For this example, consider the two-dimensional circular cylinder of Figure 7.7. Calculate and display the two-dimensional bistatic radar cross section for a cylinder of radius 3 meters at a frequency of 300 MHz.

Solution: The solution to the above example is given in the Python code *infinite_cylinder _example.py* and in the MATLAB code *infinite_cylinder_example.m*. Running the Python example code displays a GUI allowing the user to enter the radius and length of the cylinder and the operating frequency. The code then calculates and displays the two-dimensional bistatic radar cross section as a function of observation angle, as shown in Figure 7.32. The user is encouraged to vary the parameters to see how this affects the radar cross section.

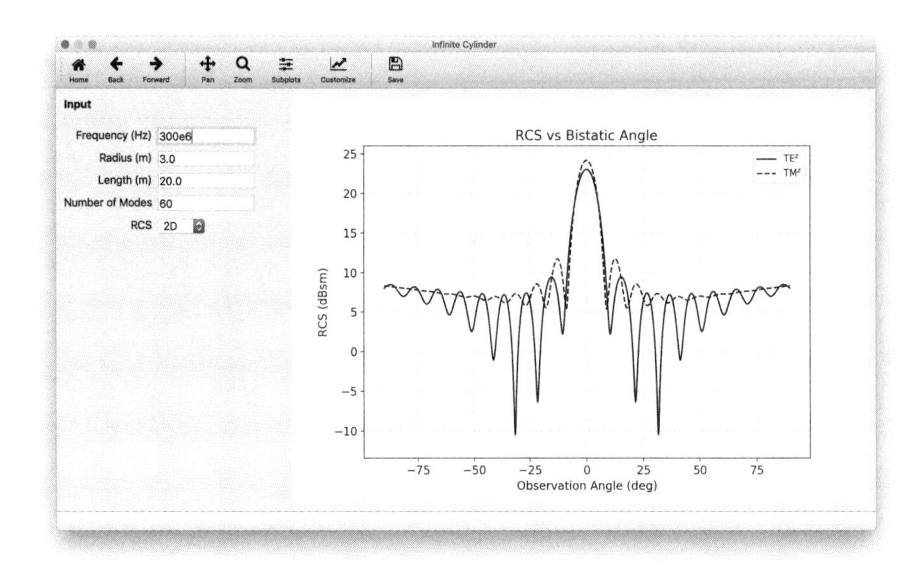

Figure 7.32 The two-dimensional bistatic radar cross section of an infinite cylinder calculated by *infinite_cylinder_example.py*.

7.6.3 Two-Dimensional Cylinder Oblique Incidence

Next, a two-dimensional circular cylinder with oblique incidence is examined, as shown in Figure 7.8. Calculate and display the two-dimensional bistatic radar cross section for a cylinder of radius 3 meters, length 20 meters, a frequency of 300 MHz, and an incident angle of 35°.

Solution: The solution to the above example is given in the Python code *infinite_cylinder_oblique_example.py* and in the MATLAB code *infinite_cylinder_oblique_example.m*. Running the Python example code displays a GUI allowing the user to enter the radius and length of the cylinder, the operating frequency, and the angle of incidence. The code then calculates and displays the two-dimensional bistatic radar cross section as a function of observation angle, as shown in Figure 7.33. The user is encouraged to vary the angle of incidence to see how this affects the radar cross section.

7.6.4 Rectangular Plate

The previous examples studied two-dimensional radar cross section. For this example, the three-dimensional radar cross section of a rectangular plate, as shown in Figure 7.9, is studied. Calculate and display the three-dimensional bistatic radar cross section for a

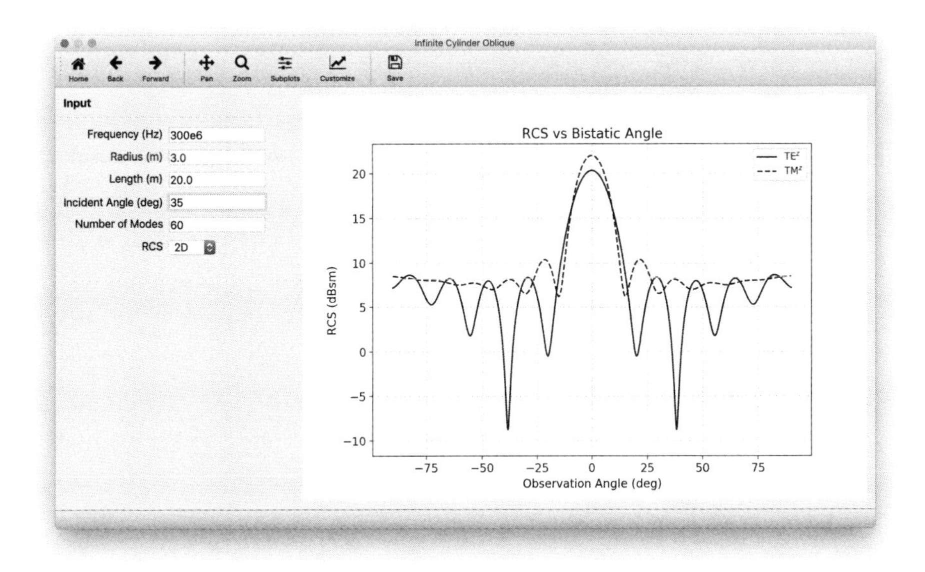

Figure 7.33 The two-dimensional bistatic radar cross section of an infinite cylinder with oblique incidence calculated by *infinite_cylinder_oblique_example.py*.

rectangular plate of width 3 meters and length 5 meters. The frequency is 300 MHz and the incident angle is 15°.

Solution: The solution to the above example is given in the Python code *rectangular_plate_example.py* and in the MATLAB code *rectangular_plate_example.m*. Running the Python example code displays a GUI allowing the user to enter the width and length of the plate, the operating frequency, and the angle of incidence. The code then calculates and displays the three-dimensional bistatic radar cross section as a function of observation angle, as shown in Figure 7.34. The user is encouraged to vary the width, length, frequency, and angle of incidence to see how these affect the radar cross section.

7.6.5 Stratified Sphere

For this example, consider a stratified sphere, as shown in Figure 7.10. Calculate and display the three-dimensional bistatic radar cross section for a stratified sphere with a perfectly conducting core, and a dielectric outer layer with a relative permittivity of 4.

Solution: The solution to the above example is given in the Python code *stratified_sphere_example.py* and in the MATLAB code *stratified_sphere_example.m*. Running

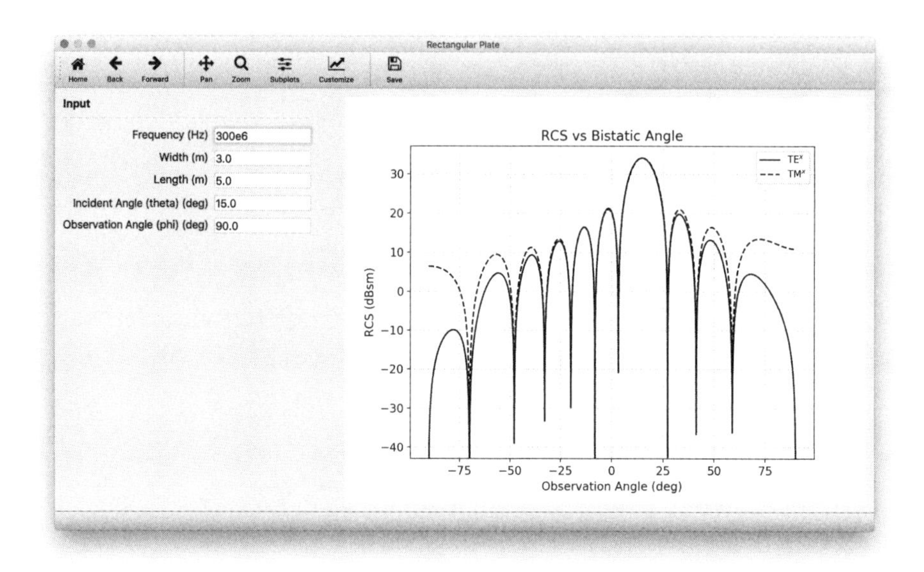

Figure 7.34 The three-dimensional bistatic radar cross section of a rectangular plate calculated by *rectangular_plate_example.py*.

the Python example code displays a GUI allowing the user to enter the relative permittivity, relative permeability, and conductivity of each layer. The GUI also allows the user to enter the frequency and select whether or not the core is conducting. The code then calculates and displays the three-dimensional bistatic radar cross section as a function of observation angle, as shown in Figure 7.35. The user is encouraged to vary the parameters of the sphere and the number of layers to see how these affect the radar cross section.

7.6.6 Circular Cone

Next, consider the right circular cone, as shown in Figure 7.11. Calculate and display the three-dimensional monostatic radar cross section for a right circular cone with a half cone angle of 15°, a base radius of 1.4 meters, at a frequency of 1 GHz.

Solution: The solution to the above example is given in the Python code *right_circular_cone_example.py* and in the MATLAB code *right_circular_cone_example.m*. Running the Python example code displays a GUI allowing the user to enter the frequency, cone half angle, and the base radius. The code then calculates and displays the three-dimensional monostatic radar cross section as a function of observation angle, as shown in Figure 7.36. The user is encouraged to vary the parameters of the cone to see how these affect the radar cross section.

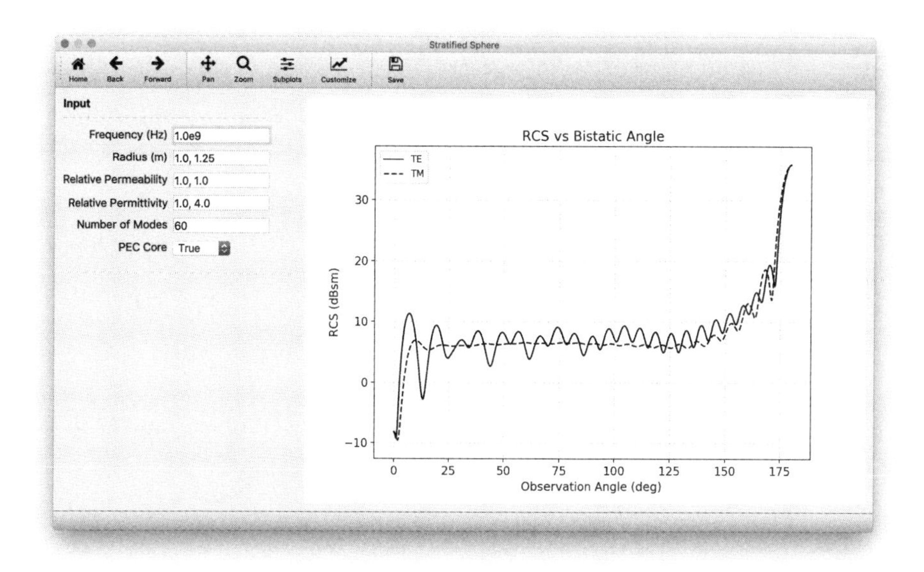

Figure 7.35 The three-dimensional bistatic radar cross section of a stratified sphere calculated by *stratified_sphere_example.py*.

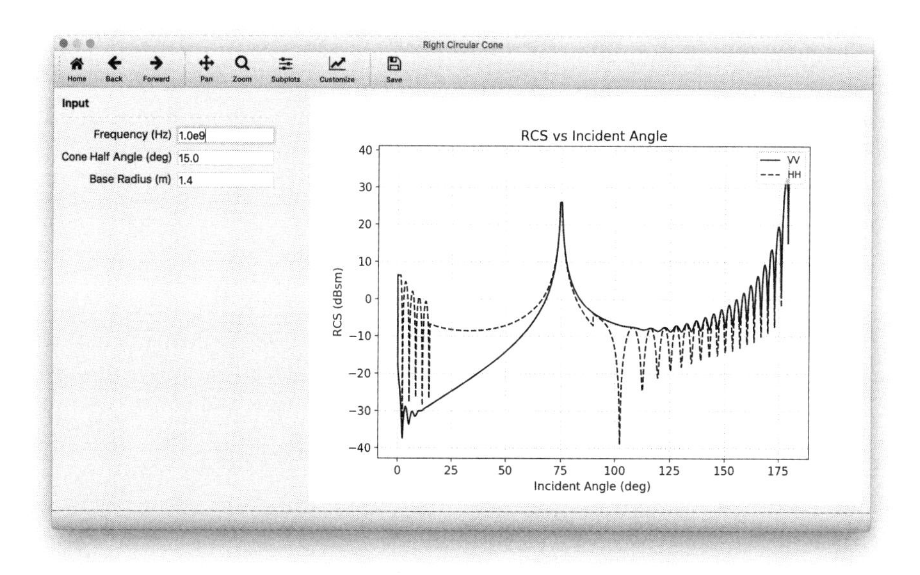

Figure 7.36 The three-dimensional monostatic radar cross section of a circular cone calculated by *right_circular_cone_example.py*.

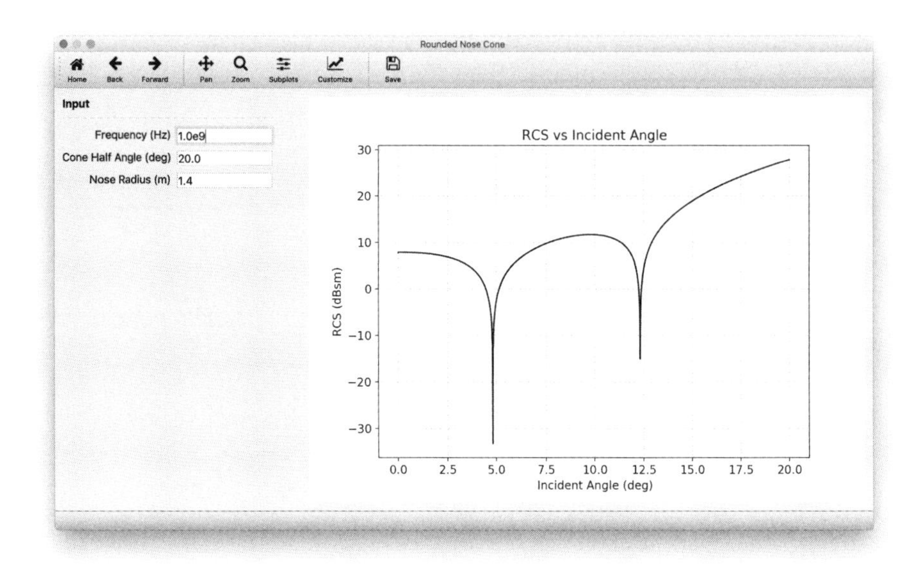

Figure 7.37 The three-dimensional monostatic radar cross section of a rounded nose cone calculated by *rounded_nose_cone_example.py*.

7.6.7 Rounded Nose Cone

For this example, consider the rounded nose cone, as shown in Figure 7.12. Calculate and display the three-dimensional monostatic radar cross section for a rounded nose cone with a half cone angle of 20°, a nose radius of 1.4 meters, at a frequency of 1 GHz.

Solution: The solution to the above example is given in the Python code *rounded_nose_cone_example.py* and in the MATLAB code *rounded_nose_cone_example.m*. Running the Python example code displays a GUI allowing the user to enter the frequency, cone half angle, and the nose radius. The code then calculates and displays the three-dimensional monostatic radar cross section as a function of observation angle, as shown in Figure 7.37. The user is encouraged to vary the parameters of the cone to see how these affect the radar cross section.

7.6.8 Frustum

In this next example, consider the frustum, as shown in Figure 7.13. Calculate and display the three-dimensional monostatic radar cross section for a frustum with a base radius of 0.2 meters, a nose radius of 0.1 meters, a length of 0.8 meters, at a frequency of 1 GHz.

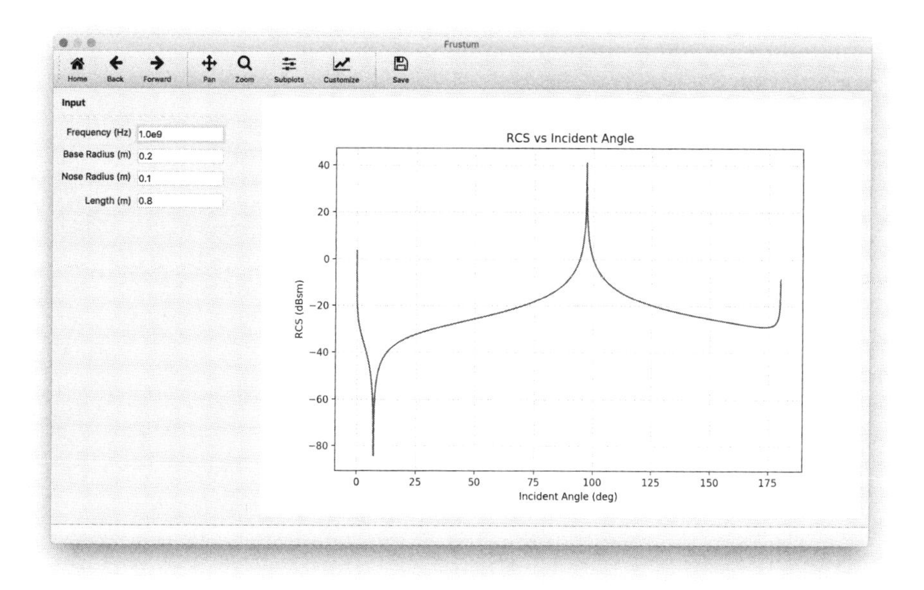

Figure 7.38 The three-dimensional monostatic radar cross section of a frustrum calculated by *frustum_example.py*.

Solution: The solution to the above example is given in the Python code *frustum_example .py* and in the MATLAB code *frustum_example.m*. Running the Python example code displays a GUI allowing the user to enter the frequency, base radius, nose radius, and length. The code then calculates and displays the three-dimensional monostatic radar cross section as a function of observation angle, as shown in Figure 7.38. As seen in the figure, this is a simple approximation and overestimates the scattering in the specular region. The user is encouraged to vary the parameters of the frustum to see how these affect the radar cross section.

7.6.9 Physical Optics

In this section, the physical optics scattering for a few targets is covered; namely, the rectangular plate, right circular cone, frustum, and double ogive. The code for this section depends on the target surfaces being discretized with triangular facets. The scattering for each facet is calculated and coherently added to the total scattering.

Solution: The solution to the examples in this section is given in the Python code *po_example.py* and in the MATLAB code *po_example.m*. Running the Python example code displays a GUI allowing the user to select the desired target and enter the frequency,

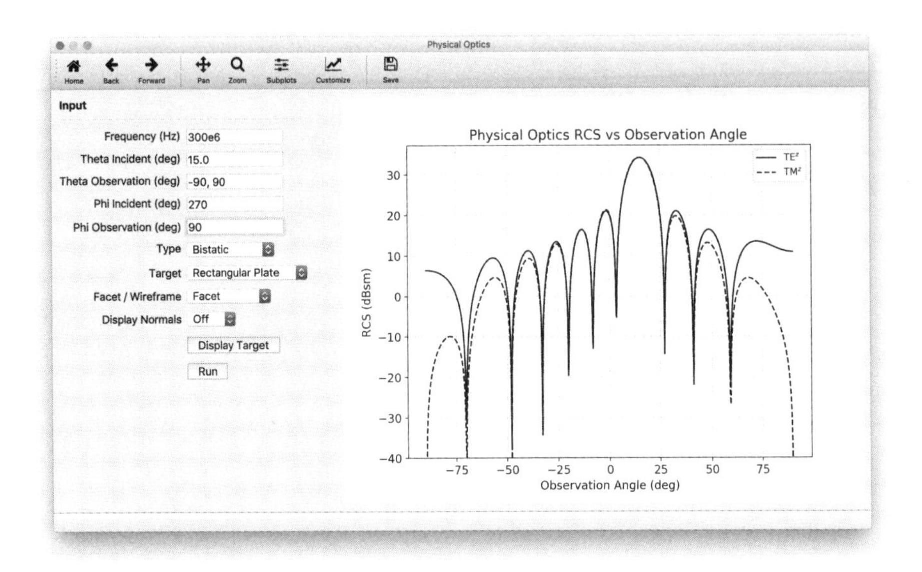

Figure 7.39 The three-dimensional bistatic radar cross section of a rectangular plate calculated by *po_example.py*.

incident angles, and observation angles. The GUI allows the user to select either monostatic or bistatic calculations. The GUI also displays the target facet model or wireframe model with or without surface normals.

Begin with the rectangular plate, as given in the example of Section 7.6.4. The solution using *po_example.py* is shown in Figure 7.39. This result matches with the result shown in Figure 7.34, which is expected since the solution given in Section 7.4.1.4 is based on physical optics.

Next, the right circular cone, as given in the example of Section 7.6.6, is examined. The solution using *po_example.py* is shown in Figure 7.40. Comparing this result to Figure 7.36 shows some differences, especially in the nose region. The intensity of the specular and base regions matches reasonably well.

The next physical optics example is the frustum, as given in the example of Section 7.6.8. The solution using *po_example.py* is shown in Figure 7.41. This result shows some large differences from the result in 7.38. The nose and base scattering matches well, but the specular region is overestimated in the approximation of (7.61). This also shows (7.60) is a simple approximation of the radar cross section as there is little structure to the scattering.

The final physical optics example is the double ogive, as defined by the Electromagnetic Code Consortium (EMCC). One function of the EMCC is the generation and

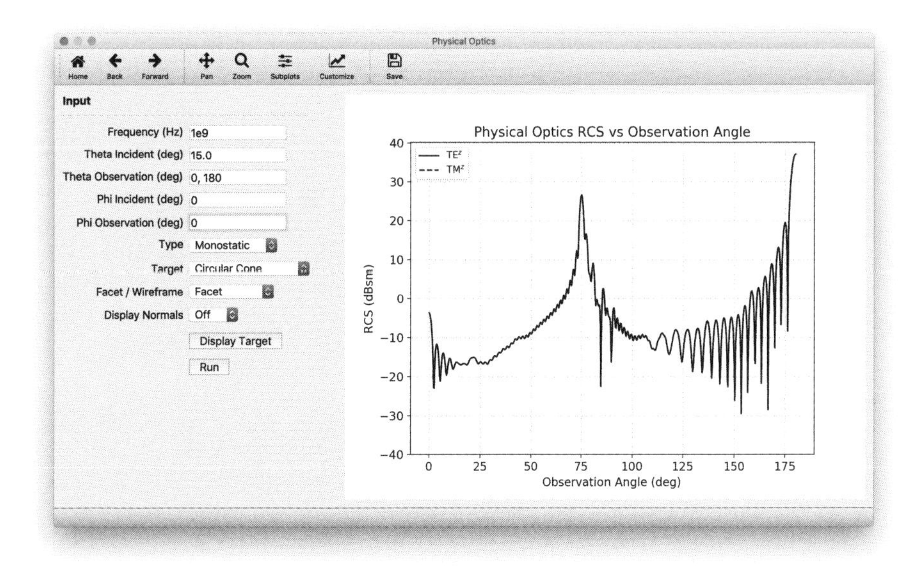

Figure 7.40 The three-dimensional monostatic radar cross section of a circular cone calculated by *po_example.py*.

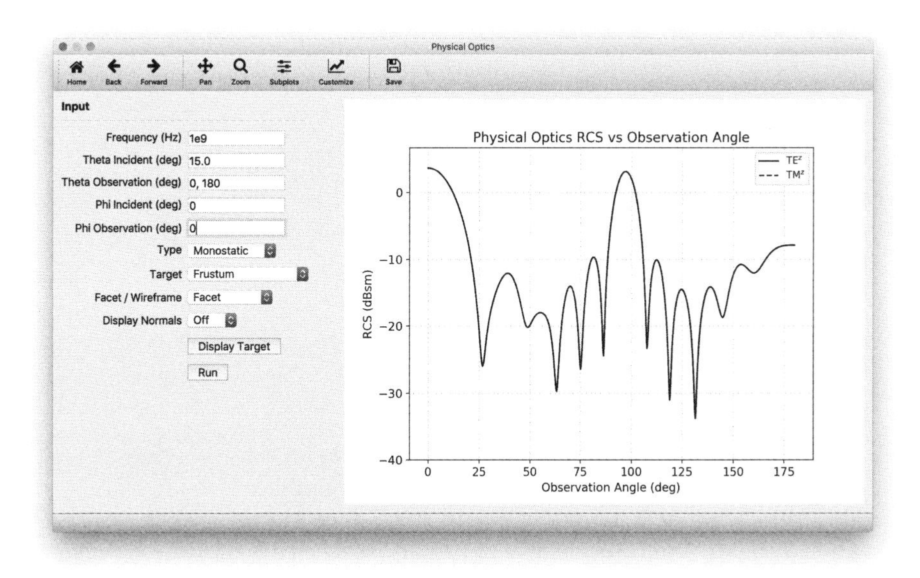

Figure 7.41 The three-dimensional monostatic radar cross section of a frustum calculated by *po_example.py*.

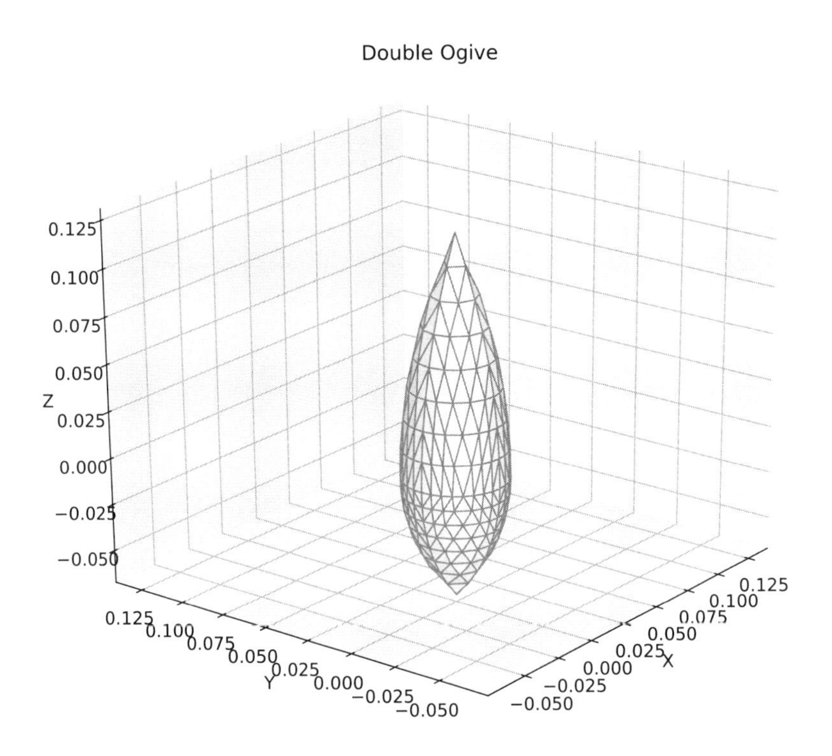

Figure 7.42 The three-dimensional monostatic radar cross section of a double ogive calculated by *po_example.py*.

management of electromagnetics code benchmark geometries and data along with associated measurement efforts [67]. The geometry for the double ogive is displayed with *po_example.py* and is shown in Figure 7.42. For details of the geometry, the reader is referred to [52, 67]. The monostatic radar cross section for the double ogive is given in Figure 7.43.

7.6.10 Finite Difference Time Domain Method

As the final Python example, consider the scattering by a two-dimensional dielectric rectangular cylinder calculated by the finite difference time domain method. Calculate and display the total electric field in the presence of a two-dimensional rectangular cylinder of dimension 2 mm by 2 mm. The number of PML layers is 10, the incident angle is 90°, and the pulse width is 10 time steps and has an amplitude of 1. The relative permittivity of the cylinder is 4.

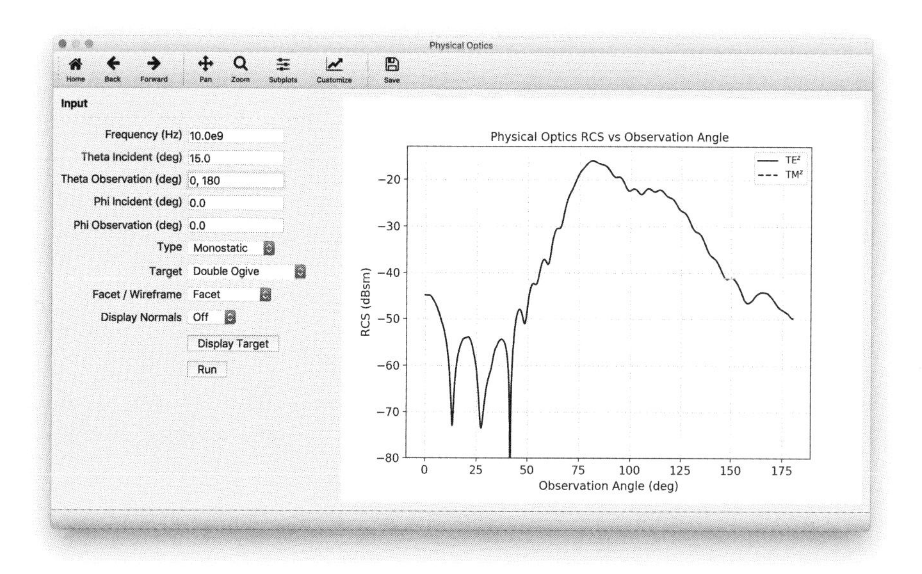

Figure 7.43 The three-dimensional monostatic radar cross section of a double ogive calculated by *po_example.py*.

Solution: The solution to the examples in this section is given in the Python code *fdtd_example.py* and in the MATLAB code *fdtd_example.m*. Running the Python example code displays a GUI allowing the user to select either TE or TM mode, the incident angle, the number of time steps, the Gaussian pulse width and amplitude, and the number of PML layers. The solution using *fdtd_example.py* is shown in Figures 7.44, 7.45, 7.46, and 7.47. In these figures, the wave can be seen propagating through the computational domain. The interaction of the incident wave with the rectangular dielectric cylinder is also demonstrated. Figure 7.44 shows the incident wave before it reaches the rectangular dielectric cylinder. Figure 7.45 illustrates the wave initially interacting with the cylinder. Figure 7.46 shows the fields as the incident wave has just passed the cylinder. Finally, Figure 7.47 is after the wave has completely passed through the computational domain.

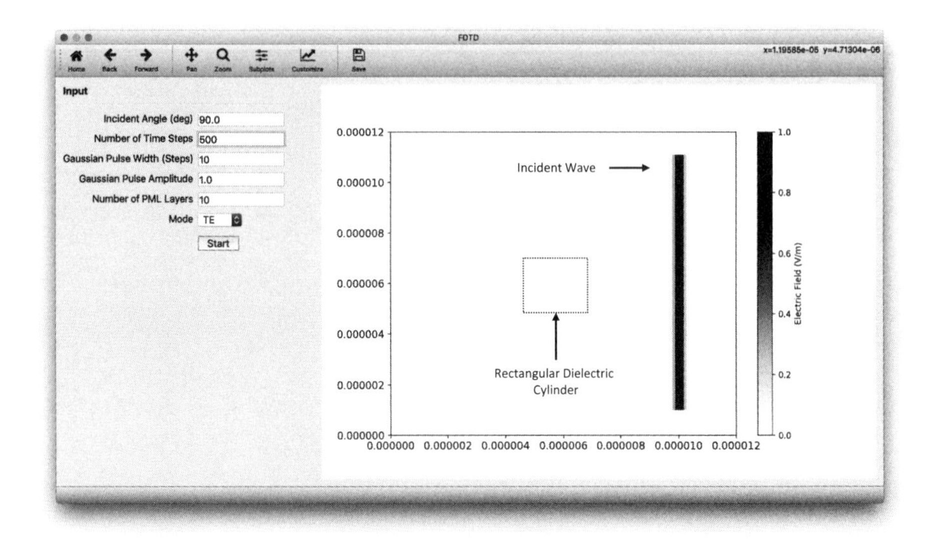

Figure 7.44 The two-dimensional scattering from a rectangular dielectric cylinder after 50 time steps calculated by *fdtd_example.py*.

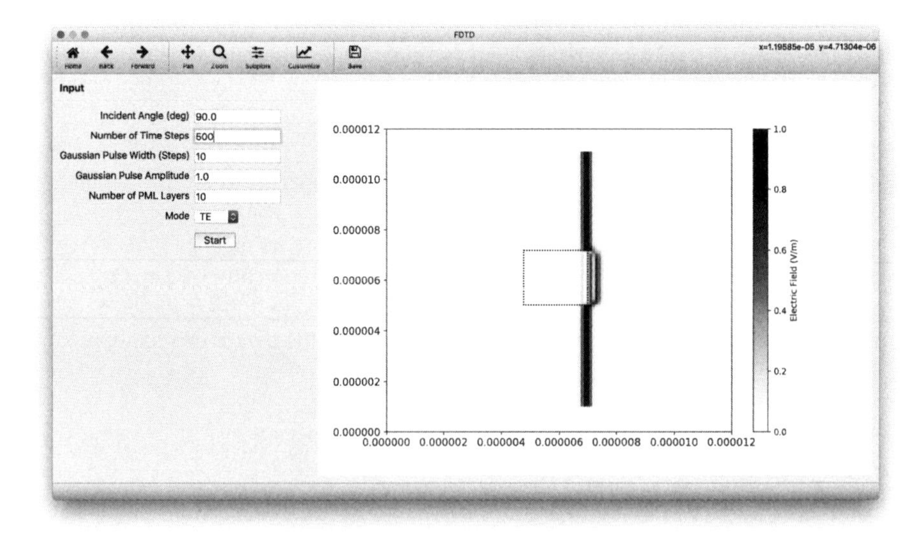

Figure 7.45 The two-dimensional scattering from a rectangular dielectric cylinder after 80 time steps calculated by *fdtd_example.py*.

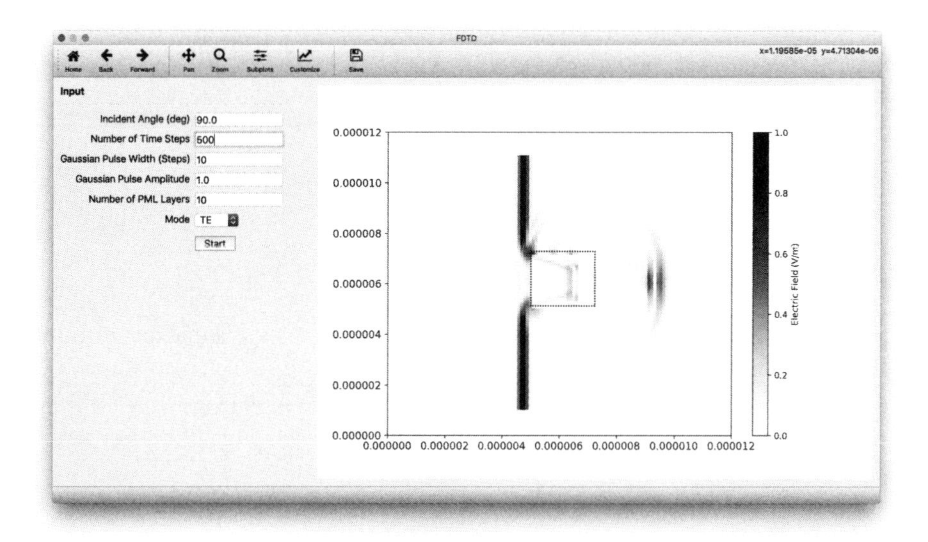

Figure 7.46 The two-dimensional scattering from a rectangular dielectric cylinder after 120 time steps calculated by *fdtd_example.py*.

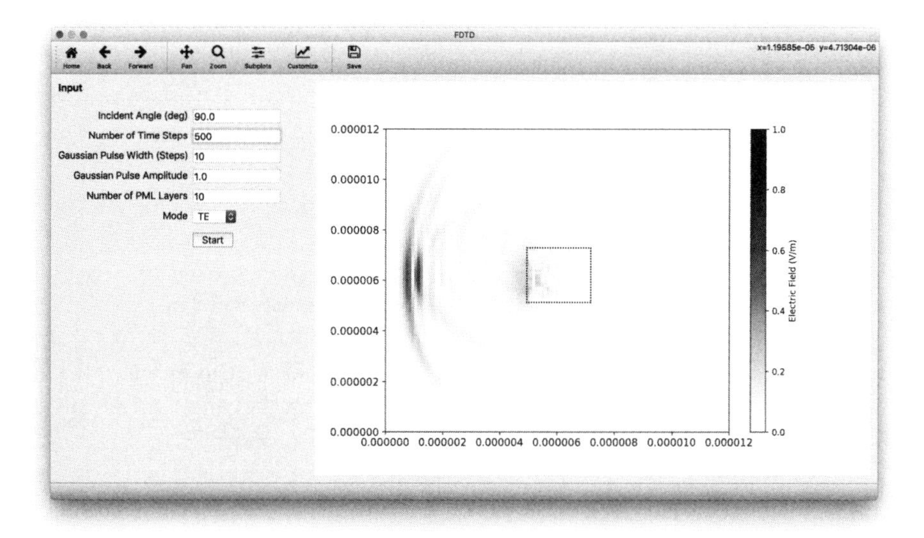

Figure 7.47 The two-dimensional scattering from a rectangular dielectric cylinder after 175 time steps calculated by *fdtd_example.py*.

PROBLEMS

7.1 For an incident electric field given by $\mathbf{E}^i = E_0\, e^{-jk\rho\cos\phi}\;\hat{\mathbf{z}}$ (V/m) and a scattered field given as

$$\mathbf{E}^s = -E_0 \sum_{n=-\infty}^{\infty} j^{-n} \frac{J_n(ka)}{H_n^{(2)}(ka)}\, H_n^{(2)}(k\rho)\, e^{jn\phi}\;\hat{\mathbf{z}} \;\text{(V/m)},$$

calculate the two-dimensional radar cross section given in (7.12).

7.2 Describe how the radar cross section of a rectangular plate varies as a function of aspect angle, frequency, and polarization.

7.3 Explain why the scattering matrix for monostatic radars is reciprocal for most targets and give an example when this is not true.

7.4 Convert the scattering matrix below from linear polarization to circular polarization.

$$\begin{bmatrix} S_{HH} & S_{HV} \\ S_{VH} & S_{VV} \end{bmatrix} = \begin{bmatrix} 30.0 + j10.0 & 0.8 - j2.0 \\ 0.8 - j2.0 & 23.0 + j17.0 \end{bmatrix}$$

7.5 Describe the difference between the forward-scattering alignment and back-scattering alignment conventions for incident and scattered waves.

7.6 Explain the main scattering mechanisms for a rectangular plate and a sphere.

7.7 What is the major drawback of analytic techniques for radar cross-section prediction?

7.8 Using the expressions in Table 7.1, compare the radar cross section of a circular reflector of radius 0.7 m and a corner reflector of dimension 0.2×0.8 m.

7.9 Compare the bistatic radar cross section of a two-dimensional strip of width 0.75m with a rectangular plate of dimension 0.6×1.2 m at a frequency of 12 GHz with an incident angle of 35°.

7.10 Calculate the bistatic radar cross section of a two-dimensional cylinder of radius 1.1 m at a frequency of 8 GHz. Compare the result to a cylinder of length 25 m with an oblique incidence angle of 15°.

7.11 For a half cone angle of 18°, base radius of 1.2 m, and frequency of 3 GHz, compare the radar cross section of right circular cone with a rounded-nose cone with a nose radius of 0.2 m.

7.12 Explain the difference between exact or low-frequency numerical methods and approximate or high frequency methods. Give examples of situations where it would be advantageous to use an approximate method rather than an exact method.

7.13 Describe the difference between tapered chambers and compact ranges and give examples of practical targets that would be suitable for measurement in each.

7.14 Calculate the power at the receiver in a compact range for a transmit power of 2.5W, feed horn gain of 0 dB, and frequency of 1 GHz. The target radar cross section is −3 dB, and the range from the feed horn to reflector is 10 m.

7.15 What are the major disadvantages of scale model radar cross section measurements?

7.16 Describe methods for reducing radar cross-section measurement errors in compact ranges.

7.17 Describe methods for radar cross-section reduction and give practical examples of each.

References

[1] M. Born and E. Wolf. *Principles of Optics: Electromagnetic Theory of Propagation, Interference and Diffraction of Light,* 7th ed. Cambridge University Press, Cambridge, UK, 1999.

[2] V. Twersky. Rayleigh scattering. *Applied Optics*, 3:1150–1162, October 1964.

[3] R. E. Kleinman. The Rayleigh region. *Proceedings of the IEEE*, pages 848–856, August 1965.

[4] P. Beckman. Depolarization of electromagnetic waves backscattered from the lunar surface. *Journal of Geophysical Research*, 73(2):649–655, January 1968.

[5] J. J. Mikulski. The scattering matrix, polarization, power and periodic bodies from the viewpoint of matrix theory. Technical Report 38, Lincoln Laboratory-MIT, March 1960.

[6] G. Bao and J. Lai. Radar cross section reduction of a cavity in the ground plane. *Global Science Preprint*, 8(3), June 2015.

[7] H. Ling, S. Lee and R. Chou. High-frequency RCS of open cavities with rectangular and circular cross sections. *IEEE Transactions on Antennas and Propagation*, 37:648–654, May 1989.

[8] H. Ammari, G. Baoand and A. W. Wood. Analysis of the electromagnetic scattering from a cavity. *Japan Journal of Industrial Applied Mathematics*, 19:301–310, 2002.

[9] G. T. Ruck, et al. *Radar Cross Section Handbook.* Plenum Press, New York, 1970.

[10] C. A. Balanis. *Advanced Engineering Electromagnetics,* 2nd ed. John Wiley and Sons, New York, 2012.

[11] M. Abramowitz and I. A. Stegun. *Handbook of Mathematical Functions: with Formulas, Graphs, and Mathematical Tables.* U.S. National Bureau of Standards, Applied Mathematics Series, Washington, DC, 1964.

[12] E. Jones, et al. SciPy: Open source scientific tools for Python, 2001–. `http://www.scipy.org/`.

[13] G. Mie. Beitrage zur optik truber medien, speziell kolloidaler metallosungen. *Annalen der Physik*, 25(330):377–445, 1908.

[14] J. B. Keller. Geometrical theory of diffraction. *Journal of the Optical Society of America*, 52:116–130, February 1962.

[15] M. E. Bechtel. Application of geometric diffraction theory to scattering from cones and disks. *Proceedings of the IEEE*, pages 877–882, August 1965.

[16] B. R. Mahafza. *Radar Systems Analysis and Design Using MATLAB,* 3rd ed. CRC Press, Boca Raton, FL, 2013.

[17] J. Song, C. C. Lu and W. C. Chew. Multilevel fast multipole algorithm for electromagnetic scattering by large complex objects. *IEEE Transactions on Antennas and Propagation*, 45(10):1488–1493, 1997.

[18] K. Zhao, M. N. Vouvakis and J. F. Lee. The adaptive cross approximation algorithm for accelerated method of moments computations of EMC problems. *IEEE Transactions on Antennas and Propagation*, 47(4):763–773, 2005.

[19] K. S. Kunz and R. J. Luebbers. *The Finite Difference Time Domain Method for Electromagnetics.* CRC Press, Boca Raton, FL, 1993.

[20] A. Taflove and S. C. Hagness. *Computational Electrodynamics: The Finite-Difference Time-Domain Method,* 3rd ed. Artech House, Norwood, MA, 2005.

[21] A. Z. Elsherbeni and V. Demir. *The Finite Difference Time Domain Method for Electromagnetics with MATLAB,* 2nd ed. SciTech Publishing, Edison, NJ, 2015.

[22] X. Chen, et al. An efficient implementation of parallel FDTD. *2007 IEEE International Symposium on Electromagnetic Compatibility*, 4:1–5, 2007.

[23] K. Yee. Numerical solution of initial boundary value problems involving Maxwell's equations in isotropic media. *IEEE Transactions on Antennas and Propagation*, 14(3):302–307, 1966.

[24] S. Benkler, N. Chavannes and N. Kuster. A new 3-D conformal PEC FDTD scheme with user defined geometric precision and derived stability criterion. *IEEE Transactions on Antennas and Propagation*, 54(6):1843–1849, 2006.

[25] F. Zhen, Z. Chen and J. Zhang. Toward the development of a three-dimensional unconditionally stable finite-difference time-domain method. *IEEE Transactions on Microwave Theory and Techniques*, 48(9):1550–1558, 2000.

[26] F. Zhen and Z. Chen. Numerical dispersion analysis of the unconditionally stable 3-D ADI-FDTD method. *IEEE Transactions on Microwave Theory and Techniques*, 49(5):1006–1009, 2001.

[27] I. Ahmed, et al. Development of the three-dimensional unconditionally stable LOD-FDTD method. *IEEE Transactions on Antennas and Propagation*, 56(11):3596–3600, 2008.

[28] J. A. Roden and S. D. Gedney. Convolution PML (CPML): An efficient FDTD implementation of the CFS-PML for arbitrary media. *Microwave and Optical Technology Letters*, 27(5):334–339, 2000.

[29] D. S. Katz, E. T. Thiele and A. Taflove. Validation and extension to three dimensions of the Berenger PML absorbing boundary condition for FDTD meshes. *IEEE Microwave and Guided Wave Letters*, 4(8):268–270, 1994.

[30] C. E. Reuter, et al. Ultrawideband absorbing boundary condition for termination of waveguiding structures in FDTD simulations. *IEEE Microwave and Guided Wave Letters*, 4(10):344–346, 1994.

[31] J. P. Berenger. A perfectly matched layer for the absorption of electromagnetic waves. *Journal of Computational Physics*, 114(2):185–200, 1994.

[32] A. Hrennikoff. Solution of problems of elasticity by the framework method. *Journal of Applied Mechanics*, 8(4):169–175, 1941.

[33] R. Courant. Variational methods for the solution of problems of equilibrium and vibrations. *Bulletin of the American Mathematical Society*, 49:1–23, 1943.

[34] NASA Langley Research Center. *NASTRAN - NASA Structural Analysis*. `http://software.nasa.gov/software/LAR-16804-GS`.

[35] R. D. Cook. *Concepts and Applications of Finite Element Analysis*. Wiley, New York, 1981.

[36] S. S. Rao. *The Finite Element Method in Engineering*. Pergamon Press, Oxford, 1982.

[37] J. Jin. *The Finite Element Method in Electromagnetics,* 2nd ed. Wiley, New York, 2002.

[38] C. Geuzaine and J. F. Remacle. *Gmsh: A three-dimensional finite element mesh generator with built-in pre- and post-processing facilities*. `http://gmsh.info`.

[39] Robert McNeel and Associates. *Rhinoceros*. `https://www.rhino3d.com`.

[40] PTC. *Pro/Engineer*. `https://www.ptc.com/en/products/cad/pro-engineer`.

[41] L. A. Harrison, et al. Error reduction using Richardson extrapolation in the finite element solution of partial differential equations using wavelet-like basis functions. *USNC/URSI National Radio Science Meeting 1998 Digest*, page 202, 1998.

[42] L. A. Harrison, et al. The use of Richardson extrapolation in the finite element solution of partial differential equations using wavelet-like basis functions. *Proceedings of the 30th Southeastern Symposium on System Theory*, pages 98–101, 1998.

[43] L. A. Harrison, et al. The numerical solution of elliptic problems using compactly supported wavelets. *Proceedings of the 30th Southeastern Symposium on System Theory*, pages 102–106, 1998.

[44] L. A. Harrison and R. K. Gordon. The use of wavelet-like basis functions in the finite element analysis of heterogeneous one-dimensional regions. *Proceedings of the IEEE Southeastcon 1996*, pages 301–304, 1996.

[45] L. A. Harrison and R. K. Gordon. Finite element analysis of boundary value problems using wavelet-like basis functions. *Proceedings of the 28th Southeastern Symposium on System Theory*, pages 103–107, 1996.

[46] L. A. Harrison and R. K. Gordon. Investigation of the properties of wavelet-like basis functions in the finite element analysis of elliptic problems. *Proceedings of The Twelfth Annual Review of Progress in Applied Computational Electromagnetics*, pages 375–382, 1996.

[47] K. K. Mei and J. Van Bladel. Scattering by perfectly conducting rectangular cylinders. *IEEE Transactions on Antennas and Propagation*, 11:185–192, March 1963.

[48] M. G. Andreasen. Scattering from parallel metallic cylinders with arbitrary cross section. *IEEE Transactions on Antennas and Propagation*, 12:746–754, November 1964.

[49] F. K. Oshiro. Source distribution techniques for the solution of general electromagnetic scattering problems. *Proceedings of the First GISAT Symposium, Mitre Corporation*, 1:83–107, 1965.

[50] R. F. Harrington. *Field Computation by Moment Methods.* Macmillan, New York, 1968.

[51] S. M. Rao, D. R. Wilton and A. W. Glisson. Electromagnetic scattering by surfaces of arbitrary shape. *IEEE Transactions on Antennas and Propagation*, 30(3):409–418, 1982.

[52] W. C. Gibson. *The Method of Moments in Electromagnetics,* 2nd ed. CRC Press, Boca Raton, FL, 2015.

[53] J. D. Jackson. *Classical Electrodynamics,* 3rd ed. John Wiley and Sons, New York, 1998.

[54] W. Panofsky and M. Phillips. *Classical Electricity and Magnetism,* 2nd ed. Addison-Wesley, Reading, MA, 1962.

[55] L. Greengard and V. Rohklin. A fast algorithm for particle simulations. *Journal of Computational Physics*, 73:325–348, 1987.

[56] M. Bebendorf. Approximation of boundary element matrices. *Numerische Mathematik*, 86(4):565–589, 2000.

[57] W. L. Stutzman and G. A. Thiele. *Antenna Theory and Design,* 3rd ed. Wiley, New Jersey, 2013.

[58] R. G. Kouyoumjian and P. H. Pathak. A UTD of diffraction for an edge in a perfectly conducting surface. *Proceedings of the IEEE*, 62(11):1448–1461, 1974.

[59] T. Eibert and V. Hansen. Calculation of unbounded field problems in free space by a 3D FEM/BEM-hybrid approach. *Journal of Electromagnetic Waves and Applications*, 10(1):61–78, 1996.

[60] J. Guan, S. Yan and J. M. Jin. An accurate and efficient finite element-boundary integral method with GPU acceleration for 3-D electromagnetic analysis. *IEEE Transactions on Antennas and Propagation*, 62(12):6325–6336, 2014.

[61] M. J. Coulombe, et al. A 585 GHz compact range for scale model RCS measurements. *Antenna Measurements and Techniques Association Proceedings*, 1993.

[62] D. W. Hess. Introduction to RCS measurements. *Antennas and Propagation Conference*, pages 37–44, 2008.

[63] Cumming Microwave Corporation. *Cumming Microwave*, 2018. `https://www.cumingmicrowave.com/`.

[64] H. Singh and R. M. Jha. *Active Radar Cross Section Reduction Theory and Applications*. Cambridge University Press, India, 2015.

[65] Aerobotix, Inc. *Robotic Coating and Paint Application Systems*. https://www.aerobotix.net.

[66] J. C. Toomay and P. J. Hannen. *Radar Principles for the Non-Specialist,* 3rd ed. SciTech Publishing, Rayleigh, NC, 2004.

[67] U.S. Air Force Virtual Distributed Laboratory. *Electromagnetic Code Consortium*. https://www.vdl.afrl.af.mil/emcc/.

Chapter 8

Pulse Compression

In order to detect closely spaced targets, a radar system must be capable of transmitting large average power while simultaneously producing good range resolution. Pulse compression techniques achieve this by transmitting waveforms with long pulsewidths, which are then processed to produce effective pulsewidths much less than the transmitted pulsewidths. This chapter begins with an overview of range resolution and simple waveforms. Next, the stepped frequency waveform, which is an interpulse compression technique, is studied. Then, the matched filter and stretch processor for intrapulse compression are examined. The ambiguity function and its associated properties are presented. This is followed by the derivation of the ambiguity function for various waveforms, including pulse trains and phase coded waveforms. The chapter concludes with several Python examples to reinforce the mechanism of pulse compression as well as the ambiguity function.

8.1 RANGE RESOLUTION

The ability of a radar system to resolve targets in the range dimension is a key feature of the system. As targets become more closely spaced, finer range resolution is required. This is achieved by decreasing the effective pulsewidth of the waveforms used by the radar. Consider the situation in Figure 8.1, where there are two targets, T_1 and T_2, located at ranges r_1 and r_2, respectively. The difference in range is $\Delta r = r_2 - r_1$. The range to the target is found from the round-trip time as

$$r = \frac{ct}{2} \qquad \text{(m)}, \tag{8.1}$$

where c is the speed of light, and t is the two-way time delay from the radar to the target and back to the radar. For the scenario in Figure 8.1, the difference in range is written as

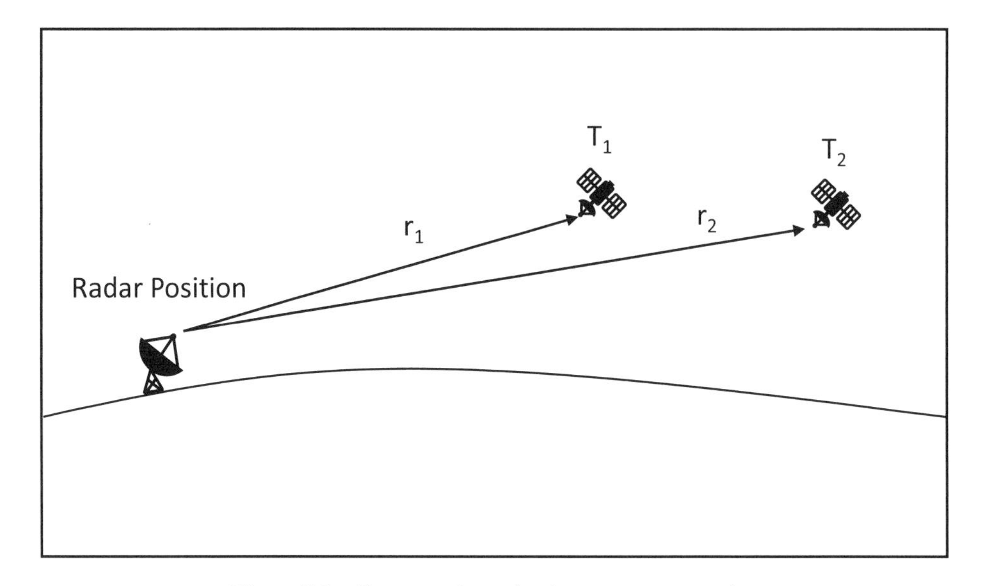

Figure 8.1 Geometry for a simple two target scenario.

$$\Delta r = \frac{c\,(t_2 - t_1)}{2} = \frac{c\,\Delta t}{2} \qquad \text{(m).} \tag{8.2}$$

The pulsewidth required to achieve this resolution is then, $\tau = \Delta t$. Therefore, to have high range resolution requires a narrow pulsewidth. However, the consequence of a narrow pulsewidth is low average power, which degrades radar performance. Ideally, a system would simultaneously have large average power and high range resolution. Pulse compression techniques produce an effective pulsewidth much less than the transmitted pulsewidth by increasing the bandwidth of the waveform. This smaller effective pulsewidth provides an increase in the range resolution. This improves the detection of closely spaced targets as well as enhancing target detection in clutter environments. Various techniques for increasing the bandwidth of the radar waveform are available and may be classified into four groups. These groups are ultrawideband (UWB), super-resolution, intrapulse compression, and interpulse compression. This chapter focuses on intrapulse and interpulse techniques.

Ultrawideband techniques achieve range resolution by transmitting extremely narrow pulses. This requires sophisticated transmitting and receiving hardware that have very large instantaneous bandwidths. Also, UWB systems transmit a low average power, resulting in lower signal-to-noise ratio and probability of detection, as shown in Chapter 6. Super-resolution techniques make use of the covariance matrix of the received signal to perform optimal estimation using an eigenstructure method [1]. Intrapulse compression methods increase the instantaneous bandwidth of the waveform through frequency and/or

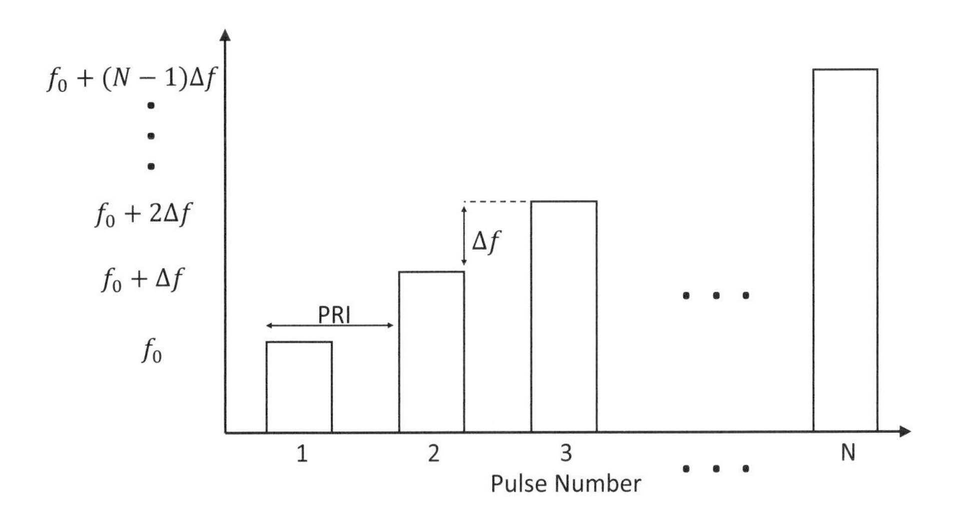

Figure 8.2 Stepped frequency waveform.

phase coding. The coded waveform is then passed through a matched filter to produce the compressed signal with the effective pulsewidth. Interpulse compression techniques vary the carrier frequency of successively transmitted pulses to synthetically increase the instantaneous bandwidth. The effective bandwidth is achieved by coherently integrating a number of pulses with differing center frequencies.

8.2 STEPPED FREQUENCY WAVEFORMS

The stepped frequency technique is an example of interpulse compression. For stepped frequency waveforms, each successive pulse is transmitted at a different center frequency, as shown in Figure 8.2. The frequency of the *i*th pulse is given by

$$f_i = f_0 + (i-1)\,\Delta f \qquad \text{(Hz)}, \tag{8.3}$$

where f_0 is the carrier frequency, and Δf is the frequency step size between successive pulses. The instantaneous bandwidth of each transmitted pulse is approximately $1/\tau$ Hz. However, the effective bandwidth is much larger, and is written as

$$B_{\mathit{effective}} = N\Delta f \qquad \text{(Hz)}, \tag{8.4}$$

where N is the number of pulses. The spectrum of the stepped frequency waveform is illustrated in Figure 8.3.

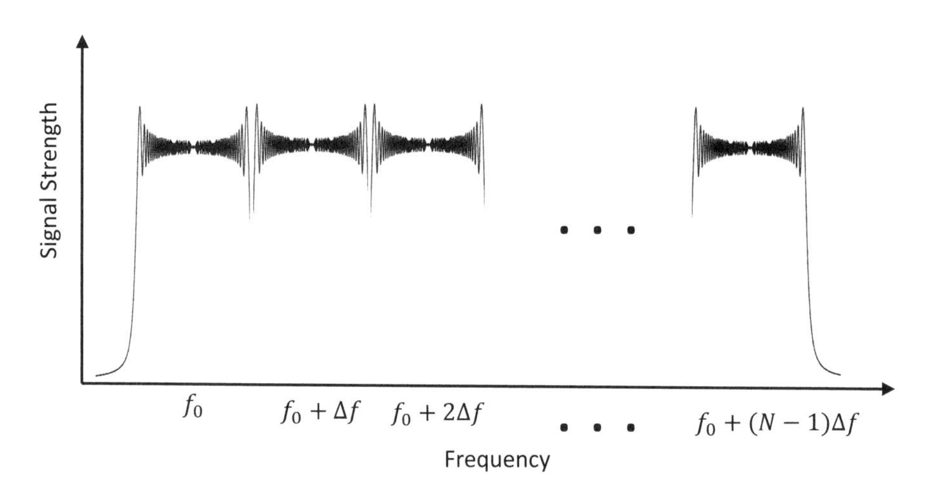

Figure 8.3 Spectrum of the stepped frequency waveform.

The range resolution for the stepped frequency waveform is then

$$\Delta r = \left(\frac{c}{2N\Delta f}\right) \qquad \text{(m)}. \tag{8.5}$$

The return signal is composed of the in-phase and quadrature components and may be expressed as $s_i(t) = A_i(t)\, e^{j\phi_i(t)}$, where the phase is $\phi_i(t) = 2\pi f_i\, t_i$ radians. The time delay to a given target is $t_i = \dfrac{2\, r_i}{c}$ seconds, where $r_i = r_0 + (i-1)\, v\, T$ meters, and

r_0 = initial range of the target (m),
v = velocity of the target in the radial direction (m/s),
T = pulse repetition interval (s).

The overall baseband return signal then becomes

$$s_i(t) = A_i(t)\, \exp\left[-j\frac{4\pi}{c}\,(f_0 + (i-1)\Delta f)\,(r_0 + (i-1)\, v\, T)\right]. \tag{8.6}$$

The expression in (8.6) is nonlinear in phase. This nonlinearity causes range-Doppler coupling [2], which impacts the range resolution and integration gain, as shown in the examples in Section 8.8. The expression in (8.6) is converted to a time domain reflectivity function, often termed *range profile*, by use of the inverse Fourier transform. The output of the stepped frequency processing is implemented in the code shown in Listing 8.1. The target is located at a range of 2 km and has a velocity of 0 m/s, and the output of this

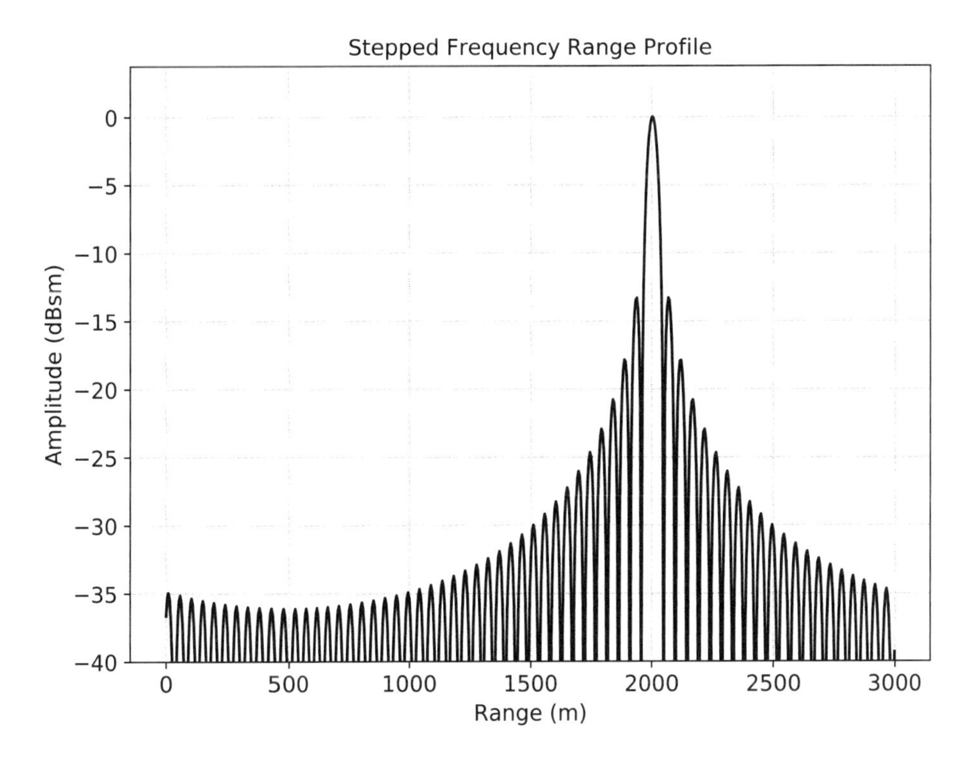

Figure 8.4 Stepped frequency output for Listing 8.1.

code is given in Figure 8.4. As seen in the figure, the peak of the output occurs at $r = 2$ km, corresponding to the target location. The inverse Fourier transform acts as a matched filter for stepped frequency waveforms. The phase shift due to radial velocity causes a mismatch in the inverse Fourier transform, which lowers the peak scatterer level, reduces the range resolution, and shifts the peak of the scatterer. This mismatch is often referred to as dispersion, and the amount of dispersion is given as

$$D = \frac{v\,N\,T}{\Delta r}. \tag{8.7}$$

For practical applications, D should be kept below 3 [3]. The number of bins the peak of the scatterer is shifted is given by

$$L = f_d\,N\,T, \tag{8.8}$$

where f_d is the Doppler frequency. This may also be written as

Listing 8.1 Calculation of the Stepped Frequency Output

```
from scipy import exp, linspace, zeros
from scipy.fftpack import ifft, next_fast_len
from scipy.constants import c, pi
from matplotlib import pyplot as plt

# Set the stepped frequency waveform parameters
number_of_pulses = 64
frequency_step = 50e3 # Hz
prf = 100 # Hz

# Set the target parameters
target_range = [2e3] # meters
target_velocity = [0] # m/s

# Initialize the return signal
s = zeros(number_of_pulses, dtype=complex)

# Calculate the return signal
for r, v in zip(target_range, target_velocity):
    s += [exp(-1j * 4.0 * pi / c * (i * frequency_step) * \
              (r - v * (i / prf)))
          for i in range(number_of_pulses)]

# Calculate the iFFT
n = next_fast_len(10 * number_of_pulses)
sf = abs(ifft(s, n))

# Set up the range window for plotting
    range_window = linspace(0, c / (2.0 * frequency_step), n)

# Plot the iFFT output
plt.figure(1)
plt.plot(range_window, sf / max(sf))
plt.title('Stepped Frequency Range Profile')
plt.xlabel('Range (m)')
plt.ylabel('Relative Amplitude')
plt.show()
```

$$L = \frac{f_c' D}{B_{effective}}, \tag{8.9}$$

where

$$f_c' = f_c + \frac{B_{effective}}{2} \qquad \text{(Hz)}. \tag{8.10}$$

If the radial velocity of the target is accurately measured, the mismatch may be reduced by multiplying the target returns by a correction factor, which is written as

$$c_i(t) = \exp\left[j\frac{4\pi}{c}\left(f_0 + (i-1)\Delta f\right)(i-1)\, v\, T\right]. \tag{8.11}$$

The major difficulty with applying this correction factor is the velocity must be known *a priori*, or accurately measured in a continuous fashion.

8.3 MATCHED FILTER

The matched filter is an intrapulse compression technique commonly used in radar systems. The matched filter is the optimal linear filter that maximizes the signal-to-noise ratio in the presence of additive random noise. The term *matched* comes about as the filter impulse response is matched to the radar transmit signal. These were first referred to as *North filters*, as the concept first appeared in a report by North in 1943 [4]. Consider the matched filter diagram in Figure 8.5, where t_0 is the time delay to the target. The input signal to the filter is composed of both the input signal, $s_i(t)$, and the input noise, $n_i(t)$, as

$$x_i(t) = s_i(t) + n_i(t), \tag{8.12}$$

where the input signal is the delayed version of the transmit signal; that is, $s_i(t) = s(t - t_0)$. If the filter response, $h(t)$, is linear, the filter output is written as

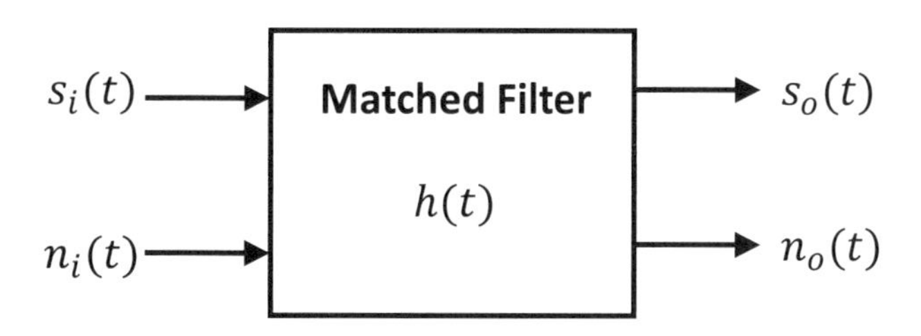

Figure 8.5 Matched filter block diagram.

$$x_o(t) = s_o(t) + n_o(t), \tag{8.13}$$

where

$$s_o(t) = s(t - t_0) * h(t), \tag{8.14}$$

$$n_o(t) = n_i(t) * h(t), \tag{8.15}$$

and the symbol $(*)$ represents linear convolution. Since the input and output noise signals are random processes, statistical methods are employed. The output noise power is assumed to be stationary. Specifically, its variance is the same regardless of the time the measurement is taken. The average noise power at the output of the matched filter is represented by

$$P_{noise} = \mathbb{E}\left\{|n_o(t)|^2\right\} \qquad \text{(W)}, \tag{8.16}$$

where $\mathbb{E}\{x\}$ is the expectation of x. The peak instantaneous signal power at the output of the matched filter is given by

$$P_{signal} = \left|s_o(t_0)\right|^2 \qquad \text{(W)}. \tag{8.17}$$

The peak signal-to-noise ratio is written as

$$SNR = \frac{P_{signal}}{P_{noise}} = \frac{\left|s_o(t_0)\right|^2}{\mathbb{E}\left\{|n_o(t)|^2\right\}}. \tag{8.18}$$

For convenience, treatment of the matched filter continues in the frequency domain. Using the Fourier transform results in

$$H(f) = \mathcal{F}\{h(t)\}, \tag{8.19}$$

$$S_i(f) = \mathcal{F}\{s_i(t)\}, \tag{8.20}$$

$$S_o(f) = \mathcal{F}\{s_o(t)\}, \tag{8.21}$$

where $\mathcal{F}\{x\}$ is the Fourier transform of x, and

$$S_o(f) = H(f)\, S_i(f). \tag{8.22}$$

Since the input and output noise are random processes, the Fourier transform of the expected value is

$$N_i(f) = \mathcal{F}\left\{\mathbb{E}[n_i(t+\tau)\, n_i^*(t)]\right\}, \tag{8.23}$$

$$N_o(f) = \mathcal{F}\{\mathbb{E}[n_o(t+\tau)\, n_o^*(t)]\}, \tag{8.24}$$

and

$$N_o(f) = |H(f)|^2\, N_i(f). \tag{8.25}$$

The expression in (8.25) is a power spectral density, and the average noise power is found by integrating over all frequencies as

$$P_{noise} = \int_{-\infty}^{\infty} N_o(f) df = \int_{-\infty}^{\infty} |H(f)|^2\, N_i(f) df. \tag{8.26}$$

The peak instantaneous power given in (8.17) may be expressed as

$$P_{signal} = \left|\mathcal{F}^{-1}\{S_0(f)\}|_{t=t_0}\right|^2 = \left|\int_{-\infty}^{\infty} S_i(f) H(f)\, e^{j2\pi f t_0}\, df\right|^2. \tag{8.27}$$

Then expression for the signal-to-noise ratio is then

$$SNR = \frac{P_{signal}}{P_{noise}} = \frac{\left|\int_{-\infty}^{\infty} S_i(f) H(f)\, e^{j2\pi f t_0}\, df\right|^2}{\int_{-\infty}^{\infty} |H(f)|^2\, N_i(f) df}. \tag{8.28}$$

Assuming $n_i(t)$ is white noise with power spectral density of kT_0FG gives

$$SNR = \frac{P_{signal}}{P_{noise}} = \frac{\left|\int_{-\infty}^{\infty} S_i(f) H(f)\, e^{j2\pi f t_0}\, df\right|^2}{kT_0FG \int_{-\infty}^{\infty} |H(f)|^2\, df}. \tag{8.29}$$

To maximize the signal-to-noise ratio, the Cauchy-Schwarz inequality is used as [5]

$$\left|\int_a^b X(f) Y(f) df\right|^2 \le \left(\int_a^b |X(f)|^2 df\right)\left(\int_a^b |Y(f)|^2 df\right). \tag{8.30}$$

The equality holds when $X(f) = KY^*(f)$, where K is an arbitrary complex constant. Making use of (8.30), (8.29) becomes

$$SNR \leq \frac{\left(\int\limits_a^b |S_i(f)|^2 df\right)\left(\int\limits_a^b |H(f)|^2 df\right)}{kT_0FG \int\limits_{-\infty}^{\infty} |H(f)|^2\, df} \leq \frac{\int\limits_a^b |S_i(f)|^2 df}{kT_0FG}. \tag{8.31}$$

The expression in (8.31) provides an upper bound for the signal-to-noise ratio for all $H(f)$. The filter response required to achieve the maximum value in (8.31) is found from the Cauchy-Schwarz inequality, and is written as

$$H(f) = KS_i^*(f)\, e^{-j2\pi f t_0}. \tag{8.32}$$

Using Parseval's theorem [6], the signal-to-noise ratio may be written as

$$SNR = \frac{E_i}{kT_0FG}, \tag{8.33}$$

where E_i is the energy in the input signal. This illustrates the maximum instantaneous signal-to-noise ratio for the matched filter only depends on the energy in the input signal and the input noise and does not depend on the waveform used. To find the filter impulse response, $h(t)$, take the inverse Fourier transform of (8.32), which gives

$$\begin{aligned} h(t) &= \mathcal{F}^{-1}\{KS_i^*(f)\, e^{-j2\pi f t_0} \\ &= \int\limits_{-\infty}^{\infty} KS_i^*(f)\, e^{-j2\pi f t_0}\, e^{j2\pi f t}\, df = Ks_i^*(t_0 - t). \end{aligned} \tag{8.34}$$

Therefore, $h(t)$, is the conjugate of a scaled, time reversed, and time shifted version of transmitted signal, $s(t)$. This correspondence of the filter response to the transmitted signal gives rise to the name *matched filter*. The matched filter may be implemented by using the Fourier transform and inverse transform from the SciPy module *fftpack*. This module includes various functions for computing discrete and fast Fourier transforms along with a number of helper functions. Employing the frequency domain form of (8.22) allows the output of the matched filter to be written as

$$s_o(t) = \mathcal{F}^{-1}\Big\{\mathcal{F}\{h(t)\} \times \mathcal{F}\{s_i(t)\}\Big\}, \tag{8.35}$$

where

$s_i(t)$	=	matched filter input signal,
$h(t)$	=	matched filter impulse response,
$s_0(t)$	=	matched filter output.

Listing 8.2 Calculation of the Matched Filter Output

```
from scipy.fftpack import fft, ifft, fftshift
from scipy import conj, linspace, exp
from scipy.constants import pi
from matplotlib import pyplot as plt

t = linspace(-1, 1, 1024) # Pulsewidth (s)

t0 = 0.2 # Time delay to the target (s)

# Set up the transmit and receive signals (s)
st = exp(1j * 10 * pi * t ** 2)
sr = exp(1j * 10 * pi * (t - t0) ** 2)

# Impulse response and matched filtering
Hf = fft(conj(st))
Si = fft(sr)
so = fftshift(ifft(Si * Hf))

# Plot the matched filter output
plt.figure(1)
plt.plot(t, abs(so) / max(abs(so)))
plt.title('Matched Filter Output')
plt.xlabel('Time Delay (s)')
plt.ylabel('Relative Amplitude')
plt.show()
```

The expression in (8.35) is implemented in the code shown in Listing 8.2. The target is located at a time delay of 0.2 seconds, and the output of this code is given in Figure 8.6. As seen in the figure, the peak of the matched filter output occurs at $t = 0.2$ seconds, corresponding to the target location.

An important aspect of pulse compression is the *time-bandwidth product*, given by $B\,\tau$, where B is the bandwidth in Hz, and τ is the pulsewidth in seconds. The time-bandwidth product is a figure of merit representing the increase in the signal-to-noise ratio at the output of the matched filter compared to the input. It is sometimes referred to as compression gain or matched filter gain. Bandwidth limited waveforms, such as a Gaussian pulse, have the lowest time-bandwidth products. The time-bandwidth product for unmodulated pulses approaches 1 as $B \approx 1/\tau$ for these waveforms. The time-bandwidth product of binary phase coded waveforms then depends on the length of the code. For example, Barker codes would be limited to a maximum time-bandwidth product of 13 [7].

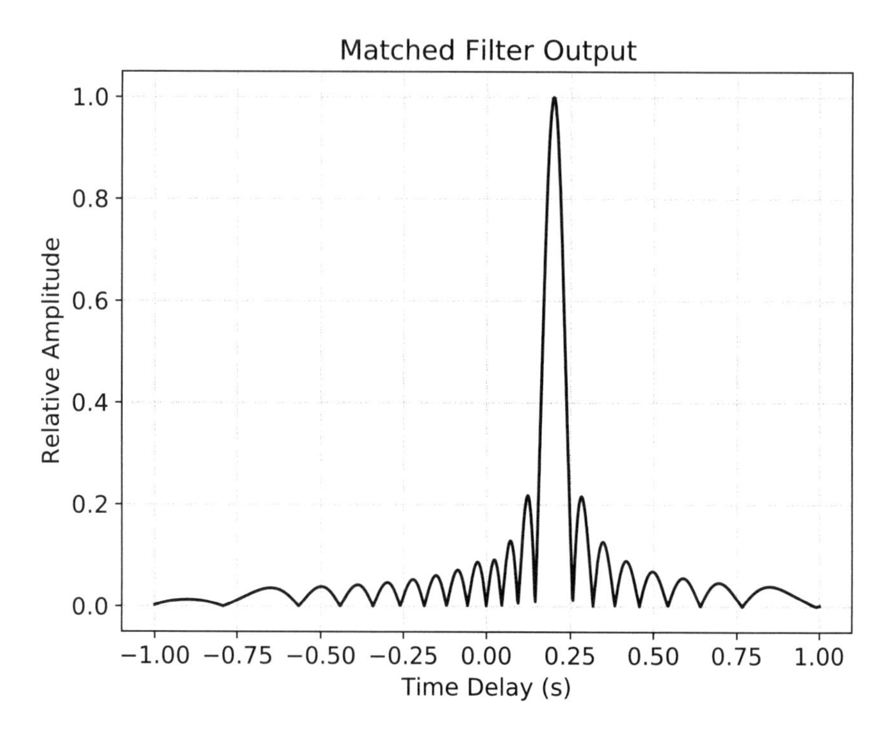

Figure 8.6 Matched filter output for Listing 8.2.

Linear frequency modulated waveforms, often referred to as chirp waveforms, may have extremely large time-bandwidth products of $100,000$ or more. Perfect matching to the transmitted waveform can be difficult, which reduces the compression gain compared to the theoretical value of $B\tau$. Also, radar systems often have stringent requirements on time sidelobes that may be achieved by using a modified version of the transmitted waveform, resulting in a lower compression gain.

8.4 STRETCH PROCESSING

Stretch processing is an intrapulse compression technique used for processing waveforms of very large bandwidth. It was originally developed to reduce the sampling requirements for large bandwidth, high range resolution systems such as synthetic aperture radar (SAR) systems [8]. While stretch processing may be applied to various forms of wide bandwidth waveforms, this discussion is focused on linear frequency modulation (LFM) waveforms. A block diagram of a stretch processor is given in Figure 8.7, and in general, stretch processing consists of the following basic steps:

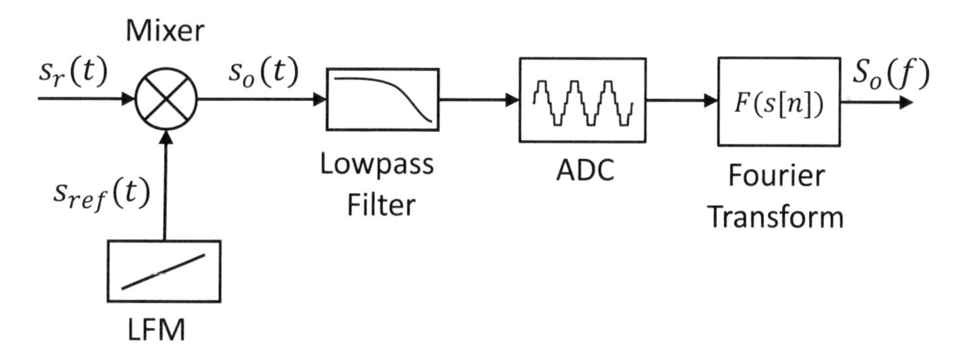

Figure 8.7 Block diagram of a stretch processor.

1. Mix the return signal with a reference signal.
2. Low-pass filter the mixed signals.
3. Analog-to-digital conversion of the low-pass filtered signals.
4. Fourier transform of the digital signals.

The first step in stretch processing is to mix the return signal with a reference signal. The reference signal is typically generated with the same frequency slope as the transmitted signal. The transmitted LFM signal may be written as

$$s(t) = \exp\left[j\,2\pi\left(f_0 t + \frac{\beta}{2}t^2\right)\right] \qquad 0 \le t \le \tau_p, \tag{8.36}$$

where

β = frequency slope, B/τ_p, (Hz/s),
f_0 = center frequency (Hz),
$s(t)$ = transmitted LFM signal.

The return signal from a point scatterer is written as

$$s_r(t) = A\,\exp\left[j\,2\pi\left(f_0(t - t_0) + \frac{\beta}{2}(t - t_0)^2\right)\right], \tag{8.37}$$

where

t_0 = time delay associated with the target, $2r_0/c$, (s),
r_0 = range to the target (m),
$s_r(t)$ = return signal from the target.

The factor, A, is a constant related to the target's radar cross section, antenna gain, path loss, and other factors. The reference signal is written as

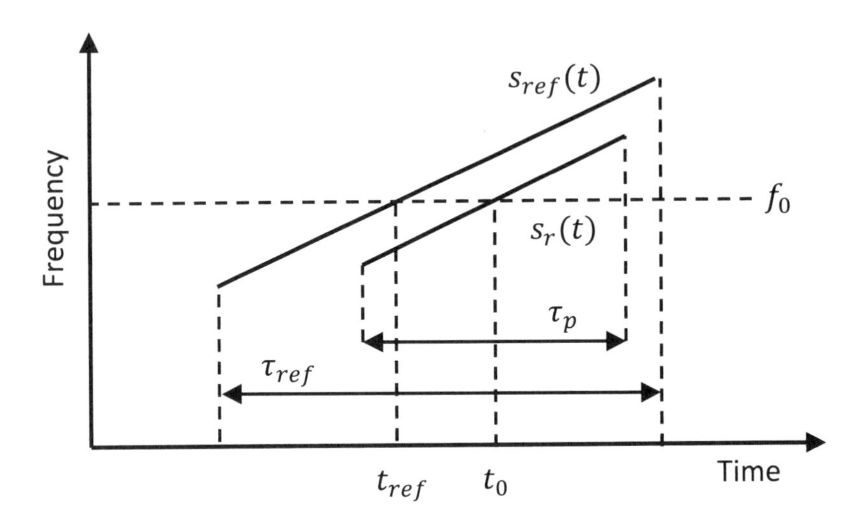

Figure 8.8 The return and reference signals for stretch processing.

$$s_{ref}(t) = \exp\left[j\,2\pi\left(f_0(t - t_{ref}) + \frac{\beta}{2}(t - t_{ref})^2\right)\right], \tag{8.38}$$

where t_{ref} is the reference time of the signal, which must closely match t_0. Figure 8.8 illustrates the relationship between the return signal, $s_r(t)$, and the reference signal, $s_{ref}(t)$. As shown in the Figure 8.8, τ_{ref} is chosen such that $s_{ref}(t)$ completely overlaps $s_r(t)$ in time. In other words, the pulsewidth of the reference signal is greater than the pulsewidth transmitted signal, $\tau_{ref} > \tau_p$, for the expected values of t_0 relative to t_{ref}. These conditions result in a signal-to-noise ratio close to that of the matched filter [2, 9]. If $s_{ref}(t)$ does not completely overlap $s_r(t)$ in time, there will be a loss in the resulting signal-to-noise ratio as well as a degradation of the range resolution. This leads to the following requirement on the pulsewidth of the reference signal,

$$\tau_{ref} \geq t_r + \tau_p \qquad \text{(s)}, \tag{8.39}$$

where t_r is the range of time over which stretch processing is to be employed, as illustrated in Figure 8.8. This is expressed as $t_r = t_{max} - t_{min}$. The second requirement is $t_{ref} - t_{min} \leq t_0 \leq t_{max} + t_{ref}$. Now that the requirements on the reference signal have been specified, the output of the mixing step is written as

$$s_o(t) = A\,\exp\left[j\,2\pi f_0(t_0 - t_{ref}) + j\,2\pi\beta(t_0 - t_{ref})t + j\,\pi\beta(t_{ref}^2 - t_0^2)\right]. \tag{8.40}$$

The instantaneous frequency is the time derivative of the phase, which is written as

$$f_i = \frac{1}{2\pi}\frac{d\phi(t)}{dt} \qquad \text{(Hz)}, \tag{8.41}$$

where $\phi(t)$ is the phase of the signal in (8.40), which is simply

$$\phi(t) = 2\pi f_0(t_0 - t_{ref}) + 2\pi\beta(t_0 - t_{ref})t + \pi\beta\left(t_{ref}^2 - t_0^2\right) \qquad \text{(rad)}. \tag{8.42}$$

The instantaneous frequency is then

$$f_i = \beta(t_0 - t_{ref}) \qquad \text{(Hz)}. \tag{8.43}$$

This allows the time delay of the target to be written as

$$t_0 = \frac{f_i}{\beta} + t_{ref} \qquad \text{(s)}. \tag{8.44}$$

Therefore, the target range is related to the instantaneous frequency as

$$r_0 = \frac{c}{2}\left(\frac{f_i}{B}\tau_p + t_{ref}\right) \qquad \text{(m)}. \tag{8.45}$$

The expression in (8.43) indicates the resulting instantaneous frequency is constant and related to the range to the target, as illustrated in Figure 8.9. Performing analog-to-digital sampling on the signal output from the mixer only requires sampling at the maximum instantaneous frequency in (8.43), rather than the full bandwidth. For example, consider a waveform with a bandwidth of $B = 1$ GHz, a pulsewidth of $\tau_p = 8$ ms, and a maximum receive window of $r_{max} = 100$ m. The number of analog-to-digital samples for stretch processing is then

$$N \geq 2\tau_p f_i|_{max} \geq 2B(t_0 - t_{ref}) = \frac{4B\, r_{max}}{c} \geq \frac{4 \cdot 10^9 \cdot 100}{3 \times 10^8} \approx 1333. \tag{8.46}$$

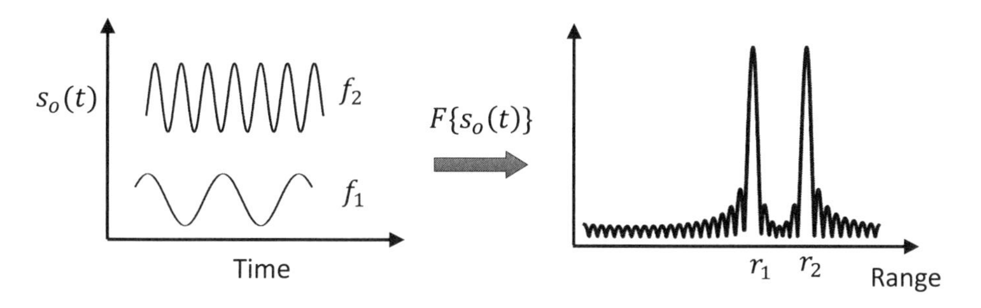

Figure 8.9 The instantaneous frequency related to the target range.

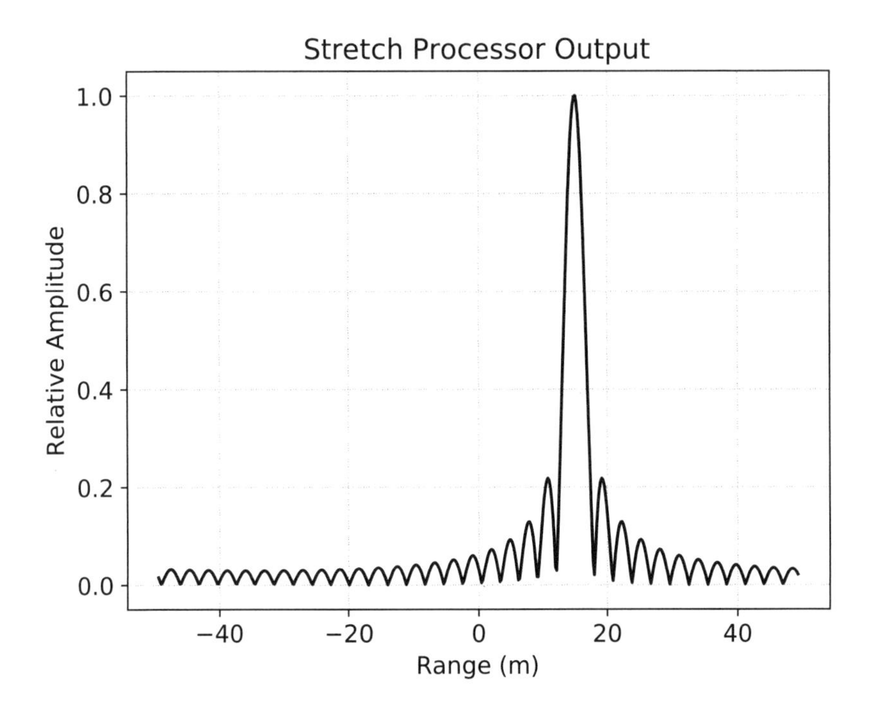

Figure 8.10 Stretch processor output for Listing 8.3.

If using a matched filter instead, the required number of samples is

$$N \geq 2B\tau = 2 \cdot 10^9 \cdot 10^{-2} = 20 \times 10^6. \tag{8.47}$$

This is a very large savings in the number of samples, ADC rate, memory, and processing. The stretch processor allows the system to utilize waveforms with bandwidths not feasible with the matched filter. Finally, performing the Fourier transform on the sampled signal results in ***sinc*** functions located at the frequency corresponding to the relative time delay of the target. Listing 8.3 illustrates this with a simple example, and the output is given in Figure 8.10. To study the range resolution associated with the stretch processor, consider the situation where there are two point scatterers present at ranges r_1 and r_2, with associated time delays of $t_1 = 2r_1/c$ and $t_2 = 2r_2/c$. Using (8.43), the instantaneous frequencies are

$$f_{i1} = \beta(t_1 - t_{ref}), \qquad f_{i2} = \beta(t_2 - t_{ref}) \qquad \text{(Hz)}. \tag{8.48}$$

The frequency resolution of the stretch processor is written as

Listing 8.3 Calculation of the Stretch Processor Output

```
from scipy.fftpack import fft, fftshift, fftfreq
from scipy import ceil, linspace, exp
from scipy.constants import pi, c
from matplotlib import pyplot as plt

bandwidth = 50e6  # Bandwidth (Hz)

pulsewidth = 1e-3  # Pulsewidth (s)

range_window_length = 50.0  # Range window length +/- (m)

target_range = 15.0  # Target range (m)
t0 = 2.0 * target_range / c  # Time delay to target (s)

# Number of samples
number_of_samples = ceil(4 * bandwidth * range_window_length / c)

# Time sampling
t, dt = linspace(-0.5 * pulsewidth, 0.5 * pulsewidth,
                 number_of_samples, retstep=True)

# Sampled signal after mixing
so = exp(1j * 2.0 * pi * bandwidth / pulsewidth * t0 * t)

# Fourier transform
so = fftshift(fft(so, 512))

# FFT frequencies
frequencies = fftshift(fftfreq(512, dt))

# Range window
range_window = 0.5 * frequencies * c * pulsewidth / bandwidth

# Plot the stretch processor output
plt.figure(1)
plt.plot(range_window, abs(so) / max(abs(so)), 'k')
plt.title('Stretch Processor Output')
plt.xlabel('Range (m)')
plt.ylabel('Relative Amplitude')
plt.show()
```

$$\Delta f = \frac{1}{\tau_p} \qquad \text{(Hz)}. \tag{8.49}$$

The two point scatterers may be resolved if their associated ranges are such that the instantaneous frequencies are greater than the frequency resolution of the stretch processor. This may be expressed as

$$f_{i2} - f_{i1} = \beta(t_2 - t_{ref}) - \beta(t_1 - t_{ref}) \geq \frac{1}{\tau_p} \qquad \text{(Hz)}. \tag{8.50}$$

This leads to the following time resolution

$$t_2 - t_1 \geq \frac{1}{B} \qquad \text{(s)}, \tag{8.51}$$

which gives a range resolution of

$$\Delta r = r_2 - r_1 = \frac{c}{2B} \qquad \text{(m)}. \tag{8.52}$$

This is the same as the range resolution for the matched filter and the effective range resolution for stepped frequency waveforms. While the stretch processor provides a reduction in sampling requirements compared to the matched filter, it is limited to a small range window relative to the pulsewidth, as shown in Figure 8.8, whereas the matched filter tests for the presence of a target over the entire pulse repetition interval. This is sometimes referred to as *all-range* processing. Clearly, stretch processing would not be very effective for radar search functions as the range window would be of limited extent. Stretch processing is sometimes used to update an established track, when more accurate range measurements are desired for a precision track. However, the range of bandwidths requiring stretch processing tends to be excessive for tracking purposes. The two most common applications for stretch processing are target discrimination and synthetic aperture radar. In each case, it is necessary to identify individual scatterers on a target or in the scene, which requires waveforms of very wide bandwidth.

8.5 WINDOWING

As shown in Chapter 3, windowing is a method for reducing sidelobes by applying weighting coefficients to the antenna elements. Similarly, weighting coefficients may be applied to the digital samples in matched filtering or stretch processing as a technique for reducing the time (range) sidelobes. Analogous to the antenna case, windowing will reduce the range resolution by broadening the main lobe, and also reduce the signal-to-noise ratio. However, these trades may be necessary in cases where time sidelobes mask lower RCS targets or even cause false alarms. Figure 8.11 illustrates a low-intensity target being masked by range sidelobes of a high-intensity target when no window is applied.

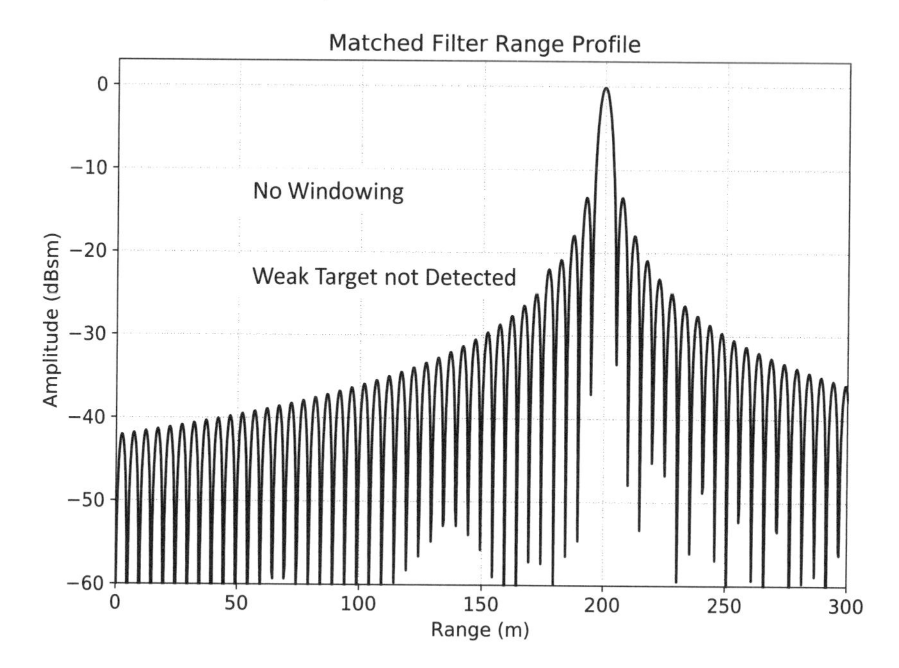

Figure 8.11 Low-intensity target masked by high-intensity target.

Figure 8.12 shows the low-intensity target detected when a Hamming window is applied to the data.

8.6 AMBIGUITY FUNCTION

As shown in Section 8.3, matched filtering results in the maximum achievable signal-to-noise ratio at the output of the receiver. However, there may be interference from targets at different ranges and velocities than the target of interest. One method for representing this interference is through the use of the ambiguity function. The ambiguity function is the output of the matched filter when the signal is received with a time delay, τ, and a Doppler shift, f, with respect to the nominal values expected by the matched filter. The return from the target of interest is located at the origin of the ambiguity function $(\tau = 0, f = 0)$, and the shape of the ambiguity function is determined by the properties of the transmitted waveform and the matched filter. Studying the resulting shape reveals the waveform's range and Doppler resolution as well as the interference from other targets. The interference from other targets results in ambiguities for the radar system. The ambiguity function provides the radar designer valuable insight into which waveforms would be best suited for various scenarios and applications. This makes the ambiguity

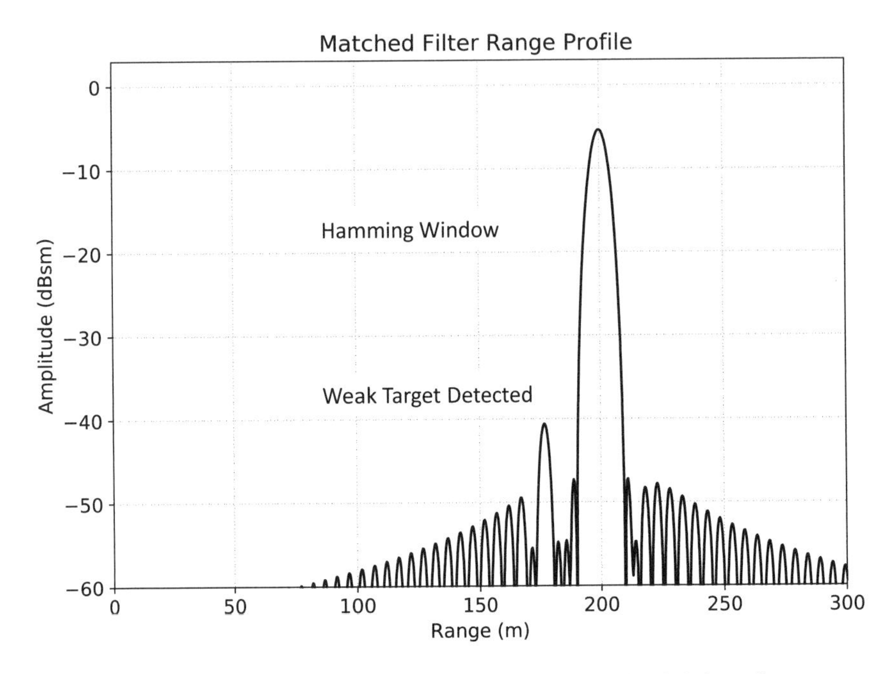

Figure 8.12 Low-intensity target detected in the presence of a high-intensity target.

function a useful tool for both analyzing and designing radar waveforms. While there are inconsistencies in the definition of the ambiguity function, the definition from [10] is used to write the ambiguity function as the magnitude squared of the output of the matched filter. This is expressed as

$$|\chi(\tau, f)|^2 = \left| \int_{-\infty}^{\infty} s(t)\; s^*(t-\tau)\; e^{j2\pi f t}\; dt \right|^2 . \tag{8.53}$$

The different definitions of the ambiguity function result in essentially the same physical meaning. However, this makes it difficult to be confident that specific properties hold for different definitions. For the definition given in (8.53), there are several useful properties for examining radar waveforms [11, 12]:

1. The value at the origin, $(\tau = 0, f = 0)$, is E^2.
2. The maximum value is located at the origin, $\left|\chi(\tau, f)\right|^2 \leq \left|\chi(0, 0)\right|^2$.
3. The volume of the ambiguity function is constant and equal to E^2,

$$\int_{-\infty}^{\infty}\int_{-\infty}^{\infty} |\chi(\tau, f)|^2 \, d\tau \, df = E^2. \tag{8.54}$$

4. The ambiguity function is symmetric, $|\chi(\tau, f)|^2 = |\chi(-\tau, -f)|^2$.
5. The time shift property states if $v(t) = s(t - \delta t)$, then

$$\chi_v(\tau, f) = e^{j2\pi f \delta t} \, \chi_s(\tau, f). \tag{8.55}$$

6. The time scaling property states if $v(t) = s(at)$, then

$$\chi_v(\tau, f) = \frac{1}{|a|} \, \chi_s(a\tau, f/a). \tag{8.56}$$

7. The modulation property states if $v(t) = s(t)\, e^{j2\pi\psi t}$, then

$$\chi_v(\tau, f) = e^{j2\pi\psi\tau} \, \chi_s(\tau, f). \tag{8.57}$$

8. The quadratic phase shift property states if $v(t) = s(t)\, e^{j\pi\beta t^2}$, then

$$\left|\chi_v(\tau, f)\right|^2 = \left|\chi_s(\tau, f + \beta\tau)\right|^2. \tag{8.58}$$

8.6.1 Single Unmodulated Pulse

As a first example, consider the ambiguity function of a single unmodulated pulse. The complex envelope of this waveform is expressed as

$$s(t) = \text{rect}\left[\frac{t}{\tau_p}\right], \tag{8.59}$$

where τ_p is the pulsewidth, and the unit rectangular function is given by

$$\text{rect}[t] = \begin{cases} 1 & \text{for} \quad 0 \le t \le 1, \\ 0 & \text{otherwise.} \end{cases} \tag{8.60}$$

Using (8.59) and (8.53) allows the ambiguity function to be written as

$$|\chi(\tau, f)|^2 = \left| \int_{-\infty}^{\infty} \text{rect}\left[\frac{t}{\tau_p}\right] \text{rect}\left[\frac{t-\tau}{\tau_p}\right] e^{j2\pi f t} \, dt \right|^2. \tag{8.61}$$

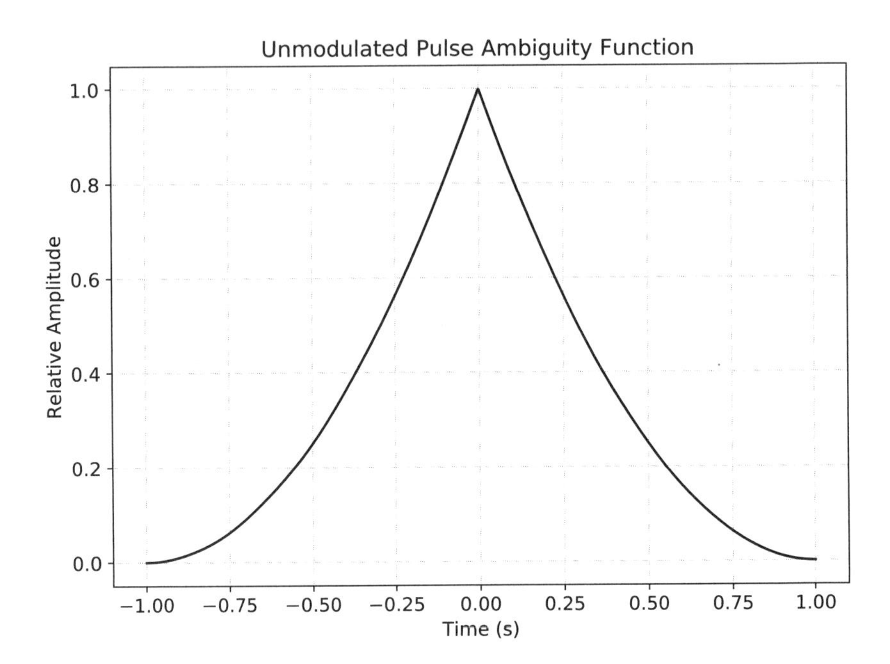

Figure 8.13 Zero-Doppler cut through the unmodulated single-pulse ambiguity function.

For $\tau > 0$, the overlap of the unit rectangular functions changes the limits of integration to (τ, τ_p). In the case $\tau < 0$, the limits become $(0, \tau_p + \tau)$. Carrying out the integration for both cases and combining the results gives

$$\left|\chi(\tau, f)\right|^2 = (\tau_p - |\tau|)^2 \left|\text{sinc}\left[f(\tau_p - |\tau|)\right]\right|^2 \quad \text{for } |\tau| \le \tau_p. \tag{8.62}$$

Figure 8.13 shows the zero-Doppler cut through the ambiguity function. This cut corresponds to $f = 0$, and is expressed as

$$\left|\chi(\tau, 0)\right|^2 = (\tau_p - |\tau|)^2 \quad \text{for } |\tau| \le \tau_p. \tag{8.63}$$

The ambiguity function along the time axis extends $(-\tau_p, \tau_p)$. Therefore, targets will be unambiguous in time if they are separated by at least τ_p seconds. This shows the range resolution for a single unmodulated pulse is $c\,\tau_p/2$ meters. Similarly, taking a cut along the zero-time axis allows for the study of the Doppler resolution and ambiguity. Figure 8.14 shows the zero-time cut through the ambiguity function. This cut corresponds to $\tau = 0$, and is expressed as

$$\left|\chi(0, f)\right|^2 = \tau_p^2 \left|\text{sinc}[f\tau_p]\right|^2. \tag{8.64}$$

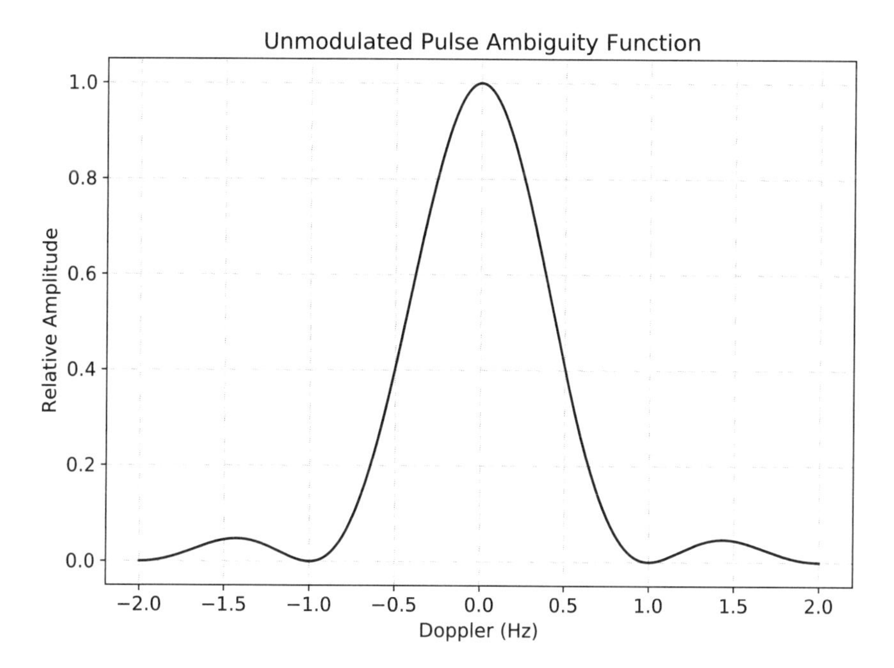

Figure 8.14 Zero-time delay cut through the unmodulated single-pulse ambiguity function.

The ambiguity function along the Doppler axis extends $(-\infty\ ,\ \infty)$, with the first null occurring at $f = \pm 1/\tau_p$ Hz. This indicates targets separated in Doppler by $1/\tau_p$ Hz can be detected without any ambiguity. The two-dimensional contour plot for the ambiguity function of the unmodulated single pulse is given in Figure 8.15. This figure demonstrates the range and Doppler resolutions are determined by the pulsewidth, τ_p.

8.6.2 Single LFM Pulse

Next, consider the ambiguity function of a single LFM pulse. The complex envelope of this waveform is expressed as

$$s(t) = \text{rect}\left[\frac{t}{\tau_p}\right] e^{j\pi\beta t^2}. \tag{8.65}$$

As before, τ_p is the pulsewidth. Using (8.65) and (8.53), the ambiguity function is written as

$$|\chi(\tau, f)|^2 = \left| \int_{-\infty}^{\infty} \text{rect}\left[\frac{t}{\tau_p}\right] \text{rect}\left[\frac{t-\tau}{\tau_p}\right] e^{j\pi\beta t^2}\, e^{-j\pi\beta(t-\tau)^2}\, e^{j2\pi f t}\, dt \right|^2. \tag{8.66}$$

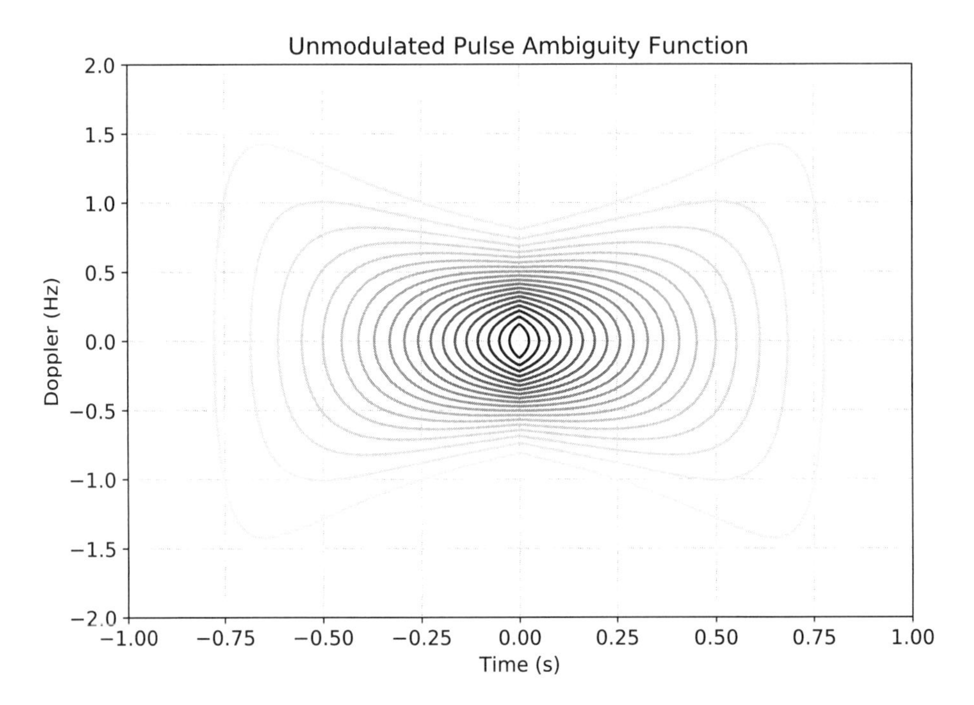

Figure 8.15 Two-dimensional contour plot of the unmodulated single-pulse ambiguity function.

As in the unmodulated single pulse case, the limits of integration due to the unit rectangular functions are (τ, τ_p) for $\tau > 0$, and $(0, \tau_p + \tau)$ for $\tau < 0$. Performing both integrations and combining the results leads to

$$\left|\chi(\tau, f)\right|^2 = \left|(\tau_p - |\tau|)\,\text{sinc}\big[\tau_p(f + \beta\tau)(\tau_p - |\tau|)\big]\right|^2 \quad \text{for } 0 \le |\tau| \le \tau_p. \tag{8.67}$$

Figure 8.16 shows the zero-Doppler cut through the ambiguity function. The ambiguity function along the time axis is a sinc function with the first null located at $\tau = \pm 1/B$ seconds. Therefore, targets will be unambiguous in time if they are separated by $1/B$ seconds. This shows the range resolution for a single LFM pulse is $c/2B$ meters. Similarly, taking a cut along the zero-time axis allows for the study of the Doppler resolution and ambiguity. Figure 8.17 shows the zero-time cut through the ambiguity function. The ambiguity function along the Doppler axis extends $(-\infty, \infty)$, with the first null occurring at $f = \pm 1/\tau_p$ Hz. This indicates targets separated in Doppler by $1/\tau_p$ Hz can be detected without any ambiguity. The two-dimensional contour plot for the ambiguity function of the single LFM pulse is given in Figure 8.18. This figure illustrates

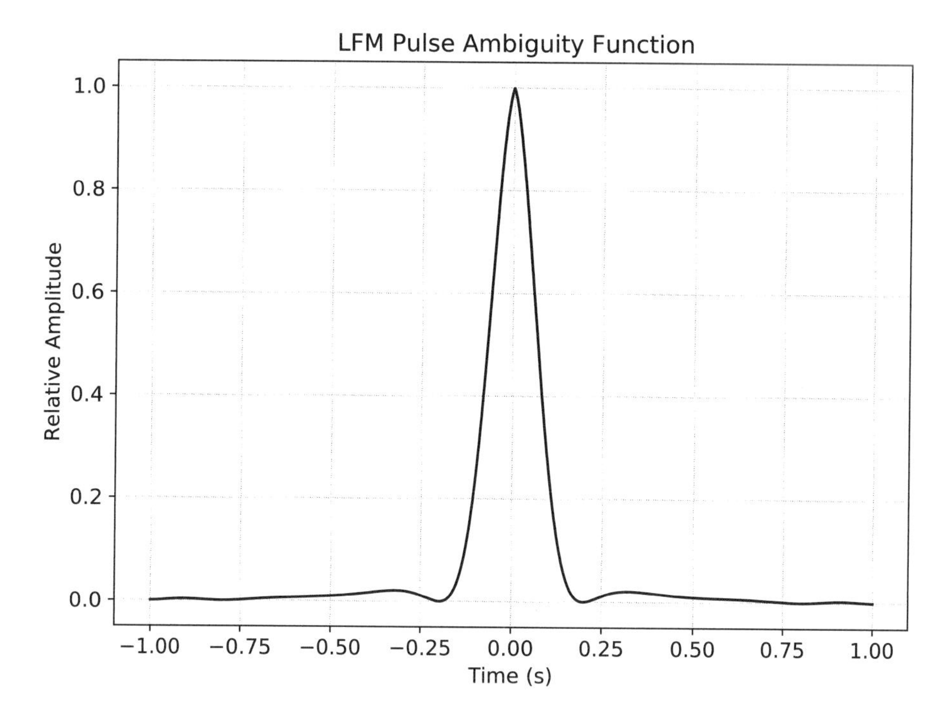

Figure 8.16 Zero-Doppler cut through the single LFM pulse ambiguity function.

the Doppler resolution is determined by the pulsewidth just as it was for the unmodulated pulse. The range resolution for the LFM pulse is determined by the bandwidth rather than the pulsewidth. The effective pulsewidth is $1/B$ seconds, and the ambiguity function is narrower than the unmodulated pulse by $\tau_p\, B$, which is the time-bandwidth product.

8.6.3 Generic Waveform Procedure

The previous sections examined the ambiguity function for simple waveforms. As waveforms become more complicated, the integral in (8.53) becomes increasingly difficult to evaluate analytically and to implement in code. Therefore, a straightforward numerical approach for calculating the ambiguity function of generic waveforms is presented. The approach is based on the frequency domain representation of the ambiguity function, which is written as

$$|\chi(\tau, f_d)|^2 = \left| \int_{-\infty}^{\infty} S(f)\, S^*(f - f_d)\, e^{-j2\pi f t}\, df \right|^2 . \tag{8.68}$$

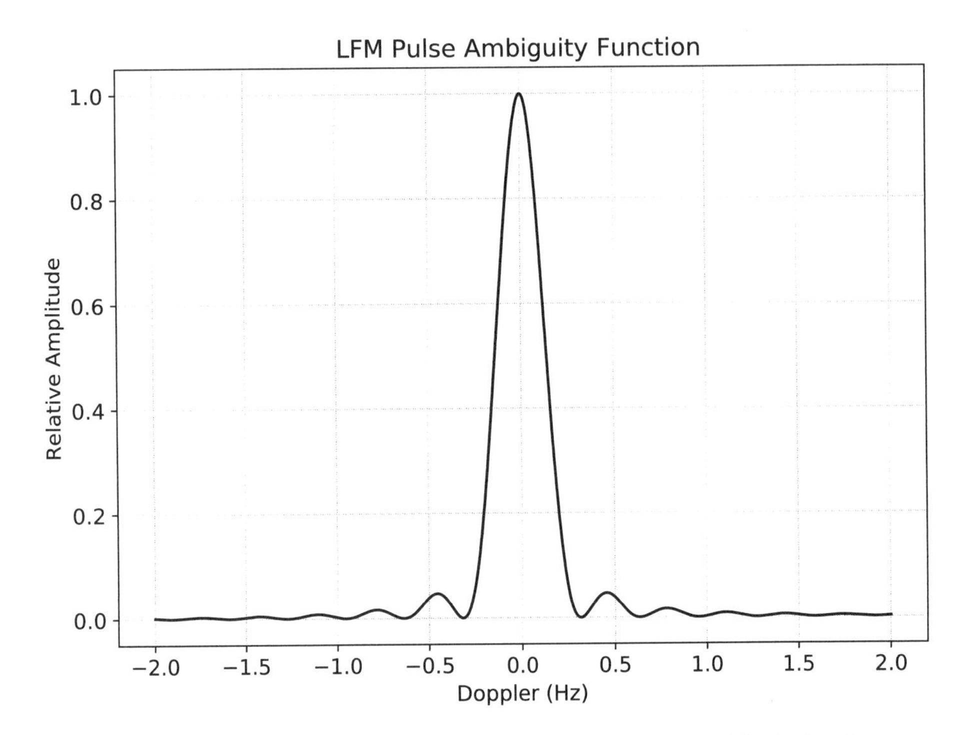

Figure 8.17 Zero-time delay cut through the single LFM pulse ambiguity function.

The following steps are used to numerically evaluate the ambiguity function:

1. Create a sampled version of $s(t)$. The number of samples must meet the Shannon requirements [13], $N > 2FT$, where F is the highest frequency in the signal and T is the time duration of the signal.

2. Compute $S(f)$ by taking the Fourier transform of $s(t)$. Performing the Fourier transform with a zero-padded version of $s(t)$ results in more detailed range cuts. SciPy's *fftpack* has efficient functions for radix $\{2, 3, 4, 5\}$, and the helper function, ***next_fast_len(N)***, returns the next composite of the prime factors 2, 3, and 5, which is greater than or equal to N. These are also known as 5-smooth numbers, regular numbers, or Hamming numbers.

3. Choose the Doppler mismatch frequency, f_d, and create a sampled version of $s(t)\,\exp(j2\pi f_d t)$. The sampling requirements are the same as above.

4. Compute $S(f - f_d)$ by taking the Fourier transform of $s(t)\,\exp(j2\pi f_d t)$.

5. Compute $\chi(\tau, f_d)$ by taking the inverse Fourier transform of $S(f)S^*(f - f_d)$.

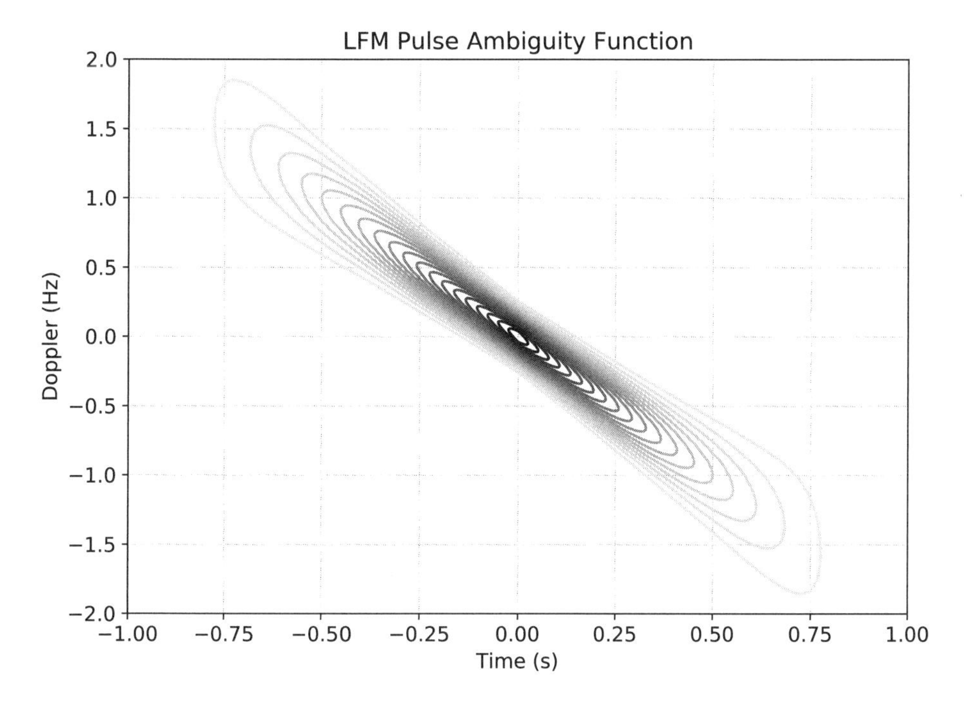

Figure 8.18 Two-dimensional contour plot of the single LFM pulse ambiguity function.

6. Steps 3-5 are repeated for various Doppler frequencies, f_d, and the results stored in a two-dimensional array.

These steps are expressed mathematically as

$$|\chi(\tau, f_d)|^2 = \left|F^{-1}\{F[s(t)]\, F[s(t)\exp(j2\pi f_d t)]^*\}\right|^2. \tag{8.69}$$

The procedure outlined above will be used in Section 8.8 to study the ambiguity function of unmodulated pulse trains, LFM pulse trains, and coded waveforms.

8.7 PHASE-CODED WAVEFORMS

Phase-coded waveforms are waveforms in which the phase is changed in discrete steps rather than continuously, as in the LFM case. A phase-coded waveform can be considered as a single pulse which is divided into subpulses, as illustrated in Figure 8.19. The subpulses are often referred to as *chips*. Phase-coded waveforms may be expressed as

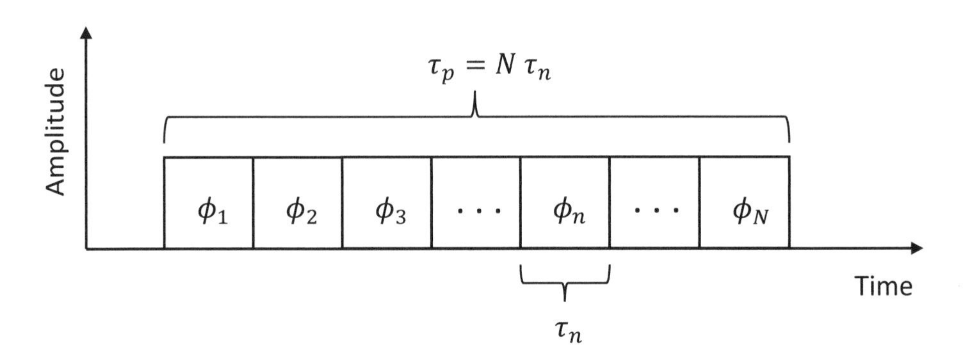

Figure 8.19 Phase-coded waveform.

$$s(t) = \sum_{n=0}^{N-1} e^{j\phi_n}\,\mathrm{rect}\left[\frac{t - n\,\tau_n}{\tau_n}\right], \tag{8.70}$$

where,

N = total number of chips,
ϕ_n = phase of each chip (rad),
τ_n = width of each chip (s).

8.7.1 Barker Codes

Binary phase coding is the case where ϕ_n takes on one of two values, 0 or π. One common set of binary phase codes for radar applications is Barker codes, which have several unique properties. The time sidelobe level of the normalized ambiguity function is $1/N$, where N is the length of the code. The ambiguity function has a length of $2N\tau_n$, the main lobe has a width of $2\tau_n$, and there are $(N-1)/2$ sidelobes on either side of the main lobe. There are only seven known Barker codes, and these are given in Table 8.1. While these are the only known Barker codes, much research has been performed to find longer binary phase codes with low time sidelobe levels. Although the time sidelobe levels are not $1/N$ as with the Barker codes, some codes have been found to have quite small sidelobe levels. The reader is referred to several excellent sources on these waveforms for further reading [14–17]. A Barker code of length 4 is illustrated in Figure 8.20. Another approach to obtain longer codes is to combine or embed one code within another. For example, a Barker code of length 5 may be used with a Barker code of length 3 to give

$$B_{MN} = [\ 0\ 0\ 0\ \pi\ 0,\ 0\ 0\ 0\ \pi\ 0,\ \pi\ \pi\ \pi\ 0\ \pi\ ]. \tag{8.71}$$

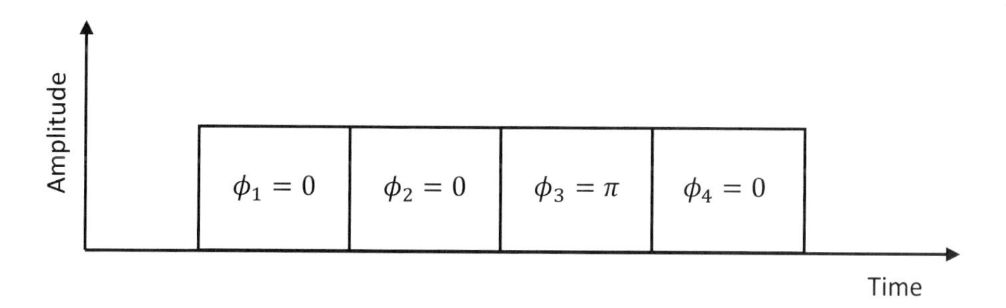

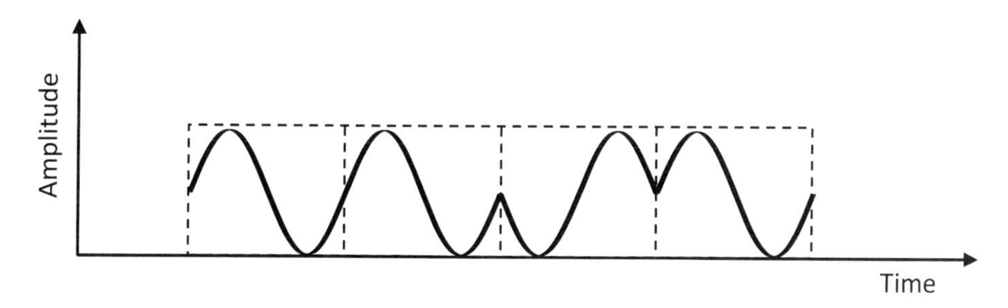

Figure 8.20 Barker phase-coded waveform of length 4.

The compression ratio of the combined Barker code is MN. However, the sidelobe level of the combined code is not $1/MN$. When using combined Barker codes, a number of sidelobes may be reduced to zero if the matched filter is followed by a linear transversal filter [10].

Table 8.1
Barker Codes

Length	*Phase Shift*	*Sidelobe Level (dB)*
2	$[0\ \pi]$ or $[0\ 0]$	−6.02
3	$[0\ 0\ \pi]$	−9.54
4	$[0\ 0\ \pi\ 0]$ or $[0\ 0\ 0\ \pi]$	−12.04
5	$[0\ 0\ 0\ \pi\ 0]$	−13.98
7	$[0\ 0\ 0\ \pi\ \pi\ 0\ \pi]$	−16.90
11	$[0\ 0\ 0\ \pi\ \pi\ \pi\ 0\ \pi\ \pi\ 0\ \pi]$	−20.83
13	$[0\ 0\ 0\ 0\ 0\ \pi\ \pi\ 0\ 0\ \pi\ 0\ \pi\ 0]$	−22.28

8.7.2 Frank Codes

Unlike binary phase coding, Frank codes are polyphase codes representing a quadratic phase shift in a discrete manner [18]. To generate a Frank code, a single pulse is divided into N groups. Next, each group is divided into N subpulses. Therefore, the length of Frank codes are perfect squares, $L = N^2$. Frank codes may be represented as a matrix of the form

$$F_N = \frac{2\pi}{N}\begin{bmatrix} 0 & 0 & 0 & \dots & 0 \\ 0 & 1 & 2 & \dots & N-1 \\ 0 & 2 & 4 & \dots & 2(N-1) \\ \vdots & \vdots & \vdots & \ddots & \vdots \\ 0 & N-1 & 2(N-1) & \dots & (N-1)^2 \end{bmatrix} \quad \text{(rad)}, \tag{8.72}$$

where each row is the phase shift associated with the subpulses in that group. Frank codes may also be expressed more compactly as

$$F_N = \frac{2\pi}{N}(m-1)(n-1) \quad \text{for} \quad m, n = 1 \dots N \quad \text{(rad)}. \tag{8.73}$$

For example, the Frank code for $N = 3$ is

$$F_3 = \frac{2\pi}{3}[\,0\ 0\ 0\ 0\ 1\ 2\ 0\ 2\ 4\,] = \left[\,0\ 0\ 0\ 0\ \frac{2\pi}{3}\ \frac{4\pi}{3}\ 0\ \frac{4\pi}{3}\ \frac{2\pi}{3}\,\right]. \tag{8.74}$$

For phase shifts larger than 2π, the phase wraps and the modulo with 2π is used. The phase shift associated with each subpulse for Frank code F_3 is illustrated in Figure 8.21. As illustrated in Section 8.8.10, the ambiguity function of Frank code waveforms is similar to that of LFM waveforms. This is expected as the Frank code is a discrete version of a quadratic phase shift. For very long Frank codes, the phase shift between subpulses becomes small. This can be problematic as phase instability of the system degrades the phase shift associated with each subpulse and reduces the performance of the waveform. Other discrete polyphase waveforms have been developed to approximate quadratic phase shifts, and the reader is referred to several references on this subject [19–21].

8.7.3 Pseudorandom Number Codes

Pseudorandom number (PRN) coded waveforms are binary phase-coded waveforms where the phase state is determined by a sequence of 0s and 1s based on a PRN sequence. PRN waveforms appear similar to noise and satisfy one or more of the standard tests for

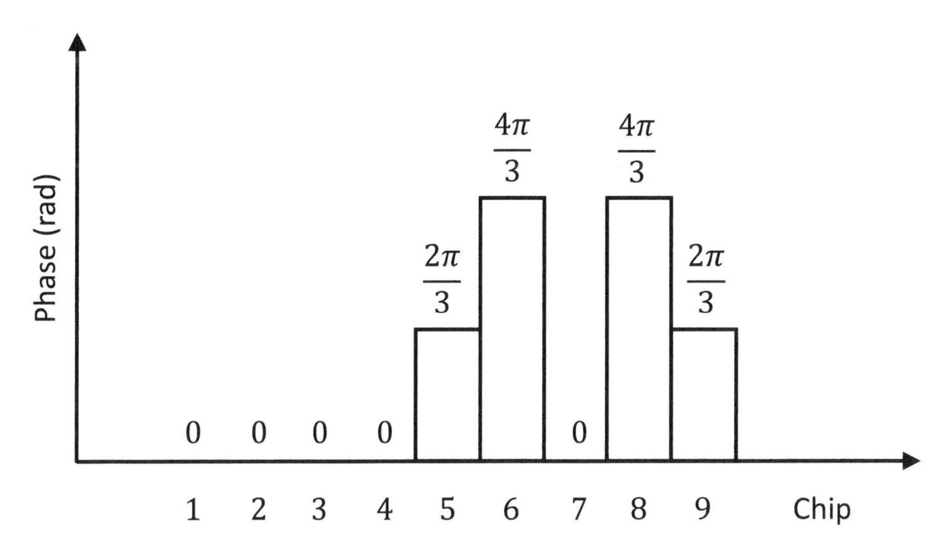

Figure 8.21 Phase shift of each subpulse for Frank code F_3.

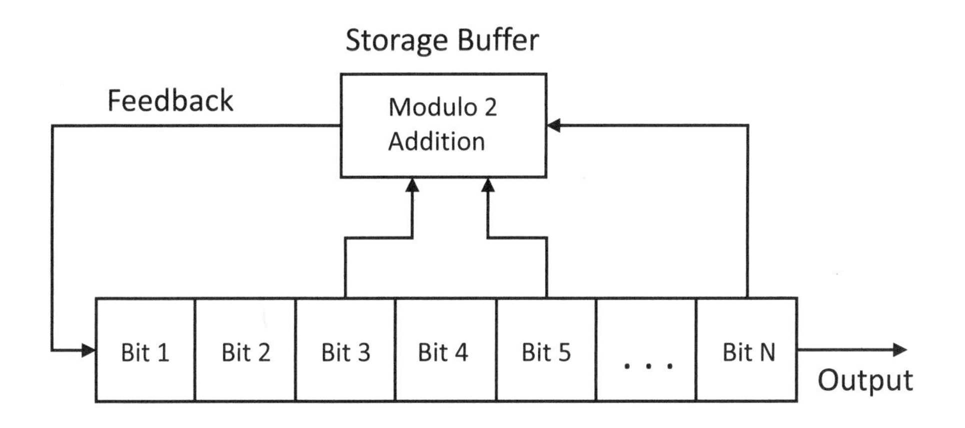

Figure 8.22 Linear feedback shift register with N bits.

statistical randomness. However, PRN codes have a deterministic sequence that repeats after its period [22, 23]. A type of PRN sequence commonly used in radar applications is the *maximum length sequence*. These are sequences generated using maximal linear feedback shift registers, as shown in Figure 8.22. The generation of a maximum length sequence consists of the following steps:

1. Initialize the register with any sequence except all 0s.

Table 8.2

Feedback Taps for Generating Maximum Length Sequences

N	*Feedback Taps*	*Characteristic Polynomial*	*Sequence Length*
3	$3, 2$	$x^3 + x^2 + 1$	7
4	$4, 3$	$x^4 + x^3 + 1$	15
5	$5, 3$	$x^5 + x^3 + 1$	31
6	$6, 5$	$x^6 + x^5 + 1$	63
7	$7, 6$	$x^7 + x^6 + 1$	127
8	$8, 6, 5, 4$	$x^8 + x^6 + x^5 + x^4 + 1$	255

2. Perform modulo 2 addition and place the result in the storage buffer.
3. Shift the contents of the register one bit to the right.
4. Store the contents of bit N in an output buffer.
5. Load the contents of the storage buffer into bit 1 of the register.
6. Repeat Steps 2-5 until the output buffer contains 2^N–1 bits.

These sequences are periodic and reproduce every binary sequence that can be represented by the shift registers. For a register of length N, the sequence is of length $2^N - 1$. The feedback connections shown in Figure 8.22 determine whether or not the sequence will be maximum length. Only certain feedback configurations lead to a maximum length sequence, as given in Table 8.2. Table 8.2 is not a comprehensive list of feedback connections and more combinations may exist for a given register. However, the number of different feedback combinations becomes very large as the number of bits in the register increases. A more thorough list is given in [24].

Maximum length shift registers are often described by their characteristic polynomial. For example, if feedback taps i, j, and k are used, then the characteristic polynomial is written as

$$x^i + x^j + x^k + 1. \tag{8.75}$$

The characteristic polynomial of a linear feedback shift register generating a maximum length sequence is a primitive polynomial. If the characteristic polynomial is known, then the associated sequence can easily be calculated. As in the shift register, the additions in (8.75) are modulo 2. The determination of the characteristic polynomial continues to be a research area for radar systems, digital communication systems, optical dielectric multilayer reflector design, and in the design of magnetic resonance imaging experiments [25]. Maximum length sequences have several interesting properties [22]. The *balance property* deals with the number of ones and zeros in a given sequence. Specifically, there are 2^{N-1} ones and $2^{N-1}-1$ zeros in a maximum length sequence. The number of zeros is one less than the number of ones since the state containing all zeros cannot occur. The *run property* is concerned with the number of consecutive ones or zeros in a given sequence, and states that $1/2$ of the runs are of length one, $1/4$ of the runs are of length two, $1/8$ of the runs are of length three, and so on. The *correlation property* shows that the circular autocorrelation of a maximum length sequence is the Kronecker delta function [26], and may be written as

$$R(n) = \sum_{m=1}^{N} s[m]\, s^*[m+n]_N = \begin{cases} N & \text{if } n = 0, \\ -1 & \text{if } 0 < n < N. \end{cases} \tag{8.76}$$

The linear autocorrelation is then an approximation to the Kronecker delta function. Therefore, the ambiguity function associated with PRN waveforms is well suited for radar applications due to their time and Doppler resolution and ambiguity properties, as shown in Section 8.8. Also, PRN waveforms are nearly orthogonal (i.e., if a PRN waveform is passed through a matched filter matched to a different PRN waveform, then the output will appear as noise). This is advantageous for multiinput multioutput (MIMO) and multistatic radar applications [27, 28].

8.8 EXAMPLES

The sections below illustrate the concepts of this chapter with several Python examples. The examples for this chapter are in the directory *pyradar\Chapter08* and the matching MATLAB examples are in the directory *mlradar\Chapter08*. The reader should consult Chapter 1 for information on how to execute the Python code associated with this book.

8.8.1 Stepped Frequency Waveform

In this example, consider a stepped frequency waveform consisting of 64 pulses transmitted at a pulse repetition frequency of 100 Hz. The frequency step between successive pulses is 50 kHz. Calculate and plot the range profile when there are three targets located at 0.5, 1.5, and 2.0 km. The radar cross section of the targets is 1, 10, and 100 m^2, respectively. Also, the targets have velocities of 0, 0, and 10 m/s.

Solution: The solution to the above example is given in the Python code *stepped_frequency_example.py* and in the MATLAB code *stepped_frequency_example.m*. Running the Python example code displays a GUI allowing the user to enter the waveform parameters and the target parameters. The GUI also allows the user to select a windowing function (Hanning, Hamming, Kaiser, or Blackman-Harris). The code then calculates and displays the range profile for the scenario, as illustrated in Figure 8.23. The figure shows the peaks in the range profile corresponding to the target locations, the range resolution,

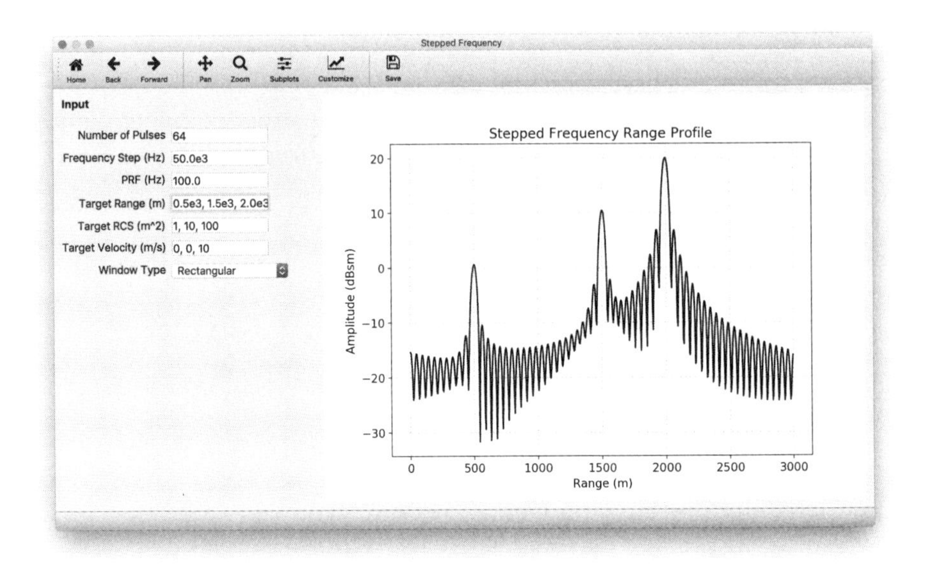

Figure 8.23 The range profile using a stepped frequency waveform calculated by *stepped_frequency_example.py*.

and the sidelobe levels. The user is encouraged to change both the waveform and target parameters to visualize how these parameters affect the range profile.

8.8.2 Matched Filter

The previous example considered an interpulse compression technique. This example focuses on an intrapulse technique using the matched filter. Suppose a radar system transmits an LFM waveform with a pulsewidth of 100 μs and a bandwidth of 20 Mhz. Calculate and plot the output of the matched filter when there are three targets in the scene. The targets are located at ranges of 100, 200, and 500m from the start of the range window and their associated radar cross section values are 1, 10, and 100 m^2.

Solution: The solution to the above example is given in the Python code *matched_filter_example.py* and in the MATLAB code *matched_filter_example.m*. Running the Python example code displays a GUI allowing the user to enter the waveform parameters and the target parameters. The GUI also allows the user to select a windowing function. The code then calculates and displays the range profile for the scenario, as illustrated in Figure 8.24. The figure shows the peaks in the range profile corresponding to the target locations, the range resolution, and the sidelobe levels. The user is encouraged to change both the waveform and target parameters to visualize how these parameters affect the range profile.

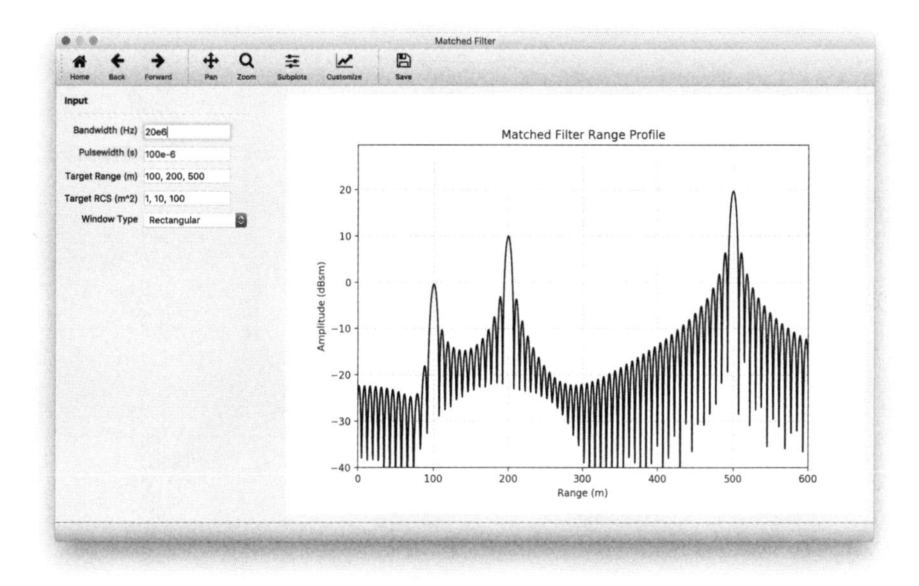

Figure 8.24 The range profile using a matched filter calculated by *matched_filter_example.py*.

8.8.3 Stretch Processor

While the matched filter may be used for narrow and medium bandwidth waveforms, the stretch processor is required for very large bandwidths due to sampling requirements. Consider a radar system transmitting an LFM waveform with a pulsewidth of 1 ms and a bandwidth of 1 GHz. Calculate and plot the output of the stretch processor with a processing window of 50m when there are two targets in the scene. The targets are located at relative ranges of 5 and 15m and their associated radar cross section values are 20 and 10 m^2.

Solution: The solution to the above example is given in the Python code *stretch_processor_example.py* and in the MATLAB code *stretch_processor_example.m*. Running the Python example code displays a GUI allowing the user to enter the waveform parameters and the target parameters. The GUI also allows the user to select a windowing function. The code then calculates and displays the range profile for the scenario, as illustrated in Figure 8.25. The figure shows the peaks in the range profile corresponding to the target locations, the range resolution, and the sidelobe levels. The user is encouraged to change both the waveform and target parameters to visualize how these parameters affect the range profile.

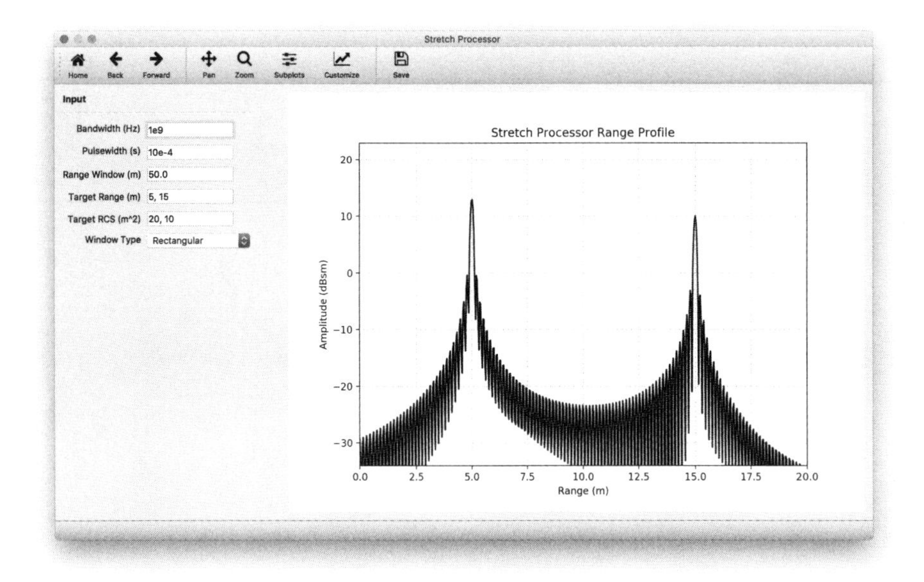

Figure 8.25 The range profile using a stretch processor calculated by *stretch_processor_example.py*.

8.8.4 Unmodulated Pulse Ambiguity

As a first example of ambiguity functions, consider the ambiguity function associated with a single unmodulated pulse. Calculate and display the ambiguity function for a single unmodulated pulse with a pulsewidth of 1 ms.

Solution: The solution to the above example is given in the Python code *single_pulse_ambiguity_example.py* and in the MATLAB code *single_pulse_ambiguity_example.m*. Running the Python example code displays a GUI allowing the user to enter the pulsewidth and select a plot type. The 2D contour, zero-Doppler, and zero-time delay plots of the ambiguity function are shown in Figures 8.26, 8.27, and 8.28. Note that the ambiguity function, and therefore the range and Doppler resolution for a single unmodulated pulse is controlled only by the pulsewidth, as discussed in Section 8.6.1.

8.8.5 LFM Pulse Ambiguity

The next ambiguity function example is of a single LFM pulse. Calculate and display the ambiguity function for single LFM pulse with a pulsewidth of 1 ms and a bandwidth of

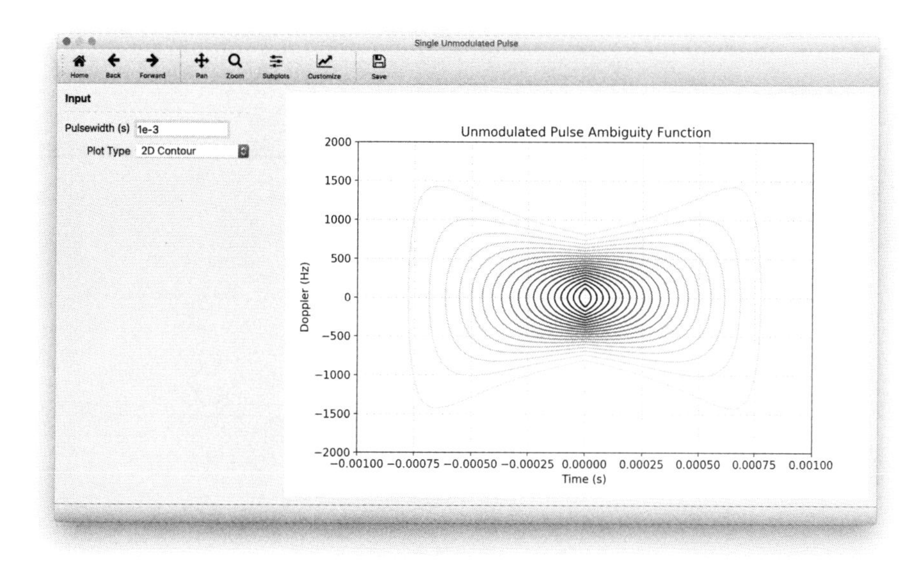

Figure 8.26 The 2D contour of the single pulse ambiguity function calculated by *single_pulse_ambiguity_example.py*.

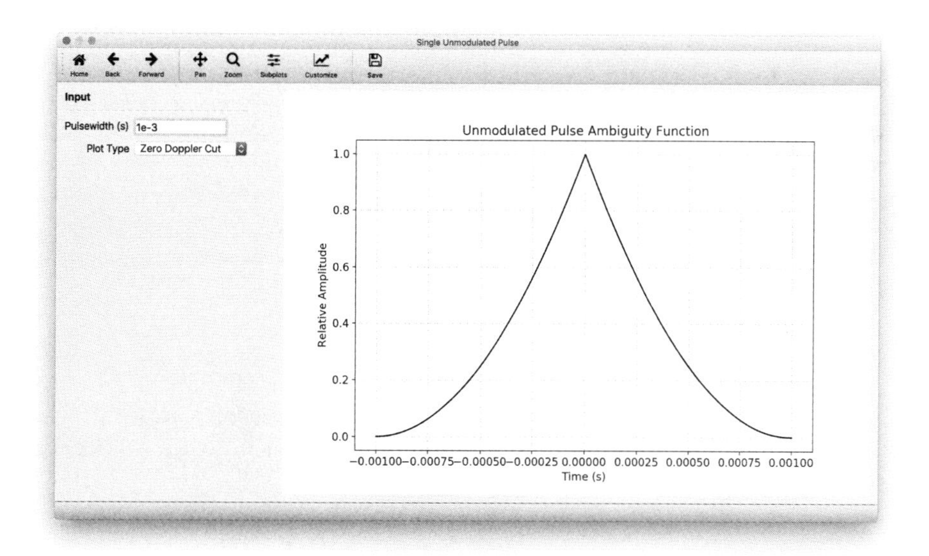

Figure 8.27 The zero-Doppler cut of the single pulse ambiguity function calculated by *single_pulse_ambiguity_example.py*.

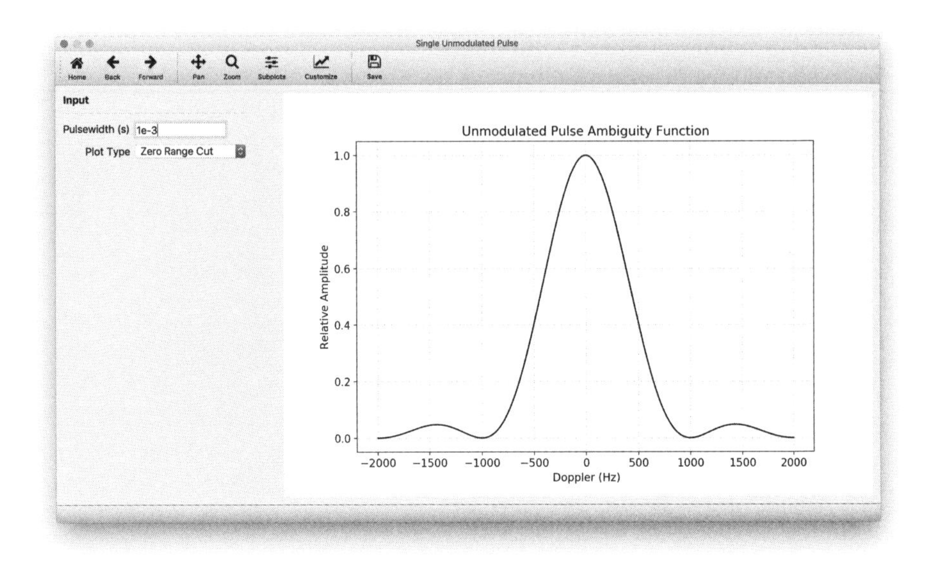

Figure 8.28 The zero-time delay cut of the single pulse ambiguity function calculated by *single_pulse_ambiguity_example.py*.

2 kHz.

Solution: The solution to the above example is given in the Python code *lfm_pulse_ ambiguity_example.py* and in the MATLAB code *lfm_pulse_ambiguity_example.m*. Running the Python example code displays a GUI allowing the user to enter the pulsewidth and bandwidth of the LFM, and to select a plot type. The 2D contour, zero-Doppler, and zero-time delay plots of the ambiguity function are shown in Figures 8.29, 8.30, and 8.31. In this case, the ambiguity function is controlled by both the pulsewidth and the bandwidth, as illustrated in Section 8.6.2

8.8.6 Coherent Pulse Train Ambiguity

Now consider the ambiguity function of a train of unmodulated coherent pulses. Calculate and display the ambiguity function for a train of 8 coherent pulses transmitted with a pulse repetition interval of 0.5s, each with a pulsewidth of 0.1s.

Solution: The solution to the above example is given in the Python code *pulse_train_ ambiguity_example.py* and in the MATLAB code *pulse_train_ambiguity_example.m*. Running the Python example code displays a GUI allowing the user to enter the number of pulses, the pulsewidth, the pulse repetition frequency, and to select a plot type. The 2D

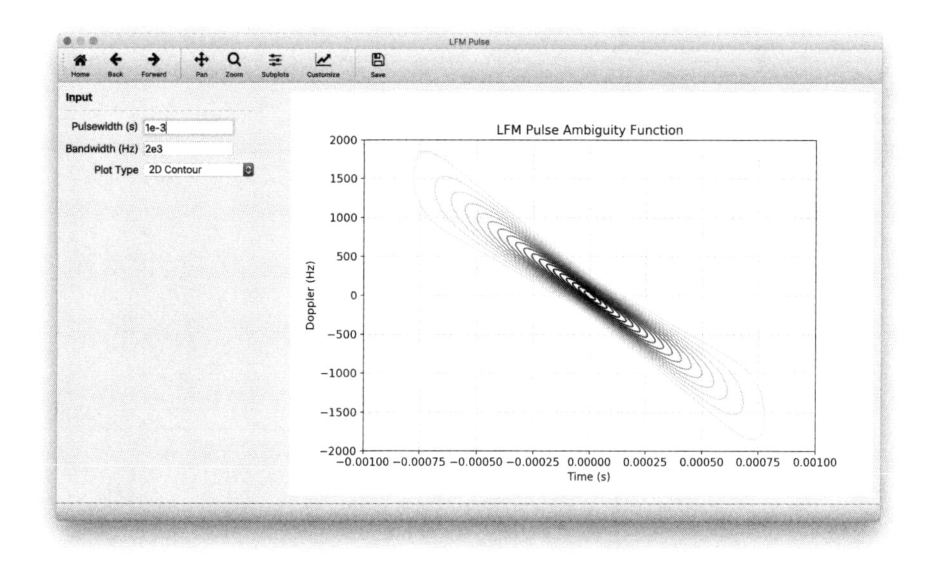

Figure 8.29 The 2D contour of the LFM ambiguity function calculated by *lfm_pulse_ambiguity_example.py*.

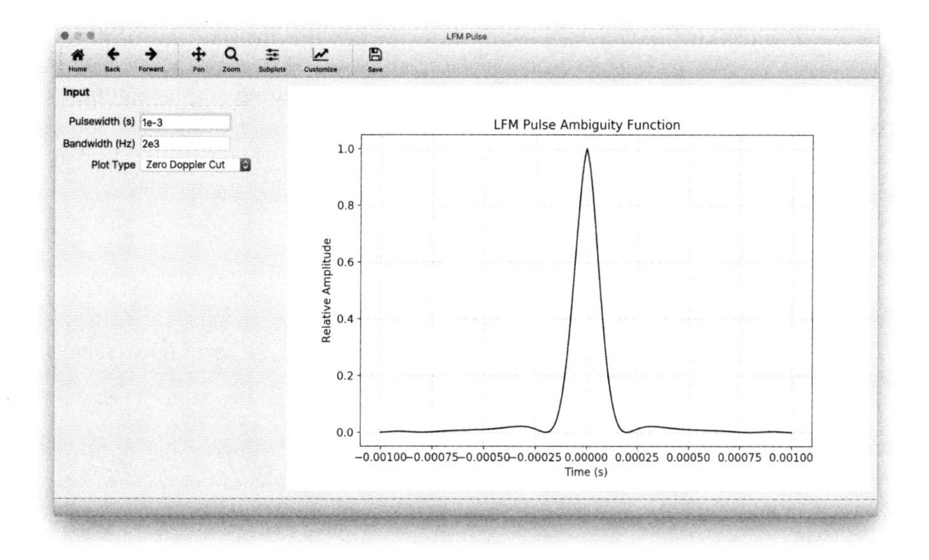

Figure 8.30 The zero-Doppler cut of the LFM ambiguity function calculated by *lfm_pulse_ambiguity_example.py*.

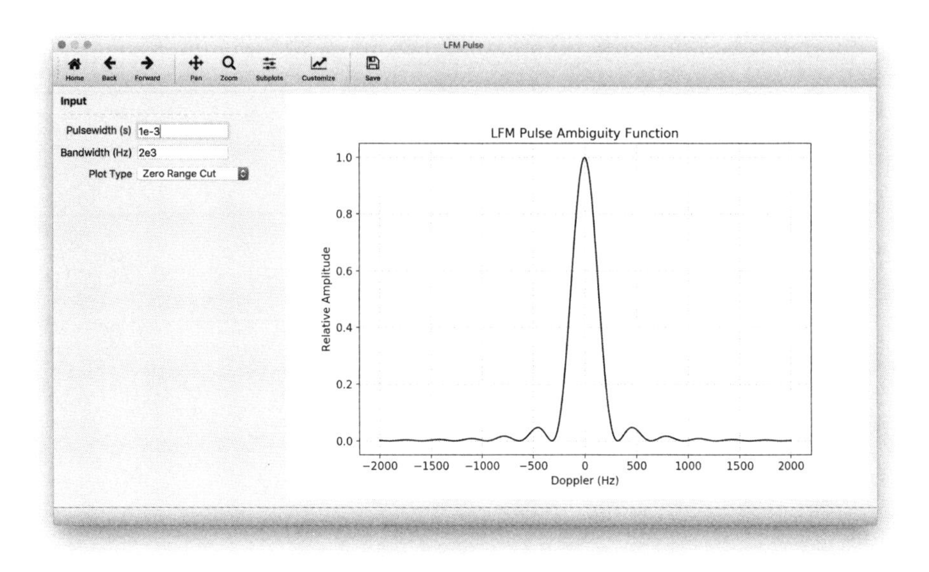

Figure 8.31 The zero-time delay cut of the LFM ambiguity function calculated by *lfm_pulse_ambiguity_example.py*.

contour, zero-Doppler, and zero-time delay plots of the ambiguity function are shown in Figures 8.32, 8.33, and 8.34. The range and Doppler resolutions are controlled by the pulsewidth as in the single pulse case. However, there are now several ambiguities in range and Doppler. Note that there are N–1 sidelobes on either side of the main lobe in the range dimension, and each sidelobe is separated by one PRI. The Doppler sidelobes are separated by $1/\tau$, where τ is the pulsewidth.

8.8.7 LFM Pulse Train Ambiguity

This example will consider a train of LFM pulses. Calculate and display the ambiguity function for a train of 6 LFM pulses transmitted with a pulse repetition interval of 1s, each with a pulsewidth of 0.4s and a bandwidth of 10 Hz.

Solution: The solution to the above example is given in the Python code *lfm_train_ ambiguity_example.py* and in the MATLAB code *lfm_train_ambiguity_example.m*. Running the Python example code displays a GUI allowing the user to enter the number of pulses, the pulsewidth, the bandwidth, the pulse repetition frequency, and to select a plot type. The 2D contour, zero-Doppler and zero-time delay plots of the ambiguity function are shown in Figures 8.35, 8.36, and 8.37. The range and Doppler resolutions are controlled by the pulsewidth and bandwidth as in the single LFM pulse case. However, there are

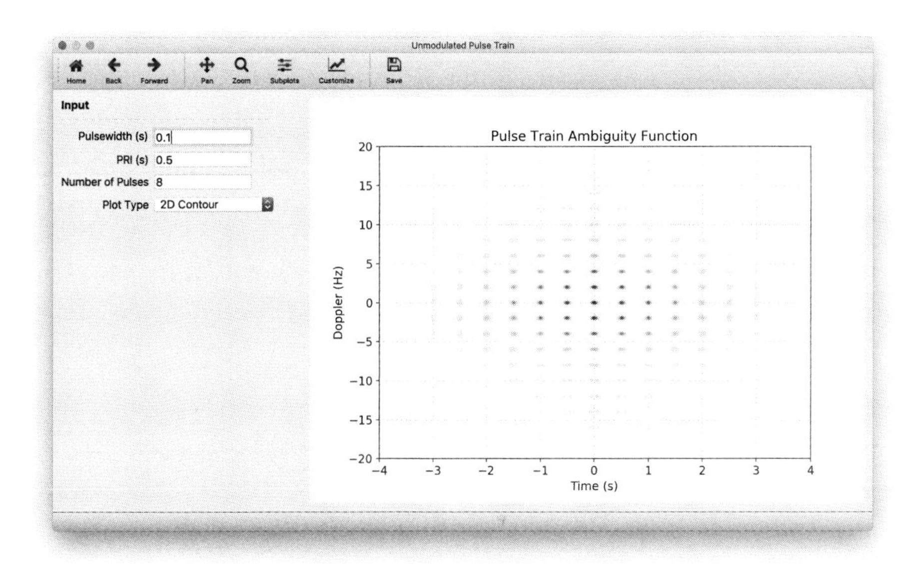

Figure 8.32 The 2D contour of the coherent pulse train ambiguity function calculated by *pulse_train_ambiguity_example.py*.

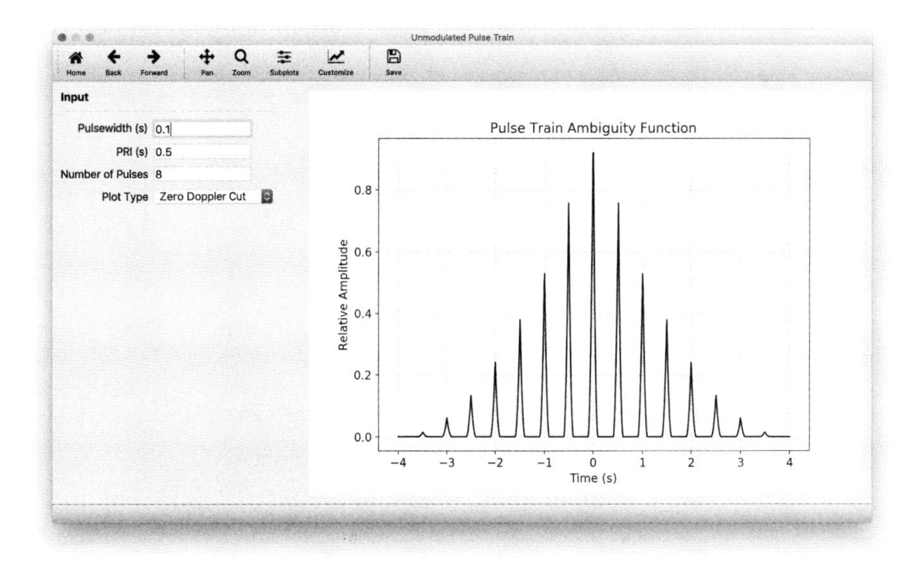

Figure 8.33 The zero-Doppler cut of the coherent pulse train ambiguity function calculated by *pulse_train_ambiguity_example.py*.

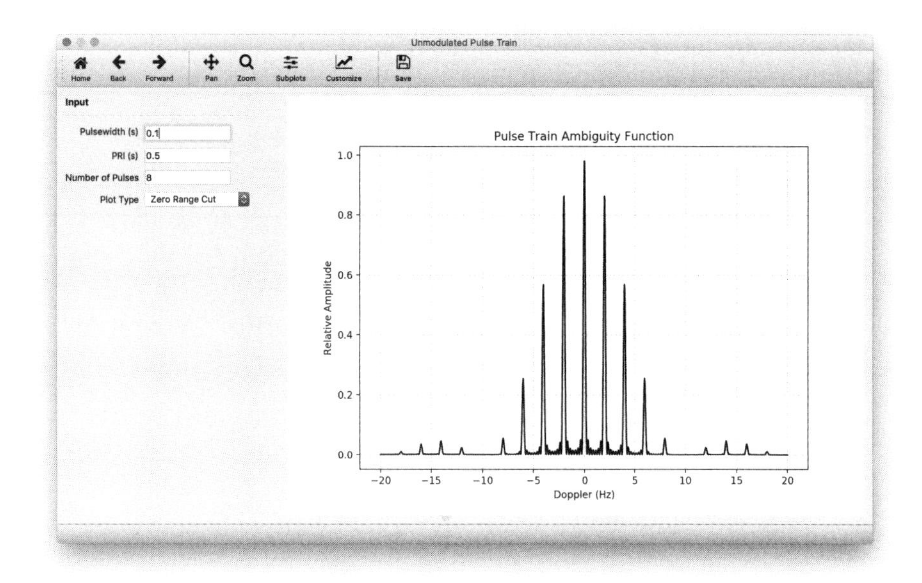

Figure 8.34 The zero-time delay cut of the coherent pulse train ambiguity function calculated by *pulse_train_ambiguity_example.py*.

now several ambiguities in range and Doppler. In comparison with the coherent pulse train case, there is also a difference in the ambiguity pattern due to the slope of the LFM.

8.8.8 Barker Code Ambiguity

The next few examples are focused on phase-coded waveforms, beginning with Barker-coded waveforms. Calculate and display the ambiguity function for a Barker-coded waveform of length 13 and a chip width of 0.1s.

Solution: The solution to the above example is given in the Python code *barker_ambiguity_example.py* and in the MATLAB code *barker_ambiguity_example.m*. Running the Python example code displays a GUI allowing the user to enter the chip width, choose a Barker code length, and to select a plot type. The 2D contour, zero-Doppler, and zero-time delay plots of the ambiguity function are shown in Figures 8.38, 8.39, and 8.40. The user is encouraged to select other Barker code lengths in the GUI to further illustrate the sidelobe structure and resolution discussed in Section 8.7.1.

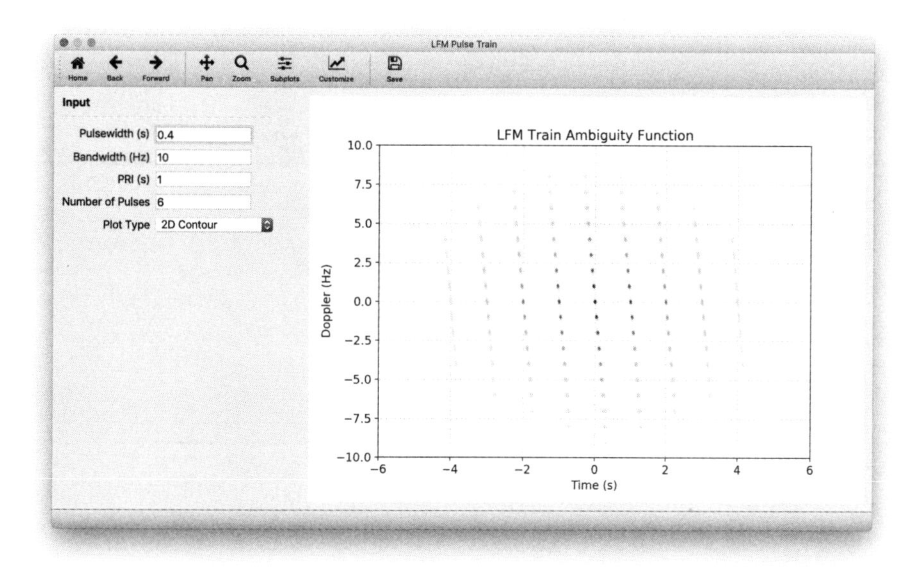

Figure 8.35 The 2D contour of the LFM pulse train ambiguity function calculated by *lfm_train_ambiguity_example.py*.

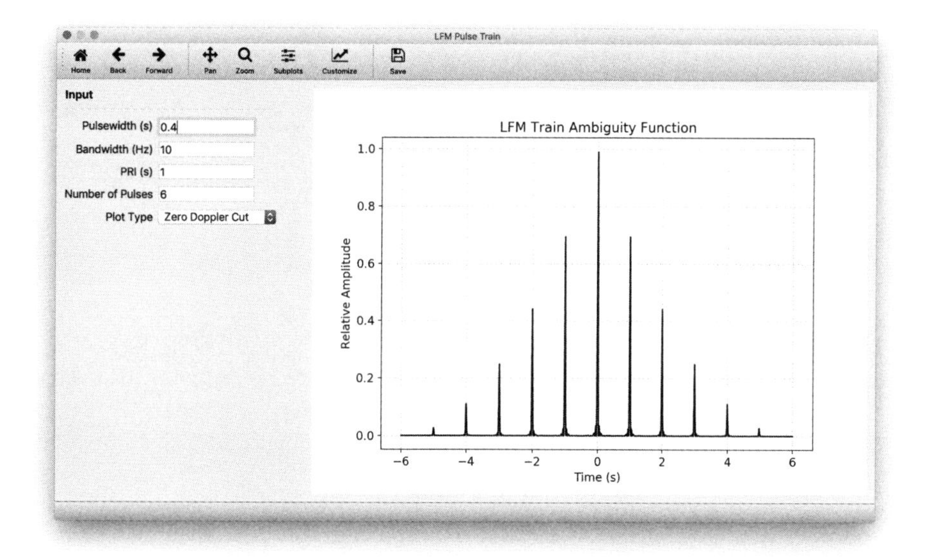

Figure 8.36 The zero-Doppler cut of the LFM pulse train ambiguity function calculated by *lfm_train_ambiguity_example.py*.

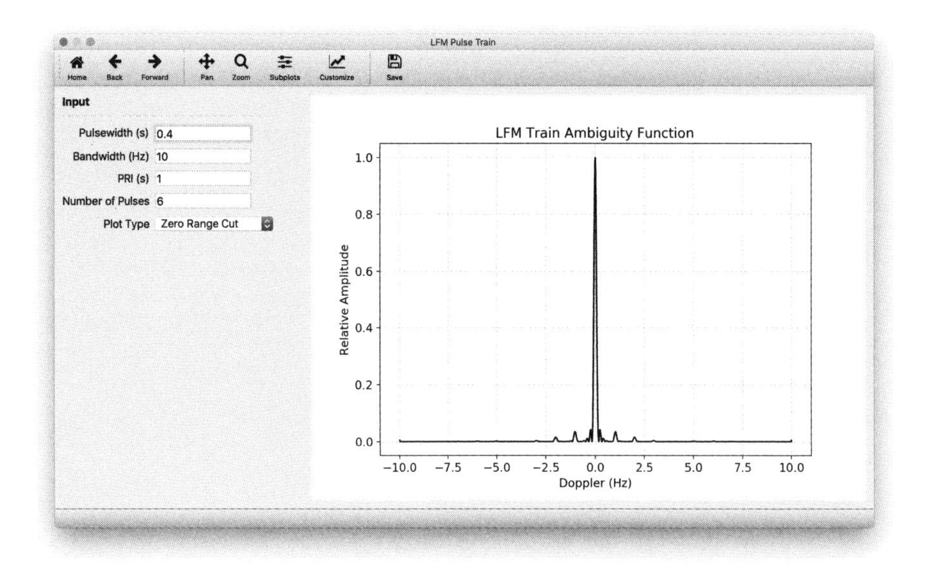

Figure 8.37 The zero-time delay cut of the LFM pulse train ambiguity function calculated by *lfm_train_ambiguity_example.py*.

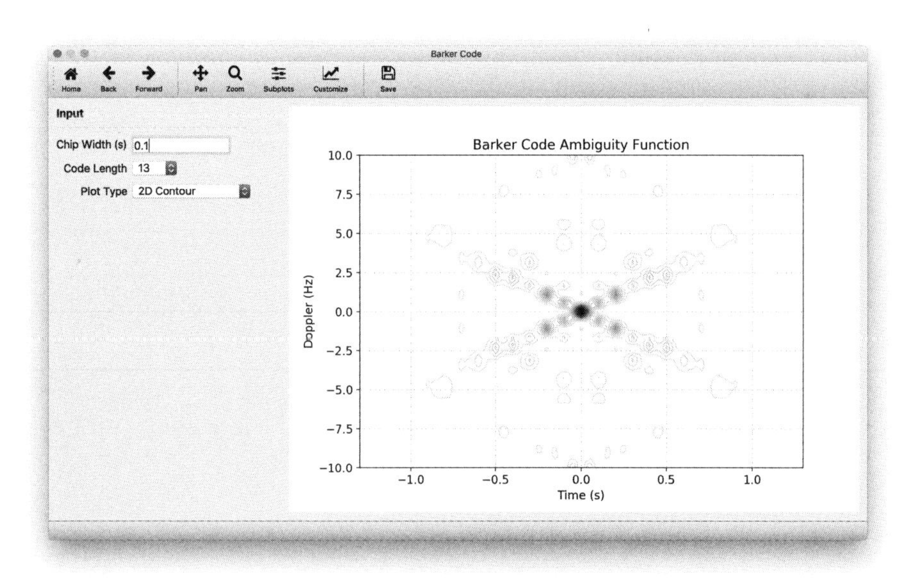

Figure 8.38 The 2D contour of the B_{13} ambiguity function calculated by *barker_ambiguity_example.py*.

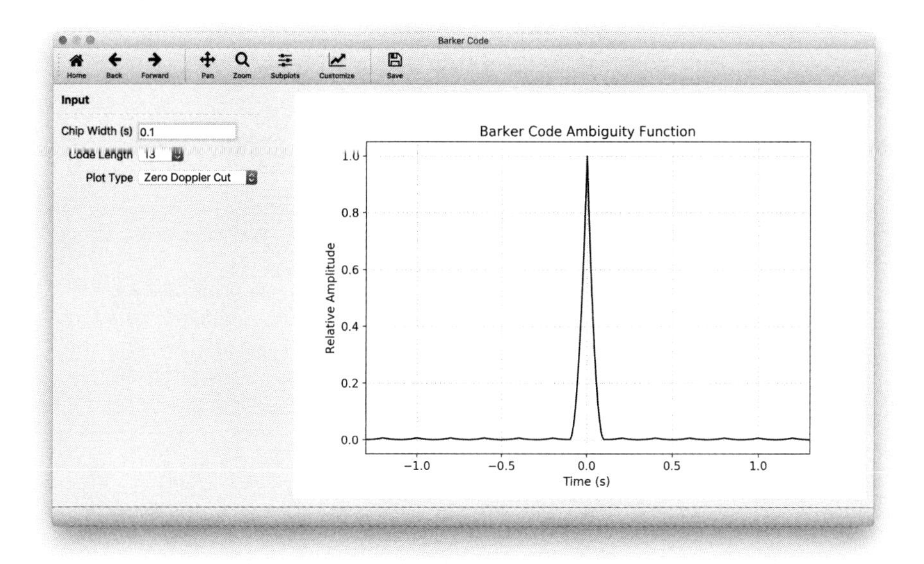

Figure 8.39 The zero-Doppler cut of the B_{13} ambiguity function calculated by *barker_ambiguity_example.py*.

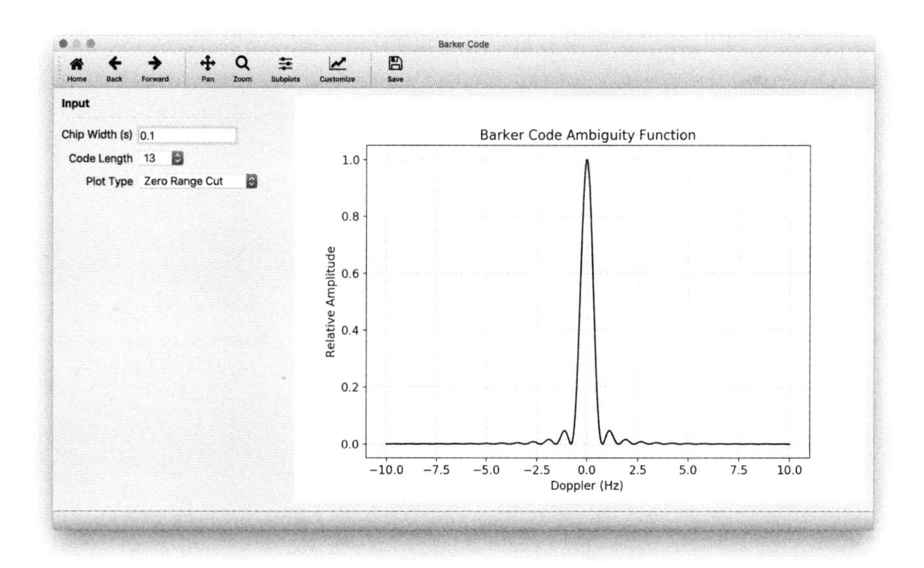

Figure 8.40 The zero-time delay cut of the B_{13} ambiguity function calculated by *barker_ambiguity_example.py*.

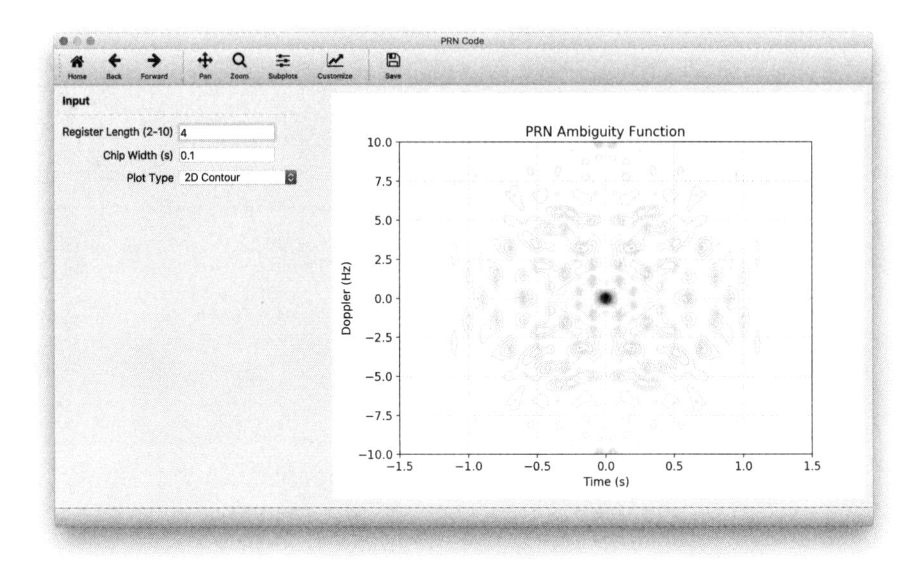

Figure 8.41 The 2D contour of a PRN$_{15}$ ambiguity function calculated by *PRN_ambiguity_example.py*.

8.8.9 PRN Code Ambiguity

The next ambiguity function example is for PRN-coded waveforms. Calculate and display the ambiguity function for a PRN-coded waveform with a chip width of 0.1s. The PRN code is generated with a register of length 4, which results in a sequence of length 15.

Solution: The solution to the above example is given in the Python code *PRN_ambiguity_example.py* and in the MATLAB code *PRN_ambiguity_example.m*. Running the Python example code displays a GUI allowing the user to enter the chip width, choose a register length, and to select a plot type. The 2D contour, zero-Doppler, and zero-time delay plots of the ambiguity function are shown in Figures 8.41, 8.42, and 8.43. Note the seemingly random pattern in the sidelobe structure. While the initial state of the linear feedback shift register has an effect on the sidelobe structure, there is no systematic method for determining the best initial state. The user is encouraged to select other register lengths in the GUI to further illustrate the sidelobe structure and resolution discussed in Section 8.7.3.

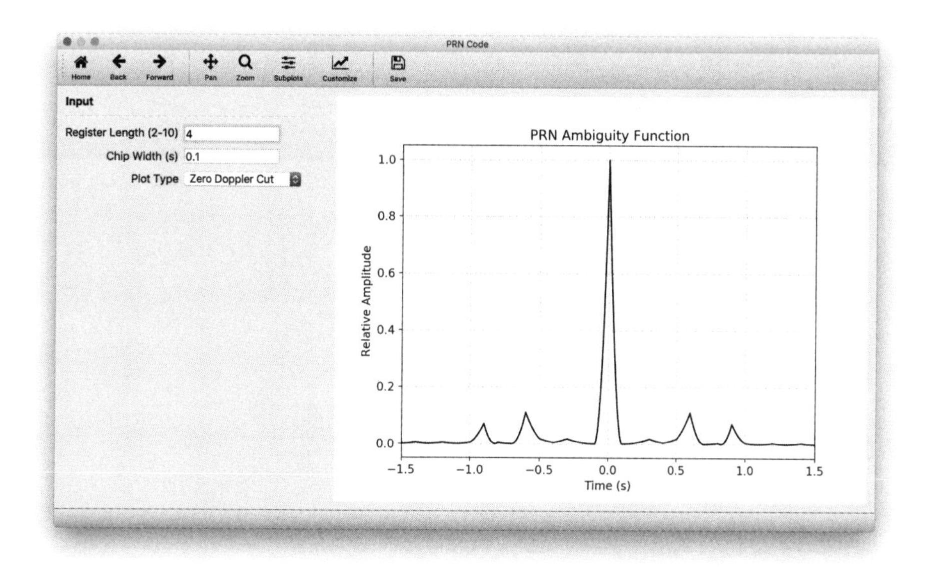

Figure 8.42 The zero-Doppler cut of a PRN_{15} ambiguity function calculated by *PRN_ambiguity_example.py*.

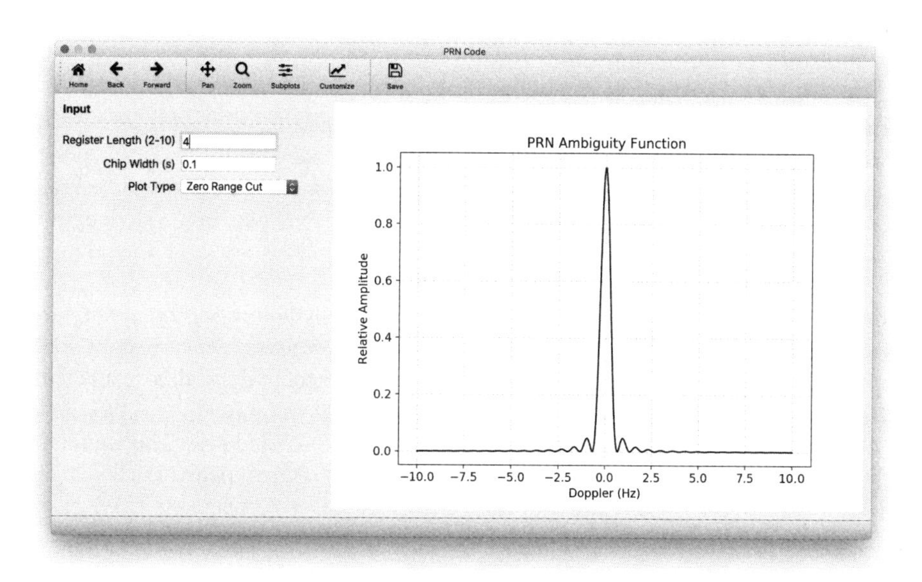

Figure 8.43 The zero-time delay cut of a PRN_{15} ambiguity function calculated by *PRN_ambiguity_example.py*.

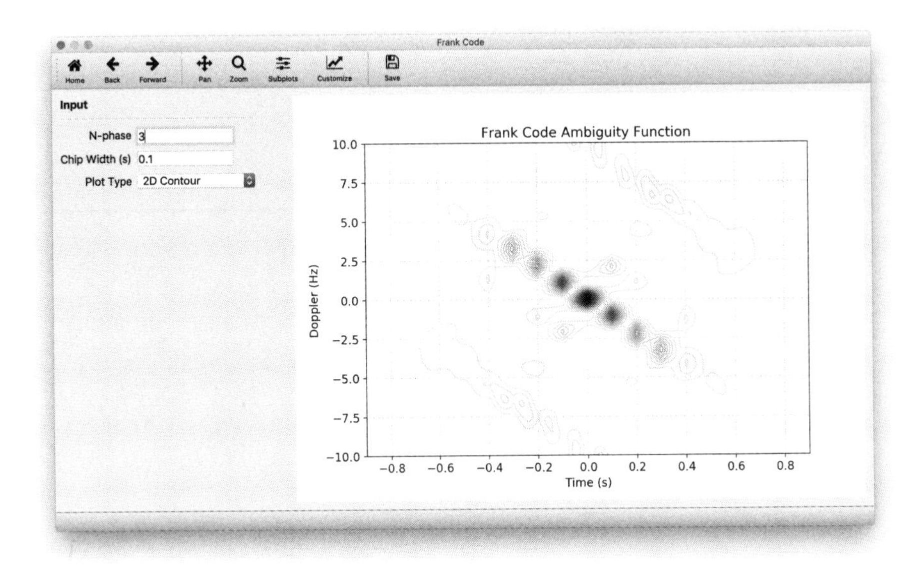

Figure 8.44 The 2D contour of the F_9 ambiguity function calculated by *frank_ambiguity_example.py*.

8.8.10 Frank Code Ambiguity

The Barker code and PRN code examples are both binary phase-coded waveforms. As a last example, consider the ambiguity function of a polyphase-coded waveform, specifically the Frank codes. Calculate and display the ambiguity function for an N-phase Frank code waveform with a chip width of 0.1s and $N = 3$, which results in a code length of $L = 9$.

Solution: The solution to the above example is given in the Python code *frank_ambiguity_example.py* and in the MATLAB code *frank_ambiguity_example.m*. Running the Python example code displays a GUI allowing the user to enter the chip width, choose a register length, and to select a plot type. The 2D contour, zero-Doppler, and zero-time delay plots of the ambiguity function are shown in Figures 8.44, 8.45, and 8.46. The ambiguity function is very similar to the LFM ambiguity function as discussed in Section 8.7.2. The user is encouraged to choose Frank codes of varying lengths to compare the LFM case.

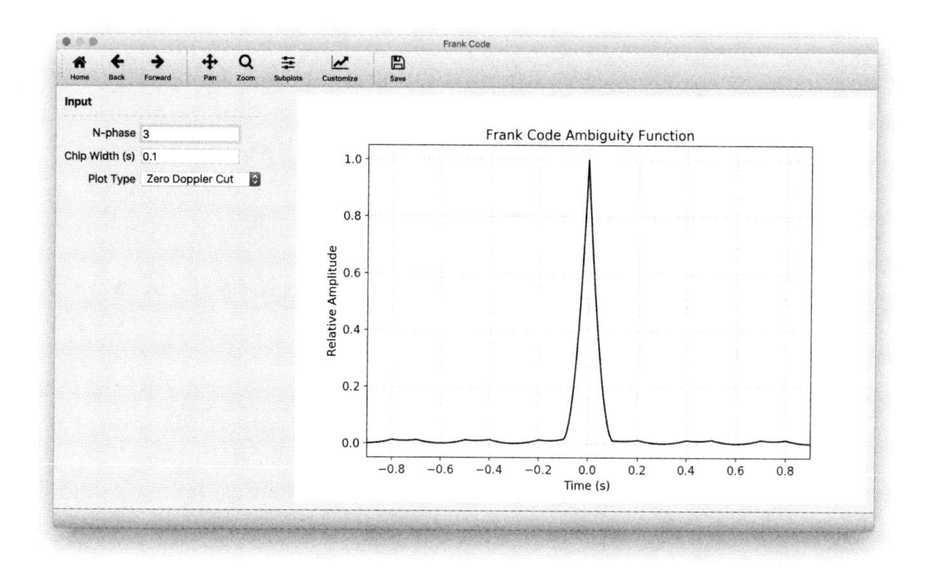

Figure 8.45 The zero-Doppler cut of the F_9 ambiguity function calculated by *frank_ambiguity_example.py*.

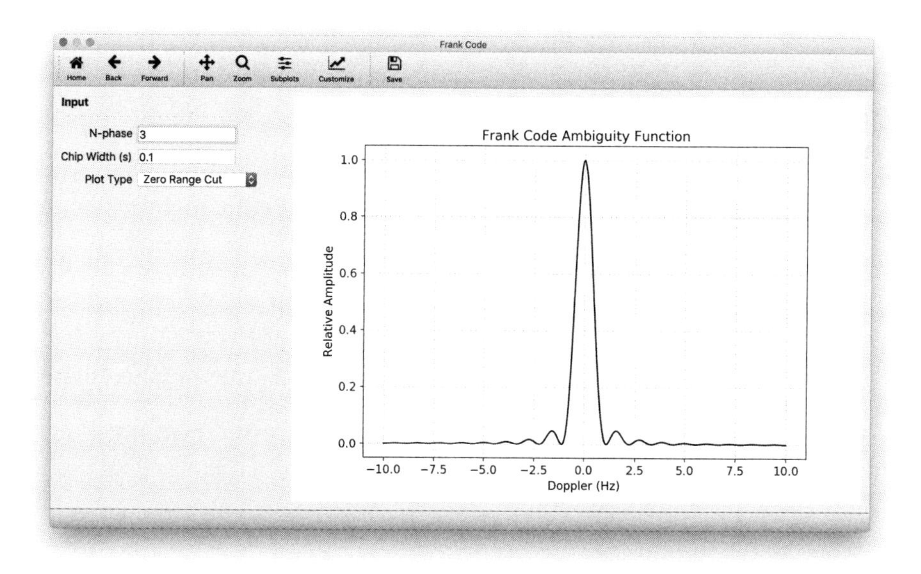

Figure 8.46 The zero-time delay cut of the F_9 ambiguity function calculated by *frank_ambiguity_example.py*.

PROBLEMS

8.1 Calculate the effective bandwidth and range resolution of a stepped frequency waveform consisting of 64 steps of 30 kHz each.

8.2 For the stepped frequency waveform given in Problem 8.1, find the dispersion for a target traveling at 50 m/s. Assume the pulse repetition interval is 0.1s.

8.3 Find the number of bins that the peak of the target scattering is shifted for the waveform and target parameters given in Problem 8.2.

8.4 Explain why the matched filter is the optimal linear filter for maximizing the signal-to-noise ratio in the presence of additive stochastic noise.

8.5 What is the increase in the signal-to-noise ratio at the output of the matched filter compared to the input for a waveform with a pulsewidth of 1 ms and bandwidth of 500 MHz?

8.6 Describe the major advantages of stretch processing and give practical examples of when stretch processing should be employed.

8.7 For a linear frequency modulated waveform with a bandwidth of 500 MHz and a pulsewidth of 0.5 ms, find the instantaneous frequency for stretch processing when the target is at a range of 100 km and the reference signal time delay is 0.6 ms.

8.8 Consider a waveform with a bandwidth of 2 GHz, a pulsewidth of 8 ms, and a maximum receive window of 100m. Compare the number of analog-to-digital samples required for stretch processing versus the matched filter.

8.9 Prove the eight properties of the ambiguity function given in Section 8.6.

8.10 Describe the difference in the ambiguity function for an unmodulated pulse and a linear frequency modulated pulse.

8.11 For a Barker phase-coded waveform of $N = 11$, what is the length of the ambiguity function, the number of sidelobes, the sidelobe level, and the main lobe width?

8.12 Explain the difference in the ambiguity function between a Frank phase-coded waveform and a linear frequency modulated waveform.

8.13 Give the phase of each chip of a Barker phase-coded waveform consisting of a Barker code of length seven embedded in a Barker code of length four.

8.14 Give the phase of each chip for a Frank phase-coded waveform for $N = 4$.

8.15 What is the sequence length of a maximal length sequence generated with a shift register of length $N = 10$?

8.16 For the maximal length sequence given in Problem 8.15, how many ones and zeros, and how many runs of length four are in the sequence?

References

[1] S. D. Blunt, K. Gerlach and T. Higgins. Aspects of radar range super-resolution. *2007 IEEE Radar Conference*, pages 683–687, April 2007.

[2] M. Skolnik. *Radar Handbook,* 3rd ed. McGraw-Hill Education, 2008.

[3] A. Paulose. High radar range resolution with the step frequency waveform. Master's thesis, Naval Post Graduate School, Monterey, CA, June 1994.

[4] D. O. North. An analysis of the factors which determine signal/noise discrimination in pulsed carrier systems. Technical Report PPR-6C, RCA Laboratories, Princeton, NJ, 1943.

[5] P. M. Vasic and D. S. Mitrinovic. *Analytic Inequalities.* Springer-Verlag, New York, 1970.

[6] W. Kaplan. *Advanced Calculus,* 4th ed. Addison-Wesley, Reading, MA, 1992.

[7] G. W. Stimson. *Stimson's Introduction to Airborne Radar,* 3rd ed. SciTech Publishing, Stevenage, UK, 2014.

[8] W. J. Caputi, Jr. Stretch: A time-transformation technique. *IEEE Transactions on Aerospace and Electronic Systems*, 7(2):269–278, March 1971.

[9] M. C. Budge and S. R. German. *Basic Radar Analysis.* Artech House, Norwood, MA, 2015.

[10] B. R. Mahafza. *Radar Systems Analysis and Design Using MATLAB,* 3rd ed. CRC Press, Boca Raton, FL, 2013.

[11] J. Mulcahy-Stanislawczyk. Properties of ambiguity functions. Master's thesis, Purdue University, West Lafayette, IN, May 2014.

[12] P. M. Woodward. *Probability and Information Theory with Applications to Radar.* Pergamon Press, London, 1953.

[13] C. E. Shannon. A mathematical theory of communication. *The Bell System Technical Journal*, 27(3):379–423, July 1948.

[14] M. N. Cohen, M. R. Fox and J. M. Baden. Minimum peak sidelobe pulse compression codes. *Proceedings of the 1990 IEEE International Radar Conference*, pages 633–638, May 1990.

[15] J. Linder. Binary sequences up to length 40 with best possible autocorrelation function. *Proceedings of the IEEE Electronic Letters*, 11(21):507, October 1975.

[16] G. Coxson and J. Russo. Efficient exhaustive search for optimal peak sidelobe binary codes. *IEEE Transactions on Aerospace and Electronic Systems*, 41(1):302–308, January 2005.

[17] A. N. Leukhin and E. N. Potekhin. Optimal peak sidelobe level sequences up to length 74. *2013 European Radar Conference*, pages 495–498, October 2013.

[18] R. L. Frank. Polyphase complementary codes. *IEEE Transactions on Information Theory*, 26(6):641–647, November 1980.

[19] S. A. Zadoff. Phase coded communication system, July 1963. United States Patent No. 3,099,796.

[20] D. C. Chu. Polyphase codes with good periodic correlation properties. *IEEE Transactions on Information Theory*, 18(4):531–532, July 1972.

[21] M. Antweiler and L. Bomer. Merit factor of Chu and Frank sequences. *Electronic Letters*, 26(25):2068–2070, December 1990.

[22] S. W. Golomb. *Shift Register Sequences*. Holden-Day, Norwood, MA, 1967.

[23] S. W. Golomb and G. Gong. *Signal Design for Good Correlation: For Wireless Communication, Cryptography, and Radar*. Cambridge University Press, Cambridge, UK, 2005.

[24] P. Koopman. *Maximal Length LFSR Feedback Terms*. `http://users.ece.cmu.edu/~koopman/lfsr/`.

[25] Wikipedia contributors. Linear-feedback shift register. `https://en.wikipedia.org/wiki/Linear-feedback_shift_register`.

[26] M. Abramowitz and I. A. Stegun. *Handbook of Mathematical Functions: with Formulas, Graphs, and Mathematical Tables*. U.S. National Bureau of Standards, Applied Mathematics Series, Washington, DC, 1964.

[27] D. J. Rabideau and P. Parker. Ubiquitous mimo multifunction digital array radar. *The Thirty-Seventh Asilomar Conference on Signals, Systems and Computers*, November 2013.

[28] M. Inggs, et al. Multistatic radar: System requirements and experimental validation. *2014 International Radar Conference*, October 2014.

Chapter 9

Target Tracking

There are many applications for tracking with a radar system, including target location, autonomous navigation, weather prediction, air traffic control, and military applications. The chapter begins with simple alpha-beta and alpha-beta-gamma tracking filters. The Kalman filter is then introduced in one dimension and later extended to the multivariate formulation. Next, adaptive filtering techniques for maneuvering targets are presented. Multiple target tracking techniques are then studied. These include global nearest neighbor, joint probabilistic data association, multiple hypothesis tracking, and random finite sets. Finally, a method for generating radar pseudomeasurements for input to tracking algorithms is given. The chapter concludes with several Python examples to demonstrate the various techniques associated with target tracking.

9.1 TRACKING FILTERS

One of the main purposes of the tracking filter is to provide an accurate and precise estimation of the position, velocity, and acceleration of a target in the presence of uncertainty. As illustrated in previous chapters, the measurement uncertainties depend on many factors such as thermal noise, atmospheric effects, quantization, and interpolation. There are also uncertainties in the model representing the motion of the target. A radar system transmits energy in the direction of the target at a given pulse repetition interval. Based on the return signal, the radar estimates the target's current position, velocity, and acceleration. Also, the radar predicts the target state for the next update. The target's predicted position is calculated using Newton's equations of motion as [1]

$$\mathbf{x} = \mathbf{x}_0 + \mathbf{v}_0 \Delta t + \frac{1}{2}\mathbf{a}\Delta t^2 \qquad \text{(m)}, \tag{9.1}$$

where

$\mathbf{x}$ = future target position (m),
$\mathbf{x}_0$ = target initial position (m),
$\mathbf{v}_0$ = target initial velocity (m/s),
$\mathbf{a}$ = target acceleration (m/s^2),
Δt = track update interval (s).

Expanding (9.1) in a three-dimensional Cartesian coordinate system results in

$$x = x_0 + v_{x0}\Delta t + \frac{1}{2}a_x\Delta t^2 \qquad \text{(m)}, \tag{9.2}$$

$$y = y_0 + v_{y0}\Delta t + \frac{1}{2}a_y\Delta t^2 \qquad \text{(m)}, \tag{9.3}$$

$$z = z_0 + v_{z0}\Delta t + \frac{1}{2}a_z\Delta t^2 \qquad \text{(m)}. \tag{9.4}$$

The target position, velocity, and acceleration $(x, y, z, v_x, v_y, v_z, a_x, a_y, a_z)$ are often referred to as the *system state*, and (9.2)–(9.4) is called the *state space model*. The current or measured state is the input to the tracking algorithm, and the predicted future state is one of the outputs of the algorithm. The prediction of the future state can be difficult to calculate as radar measurements have random errors and uncertainty, often referred to as *measurement noise*. Also, the target motion does not strictly follow (9.1) due to wind, drag, maneuvering targets, and so on. The uncertainty in the state space model is referred to as *process noise*. Both the process noise and measurement noise could result in a poor estimate of the target position. Therefore, a tracking algorithm accounting for these uncertainties is required. A few different tracking algorithms are presented in the following sections.

9.1.1 Alpha-Beta Filter

The alpha-beta filter, sometimes referred to as the g-h filter, is a simplified filter for parameter estimation and smoothing [2]. The alpha-beta filter is related to Kalman filters but does not require a detailed system model. It presumes that the system is approximated by two internal states. The first state is determined by integrating the second state over time. The radar measurements are the observations of the first model state. This is a low-order approximation and may be adequate for simple tracking problems, such as tracking a target's position where the position is found from the time integral of the velocity. Assuming the velocity remains fixed over the time interval between measurements, the position is projected forward in time to predict its value at the next sampling time as

$$\mathbf{x}_{k,k-1} = \mathbf{x}_{k-1,k-1} + \mathbf{v}_{k-1,k-1}\Delta t \qquad \text{(m)}, \tag{9.5}$$

where

$\mathbf{x}_{k-1,k-1}$ = filtered position at update $k-1$ given measurements up to and including $k-1$ (m),
$\mathbf{v}_{k-1,k-1}$ = filtered velocity at update $k-1$ given measurements up to and including $k-1$ (m/s),
Δt = measurement update interval (s),
$\mathbf{x}_{k,k-1}$ = predicted position at update k given measurements up to and including $k-1$ (m).

The velocity is assumed to be constant, so the predicted value is equal to the current value as

$$\mathbf{v}_{k,k-1} = \mathbf{v}_{k-1,k-1} \qquad \text{(m/s)}. \tag{9.6}$$

Since the alpha-beta model is a simplified dynamic model, the measurement and the predicted value are expected to deviate due to noise and other effects not included in the model. This error is often referred to as the *residual* or *innovation*, and is expressed as

$$\mathbf{r}_k = \mathbf{z}_k - \mathbf{x}_{k,k-1} \qquad \text{(m)}, \tag{9.7}$$

where

$\mathbf{z}_k$ = measurement at update k (m),
$\mathbf{x}_{k,k-1}$ = predicted position at update k (m),
$\mathbf{r}_k$ = residual at update k (m).

The alpha-beta filter uses selected constants, α and β, to adjust the position and velocity estimates as

$$\mathbf{x}_{k,k} = \mathbf{x}_{k,k-1} + \alpha\, \mathbf{r}_k \qquad \text{(m)}, \tag{9.8}$$

$$\mathbf{v}_{k,k} = \mathbf{v}_{k,k-1} + \frac{\beta}{\Delta t}\mathbf{r}_k \qquad \text{(m/s)}, \tag{9.9}$$

where

α, β = user selected constants for the filter,
$\mathbf{r}_k$ = residual at update k (m),
$\mathbf{x}_{k,k-1}$ = predicted position at update k (m),
$\mathbf{v}_{k,k-1}$ = predicted velocity at update k (m/s),
$\mathbf{x}_{k,k}$ = filtered position at update k given measurements up to and including k (m),
$\mathbf{v}_{k,k}$ = filtered velocity at update k given measurements up to and including k (m/s).

The adjustments in (9.8) and (9.9) are small steps along the estimated gradient direction. As more measurements are taken, and the adjustments accumulate, the error in the filtered state is reduced. The values of α and β are usually found through experiment. In general, larger values of α and β result in quicker response to transient changes, while smaller values of α and β reduce the noise level in the estimated state. Therefore, the values of α and β should be positive and small for good convergence and stability [3]. Typical selections for α and β are

$$0 < \alpha < 1, \tag{9.10}$$

$$0 < \beta \leq 2, \tag{9.11}$$

$$0 < 4 - 2\alpha - \beta. \tag{9.12}$$

Note that noise is amplified if $\beta \geq 1$, and reduced if $0 < \beta < 1$. The algorithm for the alpha-beta filter proceeds as follows:

1. Select values for α and β.
2. Initialize $\mathbf{x}_{0,0}$ and $\mathbf{v}_{0,0}$ with initial measurement values, or set to zero.
3. Predict state $\mathbf{x}_{k,k-1}$ and $\mathbf{v}_{k,k-1}$ using (9.5) and (9.6).
4. Calculate the residual from (9.7) and adjust the predicted state using (9.8) and (9.9).
5. Repeat steps (3)–(4) for each new measurement with $\mathbf{x}_{k,k}$ and $\mathbf{v}_{k,k}$ as the filter output.

A simple simulation for the alpha-beta filter using the steps outlined above is given in Listing 9.1. The filtered position is shown in Figure 9.1, the filtered velocity in Figure 9.2, and the residual in Figure 9.3. The measurement noise variance in this example is large, as seen in the measurement values in Figure 9.1. This leads to large residual values shown in Figure 9.3. The filtered velocity in Figure 9.2 is slowly converging and may be improved with different values of α and β. However, care must be taken as this may lead to larger noise in the filtered position. The optimal choice of α and β for a given sampling interval, process noise variance, and measurement noise variance has been researched, and the reader is referred to [4, 5] for more information.

Listing 9.1 Alpha-Beta Filter

```
from scipy import linspace, random

# Time (s)
t, dt = linspace(0, 10, 200, retstep=True)

# True position (m) and velocity (m/s)
v_true = 3.5
x_true = 2.3 + v_true * t

# Measurements (add noise)
z = x_true + 10.0 * (random.rand(200) - 0.5)

# Filter parameters
alpha = 0.1
beta = 0.001

# Initialize the position and velocity
xk_1 = 0.0
vk_1 = 0.0

# Loop over all measurements
for zk in z:
    # Predict the next state
    xk = xk_1 + vk_1 * dt
    vk = vk_1

    # Calculate the residual
    rk = zk - xk

    # Adjust the predicted state
    xk += alpha * rk
    vk += beta / dt * rk

    # Set the current state as previous
    xk_1 = xk
    vk_1 = vk
```

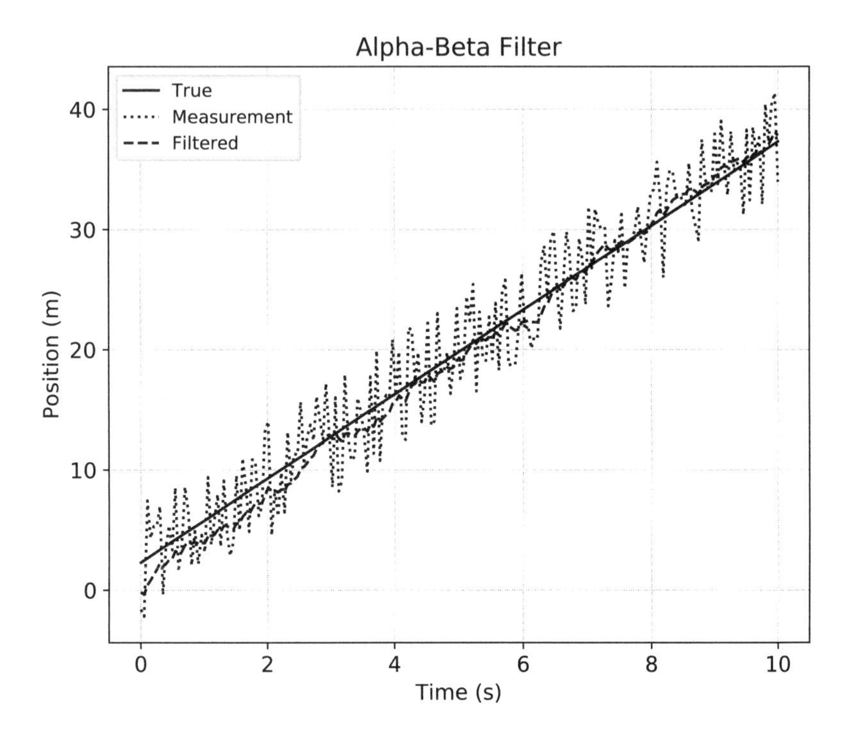

Figure 9.1 Alpha-beta filtered position for Listing 9.1.

9.1.2 Alpha-Beta-Gamma Filter

The previous section dealt with the alpha-beta filter where the velocity is assumed to remain constant over long periods of time. However, if the velocity is changing due to target acceleration, the filter needs to be extended to also include the acceleration. This leads to the alpha-beta-gamma filter, also referred to as the g-h-k filter [2]. In this extension, the velocity is obtained from integrating the acceleration. An equation for the acceleration must then be added to the state equations, and a third parameter, γ, is chosen to apply adjustments to the new state estimates. Assuming the acceleration remains fixed over the time interval between measurements, the position is projected forward in time to predict its value at the next sampling time as

$$\mathbf{x}_{k,k-1} = \mathbf{x}_{k-1,k-1} + \mathbf{v}_{k-1,k-1}\Delta t + \mathbf{a}_{k-1,k-1}\frac{\Delta t^2}{2} \qquad \text{(m)}, \tag{9.13}$$

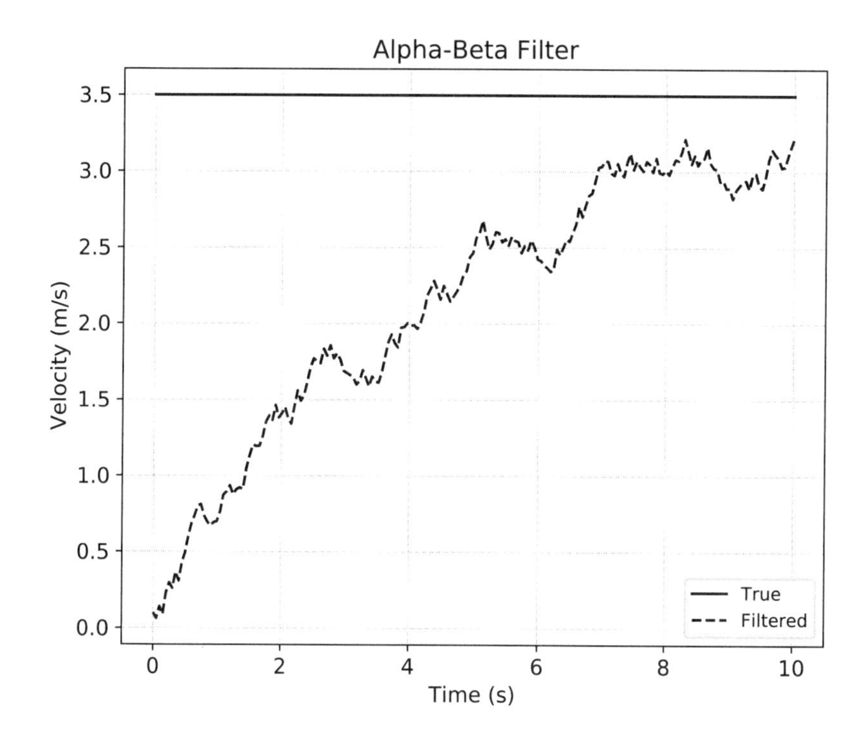

Figure 9.2 Alpha-beta filtered velocity for Listing 9.1.

where

$\mathbf{x}_{k-1,k-1}$	=	filtered position at update $k-1$ given measurements up to and including $k-1$ (m),
$\mathbf{v}_{k-1,k-1}$	=	filtered velocity at update $k-1$ given measurements up to and including $k-1$ (m/s),
$\mathbf{a}_{k-1,k-1}$	=	filtered acceleration at update $k-1$ given measurements up to and including $k-1$ (m/s^2),
Δt	=	measurement update interval (s),
$\mathbf{x}_{k,k-1}$	=	predicted position at update k given measurements up to and including $k-1$ (m).

The velocity is found from the acceleration as

$$\mathbf{v}_{k,k-1} = \mathbf{v}_{k-1,k-1} + \mathbf{a}_{k-1,k-1}\Delta t \qquad \text{(m/s)}. \tag{9.14}$$

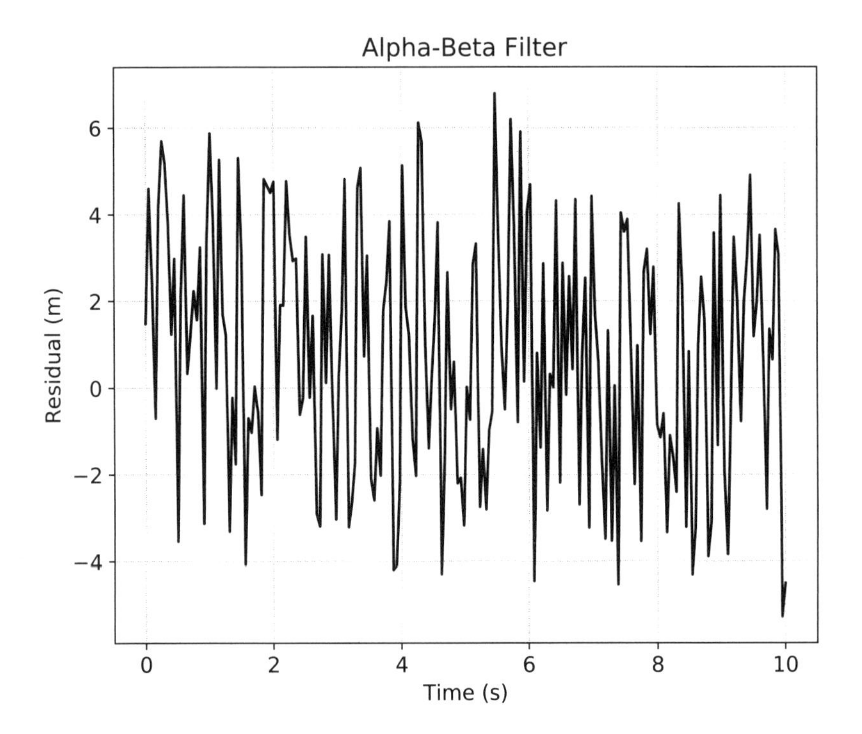

Figure 9.3 Alpha-beta residual for Listing 9.1.

The acceleration is assumed to be constant, so the predicted value is equal to the current value as

$$\mathbf{a}_{k,k-1} = \mathbf{a}_{k-1,k-1} \qquad \text{(m/s}^2\text{)}. \tag{9.15}$$

The alpha-beta-gamma filter uses the same definition for the residual as the alpha-beta filter given in (9.7). The alpha-beta-gamma filter uses selected constants, α, β, and γ, to adjust the position, velocity, and acceleration estimates as

$$\mathbf{x}_{k,k} = \mathbf{x}_{k,k-1} + \alpha\,\mathbf{r}_k \qquad \text{(m)}, \tag{9.16}$$

$$\mathbf{v}_{k,k} = \mathbf{v}_{k,k-1} + \frac{\beta}{\Delta t}\,\mathbf{r}_k \qquad \text{(m/s)}, \tag{9.17}$$

$$\mathbf{a}_{k,k} = \mathbf{a}_{k,k-1} + \frac{2\gamma}{\Delta t^2}\,\mathbf{r}_k \qquad \text{(m/s}^2\text{)}, \tag{9.18}$$

where

α, β, γ	=	selected constants for the filter,
$\mathbf{r}_k$	=	residual at update k (m),
$\mathbf{x}_{k,k-1}$	=	predicted position at update k (m),
$\mathbf{v}_{k,k-1}$	=	predicted velocity at update k (m/s),
$\mathbf{a}_{k,k-1}$	=	predicted acceleration at update k (m/s^2),
$\mathbf{x}_{k,k}$	=	filtered position at update k given measurements up to and including k (m),
$\mathbf{v}_{k,k}$	=	filtered velocity at update k given measurements up to and including k (m/s),
$\mathbf{a}_{k,k}$	=	filtered acceleration at update k given measurements up to and including k (m/s^2).

While one may extend the alpha-beta-gamma filter to higher orders, there tends to be significant interactions among multiple states. Therefore, approximating the dynamics as a simple integrator chain is less useful. Typical selections for α, β, and γ are found from [5]:

$$\alpha = 1 - s^2, \tag{9.19}$$

$$\beta = 2(1 - s)^2, \tag{9.20}$$

$$\gamma = \frac{\beta^2}{2\alpha}, \tag{9.21}$$

$$0 \le s < 1. \tag{9.22}$$

Other choices for the parameters in the alpha-beta and the alpha-beta-gamma filter may be found in [4, 5]. The algorithm for the alpha-beta-gamma filter proceeds the same as the alpha-beta filter, and a simple simulation for the alpha-beta-gamma filter is given in Listing 9.2. The filtered position, velocity, and acceleration are shown in Figures 9.4, 9.5, and 9.6, respectively. The residual is given in Figure 9.7. While the measurement noise variance for this example is large, it is not apparent in Figure 9.4. However, the residual in Figure 9.7 clearly illustrates this large variance. The convergence of the velocity and acceleration in Figures 9.5 and 9.6 is somewhat slow and may be improved with choices of the filter parameters α, β, and γ. As with the alpha-beta filter, changes in these parameters may lead to faster convergence of acceleration at the cost of more noise in the filtered position.

Listing 9.2 Alpha-Beta-Gamma Filter

```
from scipy import linspace, random

t, dt = linspace(0, 100, 1000, retstep=True) # Time (s)

# True position (m), velocity(m/s) and acceleration (m/s/s)
a_true = 3.0
v_true = 3.5 + a_true * t
x_true = 1.3 + 3.5 * t + 0.5 * a_true * t ** 2

# Measurements (add noise)
z = x_true + 10.0 * (random.rand(1000) - 0.5)

# Alpha-Beta-Gamma filter parameters
s = 0.97
alpha = 1.0 - s ** 2
beta = 2.0 * (1.0 - s) ** 2
gamma = beta ** 2 / (2.0 * alpha)

xk_1 = 0.0 # Initialize the position
vk_1 = 0.0 # Initialize the velocity
ak_1 = 0.0 # Initialize the acceleration

for zk in z: # Loop over all measurements
    # Predict the next state
    xk = xk_1 + vk_1 * dt + 0.5 * ak_1 * dt ** 2
    vk = vk_1 + ak_1 * dt
    ak = ak_1

    rk = zk - xk # Calculate the residual

    # Correct the predicted state
    xk += alpha * rk
    vk += beta / dt * rk
    ak += 2.0 * gamma / dt ** 2 * rk

    # Set the current state as previous
    xk_1 = xk
    vk_1 = vk
    ak_1 = ak
```

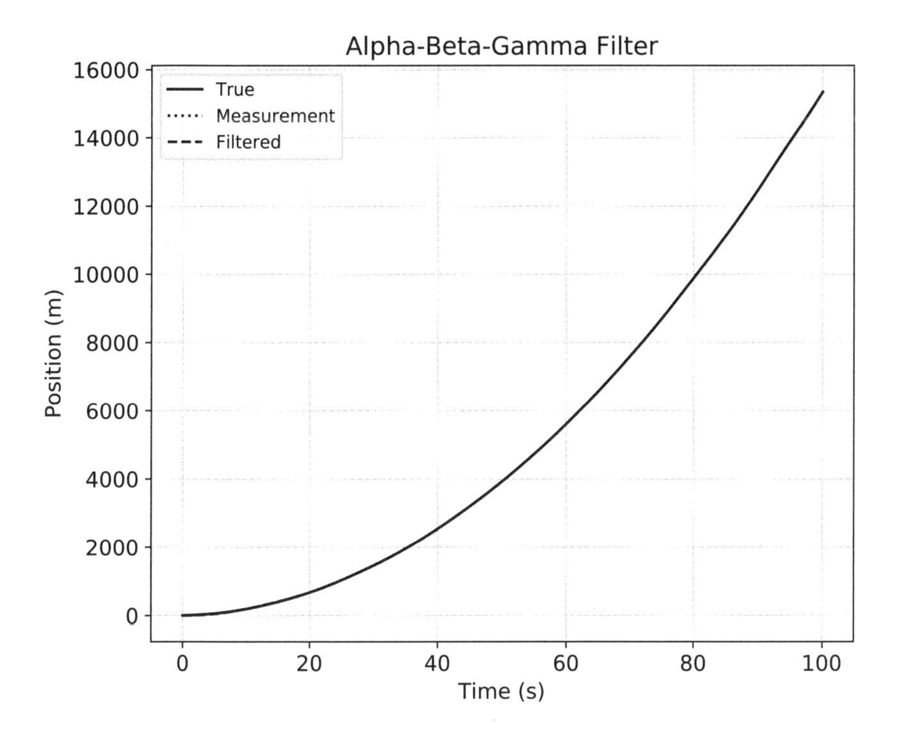

Figure 9.4 Alpha-beta-gamma filtered position for Listing 9.2.

9.1.3 Kalman Filter

The Kalman filter is named after Rudolph E. Kalman and is of the most important and often-used estimation algorithms [6]. Kalman's original paper presented a recursive solution to the discrete-data linear filtering problem. The Kalman filter follows the predict and correct construct presented in Section 9.1.1 and is optimal in the sense the estimated error uncertainty is minimized when some specific conditions are met [7]. While these conditions are not often met, the Kalman filter still performs quite well for many applications, including radar tracking, location and assisted navigation, control systems, interactive computer graphics, and multisensor fusion [8–12].

9.1.3.1 One-dimensional Kalman Filter

To simplify the equations and facilitate understanding, the Kalman filter will first be presented in one dimension. To begin, the state update equation is written as

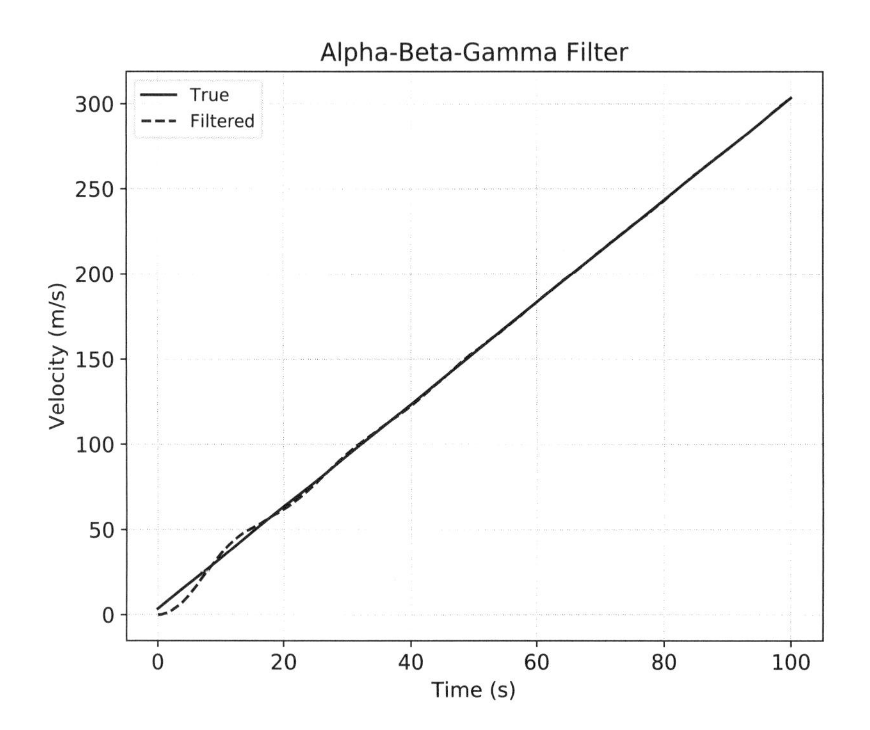

Figure 9.5 Alpha-beta-gamma filtered velocity for Listing 9.2.

$$x_{k,k} = x_{k,k-1} + K_k\,(z_k - x_{k,k-1}) = (1 - K_k)\,x_{k,k-1} + K_k\,z_k, \tag{9.23}$$

where K_k is the Kalman gain at update k. From (9.23), the Kalman gain is the weight given to the measurement, z_k, and one minus the Kalman gain is the weight given to the predicted state, $x_{k,k-1}$. In other words, the Kalman gain is an indicator of how much the estimate is changed due to a given measurement. The Kalman gain is the uncertainty in the predicted value of the state divided by the uncertainty in the predicted state, plus the uncertainty in the measurement. This is expressed as

$$K_k = \frac{P_{k,k-1}}{P_{k,k-1} + R_k}, \tag{9.24}$$

where $P_{k,k-1}$ is the predicted state uncertainty, and R_k is the measurement uncertainty. From (9.24), if the predicted state uncertainty is small and the measurement uncertainty is large, the predicted state is heavily weighted, and the measurement is given a small

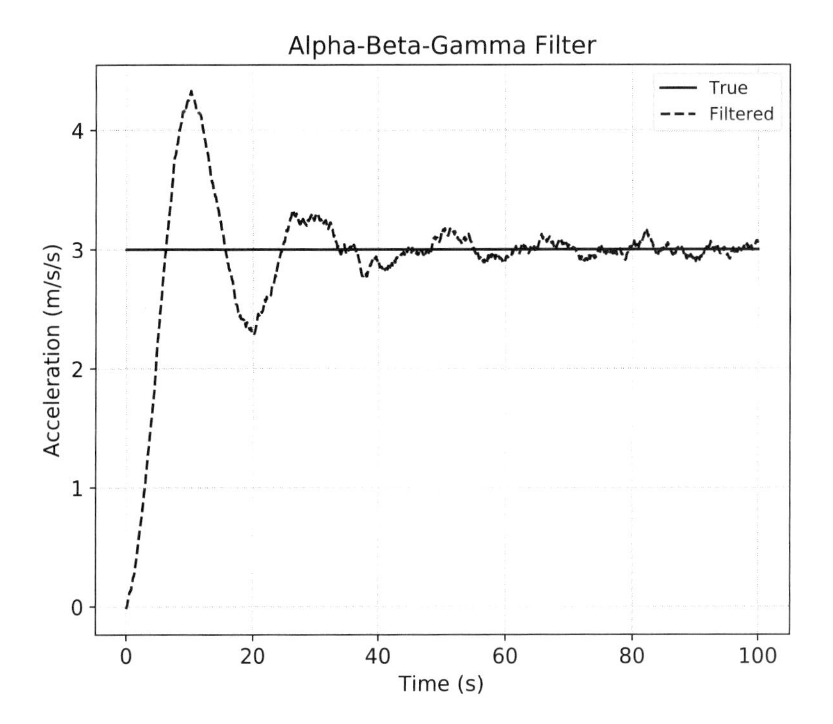

Figure 9.6 Alpha-beta-gamma filtered acceleration for Listing 9.2.

weight. However, if the predicted state uncertainty is large and the measurement uncertainty is small, the predicted state is given a small weight and the measurement is weighted heavily. Figures 9.8 and 9.9 illustrate this with the two cases of large and small Kalman gain. The state uncertainty update equation is written as

$$P_{k,k} = (1 - K_k)\ P_{k,k-1}, \tag{9.25}$$

where $P_{k,k-1}$ is the predicted state uncertainty, and $P_{k,k}$ is the updated state uncertainty. Now that the state and uncertainty update equations have been found, the predicted state and uncertainty needs to be determined. In this one-dimensional model, assuming a constant velocity, the predicted state is given by

$$x_{k,k-1} = x_{k-1,k-1} + v_{k-1,k-1}\,\Delta t \qquad \text{(m)}, \tag{9.26}$$

$$v_{k,k-1} = v_{k-1,k-1} \qquad \text{(m/s)}. \tag{9.27}$$

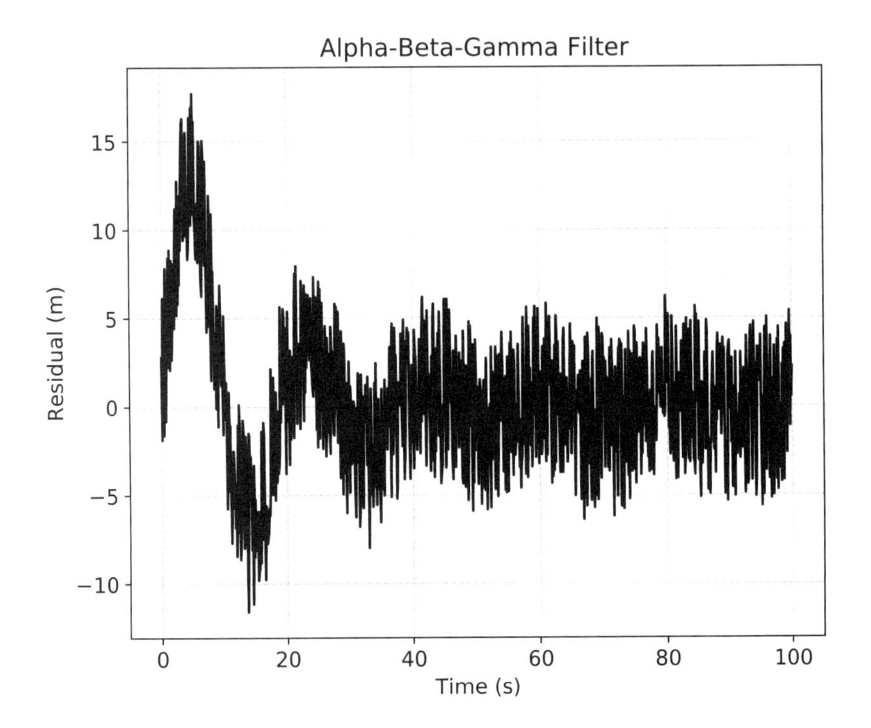

Figure 9.7 Alpha-beta-gamma residual for Listing 9.2.

The predicted state uncertainty would then be

$$P^x_{k,k-1} = P^x_{k-1,k-1} + P^v_{k-1,k-1}\,\Delta t \qquad (\text{m}^2), \tag{9.28}$$

$$P^v_{k,k-1} = P^v_{k-1,k-1} \qquad (\text{m/s})^2, \tag{9.29}$$

where P^x is the uncertainty in the position, and P^v is the uncertainty in the velocity. Practical targets of interest to radar systems do not exactly follow the dynamic model, which creates associated uncertainties. Recall that the process noise is the uncertainty of the dynamic model. For example, the dynamic model for aircraft tracking may have large uncertainties due to possible maneuvers. When tracking ballistic targets, the uncertainties may be smaller as random accelerations due to environmental factors should be small. If using a Kalman filter to estimate the position of a stationary object from radar measurements, the uncertainty associated with the dynamic model would be zero, as the

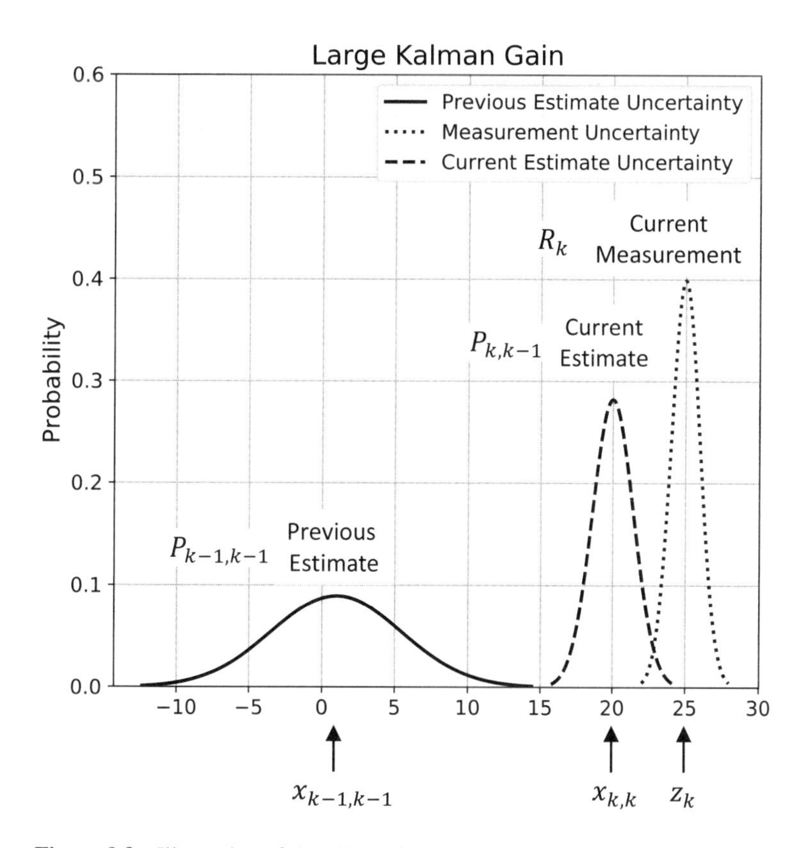

Figure 9.8 Illustration of the effect of Kalman gain on updated state estimate.

target does not move. The variance of the process noise, denoted by Q_k, is added to the predicted uncertainty as

$$P_{k,k-1} = P_{k-1,k-1} + Q_k. \tag{9.30}$$

The algorithm for a one-dimensional Kalman filter, with a constant dynamic model, is given in Listing 9.3. The output of this algorithm is shown in Figure 9.10.

9.1.3.2 Multivariate Kalman Filter

This section presents the Kalman filter equations in vector form to be used in a general framework. It follows the same iterative method of predicting the next state and uncertainty followed by an update of the current state estimate and uncertainty. The system is assumed to be modeled by the state transition equation given by

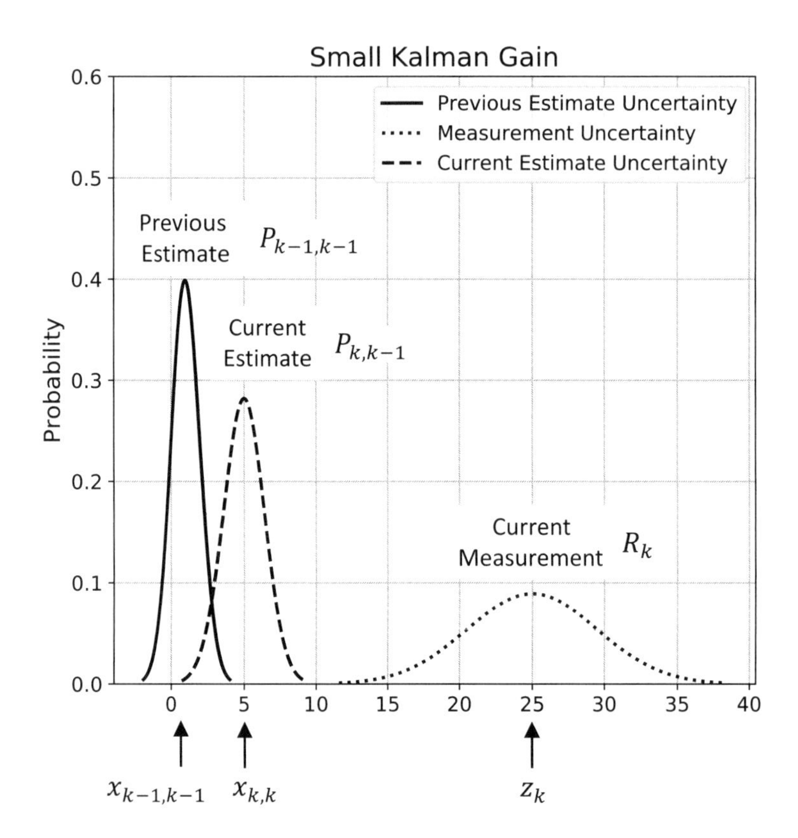

Figure 9.9 Illustration of the effect of Kalman gain on updated state estimate.

$$\mathbf{x}_k = \mathbf{A}_k \mathbf{x}_{k-1} + \mathbf{B}_k \mathbf{u}_k + \mathbf{w}_k, \tag{9.31}$$

where

$\mathbf{x}_{k-1}$	=	state at update $k-1$,
$\mathbf{A}_k$	=	state transition matrix,
$\mathbf{u}_k$	=	input control vector,
$\mathbf{B}_k$	=	relates the input control to transition matrix,
$\mathbf{w}_k$	=	process noise,
$\mathbf{x}_k$	=	state at update k.

It is also assumed the measurements of the state can be represented by a linear equation written as

$$\mathbf{z}_k = \mathbf{H}_k \mathbf{x}_k + \mathbf{v}_k, \tag{9.32}$$

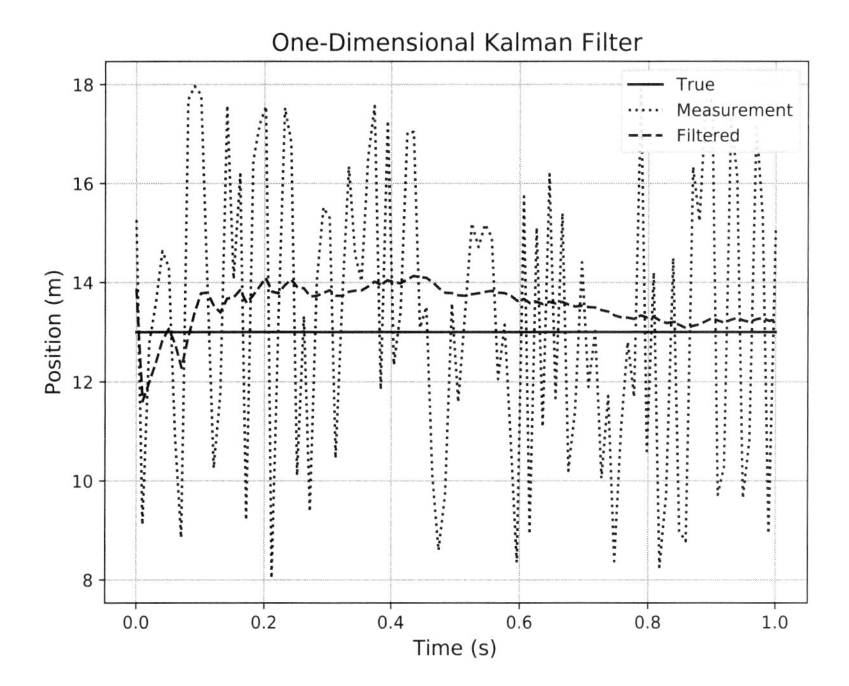

Figure 9.10 Filtered position for one-dimensional Kalman filter with constant dynamics given in Listing 9.3.

where

$$\begin{aligned}
\mathbf{x}_k &= \text{state at update } k, \\
\mathbf{H}_k &= \text{relates the state to the measurement}, \\
\mathbf{v}_k &= \text{measurement noise}, \\
\mathbf{z}_k &= \text{measurement at update } k.
\end{aligned}$$

The first step is to determine the values of **A**, **B**, and **H** that make the problem fit the model given in (9.31) and (9.32). Typically, these matrices are constant. Once these are determined, the noise parameters and initial state and uncertainty values must be determined. These may be found through a priori information or through experiment. From this point, there are two distinct sets of equations, prediction and update. Both sets of equations are applied at each time step k, for which there is a measurement. The prediction equations for the state and uncertainty are

Listing 9.3 One-Dimensional Kalman Filter

```
from scipy import linspace, random, sqrt

t, dt = linspace(0, 1, 100, retstep=True) # Time(s)

x_true = 13.0 # True position (m)

Q = 0.01 # Process noise variance (m^2)

R = 100.0 # Measurement noise variance (m^2)

# Measurements (add noise)
z = x_true + sqrt(R) * (random.rand(n) - 0.5)

# Initialize state (m) and uncertainty (m^2)
xk = 0.0
pk = 1000.0

for zk in z: # Loop over all measurements

    xk = xk # Predict the next state

    pk = pk + Q # Predict the uncertainty

    K = pk / (pk + R) # Calculate the Kalman gain

    # Update the state and uncertainty
    xk = xk + K * (zk - xk)
    pk = (1.0 - K) * pk
```

$$\mathbf{x}_{k,k-1} = \mathbf{A}_k \mathbf{x}_{k-1,k-1} + \mathbf{B}_k \mathbf{u}_k, \tag{9.33}$$

$$\mathbf{P}_{k,k-1} = \mathbf{A}_k \mathbf{P}_{k-1,k-1} \mathbf{A}_k^T + \mathbf{Q}_k, \tag{9.34}$$

where

$\mathbf{P}_{k,k-1}$	=	predicted uncertainty at update k,
$\mathbf{P}_{k-1,k-1}$	=	uncertainty estimate at update $k-1$,
$\mathbf{Q}_k$	=	process noise variance at update k.

Once predicted the state and uncertainty are calculated, the measurement value is used to update the estimated state and uncertainty. For the update, the Kalman gain is first calculated as

$$\mathbf{K}_k = \mathbf{P}_{k,k-1}\,\mathbf{H}_k^T\left(\mathbf{H}_k\,\mathbf{P}_{k,k-1}\,\mathbf{H}_k^T + \mathbf{R}_k\right)^{-1}, \tag{9.35}$$

where

$\mathbf{P}_{k,k-1}$ = predicted uncertainty at update k,
$\mathbf{H}_k$ = relates the state to the measurement,
$\mathbf{R}_k$ = measurement noise variance at update k,
$\mathbf{K}_k$ = Kalman gain at update k.

The Kalman gain in (9.35) is then used to update the current state and uncertainty as

$$\mathbf{x}_{k,k} = \mathbf{x}_{k,k-1} + \mathbf{K}_k\left(\mathbf{z}_k - \mathbf{H}_k\,\mathbf{x}_{k,k-1}\right), \tag{9.36}$$

$$\mathbf{P}_{k,k} = \left(\mathbf{I} - \mathbf{K}_k\,\mathbf{H}_k\right)\mathbf{P}_{k,k-1}, \tag{9.37}$$

where **I** is the identity matrix. Figure 9.11 illustrates the iterative process of Kalman filtering.

9.1.3.3 Adaptive Filtering

The previous sections considered the tracking of targets that closely follow the process model. Now consider the problem of tracking a maneuvering target, such as a ship,

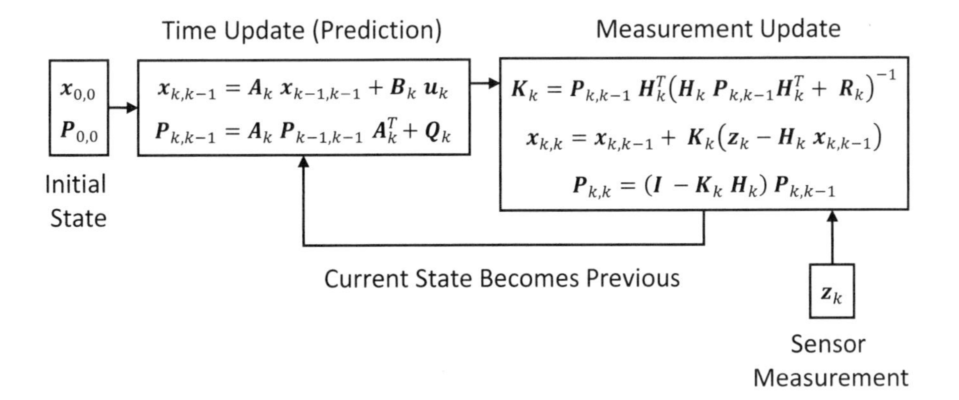

Figure 9.11 Flow diagram of the Kalman filter iterative process.

automobile, aircraft, or ballistic object experiencing atmospheric drag. While an attempt could be made to model the control inputs, there would be no way to model environmental conditions such as ocean current, wind, and temperature. There has been much research in this area, which has led to methods such as adjustable process noise [13], fading memory filters [14], multiple model and adaptive multiple model estimators [15], and interacting multiple models [16]. A simple approach to this problem would be to increase the process noise variance, $\mathbf{Q}_k$, to take into account the variance in the unknown system dynamics. While this method does give rise to a nondiverging filter, it is usually far from optimal. As seen in previous sections, a large value of process noise gives more weight to the measurement noise. A maneuvering target must undergo some type of acceleration. Therefore, implementing a constant velocity model results in filtered output that does not react quickly to maneuvers and takes a long period of time to recover, as illustrated in Figure 9.12. This is also seen in the residual given in Figure 9.13. While implementing a constant acceleration model results in filtered output that responds more quickly to maneuvers, this model amplifies the noise, as shown in Figure 9.14 and in the residual given in Figure 9.15.

Next, consider a constant velocity model with an increased process noise to account for the maneuver. This results in a filtered output that responds quickly to maneuvers but with an increased amount of noise, as shown in Figures 9.16 and 9.17. The constant velocity model does not react to accelerations quickly. However, the constant acceleration model produces a noisy output during the nonaccelerating portions of the target trajectory. The constant velocity model with large process noise also performs well with maneuvering targets. Therefore, a constant velocity model that adapts the process noise to the behavior of the target is desired. The residual is used to determine if a target is maneuvering, or in general, deviating from the process model. Two methods for adjusting the process noise are considered. The first normalizes the square of the residual as [13]

$$\epsilon_k = \mathbf{r}_k^T \, \mathbf{S}^{-1} \, \mathbf{r}_k, \tag{9.38}$$

where

$$\mathbf{S}_k = \mathbf{H}_k \, \mathbf{P}_{k,k-1} \, \mathbf{H}_k^T + \mathbf{R}_k, \tag{9.39}$$

and $\mathbf{r}_k$ is the residual. The process noise is increased when ϵ_k exceeds a set threshold, and then reduced when ϵ_k falls below this threshold. The second method, which is similar to the previous one, uses the standard deviation of the measurement and state uncertainty as [14]

$$\sigma_k = \sqrt{\mathbf{H}_k \, \mathbf{P}_{k,k-1} \, \mathbf{H}_k^T + \mathbf{R}_k}. \tag{9.40}$$

When the absolute value of the residual exceeds a multiple of the standard deviation given in (9.40), the process noise is increased by a specified amount. Figures 9.18 and 9.19 illustrate the results using the ϵ_k method, while Figures 9.20 and 9.21 show the

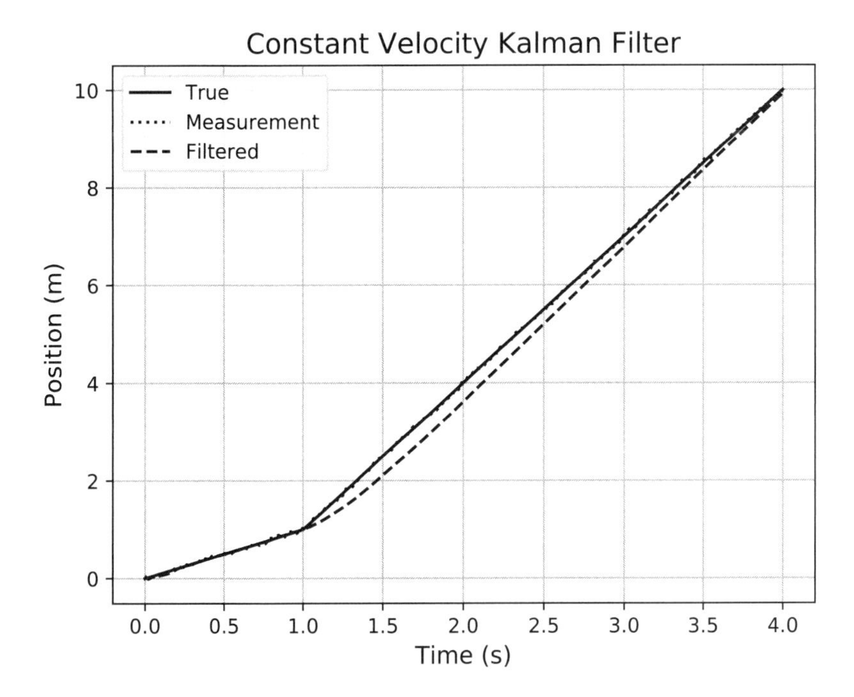

Figure 9.12 Filtered position of a constant velocity Kalman filter with a maneuvering target.

results using the σ_k method. Note that both methods respond to the maneuver quickly and have low noise during the nonmaneuvering time periods. Varying the threshold in each method changes the response to the maneuver and the noise in the output.

9.2 MULTITARGET TRACKING

To this point, tracking of only a single target has been analyzed. Now consider the problem of jointly estimating the state and uncertainty of some number of targets based on radar measurements. In addition to the measurement and process uncertainty, multitarget tracking must consider other uncertainties, including false alarms, missed detections, origin of the measurements, and targets entering and exiting the field of regard. Consider the scenario given in Figure 9.22, consisting of a single radar and four targets, T_1, T_2, T_3, and T_4. Three measurements are created $\mathbf{z}_1$ due to target T_1, $\mathbf{z}_2$ due to a false alarm, $\mathbf{z}_3$ due to T_2 and T_3 located in the same resolution cell, and no measurement for T_4. As illustrated, measurements may be due to targets, clutter, and false alarms. In addition,

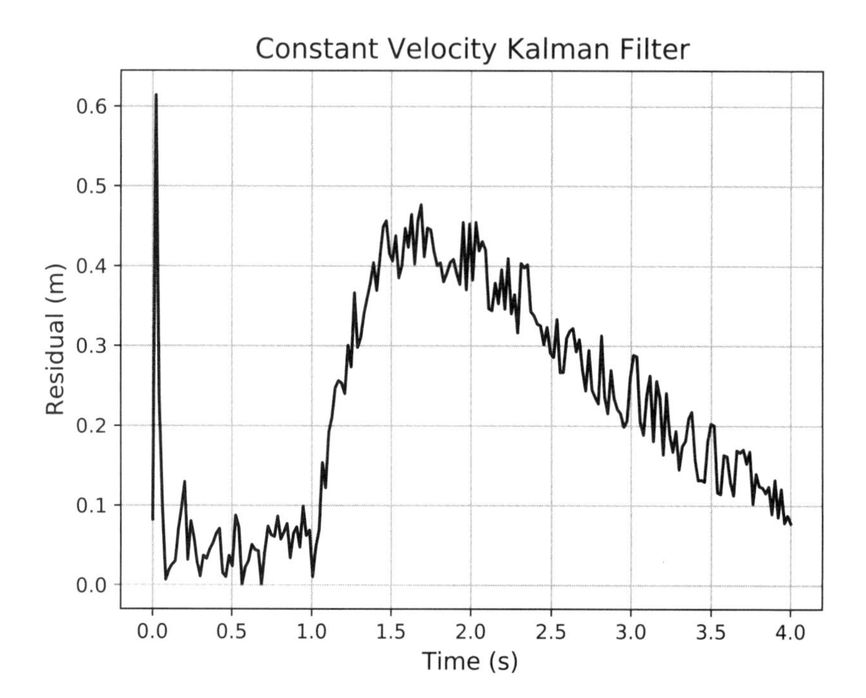

Figure 9.13 Residual of a constant velocity Kalman filter with a maneuvering target.

the number of measurements may vary from scan to scan due to noise variance, target variance, unresolved targets, extended targets, and targets entering or exiting the field of regard. For each scan, a set of measurements, $\{\mathbf{z}_1, ..., \mathbf{z}_n\}$, is sent to the tracker. The tracking filters presented in the previous sections do not apply directly to the multitarget tracking problem as there is no information as to the source of the measurement. Therefore, the two key elements of a multitarget tracker are data association, and state estimation and prediction.

Over the past several decades, numerous formulations for solving the multitarget tracking problem have emerged. However, the most commonly used techniques are the *multiple hypothesis tracker* (MHT) [17], *joint probabilistic data association filter* (JPDA) [18], and *random finite set* (RFS) [19]. The MHT and JPDA methods are widely used, and the RFS-based technique is a growing field of study. The following sections provide a summary of these methods as well as a *global nearest neighbor* (GNN) approach as an introduction to the multitarget problem.

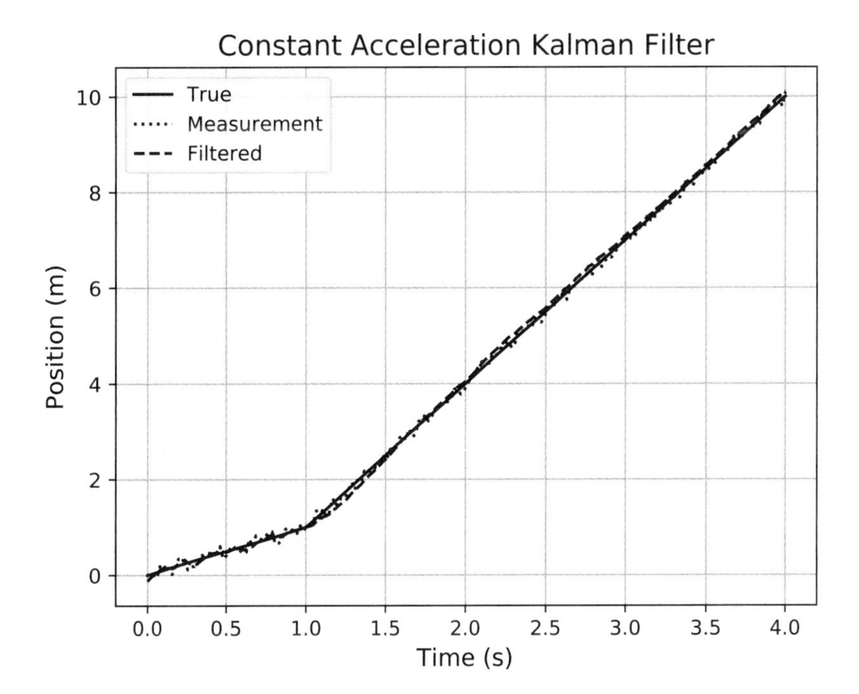

Figure 9.14 Filtered position of a constant acceleration Kalman filter with a maneuvering target.

9.2.1 Global Nearest Neighbor

As an introduction to multitarget tracking, the GNN approach is presented as it is intuitive and easy to implement. However, it has many drawbacks, including loss of track and poor performance for closely spaced targets [17]. This technique attempts to find the unique joint association of measurements to targets that minimizes some cost, such as a distance or likelihood. The GNN method then uses standard filtering techniques for each target and its associated measurement. A commonly used measure of the distance between two probability distributions is the Mahalanobis distance [20]. A cost matrix is formed using the Mahalanobis distance, and is written as

$$\mathbf{M}_k^{ij} = (\mathbf{z}_k^i - \mathbf{H}_k^j\,\mathbf{x}_k^j)^T\,\mathbf{S}_k^{-1}\,(\mathbf{z}_k^i - \mathbf{H}_k^j\,\mathbf{x}_k^j), \tag{9.41}$$

where

$$\mathbf{S}_k = \mathbf{H}_k^j\,\mathbf{P}_k^j\,\mathbf{H}_k^{j\,T} + \mathbf{R}_k^i, \tag{9.42}$$

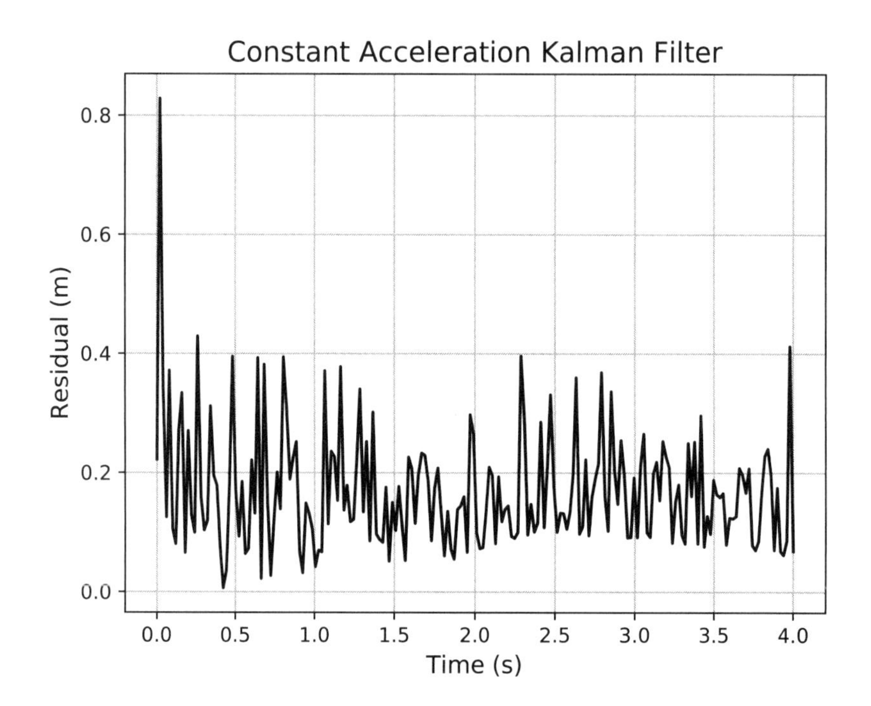

Figure 9.15 Residual of a constant acceleration Kalman filter with a maneuvering target.

and

$$\begin{aligned}
\mathbf{M}_k^{ij} &= \text{Mahalanobis distance for measurement } i \text{ and track } j, \\
\mathbf{z}_k^{i} &= \text{the } i^{th} \text{ measurement}, \\
\mathbf{x}_k^{j} &= \text{state for track } j, \\
\mathbf{H}_k^{j} &= \text{state to measurement matrix for track } j, \\
\mathbf{P}_k^{j} &= \text{uncertainty for track } j, \\
\mathbf{R}_k^{i} &= \text{uncertainty for measurement } i.
\end{aligned}$$

The cost matrix formed in (9.41) is then passed to an assignment algorithm such as Munkres [21] or Jonker-Volgenant-Castanon [22]. However, there are issues with this method as large uncertainties can lead to incorrect associations [23]. Also, GNN only considers the most likely measurement to each target. Therefore, it only performs well in situations with a low number of false alarms and widely spaced targets. Once the measurement to track associations have been made, each track is updated with the associated measurement using standard filtering techniques such as Kalman filtering.

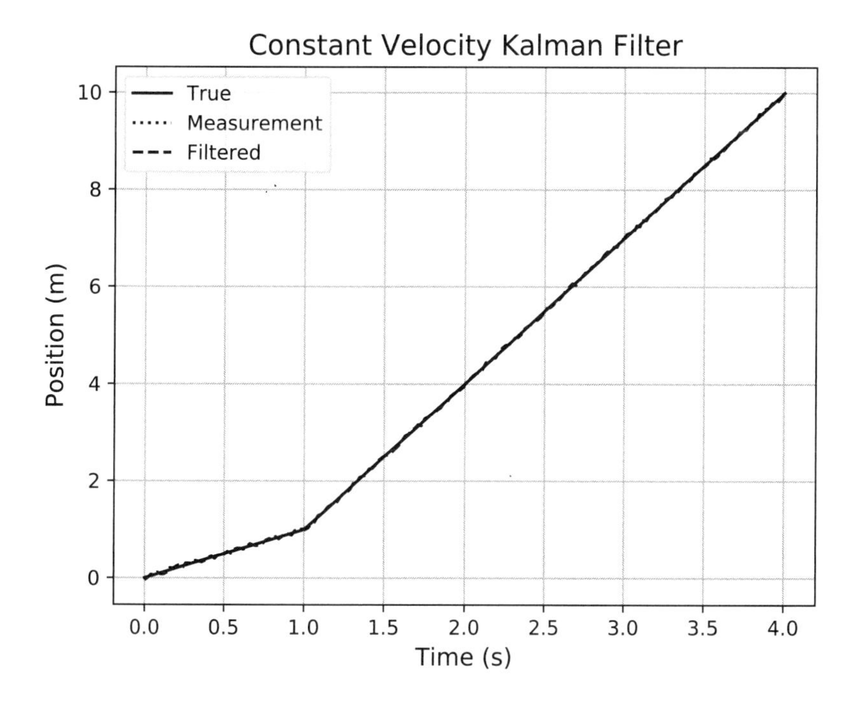

Figure 9.16 Filtered position of a constant velocity Kalman filter with a maneuvering target and large process noise.

9.2.2 Joint Probabilistic Data Association

Due to the limitations of the GNN approach, it is desirable to perform data association in a probabilistic fashion. For multitarget tracking, JPDA is the most well-known method for probabilistic data association [24]. The idea of JPDA is to enumerate all possible associations to calculate the marginal association probability [25]. To avoid conflicts in the measurement to track assignment, JPDA uses joint association events and joint association probabilities. As the number of measurements and number of targets increases, the complexity of JPDA is exponential [24]. To reduce the computational load, approximations of JPDA have been recently developed [26–28]. Moreover, the basic form of JPDA is only for a fixed and known number of targets. Several extensions of JPDA have been proposed to handle an unknown and time varying number of targets [24].

To begin, a validation matrix, $\mathbf{V}$, is created and indicates the possible sources of each measurement. The elements of the validation matrix are given as

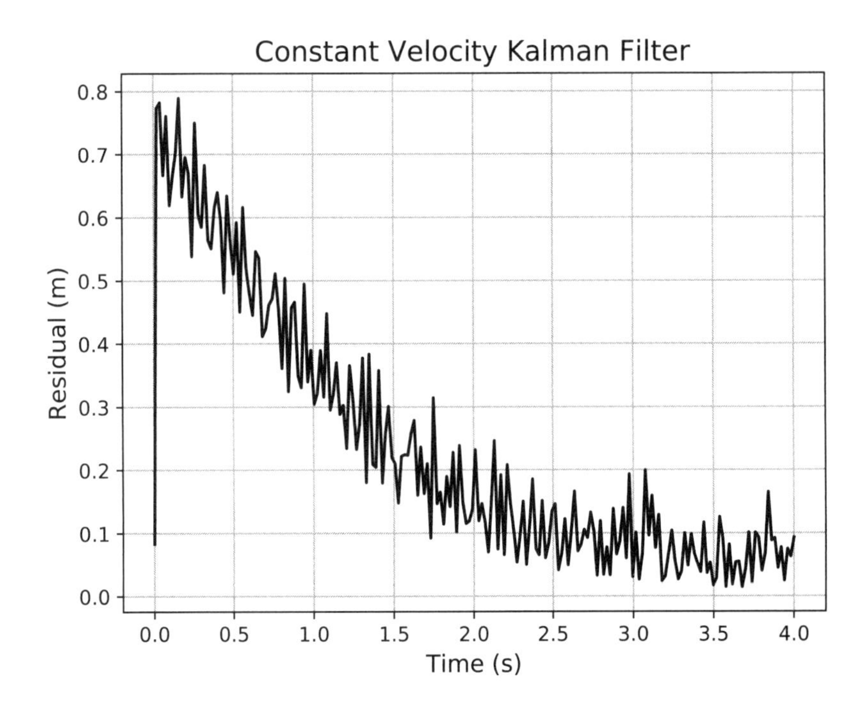

Figure 9.17 Residual of a constant velocity Kalman filter with a maneuvering target and large process noise.

$$V_{ij} = \begin{cases} 1 & \text{if measurement } i \text{ in validation region of target } j, \\ 0 & \text{otherwise,} \end{cases} \tag{9.43}$$

for $i = 1, \ldots, M$ and $j = 0, \ldots, N$. Here, M is the number of measurements and N is the number of targets. The exception is $j = 0$, which corresponds to false alarms or clutter, and is filled with all ones. Once the validation matrix in (9.43) has been formed, the feasible joint association events, **e**, are obtained by choosing one entry per row and one entry per column, except column 0, which has no limitations. This corresponds to a measurement having only one source and at most one measurement originating from a target. Next, the joint association probabilities are found from [24]

$$P\left(\mathbf{e}_k|\mathbf{Z}^k\right) = p\left(\mathbf{Z}_k|\mathbf{e}_k, M_k, \mathbf{Z}^{k-1}\right)\, P\left(\mathbf{e}_k|M_k\right), \tag{9.44}$$

where

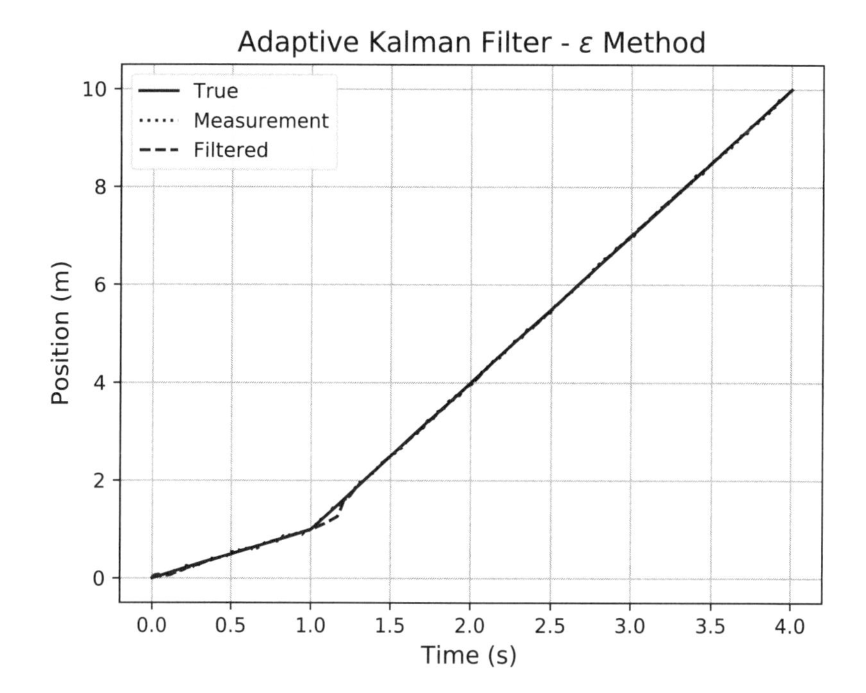

Figure 9.18 Filtered position of a constant velocity adaptive Kalman filter based on ϵ_k (9.38).

$$\begin{aligned}
\mathbf{Z}_k &= \text{set of measurements } \{\mathbf{z}_1, \ldots, \mathbf{z}_n\}, \\
\mathbf{Z}^k &= \{\mathbf{Z}_i | i = 1, \ldots, k\}, \\
\mathbf{e}_k &= \text{feasible joint association events}, \\
M_k &= \text{the number of measurements in the validation regions},
\end{aligned}$$

and p is the likelihood function given by

$$p\left(\mathbf{Z}_k | \mathbf{e}_k, M_k, \mathbf{Z}^{k-1}\right) = \prod_{i=1}^{M_k} p\left(z_k^i | \mathbf{e}_k^i, \mathbf{Z}^{k-1}\right). \tag{9.45}$$

These probabilities assume the measurements originating from targets are Gaussian distributed, false alarms are uniformly distributed, and the number of false alarms is distributed according to Poisson prior (parametric JPDA) or diffuse prior (nonparametric JPDA). To update the state and uncertainties, the marginal association probabilities need to be calculated. These are found from the joint probabilities by summing over all the joint

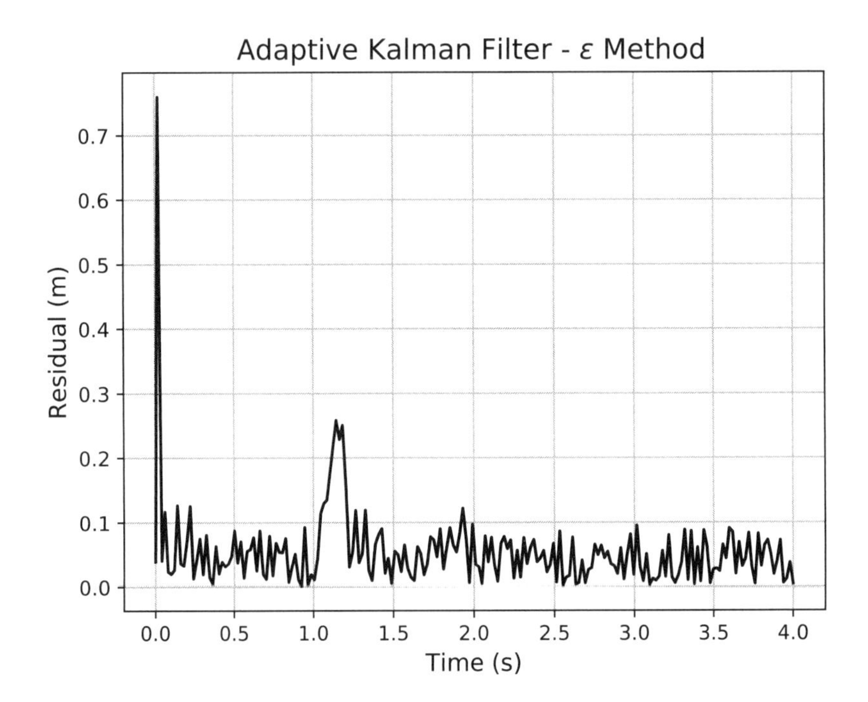

Figure 9.19 Residual of a constant velocity adaptive Kalman filter based on ϵ_k (9.38).

events in which the marginal event of interest occurs. This is expressed mathematically as [24]

$$\beta_k^{ij} = \sum_{\mathbf{e}:e_{ij} \in \mathbf{e}} P\left(\mathbf{e}|\mathbf{Z}^k\right). \tag{9.46}$$

The target state is updated similar to (9.36), where now the residual is replaced with the combined residual. The combined residual is the summation of the individual residuals weighted by the marginal association probabilities. This is written as

$$\mathbf{x}_{k,k}^j = \mathbf{x}_{k,k-1}^j + \mathbf{K}_k^j \, \mathbf{r}_k^j, \tag{9.47}$$

where the combined residual is

$$\mathbf{r}_k^j = \sum_{i=1}^{M_k} \beta_k^{ij} \left[\mathbf{z}_k^i - \mathbf{H}_k^j \mathbf{x}_{k,k-1}^j\right]. \tag{9.48}$$

The associated uncertainty is expressed as [24]

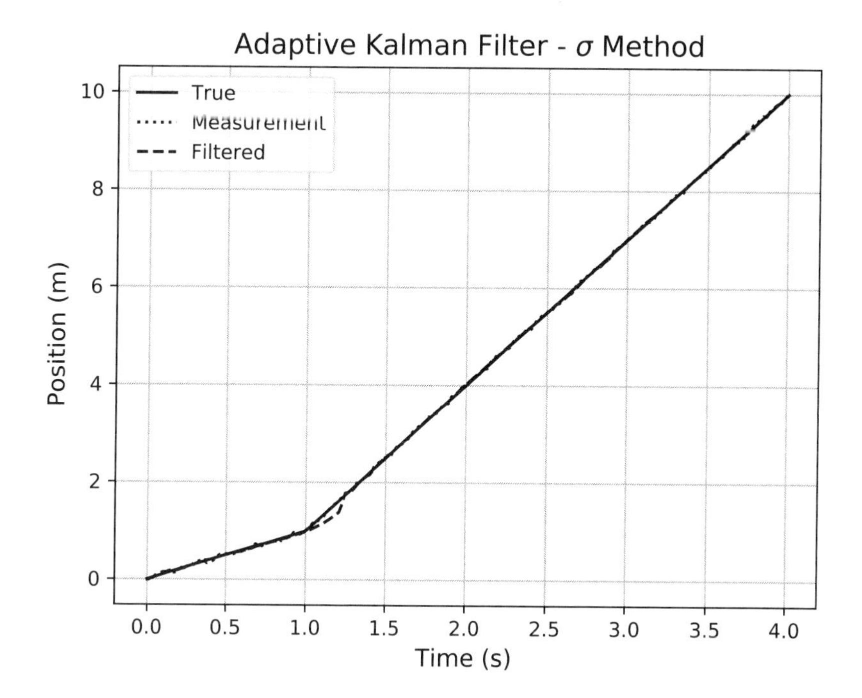

Figure 9.20 Filtered position of a constant velocity adaptive Kalman filter based on $\boldsymbol{\sigma}_k$ (9.40).

$$\mathbf{P}^j_{k,k} = \mathbf{P}^j_{k,k-1} + \left[\beta^{ij}_k - 1\right] \mathbf{K}^j_k\, \mathbf{S}^j_k\, \mathbf{K}^{j\,T}_k + \hat{\mathbf{P}}^j_k, \tag{9.49}$$

where the spread of the residual is

$$\hat{\mathbf{P}}^j_k = \mathbf{K}_k \left(\sum_{i=1}^{M_k} \beta^{ij}_k \left[\mathbf{z}^i_k - \mathbf{H}^j_k \mathbf{x}^j_{k,k-1}\right] \left[\mathbf{z}^i_k - \mathbf{H}^j_k \mathbf{x}^j_{k,k-1}\right]^T - \mathbf{r}^j_k \mathbf{r}^{j\,T}_k \right) \mathbf{K}^T_k. \tag{9.50}$$

9.2.3 Multiple Hypothesis Tracker

Due to advances in computing power, multiple hypothesis tracking has become a widely used multitarget tracking technique. Unlike JDPA, MHT is a deferred decision approach that makes use of a number of consecutive scans to solve the data association problem and is known as a maximum a posteriori estimator [24, 25]. For each scan, the MHT approach propagates a set of association hypotheses with high posterior probabilities. A new set of

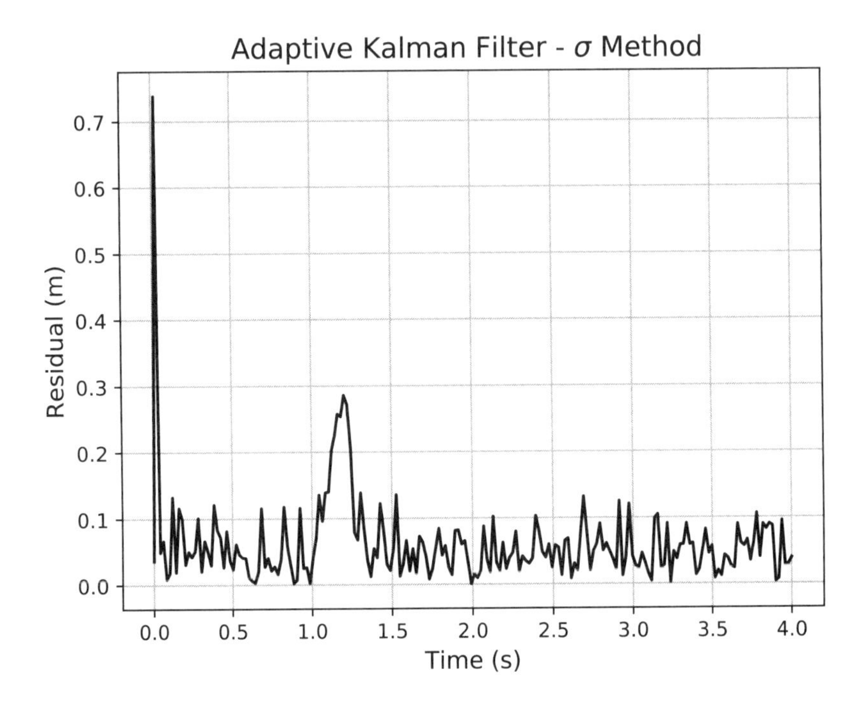

Figure 9.21 Residual of a constant velocity adaptive Kalman filter based on σ_k (9.40).

hypotheses is created from the existing hypotheses, and the posterior probabilities are updated. With this type of approach, the initiation and termination of tracks is accounted for and the unknown and time varying number of targets is handled inherently [24, 25]. Standard filtering techniques, such as Kalman or Bayes, may be used to estimate the state and uncertainty of each track using the measurements from the best hypothesis. In the MHT approach, the number of hypotheses increases exponentially with time, and some form of pruning and merging of hypotheses is required [13]. Many extensions and variations of the MHT technique have been developed in recent years to reduce this computational burden, and the reader is referred to several excellent sources on these techniques [26–28].

In general, there are two main types of MHT, hypothesis-oriented and track-oriented. Hypothesis-oriented MHT creates and grows a large number of hypotheses to solve the data association problem [25]. In contrast, a tree-based track-oriented MHT forms branches, which represent track hypotheses, and prunes these branches based on the best global hypothesis [24]. The generation of the global hypotheses is an important step in both types of MHT. For hypothesis-oriented MHT, an efficient method is to

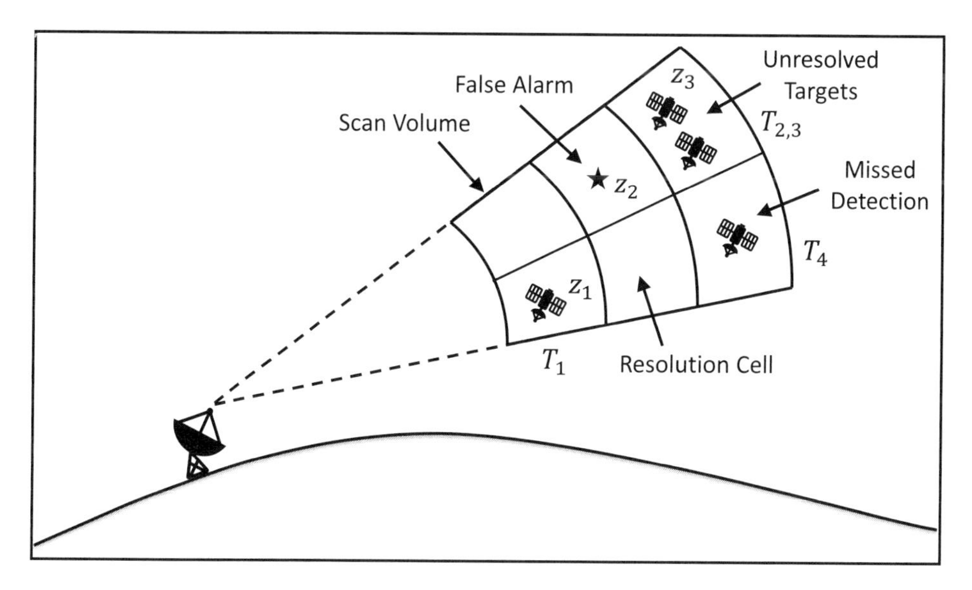

Figure 9.22 Multitarget tracking scenario.

consider only the n-best hypotheses using Murty's ranking algorithm [26]. For track-oriented MHT, multiple dimensional assignment (MDA) is the widely accepted method for implementation. Approximate methods such as Lagrangian relaxation are typically used to obtain solutions [29]. Figure 9.23 illustrates the tree-based target-oriented MHT technique. To begin, measurements from an individual scan are used to create potential measurement to track associations. This is typically performed in two steps, first with a coarse gating, followed by a method using the position and uncertainty, such as the Mahalanobis distance. For each measurement in the scan, a new track is initiated. Information about the new tracks and existing tracks is used to update the track score and find the global best hypothesis. Based on the global best, unreliable hypotheses are removed from the tree, and the remaining tracks are updated with standard filtering techniques. As an example, consider the scenario shown in Figure 9.24, where there are three frames, $k-2$, $k-1$, and k. There are two established tracks, T_1 and T_2, at scan $k-2$. During scan $k-1$, the tracker receives two measurements, $\mathbf{z}_1$ and $\mathbf{z}_2$. For scan k, a single measurement, $\mathbf{z}_3$, is received. Following the steps in the MHT process results in the formation of eight tracks, $\{t_1, \ldots, t_8\}$, as shown in Figure 9.24. Let N be the number of tracks at scan k and M be the number of resolved tracks at the root node. For this example, $N = 8$ and $M = 5$. This is a multidimensional assignment problem, and the global best hypothesis is determined by solving the binary programming problem shown in Figure 9.25, where $\mathbf{a}$ is the assignment vector. Lagrangian relaxation may be used to solve the multidimensional assignment.

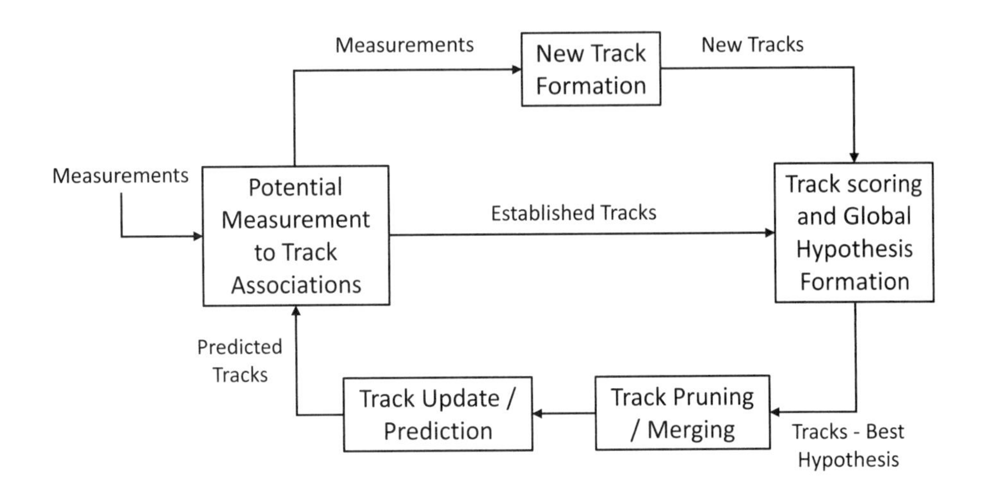

Figure 9.23 Target-oriented MHT processing steps.

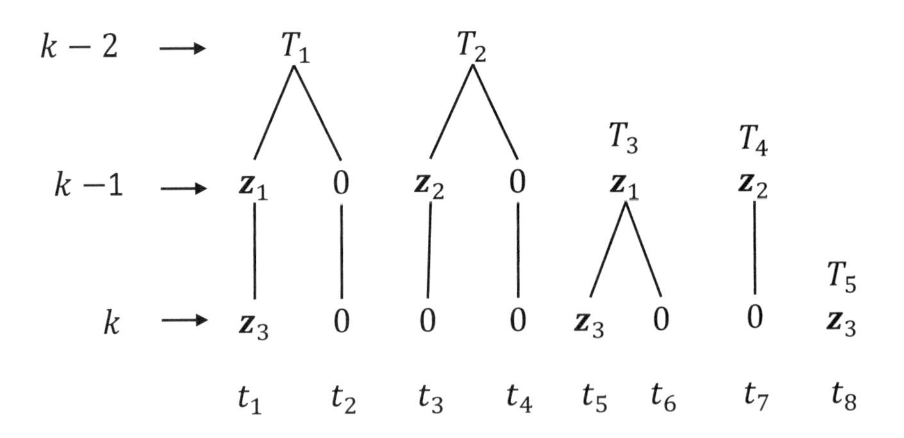

Figure 9.24 Example track tree for the MDA problem.

$$\boldsymbol{A}\boldsymbol{a} = \boldsymbol{b},$$

$$b_j = 1, \qquad j = 1, \ldots, M$$

$$a_i \in \{0, 1\}, \qquad i = 1, \ldots, N$$

$$\boldsymbol{A} = \begin{bmatrix} 1 & 1 & 0 & 0 & 0 & 0 & 0 & 0 \\ 0 & 0 & 1 & 1 & 0 & 0 & 0 & 0 \\ 1 & 0 & 0 & 0 & 1 & 0 & 0 & 0 \\ 0 & 0 & 1 & 0 & 1 & 0 & 0 & 1 \\ 1 & 0 & 0 & 0 & 1 & 0 & 0 & 1 \end{bmatrix} \begin{matrix} T_1 \\ T_2 \\ \mathbf{z}_1 \\ \mathbf{z}_2 \\ \mathbf{z}_3 \end{matrix}$$

$$t_1 \quad t_2 \quad t_3 \quad t_4 \quad t_5 \quad t_6 \quad t_7 \quad t_8$$

Figure 9.25 Example multiframe assignment problem.

9.2.4 Random Finite Set

The multitarget tracking methods presented in the previous sections make a number of approximations and have limitations when dealing with clutter, false alarms, and the initiation of new tracks. Therefore, a method for treating the tracking problem as a whole is desired. The RFS method is such an approach as it provides the full Bayes filter, which gives the probability density function for the system state, $\mathbf{x}_k$, as

$$p\left(\mathbf{x}_k|\mathbf{Z}^k\right) = \frac{p\left(\mathbf{z}_k|\mathbf{x}_k\right)\, p\left(\mathbf{x}_k|\mathbf{Z}^{k-1}\right)}{p\left(\mathbf{Z}_k|\mathbf{Z}^{k-1}\right)}. \tag{9.51}$$

The term, $p\left(\mathbf{z}_k|\mathbf{x}_k\right)$, is the probability the state, $\mathbf{x}_k$, would generate the measurement, $\mathbf{z}_k$. This requires a model for how the measurements are generated, and therefore a model of the sensor itself. The next term, $p\left(\mathbf{x}_k|\mathbf{Z}^{k-1}\right)$, is the probability of the state, $\mathbf{x}_k$, given all the previous measurements. The final term, $p\left(\mathbf{Z}^k|\mathbf{Z}^{k-1}\right)$, is for normalization and ensures the integral of the probability density function is one. Extending the expression in (9.51) to the multitarget paradigm results in

$$p\left(\mathbf{X}_k|\mathbf{Z}^k\right) = \frac{p\left(\mathbf{Z}_k|\mathbf{X}_k\right)\, p\left(\mathbf{X}_k|\mathbf{Z}^{k-1}\right)}{p\left(\mathbf{Z}_k|\mathbf{Z}^{k-1}\right)}. \tag{9.52}$$

In the RFS method, the state of the targets and the measurements are comprised of a random number of sets whose members are random variables. Suppose there are m targets with states given by $\mathbf{x}_1, \ldots, \mathbf{x}_m$, which generate n measurements expressed as $\mathbf{z}_1, \ldots, \mathbf{z}_n$. In the multitarget formulation of (9.52), $\mathbf{X}_k = \{\mathbf{x}_1, \ldots, \mathbf{x}_m\}$, $\mathbf{Z}_k = \{\mathbf{z}_1, \ldots, \mathbf{z}_n\}$, and $\mathbf{Z}^k = \{\mathbf{Z}_i | i = 1, \ldots, k\}$. To deal with (9.52), Mahler [19] defined a calculus for random finite sets. This is significant as it allows for the full Bayes filter of a wide range of measurement and dynamic system behavior, including targets entering and exiting the field of regard, targets spawning from existing targets, extended targets, unresolved targets, missed detections, false alarms, and clutter. However, the closed-form solution of Bayes theorem deals with multidimensional integrals with no known analytic solutions. Numerical techniques such as grid-based solutions and particle filters are not tractable for this problem. Therefore, some approximation methods must be employed.

The *probability hypothesis density*, $\mathbf{D}_{k,k}$, is the first moment of the of multitarget Bayes filter. $\mathbf{D}_{k,k}$ is propagated forward in time, and the integral of the probability hypothesis density over a volume is equal to the expected number of targets in the volume. Suppose the integral is calculated and gives some number of expected targets, N. The targets are localized by finding the N largest local peaks of $\mathbf{D}_{k,k}$ [30]. The *cardinalized probability hypothesis density* (CPHD) propagates the probability hypothesis density as well as the full probability distribution of the random variable for the number of targets [31]. This results in a lower variance in the calculation of the number of targets. There are two methods for practical implementation of the CPHD filter. The first method uses particles and is known as the sequential Monte Carlo method [32]. The second method uses a Gaussian mixture technique and is more computationally efficient [33]. For typical tracking scenarios, JPDA, MHT, and RFS methods show similar performance with the RFS being the least computationally complex. However, the multitarget Bayes filter is designed to deal with nonstandard tracking scenarios [34].

9.3 MEASUREMENT MODEL

While detailed radar system simulations may be used to generate measurements as input to radar tracking algorithms, a less rigorous method for generating pseudomeasurements is desired. Rather than implementing the details of the antenna elements, beam patterns, beam steering, receiver channels, signal processing, detection algorithms, complex target orientation, and radar cross section, some basic radar parameters may be used to generate sensor representative pseudomeasurements suitable for studying various tracking algorithms. The measurement model for generating range, azimuth, and elevation pseudomeasurements is expressed as [35]

$$r_{pseudo} = r_{true} + r_{noise} + r_{fixed} + r_{bias} \qquad \text{(m)}, \tag{9.53}$$

$$\theta_{pseudo} = \theta_{true} + \theta_{noise} + \theta_{fixed} + \theta_{bias} \qquad \text{(rad)}, \tag{9.54}$$

$$\phi_{pseudo} = \phi_{true} + \phi_{noise} + \phi_{fixed} + \phi_{bias} \qquad \text{(rad)}. \tag{9.55}$$

The measurement model has three terms contributing to the accuracy of the radar measurements. The noise term is dependent on the signal-to-noise ratio, the fixed term is independent of the signal-to-noise ratio, and the last term is the sensor bias.

The noise term varies with the signal-to-noise ratio and contributes to each measurement value. Therefore, a new random draw is taken for each measurement. This term is modeled as a zero mean Gaussian distribution with the standard deviation given as [35]

$$\sigma_r = \frac{r_{res}}{K_r\sqrt{2\,SNR}} \quad \text{(m)}, \tag{9.56}$$

$$\sigma_\theta = \frac{\theta_{res}}{K_\theta \cos\alpha\,\sqrt{2\,SNR}} \quad \text{(rad)}, \tag{9.57}$$

$$\sigma_\phi = \frac{\phi_{res}}{K_\phi \cos\beta\,\sqrt{2\,SNR}} \quad \text{(rad)}, \tag{9.58}$$

where

r_{res} = 3-dB range resolution (m),
θ_{res} = 3-dB θ resolution (rad),
ϕ_{res} = 3-dB ϕ resolution (rad),
K_r = sensor-specific range factor,
K_θ = sensor-specific θ factor,
K_ϕ = sensor-specific ϕ factor,
α = off-boresight angle θ direction (rad),
β = off-boresight angle ϕ direction (rad),
SNR = signal-to-noise ratio,
σ_r = standard deviation for noise range error term (m),
σ_θ = standard deviation for noise θ error term (rad),
σ_ϕ = standard deviation for noise ϕ error term (rad).

If there is no sensor specific data available for the factors K_r, K_θ, and K_ϕ, these are typically taken to be $K_r \approx 1$ and $K_\theta, K_\phi \approx 1.6$.

The fixed term is independent of the signal-to-noise ratio and contributes to each measurement value. Therefore, a new random draw is taken for each measurement value. These types of errors arise from many different sources, including quantization, interpolation, and tropospheric and ionospheric fluctuations. This term is modeled as a zero mean Gaussian distribution with the standard deviation given by

$$\sigma_r = \frac{r_{res}}{L_r} \quad \text{(m)}, \tag{9.59}$$

$$\sigma_\theta = \frac{\theta_{res}}{L_\theta \cos\alpha} \quad \text{(rad)}, \tag{9.60}$$

$$\sigma_\phi = \frac{\phi_{res}}{L_\phi \cos\beta} \qquad \text{(rad)}, \tag{9.61}$$

where

r_{res}	=	3-dB range resolution (m),
θ_{res}	=	3-dB θ resolution (rad),
ϕ_{res}	=	3-dB ϕ resolution (rad),
L_r	=	sensor-specific range factor,
L_θ	=	sensor-specific θ factor,
L_ϕ	=	sensor-specific ϕ factor,
α	=	off-boresight angle θ direction (rad),
β	=	off-boresight angle ϕ direction (rad),
σ_r	=	standard deviation for fixed range error term (m),
σ_θ	=	standard deviation for fixed θ error term (rad),
σ_ϕ	=	standard deviation for fixed ϕ error term (rad).

If there is no sensor-specific data available for the factors L_r, L_θ, and L_ϕ, these are typically taken to be ≈ 20.

True biases, which do not change over very long time frames, are removed during radar calibration. The remaining biases, which remain constant over the time frame the measurements are taken, are the ones considered in the measurement model. Since the true biases are assumed to have been removed, the remaining biases are distributed about the true measurement values with a zero mean Gaussian distribution. A single random draw from the distribution is used for all measurement values. A new random draw would be taken for each instance in a Monte Carlo simulation. The most common source of this type of bias is uncompensated atmospheric refraction.

9.4 EXAMPLES

The sections below illustrate the concepts of this chapter with several Python examples. The examples for this chapter are in the directory *pyradar\Chapter09* and the matching MATLAB examples are in the directory *mlradar\Chapter09*. The reader should consult Chapter 1 for information on how to execute the Python code associated with this book.

9.4.1 Alpha-Beta Filter

For this example, use an alpha-beta filter to track a target with an initial position of 5.0m and a velocity of 0.5 m/s. The radar has an update rate of 0.1s and a measurement variance of 2.0 m^2. The filter parameters are $\alpha = 0.1$ and $\beta = 0.001$. Track the target from 0 to 20s, and plot the resulting position, velocity, and residual compared to the true position, velocity, and measurements.

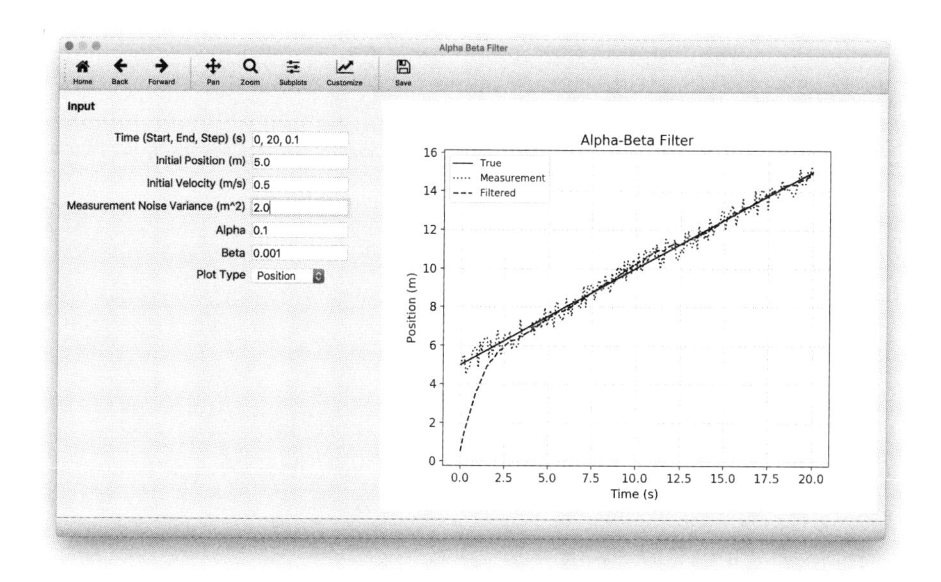

Figure 9.26 The true position, measurement position, and filtered position calculated by *alpha_beta_example.py*.

Solution: The solution to the above example is given in the Python code *alpha_beta_ example.py* and in the MATLAB code *alpha_beta_example.m*. Running the Python example code displays a GUI allowing the user to enter the filter parameters, the target trajectory parameters, the radar parameters, and to choose a plot type. The code then performs the alpha-beta filtering and displays the results. The resulting filtered position, along with the true position and measurement position, is shown in Figure 9.26. The filtered velocity and true velocity are shown in Figure 9.27. Finally, the residual for this example is shown in Figure 9.28. The reader is encouraged to vary the filter and trajectory parameters to see how these affect the tracking performance.

9.4.2 Alpha-Beta-Gamma Filter

For the next example, use an alpha-beta-gamma filter to track a target with an initial position of 1.3m, an initial velocity of 3.5 m/s, and an acceleration of 3.0 m/s^2. The radar has an update rate of 0.1s and a measurement variance of 10.0 m^2. The filter parameters are $\alpha = 0.06$, $\beta = 0.0018$, and $\gamma = 2.75 \times 10^{-5}$. Track the target from 0 to 100s, and plot the resulting position, velocity, acceleration, and residual compared to the true position, velocity, acceleration, and measurements.

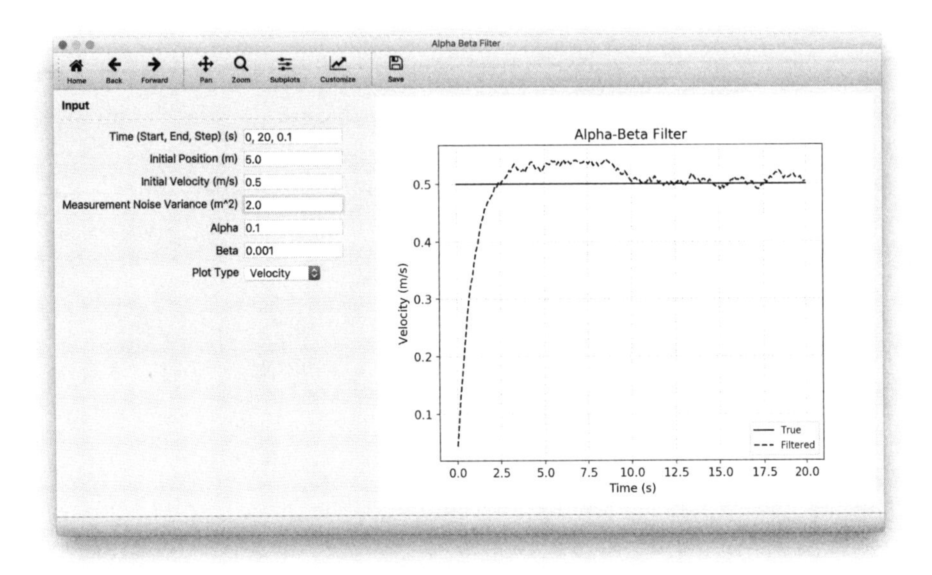

Figure 9.27 The true and filtered velocity calculated by *alpha_beta_example.py*.

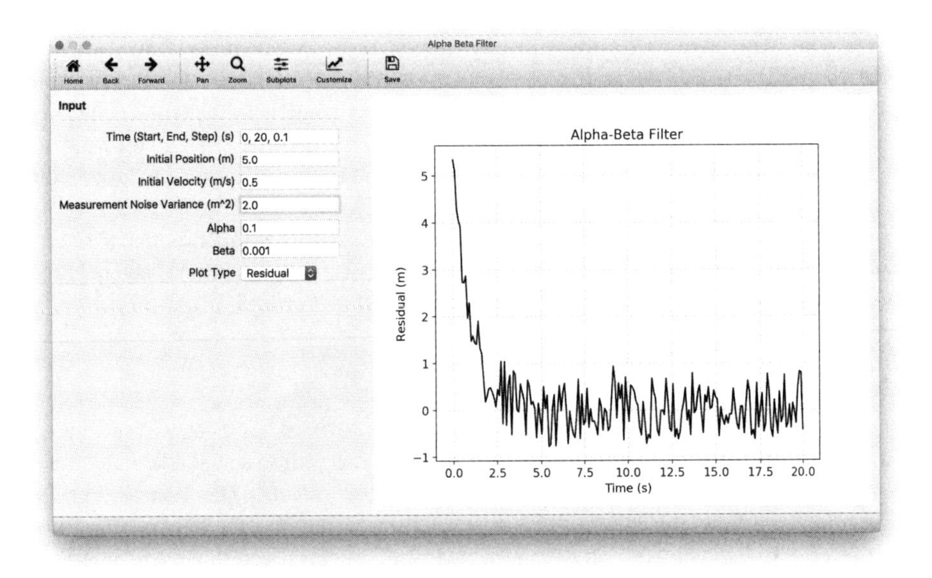

Figure 9.28 The residual calculated by *alpha_beta_example.py*.

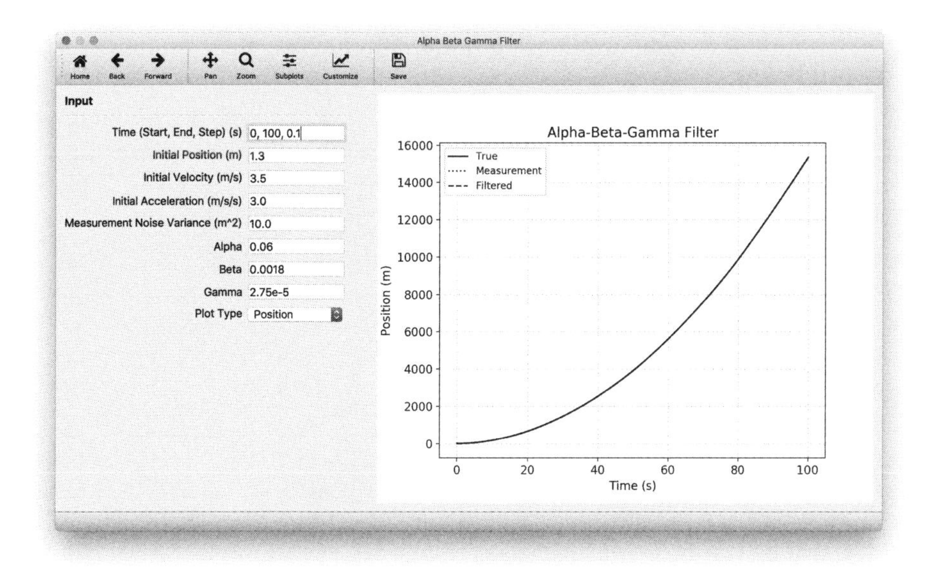

Figure 9.29 The true position, measurement position, and filtered position calculated by *alpha_beta_gamma_example.py*.

Solution: The solution to the above example is given in the Python code *alpha_beta_gamma_example.py* and in the MATLAB code *alpha_beta_gamma_example.m*. Running the Python example code displays a GUI allowing the user to enter the filter parameters, the target trajectory parameters, the radar parameters, and to choose a plot type. The code then performs the alpha-beta-gamma filtering and displays the results. The resulting filtered position, along with the true position and measurement position, is shown in Figure 9.29. The filtered velocity and true velocity are shown in Figure 9.30. The filtered acceleration and true acceleration are given in Figure 9.31. Finally, the residual for this example is shown in Figure 9.32. The reader is encouraged to vary the filter and trajectory parameters to see how these affect the tracking performance.

9.4.3 Kalman Filter: Constant Velocity

As the first Kalman filter example, use a constant velocity dynamic model to track a target with an initial position of $(1.0, 11.0, 3.2)$ m and a velocity of $(1.0, 2.0, 1.5)$ m/s. The radar has an update rate of 0.1s and a measurement variance of 10.0 m^2. Using a process noise variance of 1.0^{-6}, track the target from 0 to 40s, and plot the resulting y component of the position and velocity as well as the total residual. Compare these with

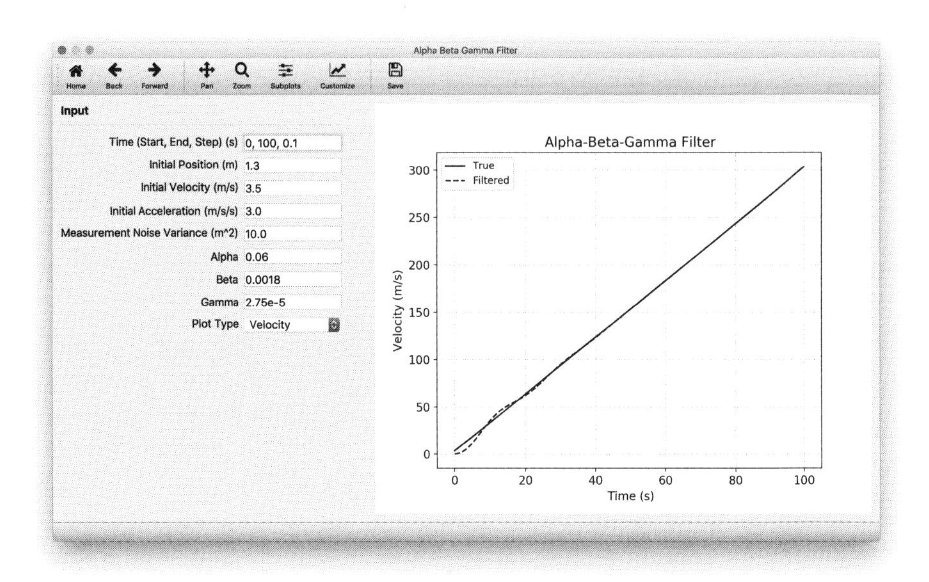

Figure 9.30 The true and filtered velocity calculated by *alpha_beta_gamma_example.py*.

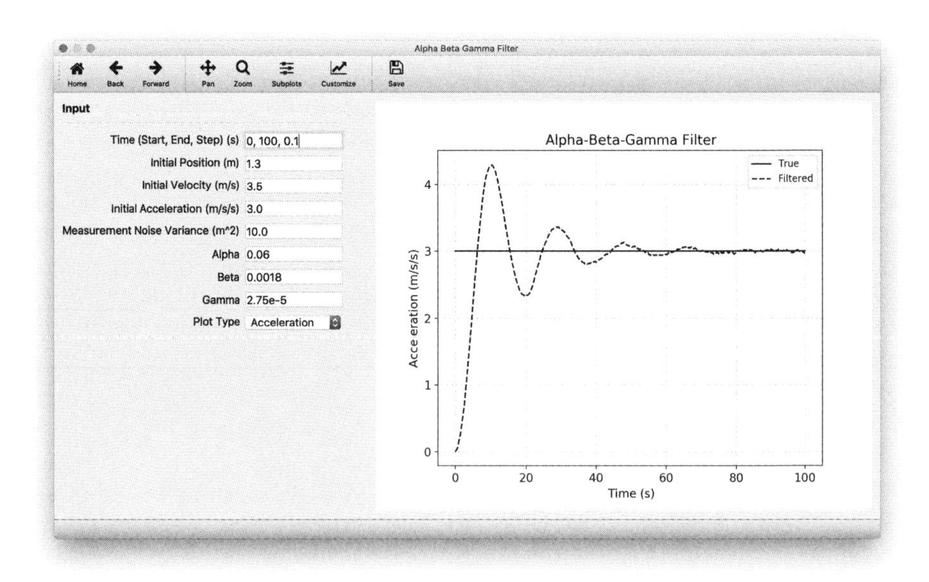

Figure 9.31 The true and filtered acceleration calculated by *alpha_beta_gamma_example.py*.

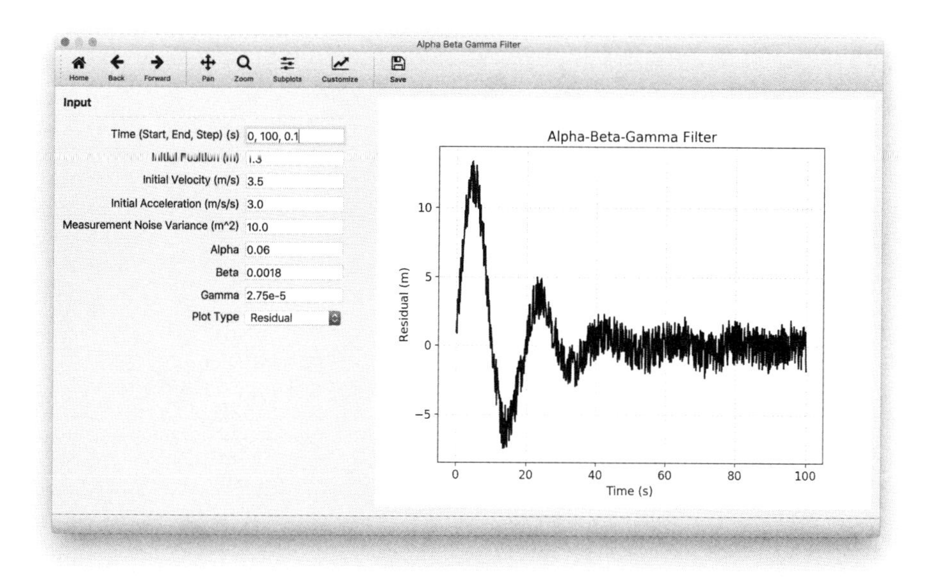

Figure 9.32 The residual calculated by *alpha_beta_gamma_example.py*.

the true y component of the position and velocity.

Solution: The solution to the above example is given in the Python code *kalman_cv_ example.py* and in the MATLAB code *kalman_cv_example.m*. Running the Python example code displays a GUI allowing the user to enter the filter parameters, the target trajectory parameters, the radar parameters, and to choose a plot type. The code then performs the Kalman filtering with a constant velocity dynamic model and displays the results. The resulting filtered position, along with the true position and measurement position, is shown in Figure 9.33. The filtered velocity and true velocity are shown in Figure 9.34. Finally, the residual for this example is shown in Figure 9.35. The reader is encouraged to vary the filter and trajectory parameters to see how these affect the tracking performance.

9.4.4 Kalman Filter: Constant Acceleration

This Kalman filter example uses a constant acceleration dynamic model to track a target with an initial position of $(7.0, 11.0, 21.0)$ m, an initial velocity of $(10.0, 20.0, 15.0)$ m/s, and an acceleration of $(0.5, 1.0, 0.75)$ m/s^2. The radar has an update rate of 0.1s and a measurement variance of 10.0 m^2. Using a process noise variance of 1.0^{-6}, track the target from 0 to 20s, and plot the resulting x component of the position, velocity, and acceleration as well as the total residual. Compare these with the true x component of the

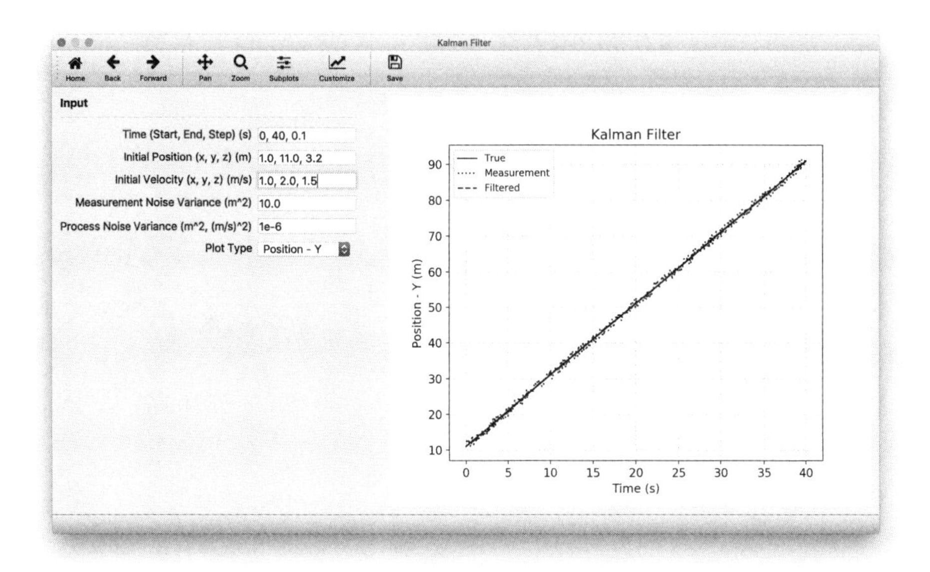

Figure 9.33 The true y position, measurement position, and filtered position calculated by *kalman_cv_example.py*.

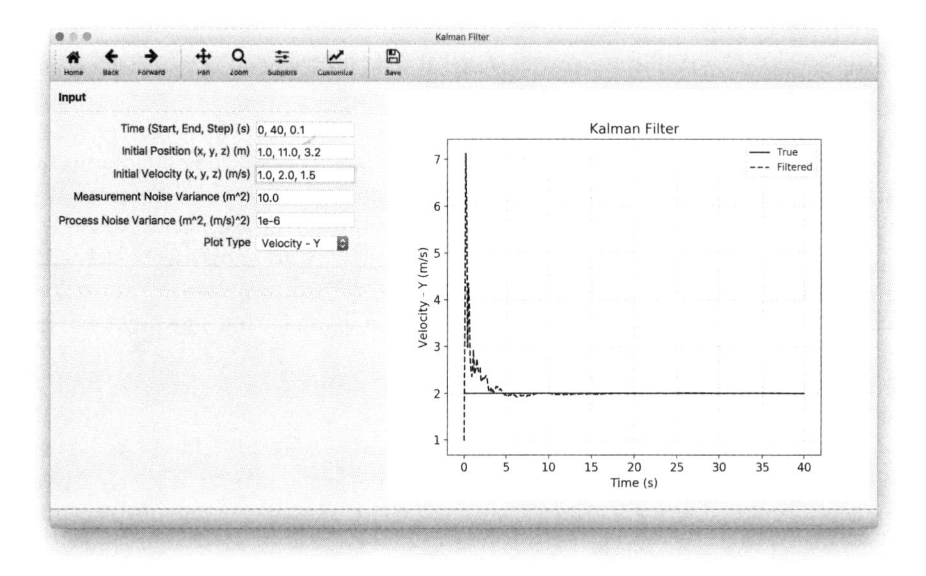

Figure 9.34 The true and filtered y velocity calculated by *kalman_cv_example.py*.

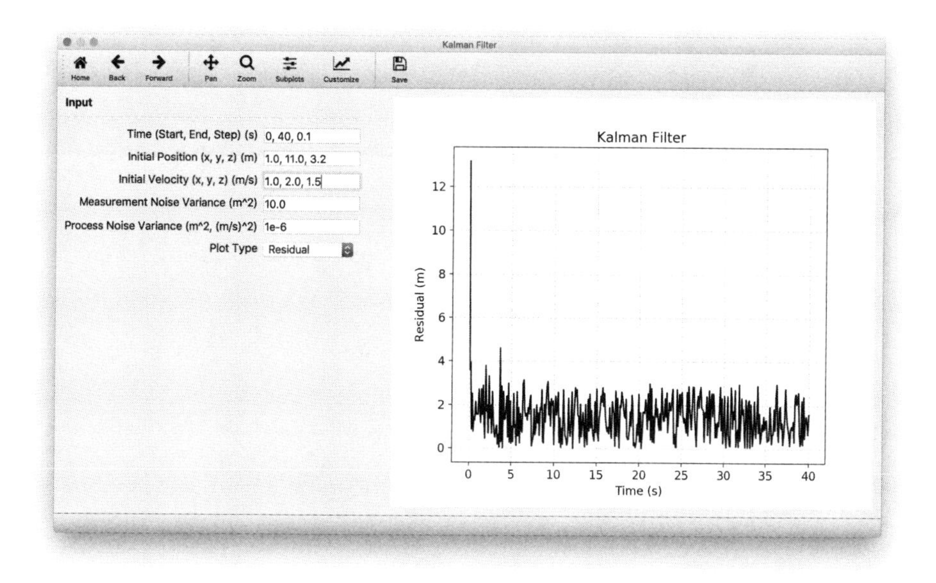

Figure 9.35 The total residual calculated by *kalman_cv_example.py*.

position, velocity, and acceleration.

Solution: The solution to the above example is given in the Python code *kalman_ca_example.py* and in the MATLAB code *kalman_ca_example.m*. Running the Python example code displays a GUI allowing the user to enter the filter parameters, the target trajectory, the radar parameters, and to choose a plot type. The code then performs the Kalman filtering with a constant acceleration dynamic model and displays the results. The resulting filtered position, along with the true position and measurement position, is shown in Figure 9.36. The filtered velocity and true velocity are shown in Figure 9.37. The filtered acceleration and true acceleration are shown in Figure 9.38. Finally, the residual for this example is shown in Figure 9.39. The reader is encouraged to vary the filter and trajectory parameters to see how these affect the tracking performance.

9.4.5 Adaptive Kalman Filter: Epsilon Method

In this example, use an adaptive Kalman filter with a constant velocity dynamic model and an adjustable process noise. The process noise will be adjusted using the ϵ_k method presented in Section 9.1.3.3. The target has an initial position of $(2.0, 1.0, 5.0)$ m and an initial velocity of $(10.0, 20.0, 15.0)$ m/s. The target makes a maneuver at 20s and the new velocity is $(100.0, 20.0, 15.0)$ m/s. The radar has an update rate of 0.1s and a

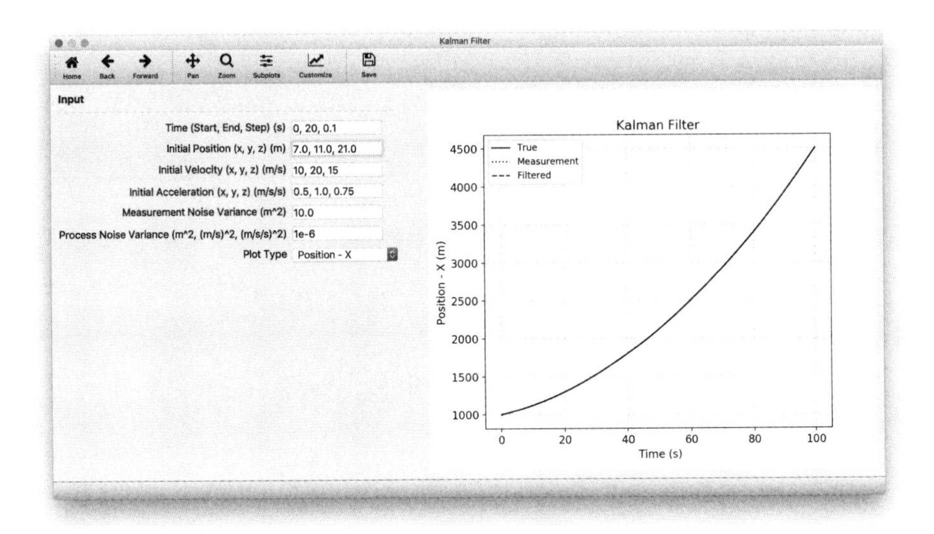

Figure 9.36 The true x position, measurement position, and filtered position calculated by *kalman_ca_example.py*.

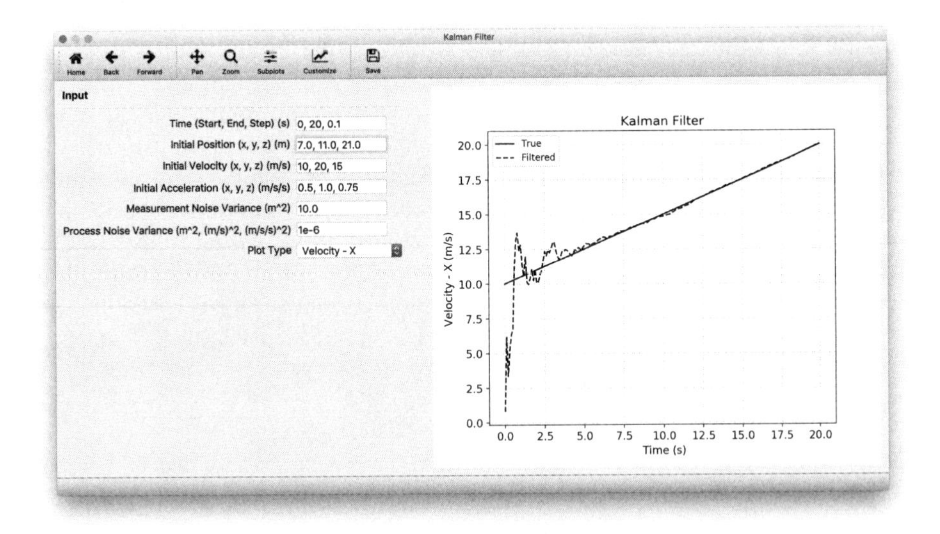

Figure 9.37 The true and filtered x velocity calculated by *kalman_ca_example.py*.

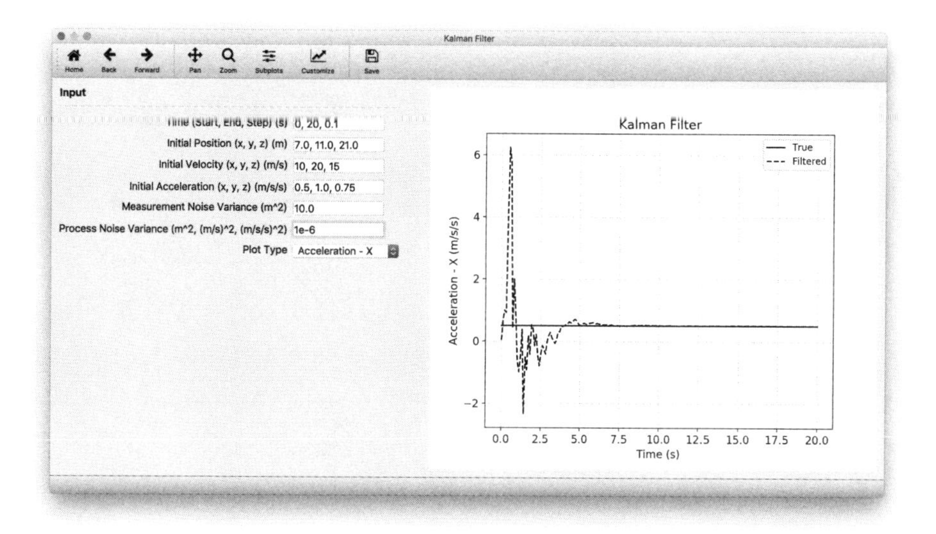

Figure 9.38 The true and filtered x acceleration calculated by *kalman_ca_example.py*.

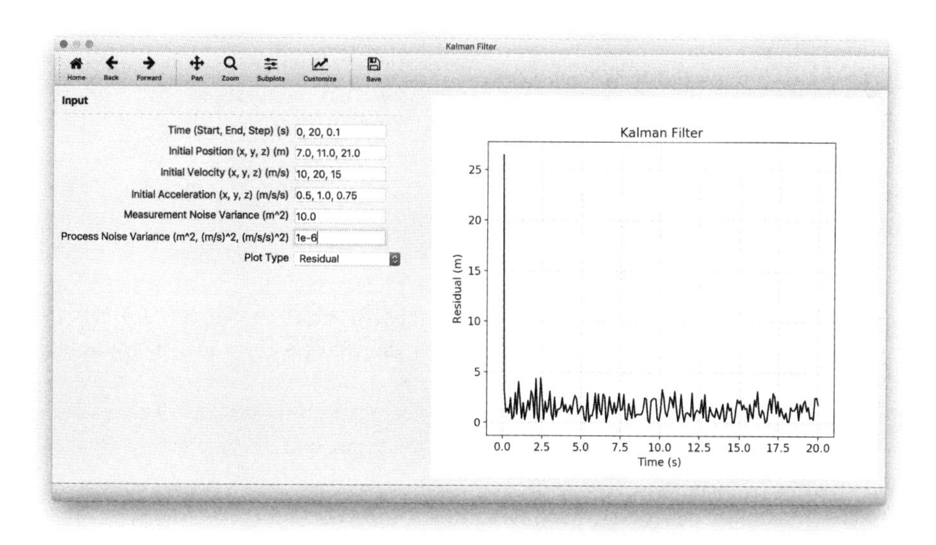

Figure 9.39 The total residual calculated by *kalman_ca_example.py*.

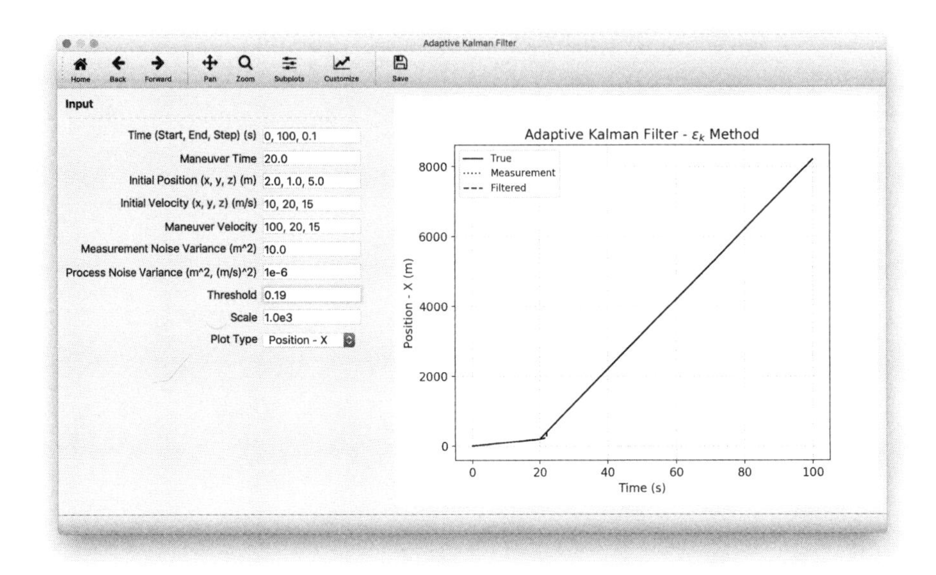

Figure 9.40 The true x position, measurement position, and filtered position calculated by *kalman_epsilon_example.py*.

measurement variance of 10.0 m^2. Using an initial process noise variance of 1.0^{-6}, a threshold of 0.19, and a scale of 1.0×10^3, track the target from 0 to 100s, and plot the resulting x component of the position and velocity as well as the total residual. Compare these with the true x component of the position and velocity.

Solution: The solution to the above example is given in the Python code *kalman_epsilon_example.py* and in the MATLAB code *kalman_epsilon_example.m*. Running the Python example code displays a GUI allowing the user to enter the filter parameters, the target trajectory, the radar parameters, and to choose a plot type. The code then performs the adaptive Kalman filtering with a constant velocity dynamic model and displays the results. The resulting filtered position, along with the true position and measurement position, is shown in Figure 9.40. The filtered velocity and true velocity are shown in Figure 9.41. Finally, the residual for this example is shown in Figure 9.42. The reader is encouraged to vary the filter and trajectory parameters to see how these affect the tracking performance.

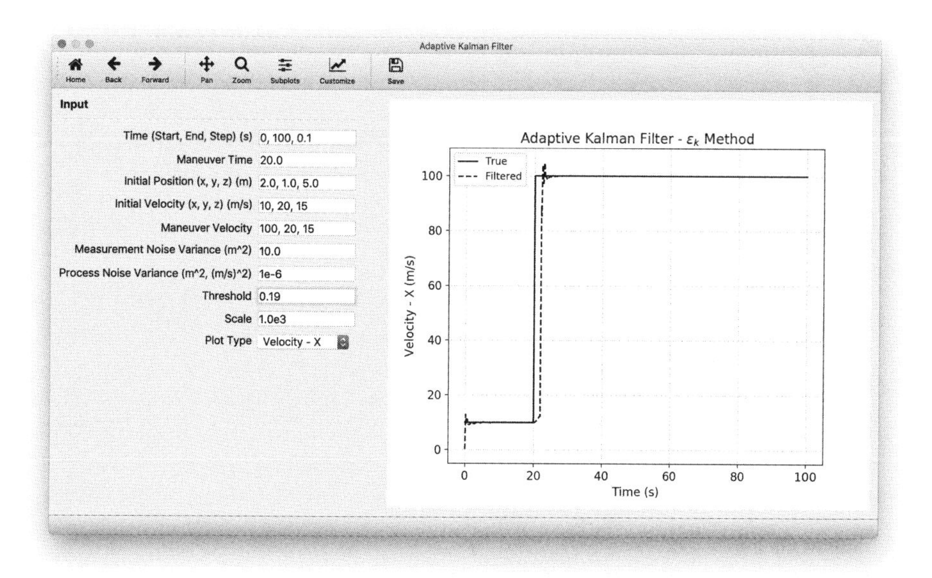

Figure 9.41 The true and filtered x velocity calculated by *kalman_epsilon_example.py*.

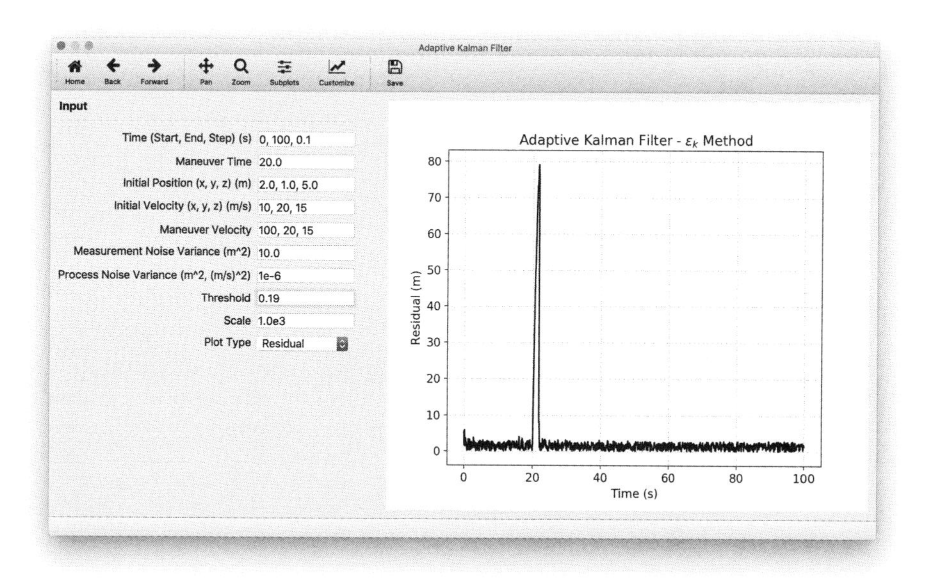

Figure 9.42 The total residual calculated by *kalman_epsilon_example.py*.

9.4.6 Adaptive Kalman Filter: Sigma Method

In this example, use an adaptive Kalman filter with a constant velocity dynamic model and an adjustable process noise. The process noise will be adjusted using the σ_k method presented in Section 9.1.3.3. The target has an initial position of $(2.0, 1.0, 5.0)$ m and an initial velocity of $(10.0, 20.0, 15.0)$ m/s. The target makes a maneuver at 20s, and the new velocity is $(100.0, 20.0, 15.0)$ m/s. The radar has an update rate of 0.1s and a measurement variance of 10.0 m^2. Using an initial process noise variance of 1.0^{-6}, a threshold of 0.08, and a scale of 1.0×10^3, track the target from 0 to 100s, and plot the resulting x component of the position and velocity as well as the total residual. Compare these with the true x component of the position and velocity.

Solution: The solution to the above example is given in the Python code *kalman_sigma_example.py* and in the MATLAB code *kalman_sigma_example.m*. Running the Python example code displays a GUI allowing the user to enter the filter parameters, the target trajectory, the radar parameters, and to choose a plot type. The code then performs the adaptive Kalman filtering with a constant velocity dynamic model and displays the results. The resulting filtered position, along with the true position and measurement position, is shown in Figure 9.43. The filtered velocity and true velocity are shown in Figure 9.44. Finally, the residual for this example is shown in Figure 9.45.

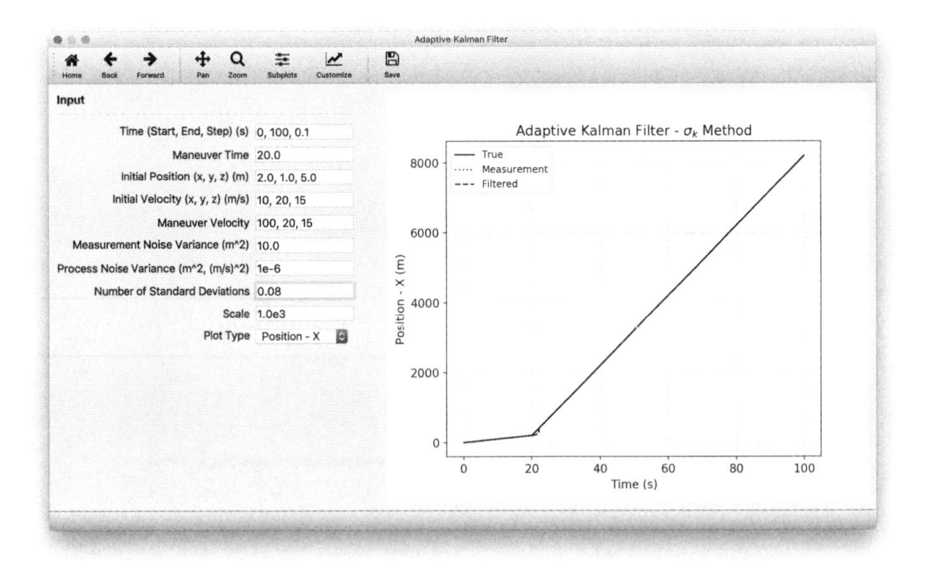

Figure 9.43 The true x position, measurement position, and filtered position calculated by *kalman_sigma_example.py*.

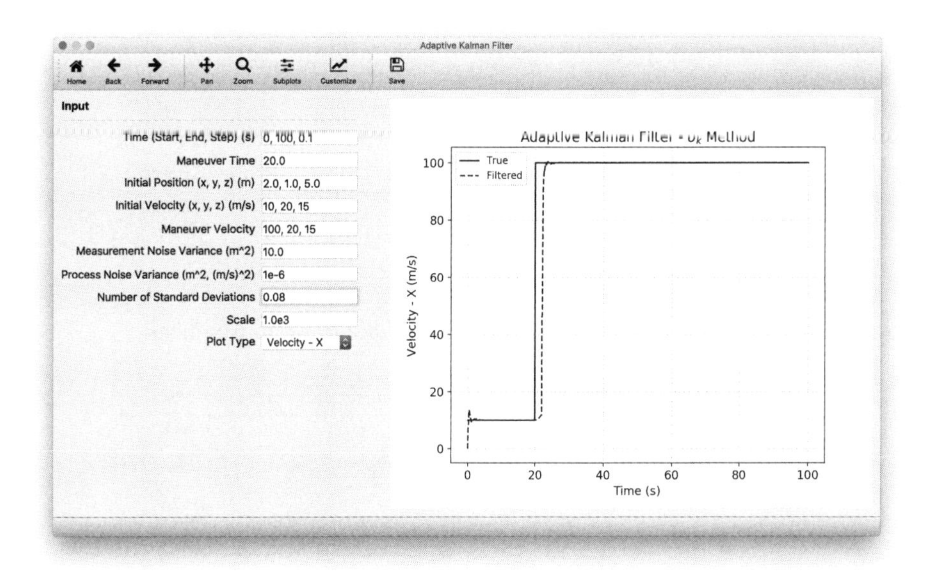

Figure 9.44 The true and filtered x velocity calculated by *kalman_sigma_example.py*.

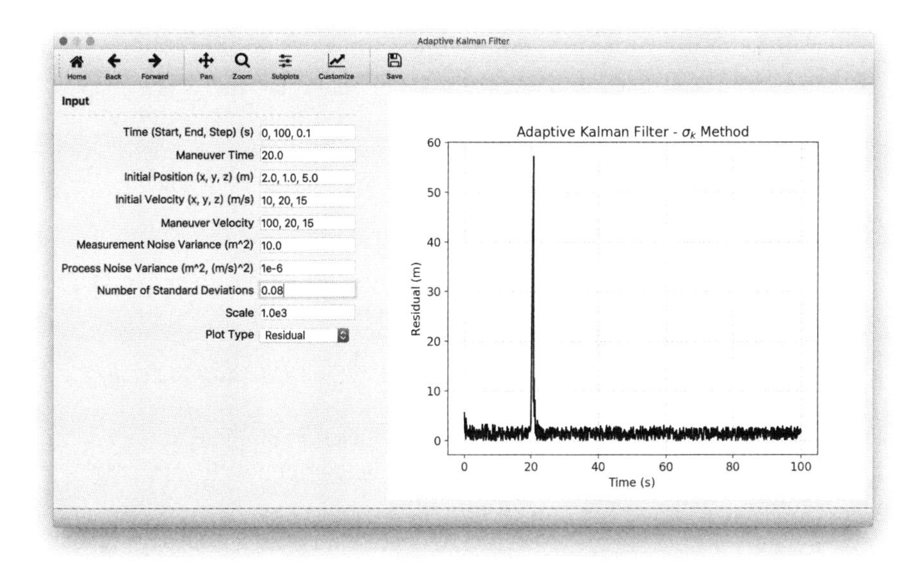

Figure 9.45 The total residual calculated by *kalman_sigma_example.py*.

PROBLEMS

9.1 Describe the purpose of tracking filters in radar systems and give practical examples of the use of tracking filters.

9.2 Explain the system model in alpha-beta filters and describe how these type of filters are related to Kalman filters.

9.3 What is the residual or innovation, and how is it used to update the filter?

9.4 How does the choice of α and β affect the performance of alpha-beta filters?

9.5 Describe the major differences between alpha-beta and alpha-beta-gamma filters.

9.6 What is the main disadvantage of extending alpha-beta-gamma to higher orders?

9.7 How does α, β, and γ affect the performance of alpha-beta-gamma filters?

9.8 What are the conditions under which the Kalman filter is optimal in the sense the estimated error uncertainty is minimized?

9.9 Describe the role of the Kalman gain in the update of the state and uncertainty. How does the state and uncertainty differ between large and small Kalman gain?

9.10 In the Kalman filter, what does the process noise represent? Give practical examples of systems with large process noise.

9.11 Explain the need for adaptive filtering and give practical examples of situations where this type of filtering could be beneficial.

9.12 Describe the difference in performance between constant acceleration and constant velocity Kalman filter models for maneuvering targets.

9.13 Characterize the differences between the σ and ϵ methods of adaptive filtering.

9.14 Describe the major disadvantages of the GNN approach to multitarget tracking and give an example scenario where this method would not perform well.

9.15 Outline the basic steps in the RFS technique for multitarget tracking and list the major advantages of JPDA and MHT.

9.16 How does the computational complexity of MHT compare to JPDA in their standard forms? Give examples of techniques to reduce the computational complexity of both methods.

9.17 In the measurement model for generating pseudomeasurements as input to tracking algorithms, explain the difference between the noise, fixed, and bias terms and how each contributes to the total measurement error.

References

[1] H. Ohanian and J. Markert. *Physics for Engineers and Scientist,* 3rd ed. W. W. Norton and Company, New York, 2007.

[2] E. Brookner. *Tracking and Kalman Filtering Made Easy*. John Wiley and Sons, New York, 1998.

[3] F. Asquith. Weight selection in first-order linear filters. Technical Report AD859332, Army Inertial Guidance and Control Laboratory Center, Redstone Arsenal, Alabama, 1969.

[4] P. Kalata. The tracking index: A generalized parameter for alpha-beta and alpha-beta-gamma target trackers. *IEEE Transactions on Aerospace and Electronic Systems*, 20(2):174–181, 1984.

[5] J. E. Gray and W. J. Murray. A derivation of an analytic expression for the tracking index for the alpha-beta-gamma filter. *IEEE Transactions on Aerospace and Electronic Systems*, 29:1064–1065, 1993.

[6] R. Kalman. A new approach to linear filtering and prediction problems. *Transaction of the ASME-Journal of Basic Engineering*, pages 35–45, March 1960.

[7] J. Crassidis and J. Junkins. *Optimal Estimation of Dynamic Systems*. CRC Press, Boca Raton, FL, 2011.

[8] R. Azuma. Improving static and dynamic registration in an optical see-through HMD. *Computer Graphics Conference Proceedings*, pages 197–204, 1994.

[9] E. Foxlin, M. Harrington and G. Pfeifer. Constellation^{TM}: A wide-range wireless motion-tracking system for augmented reality and virtual set applications. *Computer Graphics Conference Proceedings*, pages 371–378, 1998.

[10] J. Van Pabst and P. Krekel. Multi sensor data fusion of points, line segments and surface segments in 3D space. *7th International Conference on Image Analysis and Processing*, pages 174–182, 1993.

[11] A. Azarbayejani and A. Petland. Recursive estimation of motion, structure, and focal length. Technical report, Massachusetts Institute of Technology, Cambridge, MA, 1994.

[12] S. Emura and S. Tachi. Sensor fusion based measurement of human head motion. *3rd IEEE International Workshop on Robot and Human Communication*, 1994.

[13] Y. Bar-Shalom, L. Xiao-Rong and T. Kirubarajan. *Estimation with Applications to Tracking and Navigation*. John Wiley and Sons, New York, 2001.

[14] P. Zarchan and H. Musoff. *Fundamentals of Kalman Filtering: A Practical Approach,* 4th ed. American Institute of Aeronautics and Astronautics, Reston, VA, 2013.

[15] G. Marks. Multiple model adaptive estimation (MMAE) based filter banks for interception of maneuvering targets. *AIAA Guidance, Navigation and Control Conference and Exhibit*, 2007.

[16] H. Blom and Y. Bar-Shalom. The interacting multiple model algorithm for systems with Markovian switching coefficients. *IEEE Transactions on Automatic Control*, AC-8(8):780–783, 1998.

[17] A. S. Gilmour. *Design and Analysis of Modern Tracking Systems*. Artech House, Norwood, MA, 1999.

[18] Y. Bar-Shalom, P. Willett and X. Tian. *Tracking and Data Fusion: A Handbook of Algorithms*. YBS Publishing, Storrs, CT, 2011.

[19] R. Mahler. *Advances in Statistical Multisource-Multitarget Information Fusion*. Artech House, Norwood, MA, 2014.

[20] R. Altendorfer and S. Wirkert. A complete derivation of the association log-likelihood distance for multi-object tracking. *arXiv:1508.04124*, August 2015.

[21] Wikipedia Contributors. *Hungarian Algorithm*. https://en.wikipedia.org/wiki/Hungarian_algorithm.

[22] Wikipedia Contributors. *Auction Algorithm*. https://en.wikipedia.org/wiki/Auction_algorithm.

[23] Y. Bar-Shalom, X. Li and T. Kirubarajan. *Estimation with Applications to Tracking and Navigation*. Wiley, New York, 2001.

[24] B. N. Vo, et al. *Multitarget Tracking*. Wiley, New York, 2015.

[25] S. He, Hyo-Sang Shin and A. Tsourdos. Track-oriented multiple hypothesis tracking based on Tabu search and Gibbs sampling. *IEEE Sensors Journal*, 18(1):328–339, January 2018.

[26] R. J. Fitzgerald. Development of practical PDA logic for multitarget tracking by microprocessor. *American Control Conference*, pages 889–898, 1986.

[27] J. A. Roecker and G. L. Phillis. Suboptimal joint probabilistic data association. *IEEE Sensors Journal*, 29(2):510517, 1993.

[28] J. A. Roecker. A class of near optimal JPDA algorithms. *IEEE Transactions on Aerospace and Electronic Systems*, 30(2):504510, 1994.

[29] D. Bertsimas and J. Tsitsiklis. *Introduction to Linear Optimization*. Athena Scientific, 1997.

[30] O. Erdinc, P. Willett and Y. Bar-Shalom. Probability hypothesis density filter for multitarget multi-sensor tracking. *7th International Conference on Information Fusion*, page 146153, 2005.

[31] B.T. Vo, B. Vo and A. Cantoni. The cardinalized probability hypothesis density filter for linear Gaussian multi-target models. *40th Annual Conference on Information Sciences and Systems*, page 681686, 2006.

[32] B. Vo, S. Singh, and A. Doucet. Sequential Monte Carlo methods for multi-target filtering with random finite sets. *IEEE Transaction on Aerospace and Electronic Systems*, pages 1224–1245, 2005.

[33] B. Vo and W. Ma. The Gaussian mixture probability hypothesis density. *IEEE Transactions on Signal Processing*, pages 4091–4104, 2006.

[34] T. Wood. *Random Finite Sets for Multitarget Tracking with Applications*. PhD thesis, University of Oxford, 2011.

[35] D. K. Barton. *Modern Radar System Analysis*. Artech House, Norwood, MA, 1988.

Chapter 10

Tomographic Synthetic Aperture Radar

Tomography is the imaging of an object from slices or sections gathered by attenuated or reflected waves. The etymology of the word tomography is from Ancient Greek *tomos*, which means *slice*, and *grapho*, which means *to write*. This chapter begins with an introduction and history of tomography and its various applications. This is followed by a discussion of the Radon transform, which leads to a set of line integrals and projections of the object's density function. Using the Fourier slice theorem, the projection data is used to fill in the spatial frequency domain of the object's density function. Image reconstruction techniques, including the filtered backprojection algorithm, are then used to recreate the density function from the spatial frequency domain data. The tomography/SAR analogy will be shown, and a generalized three-dimensional framework will then be given. The chapter concludes with several Python examples to emphasize the concepts associated with tomographic SAR imaging.

10.1 TOMOGRAPHY

Fundamentally, tomography is the imaging of an object or scene using data collected from many different viewing directions, as illustrated in Figure 10.1. This data, referred to as projections, is typically transmission or reflection data, and may be generated by illuminating the object with various sources of energy. Tomographic imaging has been applied to many areas of science and medicine, with several examples listed in Table 10.1.

10.1.1 History

In 1917, Johann Radon demonstrated that it was possible to reconstruct an original density function from projection data [1]. In two dimensions, the *Radon transform* represents the projection of a density function defined in a plane to the space of lines in

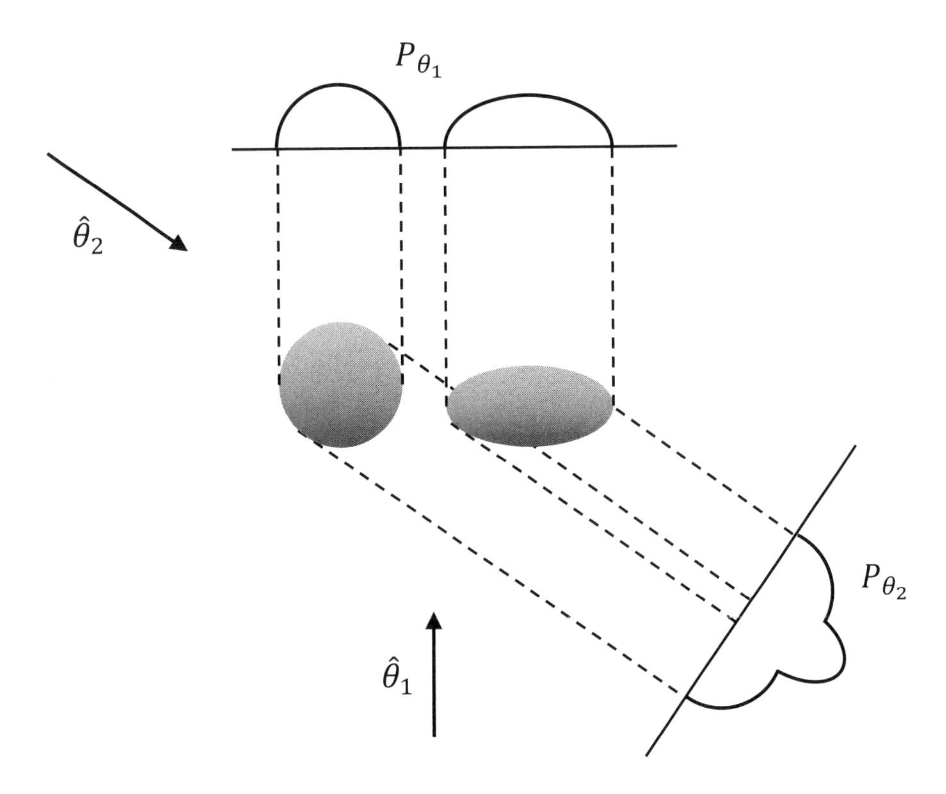

Figure 10.1 Two projections for an object consisting of a circular and elliptical cylinder.

the plane. The value of the Radon transform for a particular line is found by integrating the density function over that line. In his original work, Radon also showed the inverse transform and thus the mathematical basis for tomographic reconstruction. Radon further included formulas for the transform in three dimensions, in which the integral is taken over planes rather than lines as in two dimensions. The Radon transform has since been generalized to higher dimensional Euclidean spaces. The complex analog of the Radon transform is known as the Penrose transform [2–4].

While Radon's work was purely mathematical, Allan Cormack introduced a more practical inversion formula using the eigenfunctions from a rotational operator in 1963 and 1964 [5, 6], which provided the theoretical mathematics for a computerized reconstruction, which became known as computed tomography (CT). In 1971, Sir Godfrey Hounsfield built the X-ray computer scanner for medical work, for which Cormack and Hounsfield shared a Nobel prize [7, 8]. The device was installed at Atkinson Morely hospital. In 1972, EMI Ltd. publicly announced the EMI scanner, which was for brain scans only. Afterward, in 1975, Hounsfield built a whole-body scanner. Hounsfield's original

Table 10.1
Applications of Tomographic Imaging

Application	*Type of Tomography*
Ocean temperature and currents	Acoustic
3D Medical imaging	Positron emission, X-ray
Subsurface earth imaging	Seismic
Permittivity distribution of an object	Electrical capacitance
Flame analysis	Optical
Aquifer analysis	Hydraulic

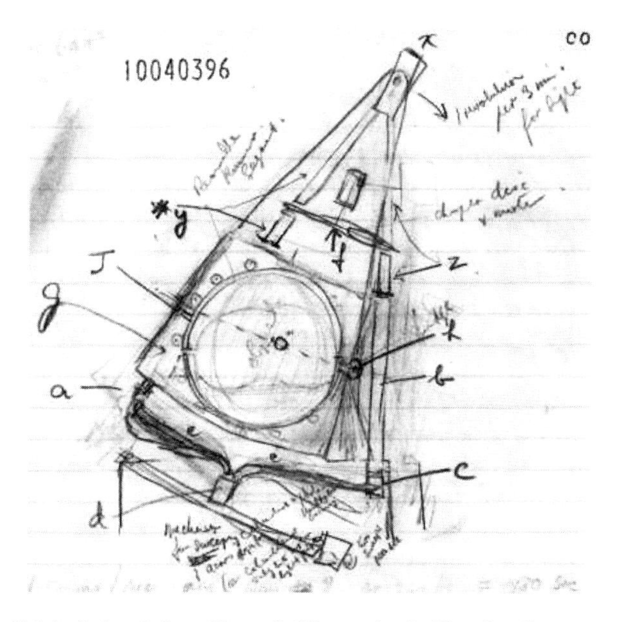

Figure 10.2 Original sketch from Hounsfield's notebook (Creative Commons Share Alike 3.0).

sketch is shown in Figure 10.2, and Figure 10.3 shows his first CT scanner prototype. CT scanning performance has grown at a remarkable rate since Hounsfield's first scanner. Table 10.2 gives some examples of today's CT scanners.

Figure 10.3 First CT scanner prototype by Hounsfield at EMI (Creative Commons Share Alike 3.0).

Table 10.2
Tomographic Scanners

Device	*Application*	*Note*
Phoenix x|aminer (GE)	High resolution inspection of electronic assemblies and components	Detail detectability down to 0.5 μm
Revolution CT (GE)	Cardiac-, neuro-, musculoskeletal-imaging for rapid trauma assessment	With GSI Xtream, provides volume spectral CT to improve small lesion detection and tissue characterization
ImagiX (North Star Imaging)	Food scans to screen for contaminates in packaging and assist in development of agricultural hybrids	Overall maximum system resolution $\approx$ 0.5 μm
Metrotom (Zeiss)	Performance testing, material analysis, and quality assurance for plastics	Standard compliant and traceable precision testing (VDI/VDE 2630) Resolution 3.5-6μm

10.1.2 Line Integrals and Projections

Consider the basic geometry for two-dimensional X-ray tomography shown in Figure 10.4, where the object's two-dimensional density function is given by $f(x, y)$. $P_\theta(t)$ is

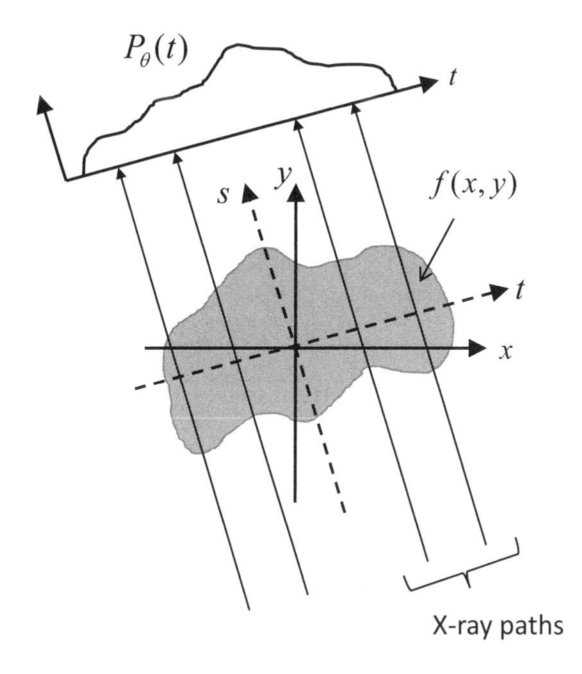

Figure 10.4 Basic geometry for two-dimensional X-ray tomography.

the value of the Radon transform of the density function, $f(x, y)$, and is found by forming a set of line integrals. Mathematically, the Radon transform is expressed as

$$P_\theta(t) = \int_\ell f(x, y)\, ds = \int_{-\infty}^{\infty}\int_{-\infty}^{\infty} f(x, y)\, \delta(x\cos\theta + y\sin\theta - t)\, dx\, dy. \qquad (10.1)$$

Once the Radon transform is found, a method for reconstructing the original density function is needed. The following sections demonstrate the reconstruction with the use of the Fourier slice theorem and the filtered backprojection algorithm.

10.1.2.1 Fourier Slice Theorem

The Fourier slice theorem, illustrated in Figure 10.5, states the one-dimensional Fourier transform of a parallel projection is equal to a slice of the two-dimensional Fourier

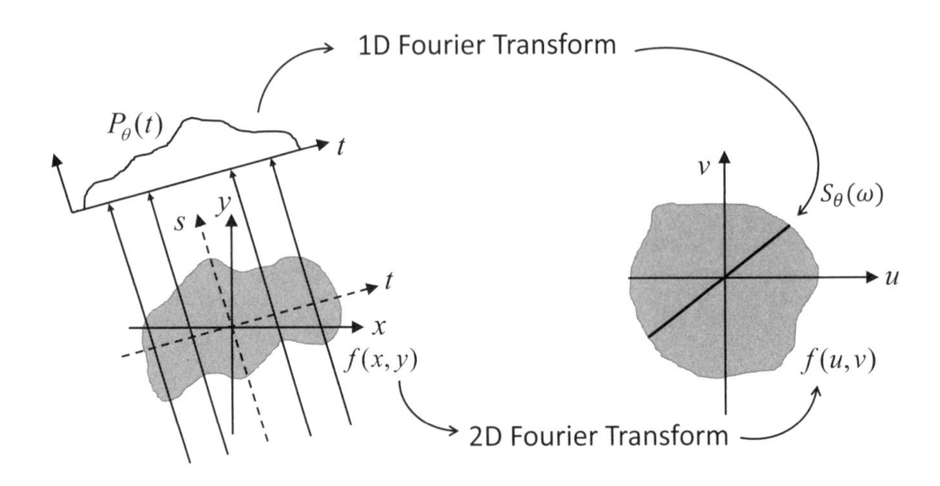

Figure 10.5 Illustration of the Fourier slice theorem.

transform of the original function. To demonstrate the Fourier slice theorem, start by defining the two-dimensional Fourier transform of the object's density function as

$$F(u,v) = \int_{-\infty}^{\infty}\int_{-\infty}^{\infty} f(x,y)\, e^{-j2\pi(ux+vy)}\, dx\, dy, \tag{10.2}$$

and the one-dimensional Fourier transform of the projection function, $P_\theta(t)$, as

$$S_\theta(\omega) = \int_{-\infty}^{\infty} P_\theta(t)\, e^{-j2\pi\omega t}\, dt. \tag{10.3}$$

Since no assumption has been made about the object's orientation, θ is set to 0 without loss of generality. This leads to $v = 0$, and allows the two-dimensional Fourier transform to be written as

$$F(u,0) = \int_{-\infty}^{\infty}\int_{-\infty}^{\infty} f(x,y)\, e^{-j2\pi ux}\, dx\, dy. \tag{10.4}$$

Note that the phase term is now a function of x only, and the double integral may be written as

$$F(u,0) = \int_{-\infty}^{\infty} \left[\int_{-\infty}^{\infty} f(x,y)\, dy \right] e^{-j2\pi ux}\, dx. \tag{10.5}$$

The term inside the brackets is an expression for the projection along lines of constant x. This may be written as

$$P_{\theta=0}(x) = \int_{-\infty}^{\infty} f(x,y)\, dy. \tag{10.6}$$

Substituting (10.6) into (10.5) results in

$$F(u,0) = \int_{-\infty}^{\infty} P_{\theta=0}(x) e^{-j2\pi ux}\, dx. \tag{10.7}$$

Recalling the definition of the one-dimensional transform in (10.3), the relationship between the one-dimensional projection function and the two-dimensional Fourier transform of the object's density function is

$$F(u,0) = S_{\theta=0}(\omega). \tag{10.8}$$

As stated earlier, no assumption was made about the object's orientation; therefore the general expression of the Fourier slice theorem is

$$F(u,v) = S_{\theta}(\omega). \tag{10.9}$$

A collection of N projections at angles $[\theta_1, \theta_2, \ldots, \theta_N]$ fills in the two-dimensional spatial frequency domain of the object's density function, as illustrated in Figure 10.6. If the number of projections is large enough, a reconstruction can be made by interpolating the values onto a rectilinear grid and then performing the inverse two-dimensional Fourier transform. This method is often referred to as *polar reformatting*. While the image may be reconstructed in this manner, the high frequency points are further apart than the low frequency points, resulting in image degradation. Therefore, other reconstruction techniques are desired.

While Radon's original work showed the mathematical procedure for reconstruction of a density function from its projections, the inverse Radon transform cannot be solved analytically. One of the earliest reconstruction approaches is the algebraic reconstruction technique (ART) [9], which was originally used by Hounsfield. Since that time, many improvements in the speed and quality of ART techniques have been developed. There are three basic versions of ART for solving the reconstruction problem. The first of these is the simultaneous iterative reconstruction technique (SIRT) [10]. In this method, pixel values are updated with an average of all weighted differences from all sensor

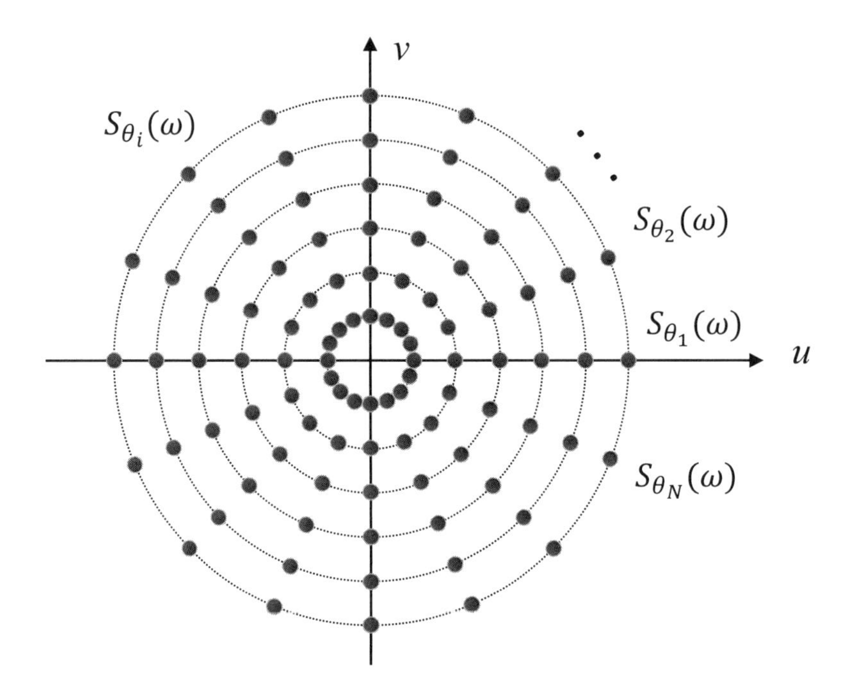

Figure 10.6 Two-dimensional spatial frequency domain from one-dimensional projection functions.

elements over all projection angles. While the SIRT reconstruction method does result in increased image quality compared to standard ART, the time to convergence is significantly increased. The second version is called simultaneous algebraic reconstruction technique (SART). This method seeks to find a middle ground between standard ART and SIRT [11] by updating pixels with the average weighted differences from all sensor elements in a single projection. The last version is the multiplicative algebraic reconstruction technique (MART) [9]. While most iterative solutions are additive in nature, MART is multiplicative. Although the MART algorithm converges faster than the additive techniques, it magnifies noise, and a pixel value of zero halts progress of the algorithm and leads to incorrect results. Researchers began using Fourier transforms in the solution, leading to a method known as filtered backprojection [12]. This is the most common image reconstruction method and will be presented in the following sections.

10.1.2.2 Filtered Backprojection

To illustrate the filtered backprojection algorithm, begin by taking the inverse Fourier transform of the object's two-dimensional spatial frequency domain representation as

$$f(x,y) = \int_{-\infty}^{\infty}\int_{-\infty}^{\infty} F(u,v)\, e^{j2\pi(xu+yv)}\, du\, dv. \tag{10.10}$$

Since the projection functions fill in the spatial frequency domain in a polar fashion, make the following substitutions

$$u = \omega \cos\theta, \tag{10.11}$$

$$v = \omega \sin\theta, \tag{10.12}$$

$$du\, dv = \omega\, d\omega\, d\theta, \tag{10.13}$$

which results in

$$f(x,y) = \int_{0}^{2\pi}\int_{0}^{\infty} F(w,\theta)\, e^{j2\pi\omega(x\cos\theta + y\sin\theta)}\, \omega\, d\omega\, d\theta. \tag{10.14}$$

Splitting (10.14) into two integrals gives

$$f(x,y) = \int_{0}^{\pi}\int_{0}^{\infty} F(w,\theta)\, e^{j2\pi\omega(x\cos\theta + y\sin\theta)}\, \omega\, d\omega\, d\theta$$
$$+ \int_{0}^{\pi}\int_{0}^{\infty} F(w,\theta+\pi)\, e^{j2\pi\omega[x\cos(\theta+\pi) + y\sin(\theta+\pi)]}\, \omega\, d\omega\, d\theta. \tag{10.15}$$

Making use of the relation, $F(\omega, \theta + \pi) = F(-\omega, \pi)$, allows (10.15) to be written as

$$f(x,y) = \int_{0}^{\pi}\int_{-\infty}^{\infty} F(w,\theta)\, e^{j2\pi\omega(x\cos\theta + y\sin\theta)}\, |\omega|\, d\omega\, d\theta. \tag{10.16}$$

The inner integral represents the convolution of the projection with a filter as

$$P_\theta(t) * h(t) = \int_{-\infty}^{\infty} F(w,\theta)\, e^{j2\pi\omega(x\cos\theta + y\sin\theta)}\, |\omega|\, d\omega. \tag{10.17}$$

The filtered projection function in (10.17) is then projected onto the image space along a line in the same direction as the projection was obtained. This is expressed as

$$f(x,y) = \int_0^{\pi} P_\theta(t) * h(t)\, d\theta. \tag{10.18}$$

The backprojection imaging process is as follows;

1. Construct the spatial domain target density array $f(x_i, y_j)$.
2. Pulse compress return signals and perform filtering.
3. Upsample the returns for better image quality. Upsample factors can be large (> 100).
4. For a given pixel point, (x_i, y_j), find the nearest upsampled time point (interpolate) and add the value of the filtered data to the array component $f(x_i, y_j)$.
5. Repeat for all pixel locations (x_i, y_j) and all pulses.

The filtered backprojection algorithm has several advantages. It is computationally efficient and employs the fast Fourier transform (FFT). It is highly parallelizable as each projection's contribution and each pixel's value may be computed separately. Also, image reconstruction may begin as soon as the first projection function is obtained, rather than waiting for all projections to be collected. Finally, the filtered backprojection algorithm does not depend on uniform angular sampling in the projection data, and multiple object rotations or platform passes can be used.

10.1.3 SAR Imaging

When processing the reflected waveform of a radar system as direct transductions of the spatial frequencies of a targets reflectivity function, a method for separating target features in both the range and cross-range dimension arises. This spatial frequency formulation makes the SAR imaging problem analogous to X-ray medical imaging [13]. Researchers at the University of Illinois were the first to formally introduce the tomography/SAR analogy in literature [14]. Table 10.3 provides an overview of a few SAR systems.

To illustrate the tomography approach to SAR imaging, consider an imaging scenario where energy from the radar illuminates an area that contains a target to be imaged, as shown in Figure 10.7. To begin, assume the target structure across the illuminated region is described by a two-dimensional reflectivity such that the radar returns from all scatterers on a constant range line are received simultaneously by the radar system. In the previous sections, the projection , $P_\theta(t)$, represented a projection of the X-ray attenuation function of the object. For SAR imaging, the projection represents a projection of the reflectivity function of the target. The projections may now be collected at a number of angles, as illustrated in Figure 10.8. Using the Fourier slice

Table 10.3
Overview of SAR Systems

Sensor	*Dates*	*Frequency (Polarization)*	*Comments*
Seasat (NASA/JPL)	1978	L (HH)	First civilian SAR satellite
Radarsat-1 (CSA)	1995–Present	C (HH)	First Canadian SAR satellite
Radarsat-2 (CSA)	2007–Present	C (Full)	Resolution up to 1m x 3m
TerraSAR-X (DLR)	2007–Present	X (Full)	First bistatic radar in space
COSMO (ASI / MiD)	2007–Present	X (Dual)	Constellation of four satellites
F-SAR (DLR)	2006–Present	Modular X, C, S, L, and P (Full)	Single-pass polarimetric interferometry capability in X- and S-bands
UAVSAR (NASA/JPL)	1988–Present	L, P, and Ka (Full)	Airborne SAR for Earth sciences

theorem as before, the projection data is used to fill in the reflectivity function's two-dimensional spatial frequency domain. The filtered backprojection algorithm is then used to reconstruct the reflectivity function from this two-dimensional spatial frequency data. As a simple example, consider a SAR imaging scenario consisting of three point scatterers located at $(0, 0)$, $(-3, -3)$, and $(3, 3)$ m relative to the center of the area to be imaged, as illustrated in Figure 10.9. The targets have radar cross-section values of 10, 5, and 20 m^2, respectively. The azimuth angular span is 3° and the bandwidth is 300 MHz. Figure 10.10 shows the image after the backprojection process with a single pulse. The resolution in the range dimension is $c/2B$, and as expected, the cross-range resolution is very poor and limited only by the beamwidth of the radar antenna. Figure 10.11 shows the image after 10 pulses have been backprojected. The range resolution is the same, and there is quite an improvement in the cross-range resolution. Figure 10.12 shows the reconstructed image after 100 pulses have been backprojected. Finally, Figure 10.13 illustrates the image after two pulses, separated by a wide angular span, have been backprojected. Aliasing is clearly present in this image and is caused by the projection functions being collected at angles with too large of a separation. The sampling requirement in the angular domain is [13, 15]

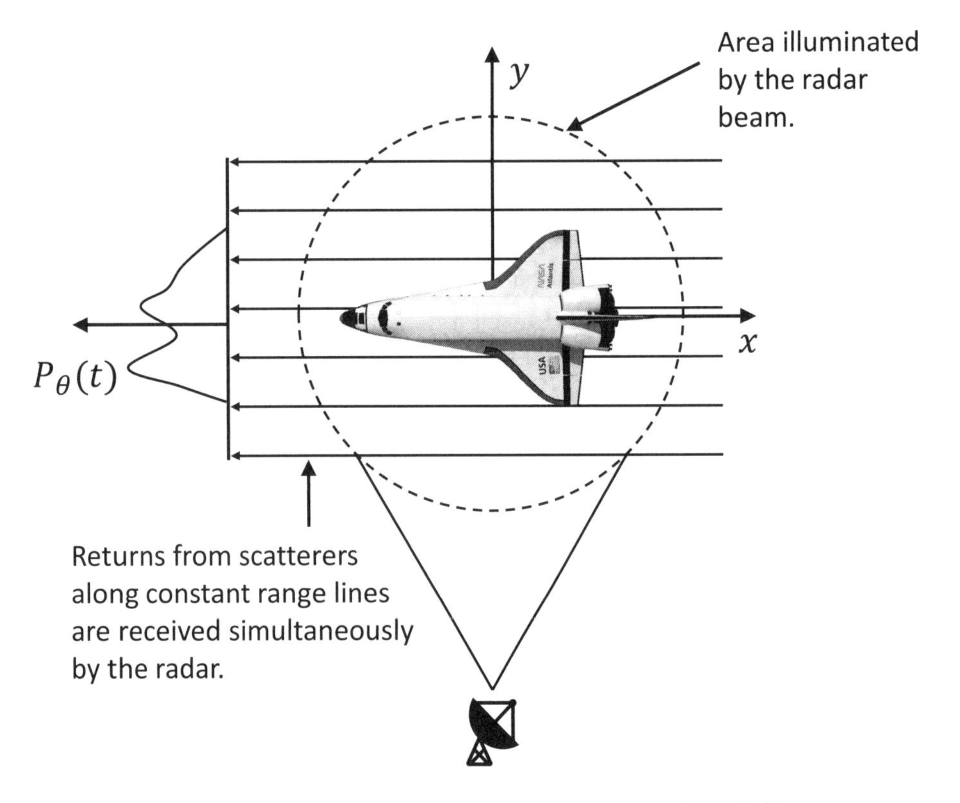

Figure 10.7 Basic scenario for tomographic SAR imaging (NASA).

$$\Delta\theta \leq \frac{2\pi}{k'_{max} - k'_{min}}, \tag{10.19}$$

where k' is the instantaneous angular frequency.

10.1.4 Three-Dimensional Tomography

When extending the Fourier slice theorem to three dimensions, two different versions may be obtained. The first is the *planar slice* version, which relates planar projection functions to planar Fourier transform slices. This version is commonly used in optical and infrared imaging applications. The second is the *linear trace* version, which relates linear projection functions to traces of the three-dimensional Fourier transform. This is the version used for radar applications as it deals with pulse-compressed radar returns.

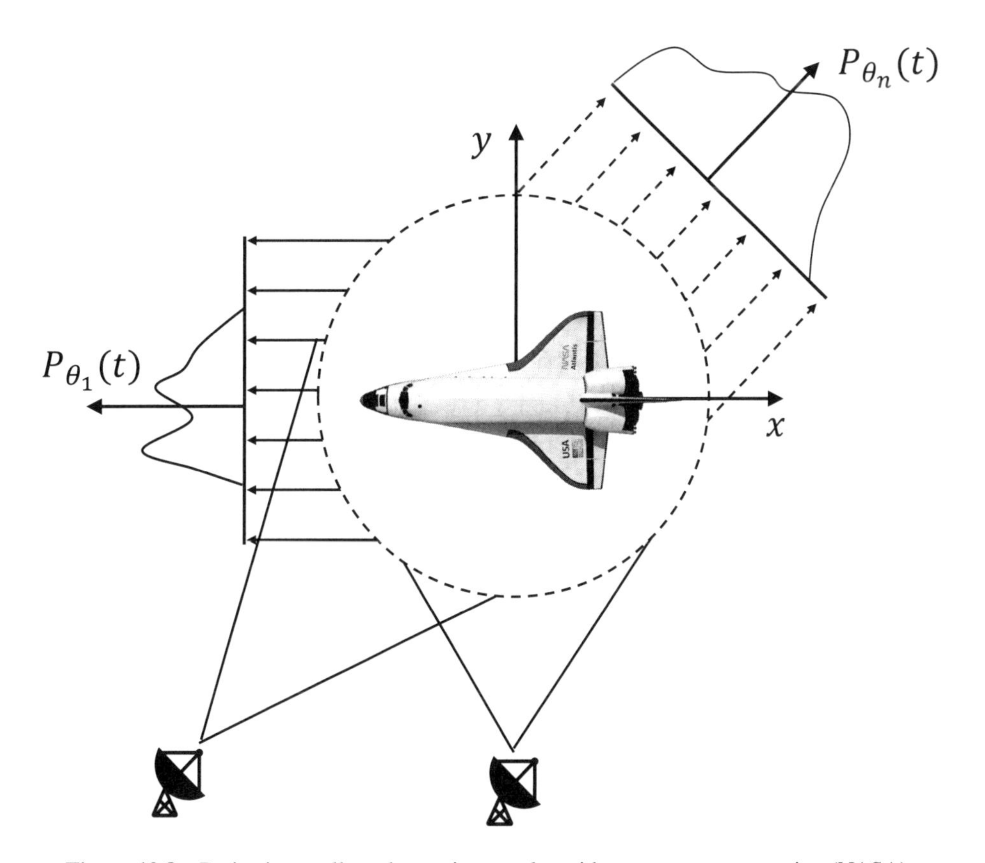

Figure 10.8 Projections collected at various angles with sensor or target motion (NASA).

10.1.4.1 Linear Trace Theorem

To show the linear trace theorem, start by defining the three-dimensional Fourier transform of the target's reflectivity function as

$$F(u, v, w) = \int_{-\infty}^{\infty}\int_{-\infty}^{\infty}\int_{-\infty}^{\infty} f(x, y, z)\, e^{-j2\pi(ux+vy+wz)}\, dx\, dy\, dz, \tag{10.20}$$

and the one-dimensional Fourier transform of the projection, $P_{\theta,\psi}(t)$, as

$$S_{\theta,\psi}(\omega) = \int_{-\infty}^{\infty} P_{\theta,\psi}(t)\, e^{-j2\pi\omega t}\, dt. \tag{10.21}$$

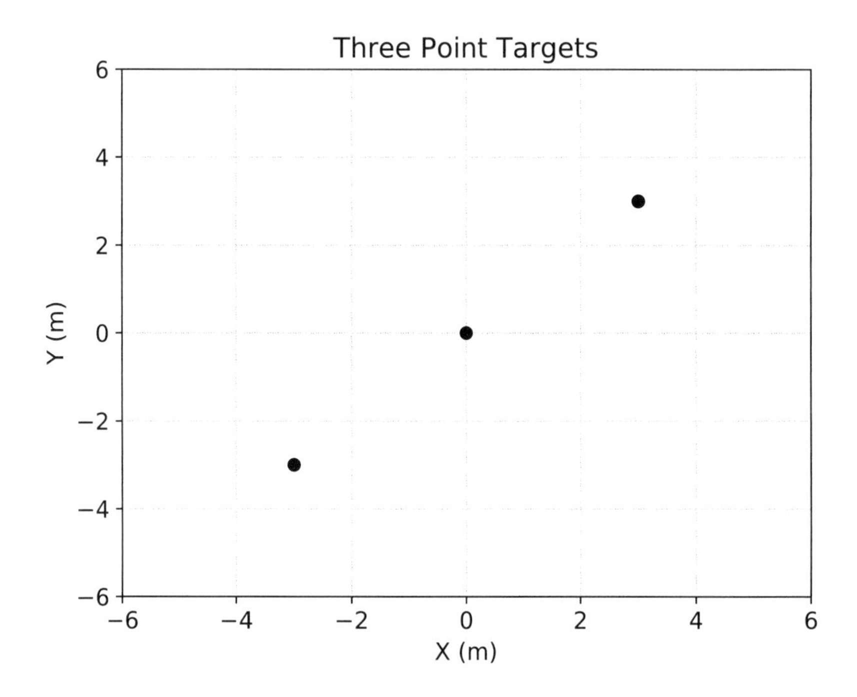

Figure 10.9 Simple SAR imaging scenario consisting of three point scatterers.

As in the two-dimensional case, no assumption has been made about the target's orientation, allowing $\theta = 0$ and $\psi = \pi/2$ without loss of generality. This leads to $v = 0$ and $w = 0$, and the three-dimensional Fourier transform is

$$F(u, 0, 0) = \int_{-\infty}^{\infty}\int_{-\infty}^{\infty}\int_{-\infty}^{\infty} f(x, y, z)\, e^{-j2\pi ux}\, dx\, dy\, dz. \tag{10.22}$$

Note that the phase factor is now a function of x only, and the triple integral may be written as

$$F(u, 0, 0) = \int_{-\infty}^{\infty}\left[\int_{-\infty}^{\infty}\int_{-\infty}^{\infty} f(x, y, z)\, dy\, dz\right] e^{-j2\pi ux}\, dx. \tag{10.23}$$

The term inside the brackets is an expression for the projection along planes of constant x. This may be expressed as

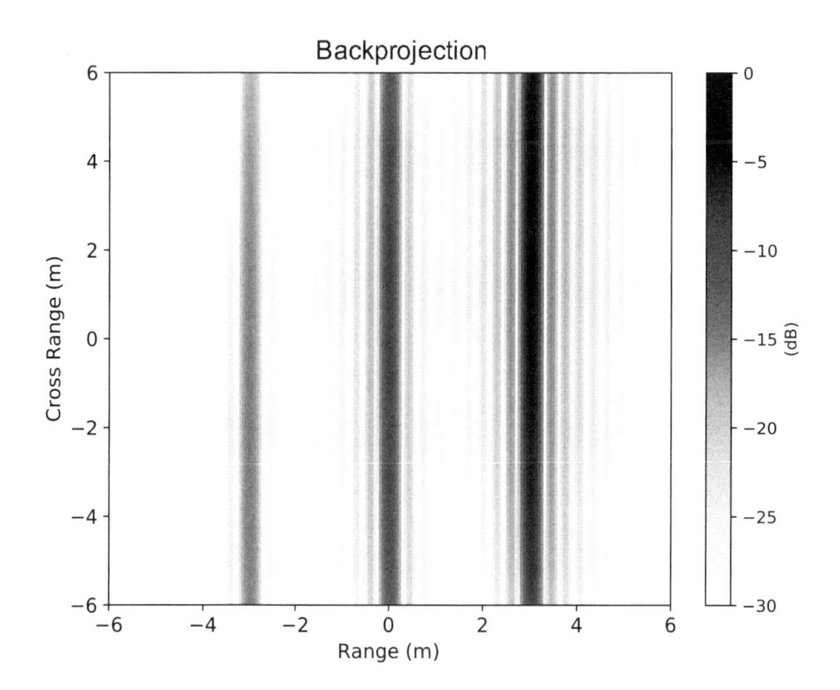

Figure 10.10 Simple SAR imaging scenario after backprojecting a single pulse.

$$P_{\theta=0,\psi=\pi/2}(x) = \int_{-\infty}^{\infty}\int_{-\infty}^{\infty} f(x,y,z)\,dy\,dz. \tag{10.24}$$

Substituting (10.24) into (10.23) results in

$$F(u,0,0) = \int_{-\infty}^{\infty} P_{\theta=0,\psi=\pi/2}(x)\,e^{-j2\pi ux}\,dx. \tag{10.25}$$

Recalling the definition of the one-dimensional transform in (10.21), the relationship between the one-dimensional projection function and the three-dimensional Fourier transform of the object's density function is

$$F(u,0,0) = S_{\theta=0,\psi=\pi/2}(\omega). \tag{10.26}$$

As before, no assumption has been made about the object's orientation; therefore the general expression of the linear trace theorem is

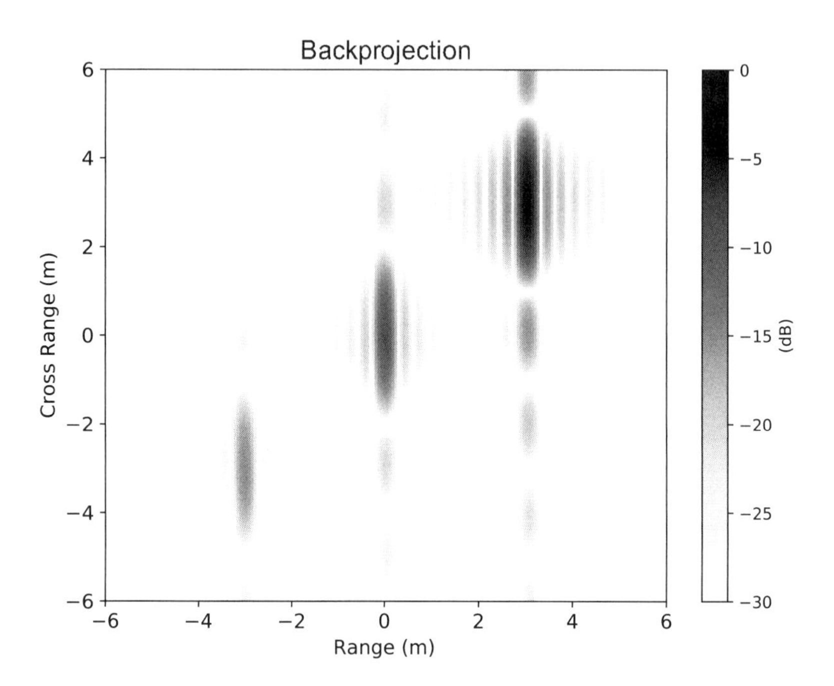

Figure 10.11 Simple SAR imaging scenario after backprojecting 10 pulses.

$$F(u, v, w) = S_{\theta,\psi}(\omega). \tag{10.27}$$

A collection of projections at pairs of angles $[(\theta_1, \psi_1), (\theta_2, \psi_2) \ldots, (\theta_N, \psi_M)]$ fills in the three-dimensional spatial frequency domain of the object's density function, as illustrated in Figure 10.14, and the image may be constructed with the backprojection algorithm.

10.1.4.2 Filtered Backprojection

For the three-dimensional filtered backprojection algorithm, begin by taking the inverse Fourier transform of the object's three-dimensional spatial frequency domain representation as

$$f(x, y, z) = \int_{-\infty}^{\infty}\int_{-\infty}^{\infty}\int_{-\infty}^{\infty} F(u, v, w)\, e^{j2\pi(xu+yv+zw)}\, du\, dv\, dw. \tag{10.28}$$

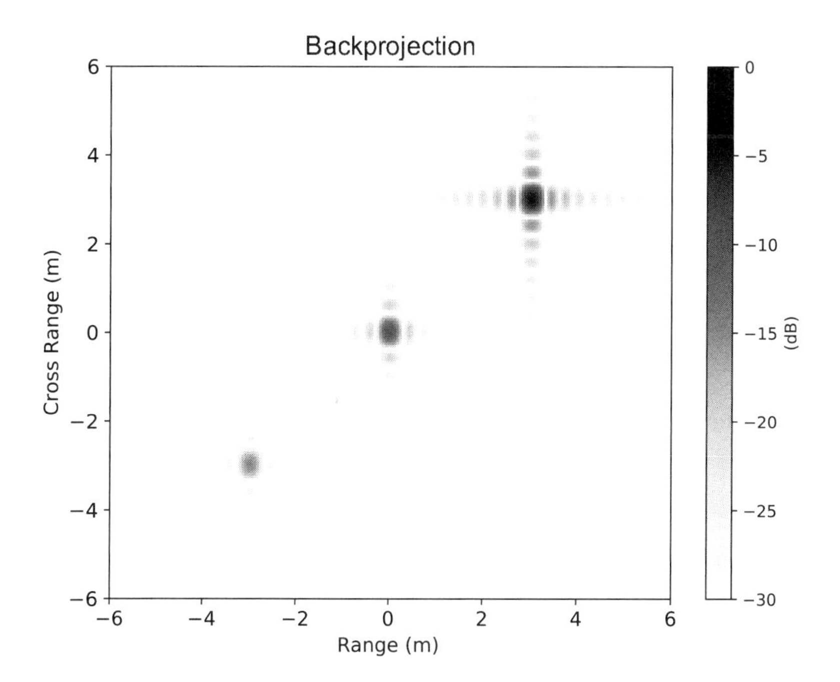

Figure 10.12 Simple SAR imaging scenario after backprojecting 100 pulses.

Since the projection functions fill in the spatial frequency domain in a spherical fashion, make the following substitutions

$$u = \omega \cos\theta \sin\psi, \tag{10.29}$$

$$v = \omega \sin\theta \sin\psi, \tag{10.30}$$

$$w = \omega \cos\psi, \tag{10.31}$$

$$du\, dv\, dw = \omega^2 \sin\psi\, d\omega\, d\theta\, d\psi, \tag{10.32}$$

which results in

$$f(x, y, z) = \int_0^{\pi} \int_0^{2\pi} \int_0^{\infty} \left[F(\omega, \theta, \psi)\, e^{j2\pi\omega(x\cos\theta\sin\psi + y\sin\theta\sin\psi + z\cos\psi)} \right] \times \omega^2\, \sin\psi\, d\omega\, d\theta\, d\psi. \tag{10.33}$$

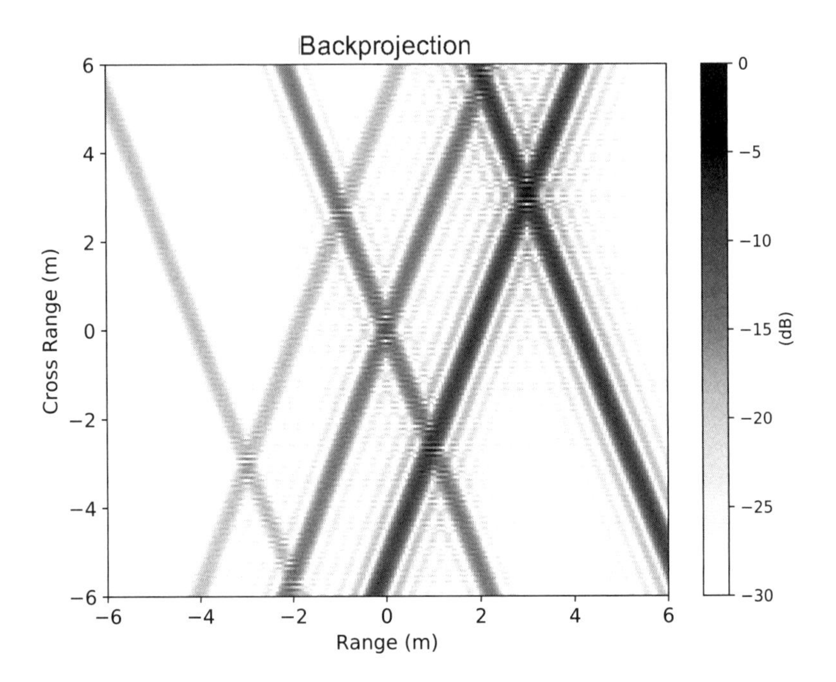

Figure 10.13 Simple SAR imaging scenario illustrating aliasing.

This may be written as

$$f(x,y,z) = \int_0^{\pi}\int_0^{2\pi}\left[\int_0^{\infty} F(\omega,\theta,\psi)\, e^{j2\pi\omega t}\,\omega^2\, d\omega\right]\sin\psi\, d\theta\, d\psi, \tag{10.34}$$

where $t = x\cos\theta\sin\psi + y\sin\theta\sin\psi + z\cos\psi$. The inner integral represents the convolution of the projection function with a filter as

$$P_{\theta,\psi}(t) * h(t) = \int_0^{\infty} F(\omega,\theta,\psi)\, e^{j2\pi\omega t}\,\omega^2\, d\omega. \tag{10.35}$$

The filtered projection function in (10.35) is then projected onto the image space along a line in the same direction as the projection was obtained. This is expressed as

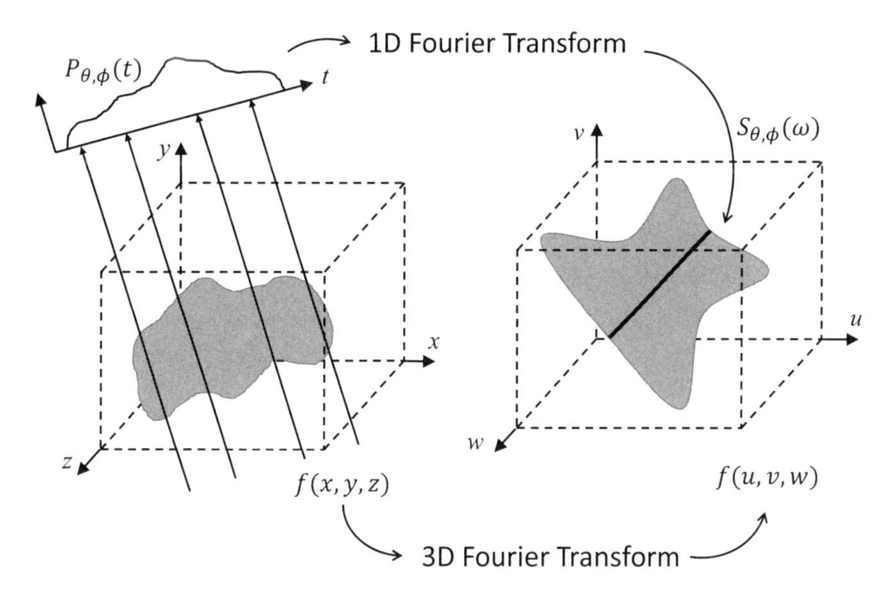

Figure 10.14 Three-dimensional spatial frequency domain from one-dimensional projection functions.

$$f(x, y, z) = \int_0^{\pi} \int_0^{2\pi} P_{\theta,\psi}(t) * h(t) \sin\psi \, d\theta \, d\psi. \tag{10.36}$$

The backprojection imaging process proceeds as in the two-dimensional case, except now the spatial domain target reflectivity array, $f(x_i, y_j, z_k)$, is three-dimensional, and the time point is based on the three-dimensional pixel point, (x_i, y_j, z_k).

10.2 EXAMPLES

The sections below illustrate the concepts of this chapter with several Python examples. The examples for this chapter are in the directory *pyradar\Chapter10* and the matching MATLAB examples are in the directory *mlradar\Chapter10*. The reader should consult Chapter 1 for information on how to execute the Python code associated with this book.

10.2.1 Two-Dimensional

This section presents examples of two-dimensional SAR imaging using the backprojection method, beginning with simple point targets, and then moving on to vehicles. The effects of elevation, polarization, and angular span on the resulting images is shown.

10.2.1.1 Point Targets

As a first example, consider an imaging scenario consisting of five point scatterers located at $(3, 3)$, $(3, -3)$, $(0, 0)$, $(-3, -3)$, and $(-3, 3)$ meters relative to the center of the area to be imaged. The targets have radar cross-section values of 10, 20, 30, 40, and 50 m^2, respectively. The azimuth angular span is -10° to 10°, the starting frequency is 1 GHz, and the bandwidth is 500 MHz. Calculate and display the two-dimensional SAR image for a 10m $\times$ 10 m image with 1000 $\times$ 1000 bins, a dynamic range of 40 dB, and use a Hanning window.

Solution: The solution to the above example is given in the Python code *back_projection_example.py* and in the MATLAB code *back_projection_example.m*. Running the Python example code displays a GUI allowing the user to enter the parameters for the point targets as well as the imaging parameters. The code then calculates and creates a plot of the backprojected image, as shown in Figure 10.15. The location and relative scattering strength of each point target is easily seen in this figure. The user is encouraged to change the target and imaging to parameters to see how each one affects the resulting image.

10.2.1.2 Toyota Camry

For this example, consider the two-dimensional SAR imaging of a Toyota Camry. The data was simulated at X-band with a high-frequency electromagnetic scattering code and provided by the U.S. Air Force Sensor Data Management System (SDMS) [16]. Calculate and display the two-dimensional SAR image of the automobile for two cases. The first is for vertical polarization, an elevation of 30°, and an azimuth span of 0° to 90°. The second case is also for vertical polarization, an elevation of 30°, and the azimuth span is now 0° to 360°. The image should span 10m $\times$ 10m and have 1000 $\times$ 1000 bins. Display the image with a dynamic range of 50 dB and use a rectangular window.

Solution: The solution to the above example is given in the Python code *back_projection_cv_example.py* and in the MATLAB code *back_projection_cv_example.m*. Running the Python example code displays a GUI allowing the user to enter the imaging parameters and choose the data set to be used. The code then calculates and creates a plot of the backprojected image. The image for the 90° azimuth span is given in Figure 10.16, and the image for the 360° azimuth span is shown in Figure 10.17. As expected, the image for the full 360° azimuth span gives much more detail than the 90° span. The user is

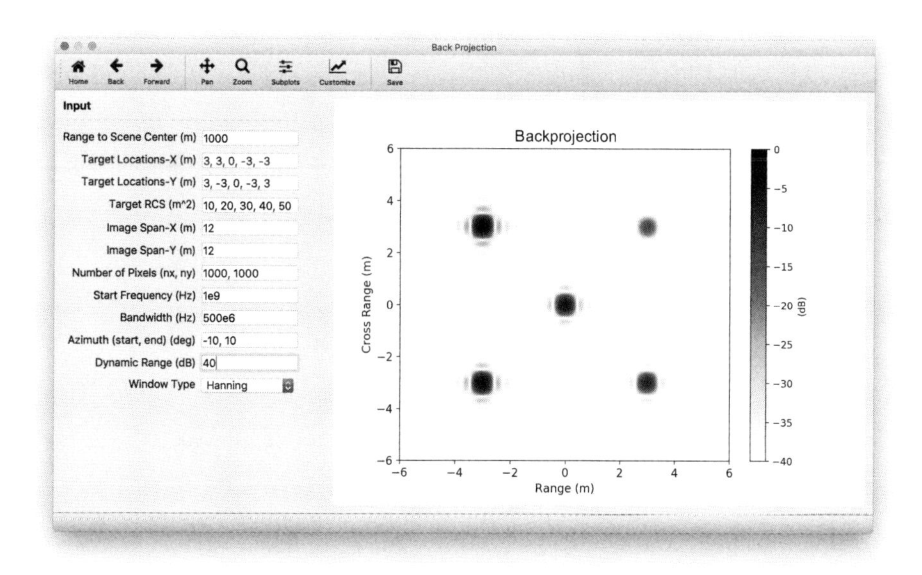

Figure 10.15 Two-dimensional backprojected image of point targets calculated by *back_projection_example.py*.

encouraged to change the imaging to parameters to see how each one affects the resulting image.

10.2.1.3 Jeep

The previous example demonstrated the effect of changing the azimuth span. In this example, the elevation angle will be varied to show how this affects the resulting image. The data set is again from the SDMS and is for a 1999 Jeep. Calculate and display the two-dimensional SAR image of the automobile for two cases. The first is for vertical polarization, an elevation of 30°, and an azimuth span of 0° to 360°. The second case is also for vertical polarization, an azimuth span of 0° to 360°, and now the elevation is 40°. The image should span 10m × 10m and have 1000 × 1000 bins. Display the image with a dynamic range of 50 dB and use a rectangular window.

Solution: The solution to the above example is given in the Python code *back_projection_cv_example.py* and in the MATLAB code *back_projection_cv_example.m*. Running the Python example code displays a GUI allowing the user to enter the imaging parameters and choose the data set to be used. The code then calculates and creates a plot of the backprojected image. The image for the 30° elevation is given in Figure 10.18, and the image for the 40° elevation is shown in Figure 10.19. The image for the 40° elevation shows the Jeep covering more of the image space. This is the expected difference between

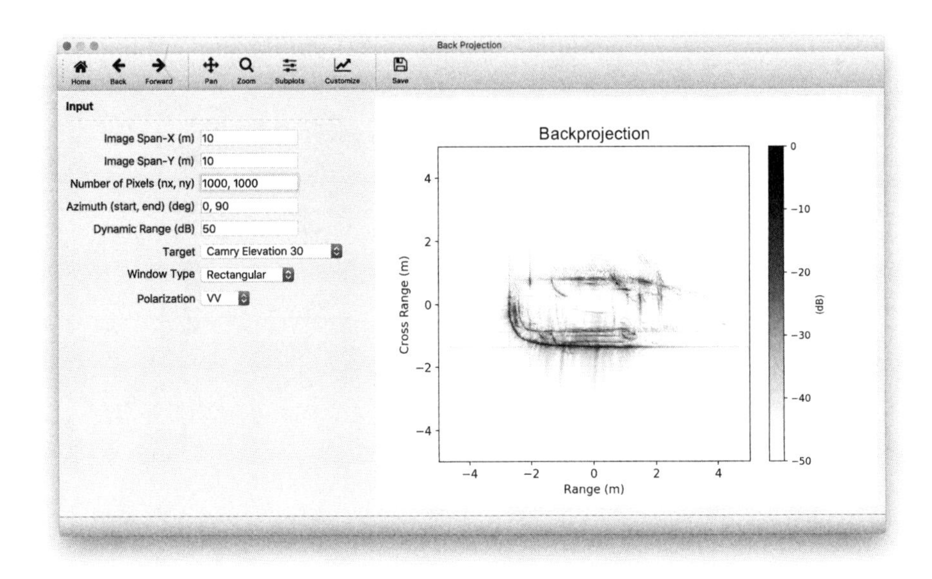

Figure 10.16 Two-dimensional backprojected image of a Toyota Camry with an azimuth span of 90° calculated by *back_projection_cv_example.py*.

slant plane images, in which distances are measured between the sensor and the target, and ground plane images, where distances are measured between the sensor ground track and the target [13]. The user is encouraged to select other elevation angles and imaging to parameters to see how these affect the resulting image.

10.2.1.4 Toyota Tacoma

This example will study the effect of polarization on the resulting SAR images. The data set is for a Toyota Tacoma, which includes full polarization information, and was provided by the SDMS. Calculate and display the two-dimensional SAR image of this automobile for three polarizations, VV, HH, and HV. Recall from Chapter 7 that VV is for transmit vertical and receive vertical, HH is transmit horizontal and receive horizontal, and HV is transmit vertical and receive horizontal. All three cases use an elevation of 30° and an azimuth span of 0° to 360°. The image should span 10m × 10m and have 1000 × 1000 bins. Display the image with a dynamic range of 50 dB and use a rectangular window.

Solution: The solution to the above example is given in the Python code *back_projection_cv_example.py* and in the MATLAB code *back_projection_cv_example.m*. Running the Python example code displays a GUI allowing the user to enter the imaging parameters

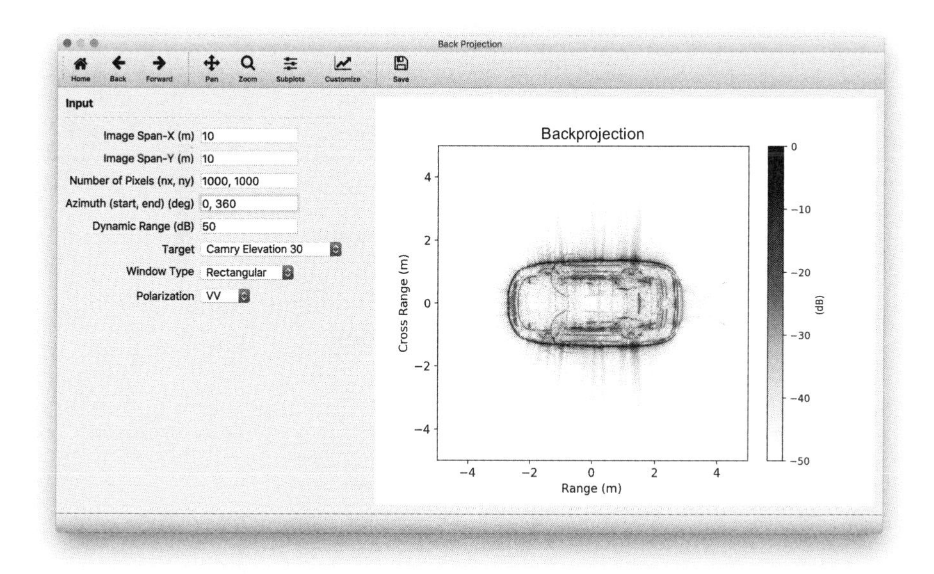

Figure 10.17 Two-dimensional backprojected image of a Toyota Camry with an azimuth span of 360° calculated by *back_projection_cv_example.py*.

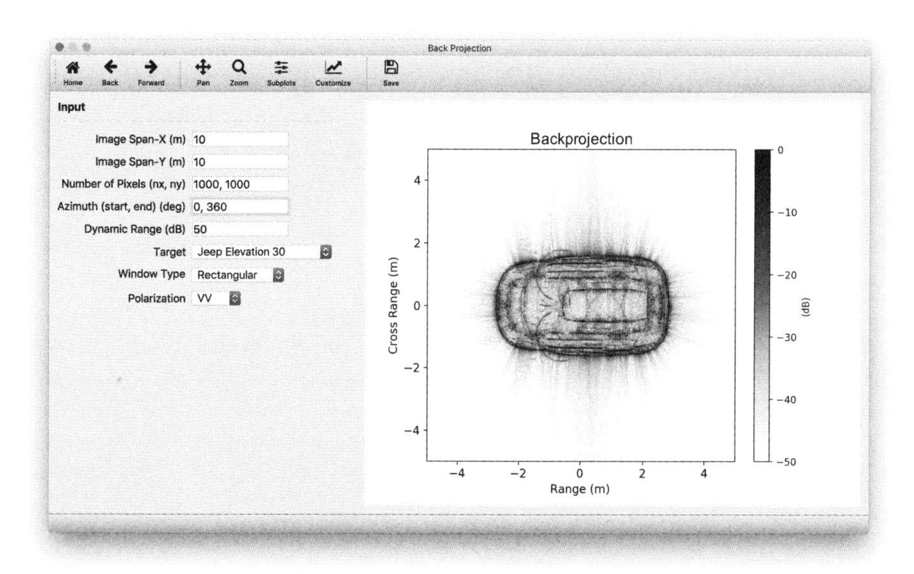

Figure 10.18 Two-dimensional backprojected image of a 1999 Jeep with an elevation of 30° calculated by *back_projection_cv_example.py*.

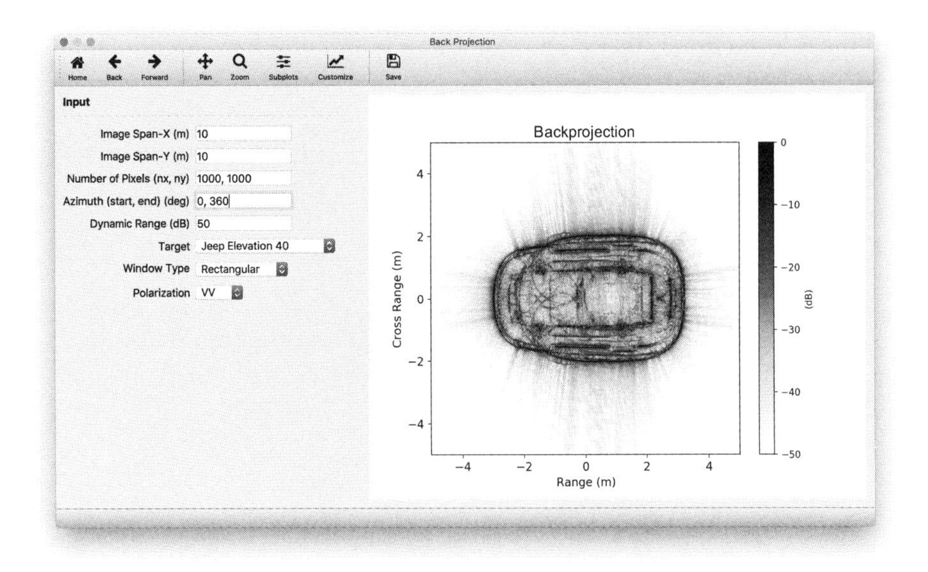

Figure 10.19 Two-dimensional backprojected image of a 1999 Jeep with an elevation of 40° calculated by *back_projection_cv_example.py*.

and choose the data set to be used. The code then calculates and creates a plot of the backprojected image. The images for VV, HH, and HV polarizations are given in Figures 10.20, 10.21 and 10.22, respectively. The HH polarization provides slightly stronger returns at this elevation angle, and the HV polarization is the weakest. Chapter 2 and Chapter 7 showed some of the conditions for strong cross-polarization scattering, which are limited for this target.

10.2.2 Three-Dimensional

This section deals with examples of three-dimensional SAR imaging using the back-projection method, beginning with simple point targets and the more complex case of a backhoe.

10.2.2.1 Point Targets

For the first three-dimensional example, consider an imaging scenario consisting of four point scatterers located at $(3, 3, -3)$, $(-3, -3, -3)$, $(3, -3, 3)$, and $(-3, 3, 3)$ meters relative to the center of the area to be imaged. The targets have radar cross-section values of 10, 10, 20, and 20 m^2 respectively. The azimuth angular span is 0° to 10°, the elevation angular span is 0° to 10°, the starting frequency is 5 GHz, and the bandwidth is 1 GHz.

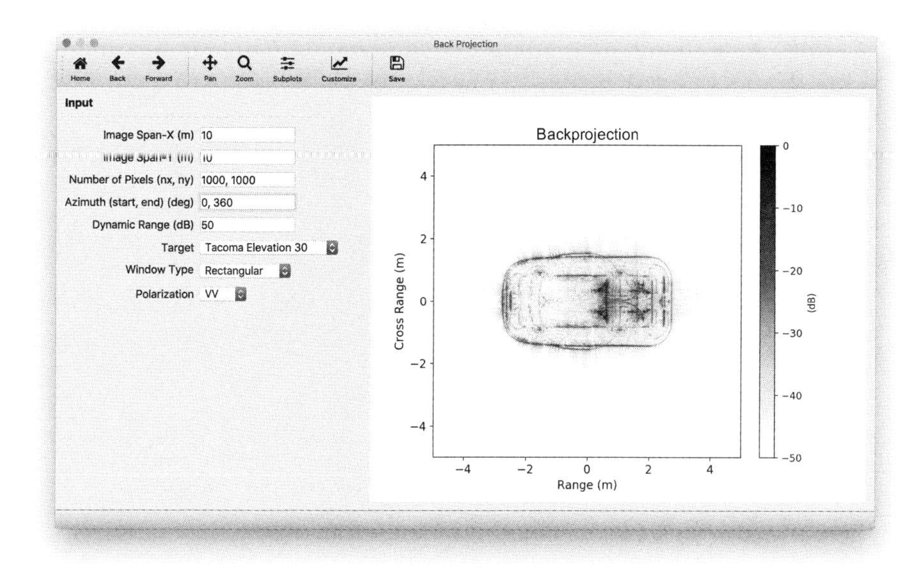

Figure 10.20 Two-dimensional backprojected image of a Toyota Tacoma for VV polarization calculated by *back_projection_cv_example.py*.

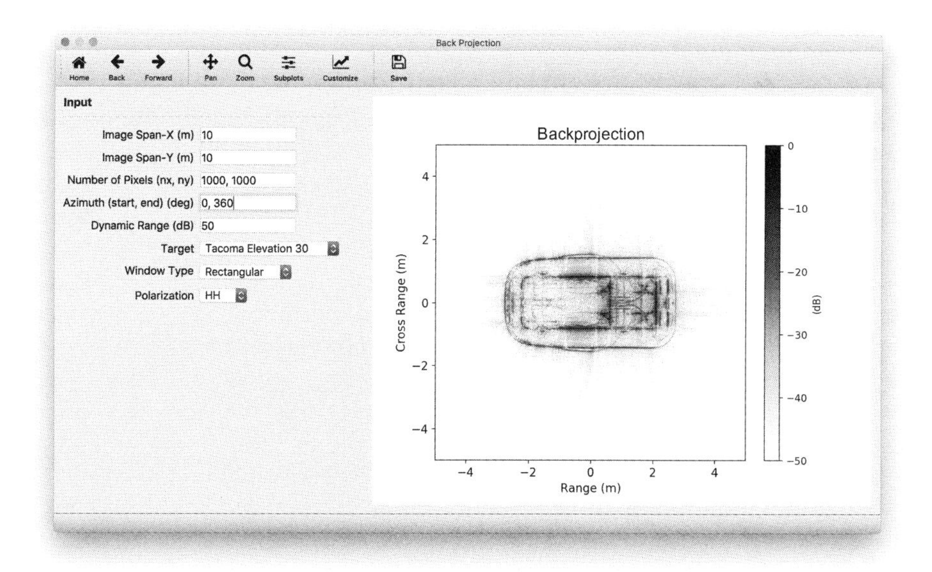

Figure 10.21 Two-dimensional backprojected image of a Toyota Tacoma for HH polarization calculated by *back_projection_cv_example.py*.

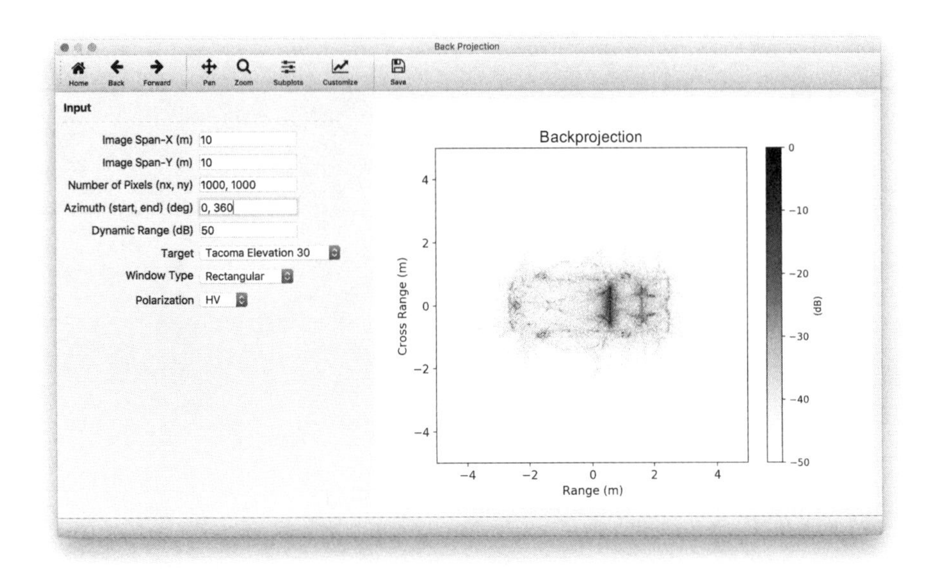

Figure 10.22 Two-dimensional backprojected image of a Toyota Tacoma for HV polarization calculated by *back_projection_cv_example.py*.

Calculate and display the two-dimensional SAR image for a 10m × 10m × 10m image with 100 × 100 × 100 bins, a dynamic range of 30 dB, and use a Hanning window.
Solution: The solution to the above example is given in the Python code *back_projection_3d_example.py* and in the MATLAB code *back_projection_3d_example.m*. Running the Python example code displays a GUI allowing the user to enter the parameters for the point targets as well as the imaging parameters. The code then calculates and creates a plot of the backprojected image, as shown in Figure 10.23. The location and relative scattering strength of each point target is easily seen in this figure. Note the larger sidelobe levels on the 20 m^2 targets versus the 10 m^2 targets. The user is encouraged to change the target and imaging to parameters to see how each one affects the resulting image.

10.2.2.2 Backhoe

This example is for the three-dimensional SAR imaging for a backhoe. As with the previous examples, this data set is from the SDMS. The data set covers an elevation span from 18° to 42° and an azimuth span of 66° to 114°. Display the three-dimensional SAR image of the backhoe for the VV polarization and a dynamic range of 45 dB. For reference, the facet model for the backhoe is given in Figure 10.24.

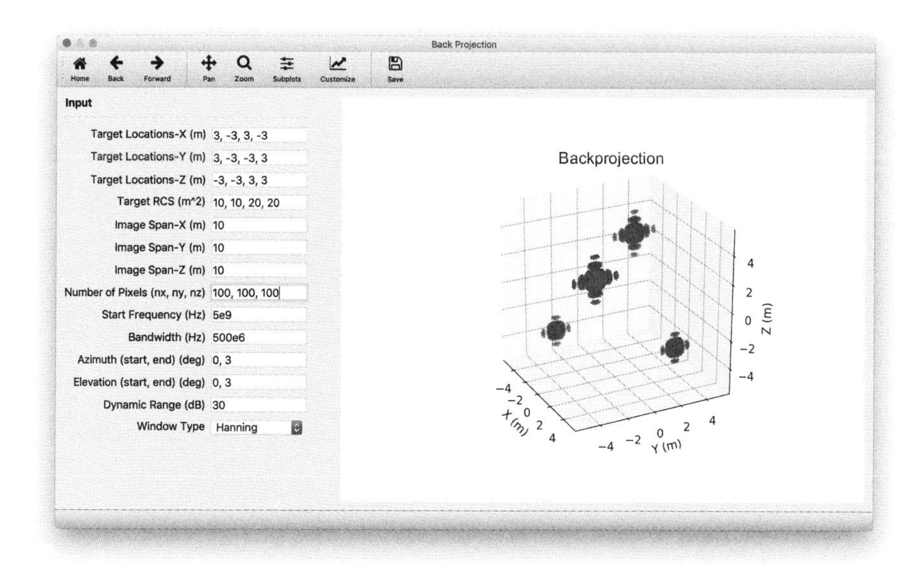

Figure 10.23 Three-dimensional backprojected image of point targets calculated by *back_projection_3d_example.py*.

Solution: The solution to the above example is given in the Python code *back_projection_backhoe_example.py* and in the MATLAB code *back_projection_backhoe_example.m*. Running the Python example code displays a GUI allowing the user to enter the imaging parameters. The code then creates a plot of the backprojected image, as shown in Figure 10.25.

Figure 10.24 Facet model of the backhoe.

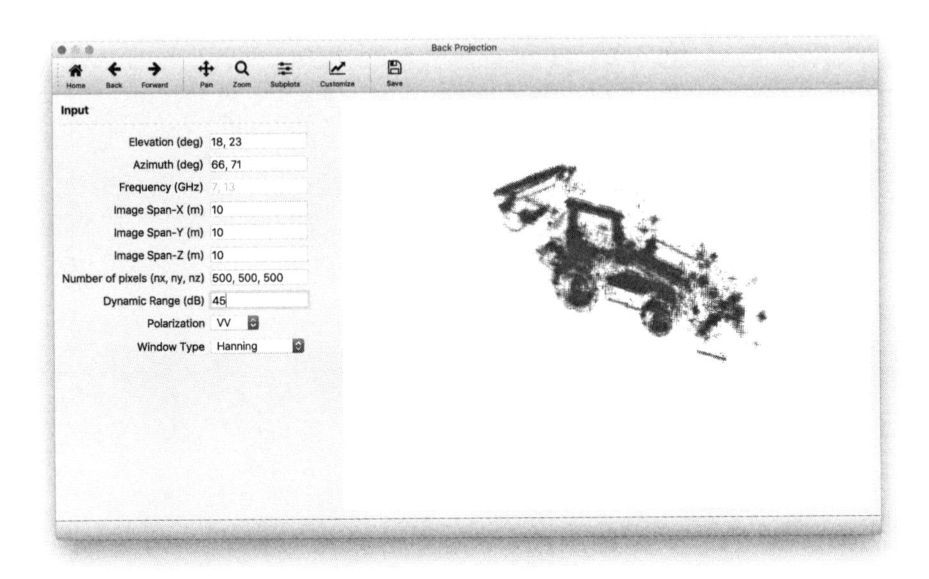

Figure 10.25 Three-dimensional backprojected image of the backhoe for VV polarization as plotted by *back_projection_backhoe_example.py*.

PROBLEMS

10.1 Explain the significance and outline the basic steps in tomography, and give practical examples not listed in Table 10.1.

10.2 Describe the Radon transform and its role in the development of modern systems employing tomographic imaging.

10.3 What is a major advantage and disadvantage of polar reformatting?

10.4 Explain the differences between SIRT, SART, and MART. Give examples of situations where one may outperform the others.

10.5 Describe the major advantages of the filtered backprojection algorithm.

10.6 What are the differences between tomographic SAR imaging and tomographic X-ray imaging?

10.7 Explain the differences between the planar slice theorem and the linear trace theorem for three-dimensional SAR imaging.

10.8 Why is the linear trace version used for radar imaging? Give examples of other types of imaging that employ planar slice.

10.9 Show the three-dimensional planar slice theorem and the backprojection method for this version.

References

[1] J. Radon, P. C. Parks (translator). On the determination of functions from their integral values along certain manifolds. *IEEE Transactions on Medical Imaging*, 5(4):170–176, 1986.

[2] R. Penrose. Twistor algebra. *Journal of Mathematical Physics*, 1(8):345–366, 1967.

[3] R. Penrose. Twistor quantisation and curved space-time. *International Journal of Theoretical Physics*, 1:61–99, 1968.

[4] R. Penrose. Solutions of the zero-rest-mass equations. *Journal of Mathematical Physics*, 10(1):38–39, 1969.

[5] A. Cormack. Representation of a function by its line integrals with some radiological implications I. *Journal of Applied Physics*, 34:2722–2727, September 1963.

[6] A. Cormack. Representation of a function by its line integrals with some radiological implications II. *Journal of Applied Physics*, 35:2908–2913, October 1964.

[7] G. Hounsfield. Computerized transverse axial scanning (tomography): Part 1. description of system. *British Journal of Radiology*, 46:1016–1022, 1973.

[8] The Nobel prize in physiology or medicine 1979. `http://www.NobelPrize.org/`.

[9] R. Gordon, R. Bender and G. Herman. Algebraic reconstruction techniques (ART) for three-dimensional electron microscopy and X-ray photography. *Journal of Theoretical Biology*, 29(3):471–481, December 1970.

[10] P. Gilbert. Iterative methods for the three-dimensional reconstruction of an object from projections. *Journal of Theoretical Biology*, 36(1):105–117, 1972.

[11] A. Andersen and A. Kak. Simultaneous algebraic reconstruction technique (SART): A superior implementation of the art algorithm. *Ultrasonic Imaging*, 6(1):81–94, 1984.

[12] R. Bates and T. Peters. Towards improvements in tomography. *New Zealand Journal of Science*, 14:883–896, 1971.

[13] C. Jakowatz, et al. *Spotlight-Mode Synthetic Aperture Radar: A Signal Processing Approach*. Kluwer Academic Publishers, Boston, 1996.

[14] D. Munson, J. O'Brien and W. Jenkins. A tomographic formulation of spotlight-mode synthetic aperture radar. *Proceedings of the IEEE*, 71(8):917–925, 1983.

[15] M. Soumekh. *Synthetic Aperture Radar Signal Processing with MATLAB Algorithms*. John Wiley and Sons, New York, 1999.

[16] U.S. Air Force. *Sensor Data Management System (SDMS)*. `https://www.sdms.afrl.af.mil`.

Chapter 11

Countermeasures

The concept of countermeasures to hinder the ability of radar systems to detect, track, and discriminate targets dates back to the early days of World War II, when the United Kingdom and Germany both engaged in the development of various countermeasure techniques. Today, there are numerous methods for deceiving and interfering with a radar system's ability to perform key functions. The chapter begins with passive jamming techniques. Next, active jamming methods, including continuous noise and deceptive jamming, are examined. Methods for mitigating these jamming techniques, such as space-time adaptive processing and moving target indication are presented. Finally, a sophisticated jamming technique known as digital radio frequency memory is discussed. The chapter concludes with several Python examples to demonstrate the applications of countermeasure techniques.

11.1 PASSIVE JAMMING

Passive jammers do not use electronic means to modify the incoming radar signal or to generate a jamming signal. Passive jammers include chaff, reflectors, and decoys. The intent is to reflect the radar signal in such a fashion as to mask the target, produce false targets, disrupt the radar signal, or otherwise deceive the radar system. Passive jamming techniques are still in use today and can be effective when employed properly.

11.1.1 Chaff

Dispersing large quantities of chaff into the atmosphere is one of the most common forms of passive jamming. Chaff is comprised of small thin plastic or fiberglass particles with a thin coating of zinc or aluminum, which have a resonant frequency close to the center frequency of the victim radar. If the number of chaff particles in the radar volume is large enough, the reflected signal will be larger than the return from the target within the chaff cloud. There are several basic properties used to characterize chaff, including effective scatter area, bandwidth, dispersion method, cloud size and growth rate, and spectra of scattered signal. While chaff is a simple idea, techniques and characteristics of this type of jamming continue to be improved [1–3]. For chaff comprised of half-wavelength dipoles, the radar cross section of a single particle is given by

$$\sigma = 0.86\lambda^2 \cos^4\theta \qquad (\mathrm{m}^2), \tag{11.1}$$

where λ is the wavelength, and θ is the angle between the axis of the dipole and the incident energy. Chaff is typically dispersed by aircraft, and due to air turbulence, the orientation of the dipoles is random. For approximate calculations, assume all possible orientations are equally likely, and consider the expectation of the cross section, which is

$$\sigma_{avg} = 0.17\lambda^2 \qquad (\mathrm{m}^2). \tag{11.2}$$

This value is then used to determine the necessary quantity of chaff particles to be deployed. Since the fields produced by different particles in the cloud are incoherent, the expression for a cloud of dipoles is

$$\sigma_c = 0.17\lambda^2 N\,\eta \qquad (\mathrm{m}^2), \tag{11.3}$$

where N is the number of particles in the cloud, and η is a reduction factor for dipoles breaking, sticking, and so on. This gives only an estimate of the average cross section of a chaff cloud. Also, the previous discussion assumes the transmitting and receiving antennas are collocated and have identical polarization.

A major drawback of a dipole chaff cloud as a countermeasure is its relatively narrow bandwidth. If the wavelength of the incoming radar signal is larger than the resonant length, the radar cross section drops sharply. If the wavelength is less than the resonant length, the radar cross section drops more slowly. As the length of the dipole approaches one wavelength, the radar cross section becomes large again. However, the weight increase associated with this increase in length hinders any practical application. One method for overcoming the bandwidth issue is to package dipoles of differing lengths to be dispersed. However, increasing the bandwidth in this fashion leads to an increase in weight to achieve the necessary number of particles in a given volume. Another technique to solve the bandwidth issue is to have equipment on the deploying aircraft to cut the dipoles to a specified length. This requires long bundles of raw material to be cut, along with sensors to determine the radar operating band.

As chaff is dispersed from an aircraft, it is subject to high turbulence. The assumption that all dipole orientations are equally probable is only valid for a short time immediately after deployment. After the influence of the aircraft slip stream has become negligible, the dipoles are oriented randomly, the cloud continues to grow, and the density of dipoles begins to decrease. From studies, dipoles oriented in the vertical direction descend faster than those oriented horizontally. Therefore, the chaff cloud becomes stratified into regions of similarly oriented dipoles.

11.1.1.1 Moving Target Indication

Even though chaff may stay airborne for several hours, radar systems employing *moving target indication* (MTI) are less affected by this countermeasure technique. MTI systems are designed to reject signals from fixed or slow-moving targets such as trees, oceans, and buildings, and retain the signals from moving targets of interest such as aircraft [4]. Chaff particles have very large aerodynamic drag, and therefore the forward velocity quickly vanishes. This makes MTI a very appealing approach for mitigating the effects of chaff. The spectrum of the signal reflected by a chaff cloud is in general different from the spectrum of the signal transmitted by the radar. This is due to relative motion between the cloud and the radar. The spectral density of the signal reflected by a chaff cloud is usually represented by a Gaussian distribution of the form

$$G(f) = G_0 e^{-f^2/2\sigma_f^2}, \tag{11.4}$$

where G_0 is the value $G(f = 0)$, and $\sigma_f = 2\sigma_v/\lambda$ is the mean square of the Doppler frequency spread. Therefore, if the Doppler spread of the chaff can be estimated from radar measurements, an MTI filter may be designed to reject the portion of the spectrum related to chaff, as shown in Figure 11.1. Such an approach using multiple MTIs for target tracking in highly clutter environments is given in [5].

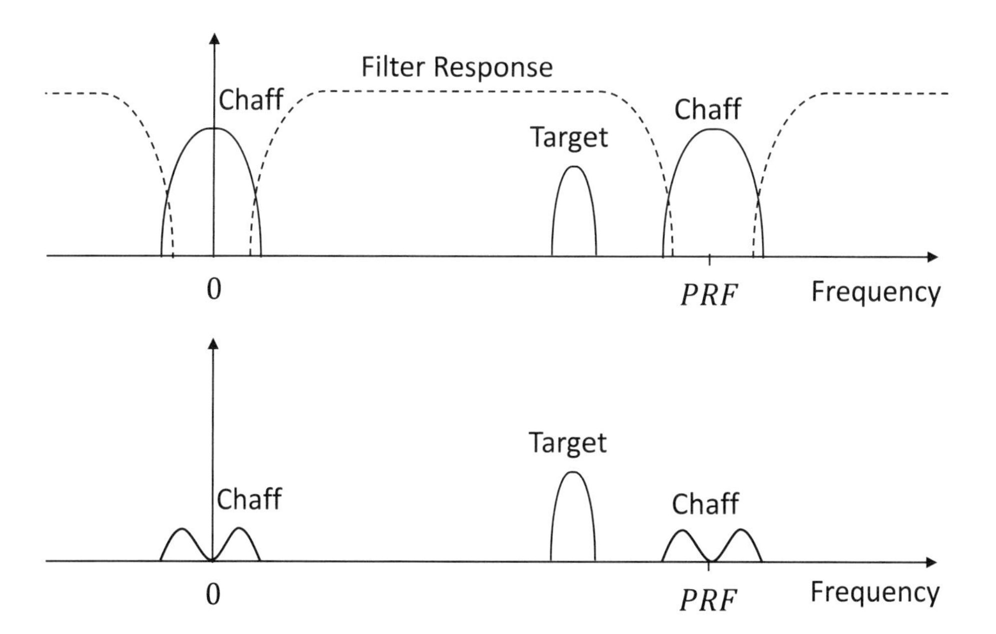

Figure 11.1 Example spectrum of chaff and MTI filtering.

Since moving targets induce a Doppler shift on the reflected signal, MTI takes advantage of this Doppler shift to separate moving targets from stationary objects. For pulsed radar systems, the Doppler shift manifests itself as a phase change between consecutive pulses. For example, the return signal from consecutive pulses incident on a stationary target will incur the same time delay, and thus have the same phase relative to the local oscillator. On the other hand, the return signal for a moving target will have slightly different time delays for consecutive pulses, and therefore a different relative phase. As an example, consider a delay line canceller shown in Figure 11.2. As a pulse is received it is delayed by the pulse repetition interval and then subtracted from the next received pulse. The frequency response of the delay line canceller is given by [6]

$$|H(f)|^2 = 4\,\sin^2(\pi f)\,, \tag{11.5}$$

where f is the normalized frequency. This frequency response is illustrated in Figure 11.3. As seen in the figure, a null is placed about $f = 0$, which corresponds to chaff. However, there are also nulls placed at multiples of the pulse repetition frequency. These are known as *blind speeds*, as targets moving with those velocities are also filtered out. These blind speeds may be alleviated with the use of varying pulse repetition intervals [4, 7].

11.1.2 Passive Deception

Passive deception jamming consists mainly of decoys, which are built to imitate real targets in shape and scattering [2, 3]. The intent is to confuse radar operators by creating patterns of targets, deceive radar signal processing algorithms, and divert weapons from the true targets. Other passive scattering mechanisms, such as corner reflectors and lens reflectors, can be added to decoys to increase scattering and make them appear larger. Airborne decoys may be towed, dropped, or have their own propulsion system such as the BriteCloud Expendable Missile Decoy [8]. More advanced decoys also have the ability to perform electronic jamming or drop chaff. Ground-based decoys are often lightweight inflatable designs with radar reflective components as well as thermal signatures [9, 10].

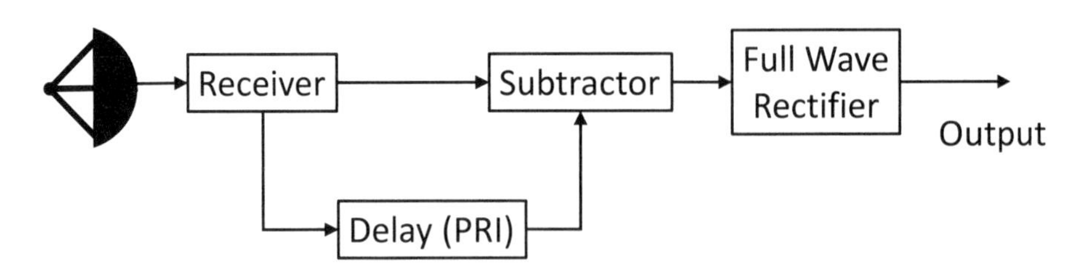

Figure 11.2 Delay line canceller.

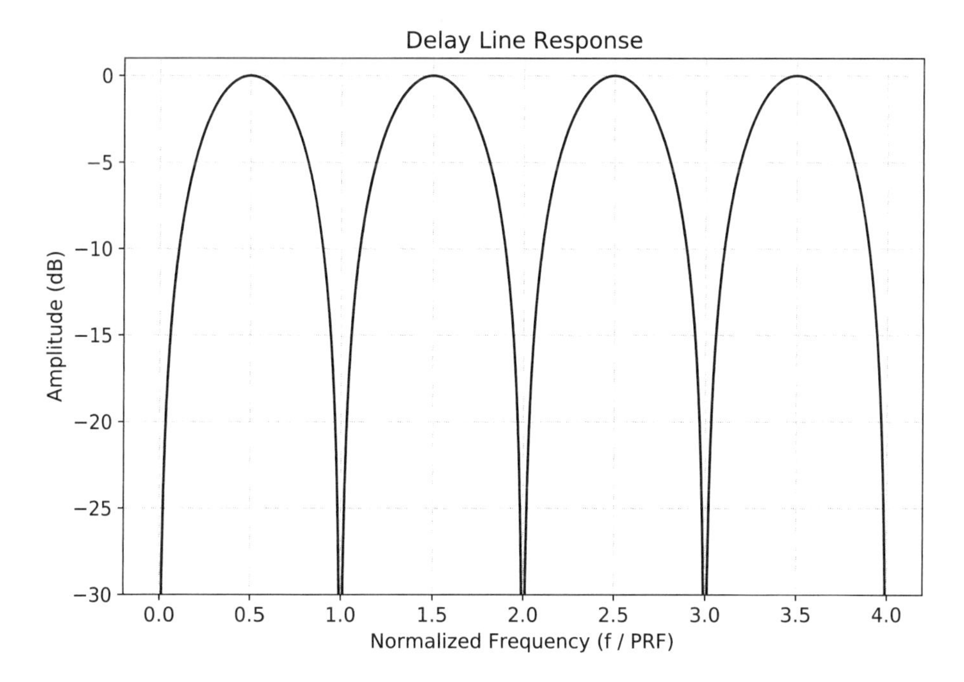

Figure 11.3 Frequency response of the delay line canceller.

11.2 ACTIVE JAMMING

Active jamming is the transmission of signals intended to interfere with or preclude the normal operations of a radar system. Active jamming produces signals at the input of the victim radar system that impede the detection and recognition of useful signals against the background environment. This results in reduced probability of detection, reduced measurement accuracy, and increased probability of false alarm. In this book, active jamming will be divided into two categories, continuous noise and deceptive jamming.

11.2.1 Continuous Noise

Continuous noise jamming, often referred to as barrage jamming, is one of the oldest methods and has been used successfully against a wide range of radar systems operating in various modes [11]. The interfering signal is transmitted continuously, where one or more of the signal parameters, including amplitude, phase, and frequency, may change randomly. It is assumed the basic jamming system parameters, such as radiated power, bandwidth, and polarization remain constant for a selected mode. As a first-order approximation, it is assumed the voltage of the jamming signal is a stationary narrowband random process in which the instantaneous values have a normal distribution and the frequency spectrum is uniform within the operating bandwidth of the jammer. This is known as band-limited white Gaussian noise [12, 13]. A stationary process is a random process whose unconditional joint probability distribution does not change when shifted in time [14]. Consequently, the statistical properties of the process, such as mean and variance, do not change over time. Narrowband processes have a spectral width many times smaller than the center frequency, which is expressed as

$$B_j \ll f_0 \qquad \text{(Hz)}, \tag{11.6}$$

where B_j is the bandwidth of the process, or in this case the jammer, and f_0 is the center frequency. Recalling the probability density function for a Gaussian process given in (6.1), the jammer power at the input to the receiver is σ_j^2, and the spectral density is

$$G_j = \sigma_j^2 / B_j \qquad \text{(W/Hz)}. \tag{11.7}$$

If the spectral density is symmetric about the center frequency, the correlation function is [12]

$$R(\tau) = \sigma_j^2\, p(\tau) \cos(2\pi f_0 \tau) = \cos(2\pi f_0 \tau) \int_{-\infty}^{\infty} G_j(f_0 - t) \cos(2\pi t \tau) dt, \tag{11.8}$$

where $p(\tau)$ is the correlation coefficient. If the spectral density is constant within the band, B_j, then

$$R(\tau) = G_j\, B_j\, \text{sinc}(B_j \tau) \cos(2\pi f_0 \tau). \tag{11.9}$$

The correlation time is given by

$$\tau_c = \int_0^{\infty} p(\tau) d\tau \qquad \text{(s)}. \tag{11.10}$$

In practice, the correlation is assumed to vanish outside the correlation time. If the process has a uniform spectral width, B_j, the correlation time is

$$\tau_c = \frac{1}{2B_j} \qquad \text{(s)}. \tag{11.11}$$

The sample functions of narrowband processes are similar to a modulated sinusoidal signal. Therefore, the usual approximation of such a process is

$$x_j(t) = X_j(t) \cos(2\pi f_0 t - \phi(t)), \tag{11.12}$$

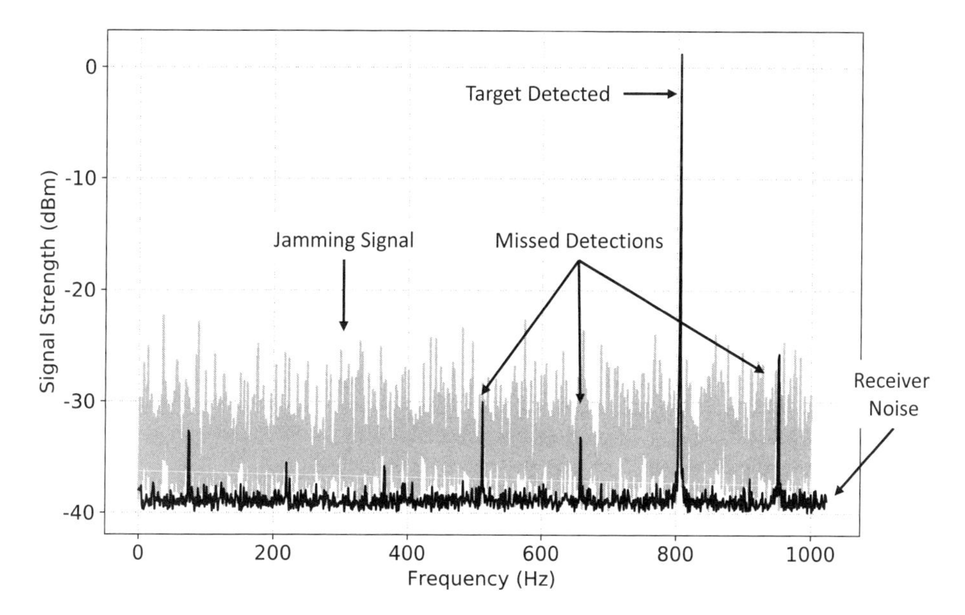

Figure 11.4 Continuous noise jamming.

where the envelope and phase functions, $X_j(t)$ and $\phi(t)$, are slowly varying in comparison to $\cos(2\pi f_0 t)$. In addition, the rate of change of $X_j(t)$ and $\phi(t)$ is inversely proportional to the spectral width. Gaussian processes are more effective than uniform continuous jamming. Figure 11.4 illustrates the concept of continuous noise jamming, which may be implemented as direct noise jamming, amplitude modulated noise jamming, or frequency modulated noise jamming. Direct noise jamming most closely represents Gaussian noise, and the reader is referred to excellent references for more information on these methods [1, 15, 16].

The effectiveness of continuous noise jamming is often characterized by the jamming power to signal power ratio in the passband, B_r, of the receiver [1, 17]. The power received by the radar from the jammer is [7]

$$J = \frac{P_j\, G_j\, \lambda^2\, G_r}{(4\,\pi\, r_j)^2\, B_j\, L_j}, \tag{11.13}$$

where,

P_j = jammer transmit power (W),
G_j = jammer antenna gain in the direction of the radar,
λ = wavelength (m),
G_r = radar receive antenna gain in the direction of the jammer,
r_j = range from the jammer to the radar (m),
B_j = jammer bandwidth (Hz),
L_j = jammer losses.

Writing (4.40) in a slightly different form for jamming applications gives the power at the radar as

$$S = \frac{P_t\, G_t^2\, \lambda^2\, \sigma}{(4\,\pi)^3\; r_t^4\, B_r\, L_r} \tag{11.14}$$

P_t = radar transmit power (W),
G_t = radar antenna gain in the direction of the target,
λ = wavelength (m),
σ = radar cross section of the target (m^2),
G_r = radar receive antenna gain in the direction of the jammer,
r_t = range from the radar to the target (m),
B_r = radar receiver bandwidth (Hz),
L_r = radar losses.

The jammer-to-signal ratio is then

$$\frac{J}{S} = \frac{P_j\, G_j\, G_r\, 4\,\pi\, r_t^4\, B_r\, L_r}{P_t\, G_t^2\, \sigma\, r_j^2\, B_j\, L_j}. \tag{11.15}$$

The jammer-to-signal ratio is often written in terms of the jammer's *effective radiated power* (ERP), which is given by

$$ERP = \frac{P_j\, G_j}{L_j} \qquad \text{(W)}. \tag{11.16}$$

This allows (11.15) to be written as

$$\frac{J}{S} = \frac{ERP\;\, G_r\, 4\,\pi\, r_t^4\, B_r\, L_r}{P_t\, G_t^2\, \sigma\, r_j^2\, B_j}. \tag{11.17}$$

Note that the expressions in (11.15) and (11.17) may be used for either self-screening or escort jamming and are valid for $B_j \le B_r$. In self-screening jamming, the target carries the jamming equipment, whereas in escort jamming the equipment is carried by an accompanying aircraft, and may be either stand-off or stand-in jamming, as shown in Figure 11.5. For stand-off jamming, the platform carrying the jamming equipment maintains a path at long ranges from the victim radar. Stand-in jamming usually consists

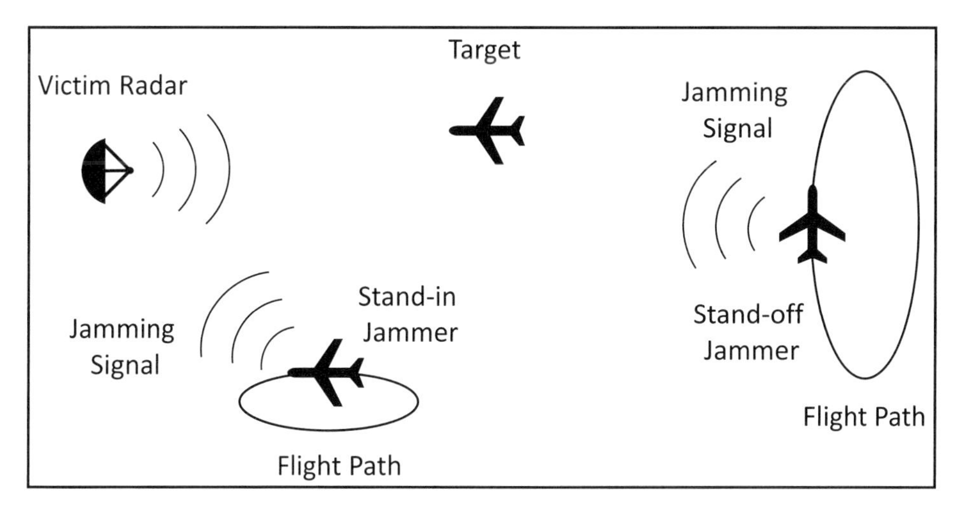

Figure 11.5 Stand-off and stand-in jamming.

of a remote vehicle very close to the victim radar. Typically, self-screening jamming is deceptive jamming, and escort jamming is noise jamming [2].

When the target of interest is located at long distances from the radar, the jamming power is typically larger than the total signal power. This is due to the jamming power being a one-way transmission inversely proportional to r^2, and the target return is a two-way transmission inversely proportional to r^4. However, as the target approaches the radar, there is a range at which the received power from the jammer and the received power from the target are equal, and the jammer-to-signal ratio becomes unity. This range is known as the *crossover range* [2], and is given by

$$r_c = \left[\frac{P_t\, G_t^2\, \sigma\, r_j^2\, B_j}{ERP\ G_r\, 4\,\pi\, B_r\, L_r}\right]^{1/4} \qquad \text{(m)}, \qquad (11.18)$$

for escort jamming, and

$$r_c = \sqrt{\frac{P_t\, G_t^2\, \sigma\, B_j}{ERP\ G_r\, 4\,\pi\, B_r\, L_r}} \qquad \text{(m)}, \qquad (11.19)$$

for self-screening jamming. As with the jammer-to-signal ratio, the expressions for the crossover range are valid for $B_j \leq B_r$. While (11.18) and (11.19) give the range for the jammer-to-signal ratio equal to one, this is not sufficient to effectively perform many radar functions such as detection, tracking, and discrimination. The range at which the radar can perform these functions based on the required jammer-to-signal ratio is referred to as the *burn-through range*, and is given by [2]

$$r_b = \left[JSR_0 \, \frac{P_t \, G_t^2 \, \sigma \, r_j^2 \, B_j}{ERP \; G_r \, 4 \, \pi \, B_r \, L_r} \right]^{1/4} \quad \text{(m)}, \tag{11.20}$$

for escort jamming, and

$$r_b = \sqrt{JSR_0 \, \frac{P_t \, G_t^2 \, \sigma \, B_j}{ERP \; G_r \, 4 \, \pi \, B_r \, L_r}} \quad \text{(m)}, \tag{11.21}$$

for self-screening jamming, where JSR_0 is the jammer-to-signal ratio required to adequately perform the necessary radar functions. The expressions in (11.20) and (11.21) are valid for $B_j \le B_r$.

11.2.1.1 Space-Time Adaptive Processing

One mitigation technique for continuous noise jamming is space-time adaptive processing (STAP), where a two-dimensional filter is designed in space and time to maximize the output of the signal-to-noise ratio and thereby selectively null jammer returns [18]. STAP uses a phased-array antenna with multiple spatial channels coupled with pulse Doppler waveforms, as illustrated in Figure 11.6. This results in an $M \times N$ matrix at each range bin, as shown in Figure 11.7. An adaptive weight vector is formed by applying the statistics of the interference environment [19, 20]. The goal is to determine the optimal weight vector and apply these weights to the received radar samples. The optimal weights are determined from

$$\mathbf{w} = \mathbf{R}^{-1} \, \mathbf{S}(\theta_t), \tag{11.22}$$

where $\mathbf{R}$ is the interference covariance matrix at each range bin and $\mathbf{S}(\theta_t)$ is the steering vector for each target in the scene. The first step in computing the covariance matrix, $\mathbf{R}$, is shown in Figure 11.7. Once the vector, $\mathbf{y}$, is found, the covariance matrix is determined from the vector cross product as

$$\mathbf{R} = \mathbf{y}^* . \mathbf{y}^T . \tag{11.23}$$

Note that $\mathbf{y}$ is of length MN, thus $\mathbf{R}$ is of dimension $MN \times MN$. The steering vector is also of length MN, and is calculated as

$$\mathbf{S}(\theta_t) = \begin{bmatrix} 1 \\ e^{-j2\pi n f_d} \\ \vdots \\ e^{-j2\pi (N-1) f_d} \end{bmatrix}, \tag{11.24}$$

where f_d is the Doppler shift, and

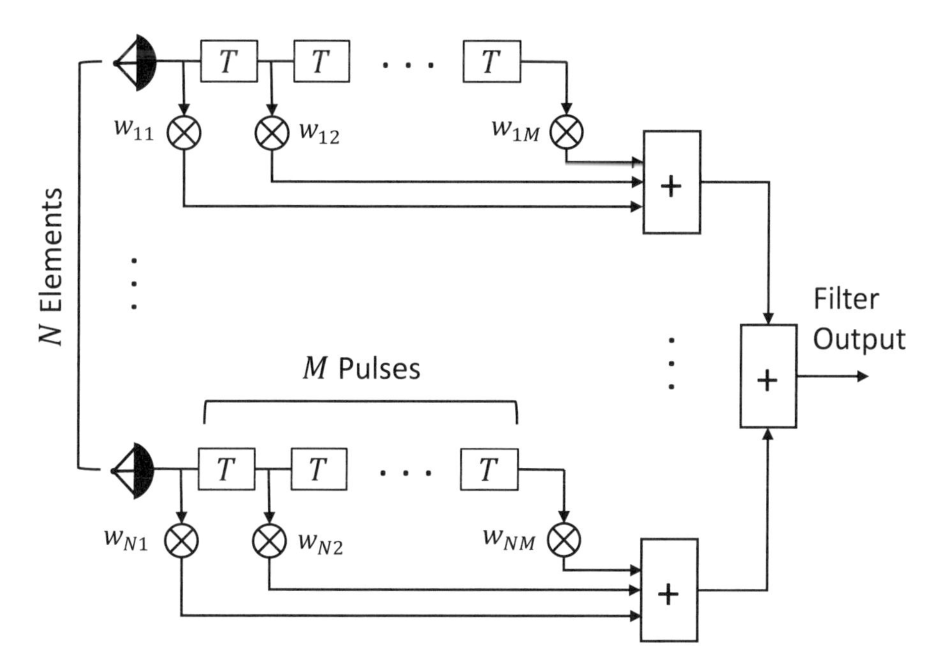

Figure 11.6 Two-dimensional STAP arrangement.

$$\mathbf{A}_\theta = e^{-j2\pi dm \sin(\theta/\lambda)} \quad \text{for } m = 0, \cdots, M-1. \tag{11.25}$$

In general, the procedure above would need to be performed at each unambiguous range bin for each target of interest. For many systems, the computational complexity of STAP is intractable, and various approaches for reducing this burden have been developed [18, 19, 21, 22].

11.2.2 Active Deception

Active deception jamming seeks to inject false target-like information into the victim radar to deny critical information about the target of interest [23]. Deceptive jammers have receivers to analyze incoming radar signal and send back simulated target echoes at various ranges and angles, hindering the ability of the radar to identity true targets from false targets. Deceptive jammers do not jam the entire bandwidth and are therefore more power-efficient.

One form of deceptive jamming, known as *range gate pull-off* (RGPO), seeks to break radar lock by stealing range gates [2]. Typically, a radar initiates a new track on a target and places range gates on either side. Signals in these range gates are zeroed out, thereby increasing the signal-to-noise ratio on the target being tracked, and to prevent

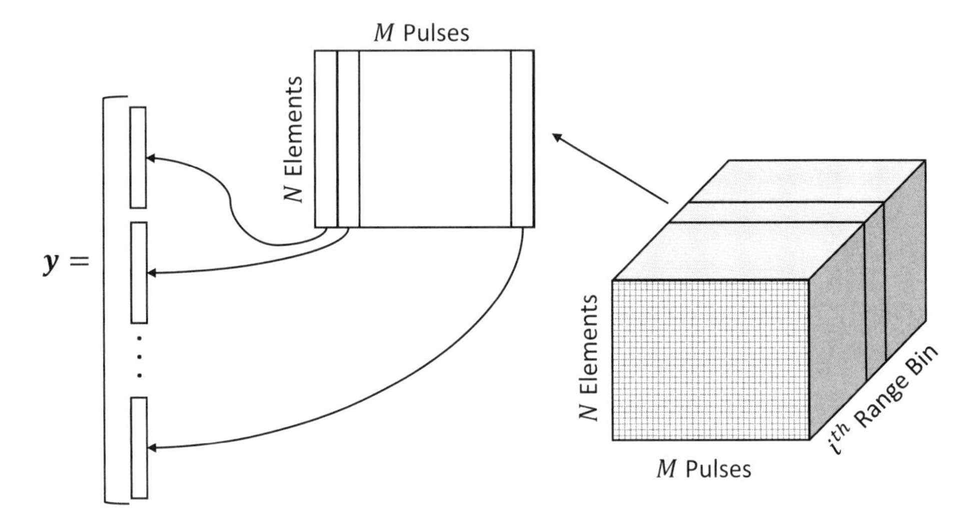

Figure 11.7 STAP data matrix at each range bin.

against asynchronous jamming. Once a radar has entered this tracking state, it is usually referred to as *locked on*. RGPO begins by sampling the radar signal, and then transmitting a target-like echo to the victim radar with as little time delay as possible, as illustrated in Figure 11.8. The power of the jamming signal is increased, and the target-like echo becomes much larger than the echo from the target. As discussed in Section 5.5, the radar sensitivity is reduced to prevent overload of the receiver components. As a result, the target's return falls below the noise floor and the jammer has effectively seized the radar's automatic gain control. The delay in the jamming signal is slowly increased and the tracking window follows this false target-like echo. Once the false target is some distance from the true target return, the jamming signal is turned off and only noise remains in the tracking window between the range gates. The jammer has accomplished its task of breaking lock, and the victim radar often goes into search or acquisition mode. In the discussion above, RGPO only produces false targets at ranges greater than the true target return. If the radar's pulse repetition frequency can be determined, then the false targets can be placed at closer ranges by calculating the time of incidence of the subsequent radar pulses.

Similar to RGPO is *velocity gate pull-off* (VGPO). This method captures velocity gates and moves those away from the true target return. VGPO by itself is effective against CW and Doppler velocity tracking systems. This is accomplished by first capturing the CW or pulse Doppler frequency, then transmitting the jamming signal to create the false target. Next, the phase (frequency) is altered to give the apparent velocity change or Doppler shift. RGPO and VGPO may be used in conjunction to provide a more realistic target motion and further deceive radar tracking systems.

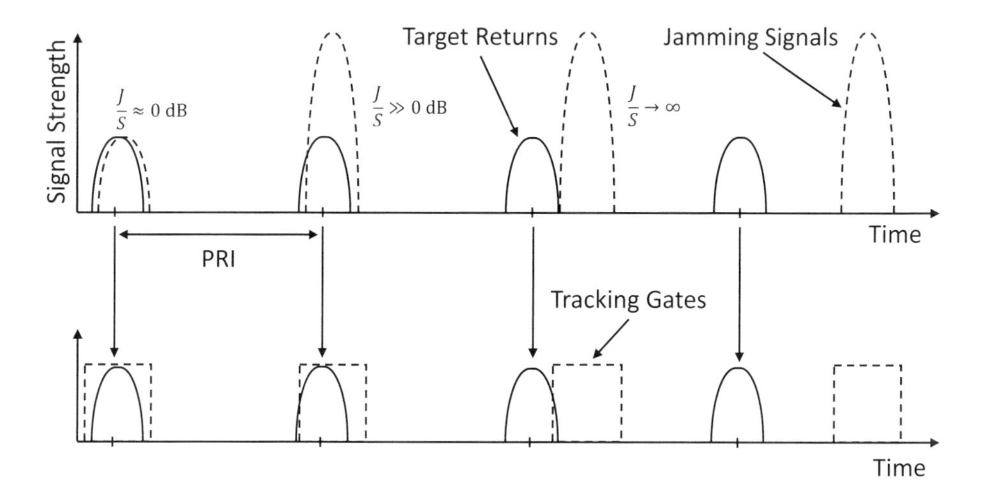

Figure 11.8 Target returns and jamming pulses for range gate pull-off.

Another form of deceptive jamming is *inverse gain* jamming. In this method, a number of false targets is created for varying scan angles. The strength of the jamming signal is inversely proportional to the strength of the incoming radar signal [24]. In this fashion, the radar receives false target-like echoes for every scan angle. Depending on strength of the signals, this method results in an error signal that quickly drives the radar away from the target on a smoothed curve. If the jamming signal is very closely matched to the inverse of the radar signal strength, the radar detects the target no matter where the main beam is pointed, as shown in Figure 11.9. This type of angle deception requires the jamming system to have large dynamic range and sensitive receiving hardware.

A different form of angle deception is wavefront distortion, often referred to as *cross-eye* jamming. This method is effective against monopulse tracking systems and employs two spatially separated jamming devices. Each device samples the incoming radar signal and transmits the same jamming signal at the same time. If the signal arrives at the radar approximately 180° out of phase, wavefront distortion occurs [25]. In monopulse tracking systems, the source of the return signal is presumed to lie along the normal to the wavefront. Cross-eye jamming distorts the wavefront and creates error in the estimation of the direction of the target. To maintain a 180° phase shift between the two separated jamming devices, the signal received by one device is transmitted by the other, and vice versa. This ensures the two propagation paths from the radar to the jamming devices and back are identical. Finally, a 180° phase shift is added to one of the paths, which then creates the wavefront distortion.

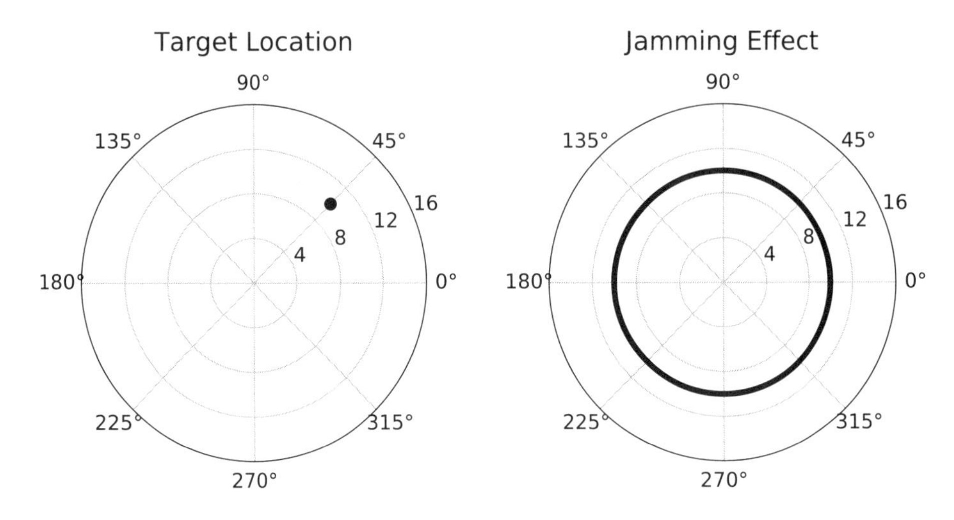

Figure 11.9 Inverse gain jamming technique.

There are numerous active deception techniques, including pseudorandom noise, spot jamming, scintillation, cross polarization, swept amplitude, track-while-scan jamming, and others. The reader is referred to several sources for further information on these techniques [2, 3, 24, 26].

11.3 DIGITAL RADIO FREQUENCY MEMORY

Digital radio frequency memory (DRFM) is a method of digitizing an incoming RF signal and transmitting a modified version of the signal. The earliest reference to a digital method for storage of an RF pulse is from Spector in 1975 [27]. Since these systems create a digital copy of the incoming radar signal, the jamming signal is coherent with the radar system. This is a significant problem as the radar processes these signals as actual targets. Not only can DRFM-based systems can create a number of false targets with varying ranges, velocities, and angles, also possible are custom signatures to deceive classification and discrimination algorithms. DRFM-based systems have also proved to be effective against communications systems. Due to the capability of DRFM systems, there has been increasing use in hardware in the loop simulations to aid in the design, development, and testing of radar systems and algorithms, especially in the areas of electronic attack protection [28–31]. This also provides a means of training radar operators, reduces cost for testing of key components earlier in the design cycle, and cuts down on the need for flight tests. Figure 11.10 illustrates the components of a basic DRFM-based system. Note that this is a block diagram, and many modern systems are implemented

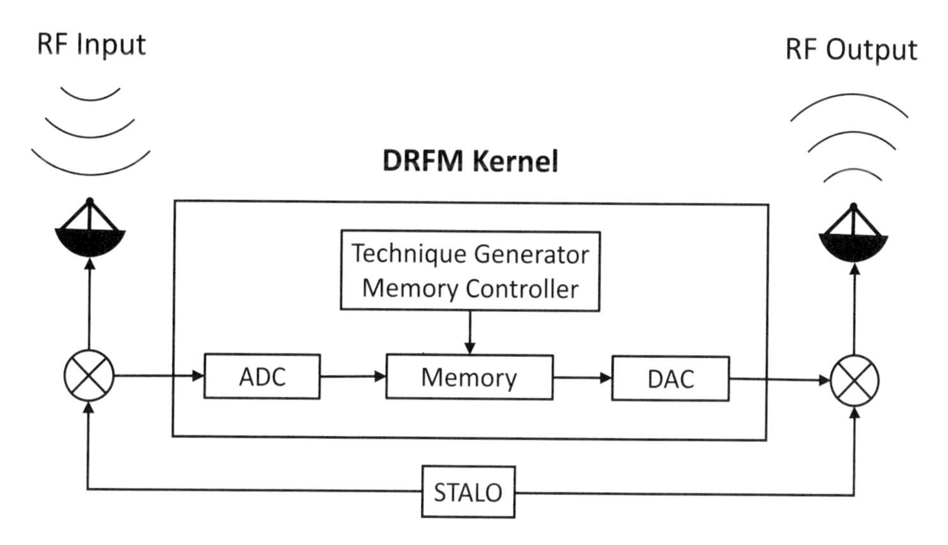

Figure 11.10 Basic DRFM components.

as field-programmable gate arrays (FPGAs). To begin, the radar signal is received, downconverted, and sampled digitally. Next, the desired electronic attack techniques, such RGPO, VGPO, or noise waveforms, are applied. The digital jamming signal is then converted to an analog signal, upconverted, and transmitted back to the victim radar. Advances in computing power and electronic manufacturing techniques have resulted in small, lightweight, low-power DRFM systems capable of being deployed on a wide range of platforms, including aircraft, unmanned vehicles, ships, and others [31–33]. Table 11.1 illustrates the advances in DRFM capability by generation [32, 34, 35].

11.4 EXAMPLES

The examples in the following sections illustrate the concepts of this chapter with several Python examples. The examples for this chapter are in the directory *pyradar\Chapter11* and the matching MATLAB examples are in the directory *mlradar\Chapter11*. The reader should consult Chapter 1 for information on how to execute the Python code associated with this book.

11.4.1 Jammer to Signal: Self-Screening

For this example, calculate and display the jammer-to-signal ratio for a self-screening continuous noise jammer with an effective radiated power of 15 dBW and a bandwidth of 10 MHz. The victim radar has a transmit power of 100 kW, an antenna gain of 20 dB, a

Table 11.1
DRFM Capability by Generation

Parameter	*2nd Generation* (1999–2003)	*3rd Generation* (2004–2006)	*4th Generation* (2007–2011)	*5th Generation* (2012–)
Sampling rate (GSPS)	1	1.2	2	5
Resolution (bits)	8	10	ADC - 10 DAC - 12	ADC - 12 DAC - 16
Instantaneous BW (MHz)	400	500	800	2000
Delay resolution (ns)	16	3.3	0.5	0.2
Memory depth (ms)	1	1.7	8.3	200
Spurious free dynamic range (dBc)	30	36	45	> 50
Digital instantaneous frequency measurement	No	No	Yes	Yes
Arbitrary modulation	No	No	Yes	Yes
Range update phase correction	No	No	Yes	Yes

bandwidth of 100 MHz, and losses of 3 dB. The target has a radar cross section of 3 dBsm.

Solution: The solution to the above example is given in the Python code *jammer_to_signal _example.py* and in the MATLAB code *jammer_to_signal_example.m*. Running the Python example code displays a GUI allowing the user to enter the parameters for the jammer, the target, and the victim radar. The GUI also allows the user to choose between self-screening and escort jamming. The code then calculates and creates a plot of the jammer-to-signal ratio, as shown in Figure 11.11.

11.4.2 Jammer to Signal: Escort

As another example of the jammer-to-signal ratio, calculate and display the jammer-to-signal ratio for an escort continuous noise jammer with an effective radiated power of 20 dBW, a bandwidth of 100 MHz, and is located 100 km from the radar. The victim radar has a transmit power of 100 kW, an antenna gain of 20 dB, an average sidelobe level of −20 dB, a bandwidth of 100 MHz, and losses of 5 dB. The target has a radar cross section

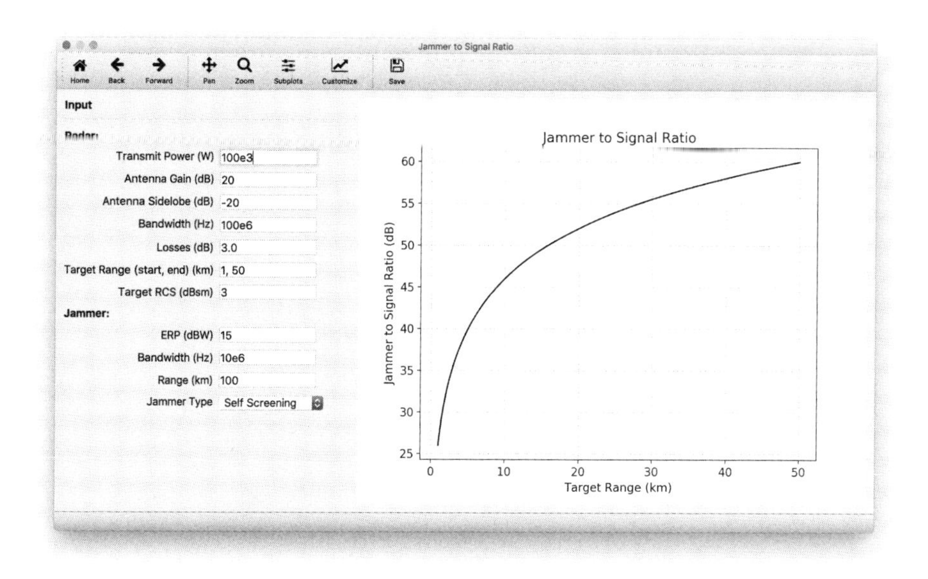

Figure 11.11 The jammer-to-signal ratio for a self-screening jammer calculated by *jammer_to_signal_example.py*.

of 3 dBsm.

Solution: The solution to the above example is given in the Python code *jammer_to_signal_example.py* and in the MATLAB code *jammer_to_signal_example.m*. Running the Python example code displays a GUI allowing the user to enter the parameters for the jammer, the target, and the victim radar. The GUI also allows the user to choose between self-screening and escort jamming. The code then calculates and creates a plot of the jammer-to-signal ratio, as shown in Figure 11.12.

11.4.3 Crossover Range: Self-Screening

In this example, calculate and display the crossover range for the case of a self-screening continuous noise jammer as a function of effective radiated power. The victim radar has a transmit power of 1 MW, an antenna gain of 40 dB, a bandwidth of 100 MHz, and losses of 3 dB. The target has a radar cross section of 3 dBsm.

Solution: The solution to the above example is given in the Python code *crossover_range_example.py* and in the MATLAB code *crossover_range_example.m*. Running the Python example code displays a GUI allowing the user to enter the parameters for the jammer,

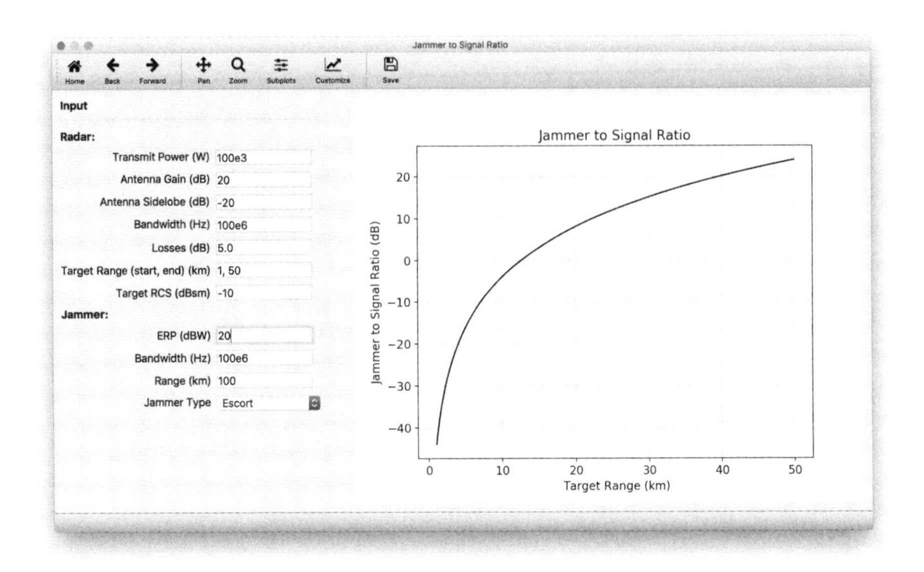

Figure 11.12 The jammer-to-signal ratio for an escort jammer calculated by *jammer_to_signal_example.py*.

the target, and the victim radar. The GUI also allows the user to choose between self-screening and escort jamming. The code then calculates and creates a plot of the crossover range, as shown in Figure 11.13.

11.4.4 Crossover Range: Escort

For the next example, calculate and display the crossover range for the case of an escort continuous noise jammer as a function of effective radiated power. The jammer is located at a range of 5 km from the radar and has a bandwidth of 200 MHz. The victim radar has a transmit power of 60 kW, an antenna gain of 40 dB, an average sidelobe level of -20 dB, a bandwidth of 100 MHz, and losses of 6 dB. The target has a radar cross section of -20 dBsm.

Solution: The solution to the above example is given in the Python code *crossover_range_example.py* and in the MATLAB code *crossover_range_example.m*. Running the Python example code displays a GUI allowing the user to enter the parameters for the jammer, the target, and the victim radar. The GUI also allows the user to choose between self-screening and escort jamming. The code then calculates and creates a plot of the crossover range, as shown in Figure 11.14.

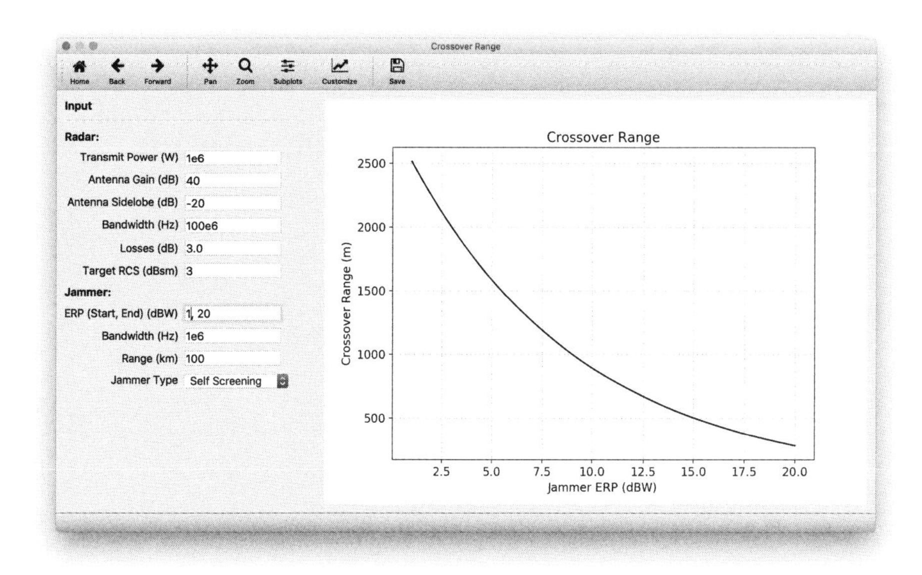

Figure 11.13 The crossover range for a self-screening jammer calculated by *crossover_range_example.py*.

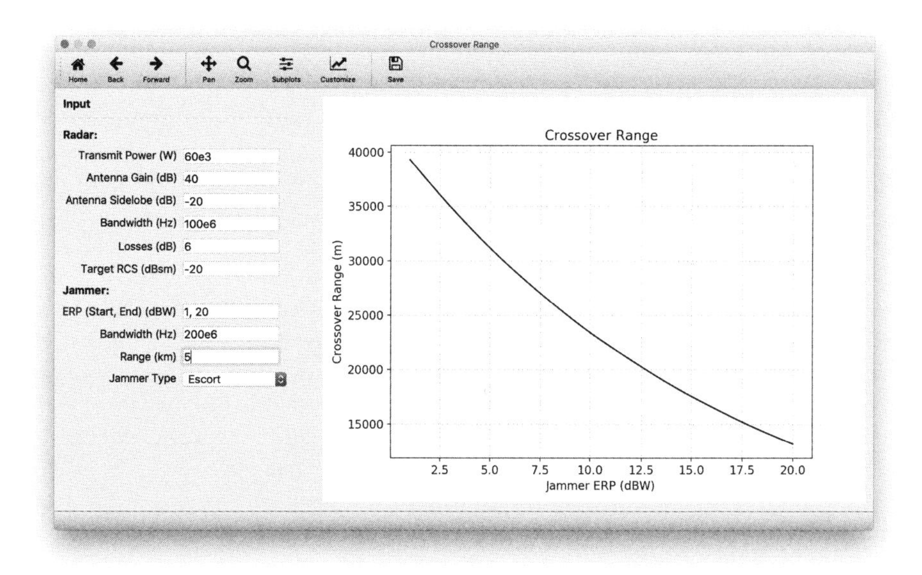

Figure 11.14 The crossover range for an escort jammer calculated by *crossover_range_example.py*.

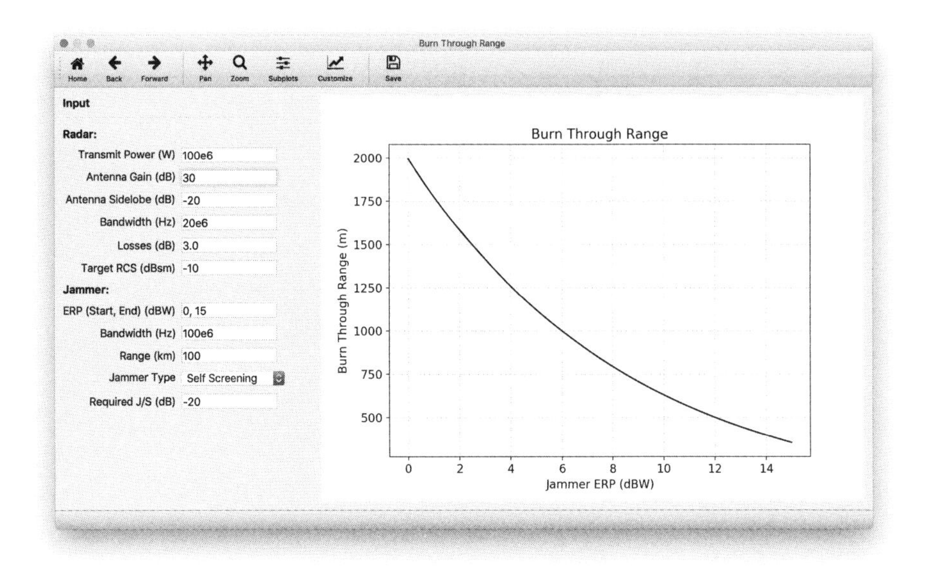

Figure 11.15 The burn-through range for a self-screening jammer calculated by *burnthrough_range_example.py*.

11.4.5 Burn-Through Range: Self-Screening

For this example, calculate and display the burn-through range for the case of a self-screening continuous noise jammer as a function of effective radiated power. The victim radar has a transmit power of 100 MW, an antenna gain of 30 dB, a bandwidth of 20 MHz, and losses of 3 dB. The target has a radar cross section of -10 dBsm. The required jammer-to-signal ratio is -20 dB.

Solution: The solution to the above example is given in the Python code *burnthrough_range_ example.py* and in the MATLAB code *burnthrough_range_example.m*. Running the Python example code displays a GUI allowing the user to enter the parameters for the jammer, the target, and the victim radar. The GUI also allows the user to choose between self-screening and escort jamming. The code then calculates and creates a plot of the burn-through range, as shown in Figure 11.15.

11.4.6 Burn-Through Range: Escort

In this example, calculate and display the burn-through range for the case of an escort continuous noise jammer as a function of effective radiated power. The jammer is located at a range of 10 km from the radar and has a bandwidth of 100 MHz. The victim radar

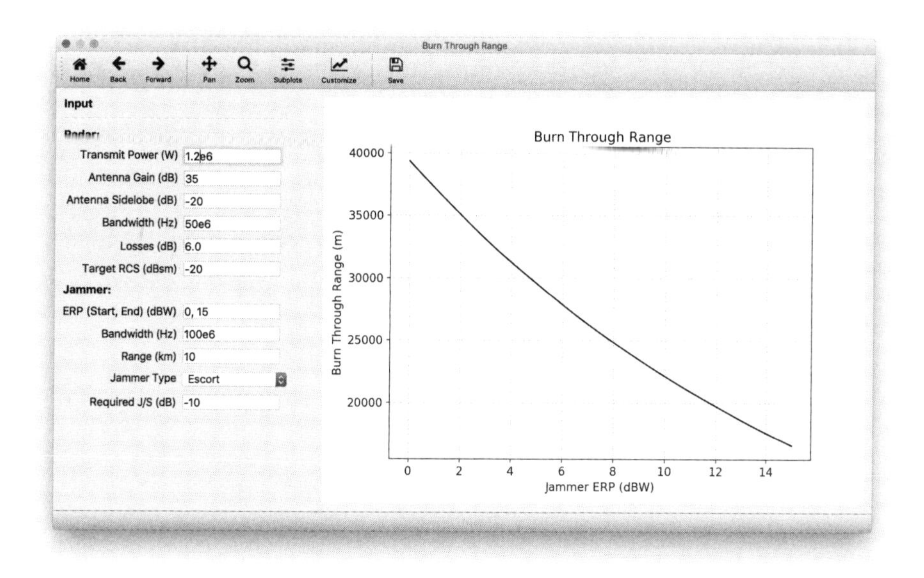

Figure 11.16 The burn-through range for an escort jammer calculated by *burnthrough_range_example.py*.

has a transmit power of 1.2 MW, an antenna gain of 35 dB, an average sidelobe level of −20 dB, a bandwidth of 50 MHz, and losses of 6 dB. The target has a radar cross section of −20 dBsm.

Solution: The solution to the above example is given in the Python code *burnthrough_range_ example.py* and in the MATLAB code *burnthrough_range_example.m*. Running the Python example code displays a GUI allowing the user to enter the parameters for the jammer, the target, and the victim radar. The GUI also allows the user to choose between self-screening and escort jamming. The code then calculates and creates a plot of the crossover range, as shown in Figure 11.16.

11.4.7 Moving Target Indication

As a last example, calculate and display the MTI filter response for the multiple pulse repetition frequency case when the second pulse repetition frequency is $5/4$ of the first pulse repetition frequency.

Solution: The solution to the above example is given in the Python code *delay_line_example.py* and in the MATLAB code *delay_line_example.m*. Running the Python example code displays a GUI allowing the user to choose single or staggered pulse repetition

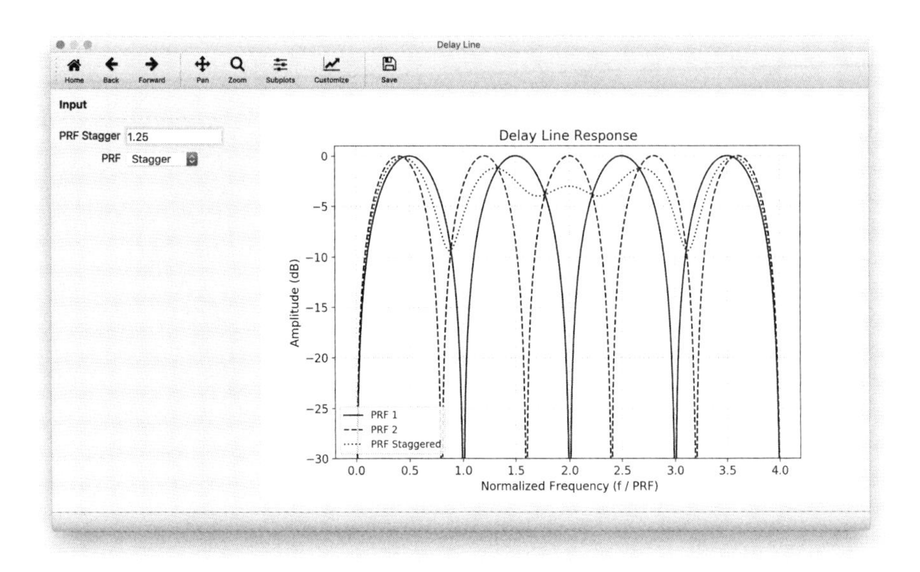

Figure 11.17 MTI filter response for staggered PRFs calculated by *delay_line_example.py*.

frequencies, and to enter the value of the stagger. The code then calculates and creates a plot of the MTI filter response, as shown in Figure 11.17.

PROBLEMS

11.1 Describe the major differences between active and passive jamming and give examples of each.

11.2 For continuous noise jamming, explain how the jammer bandwidth and radar receiver bandwidth influence the effectiveness of the technique.

11.3 Calculate the effective radiated power of a jammer with a transmit power of 10 dBW, an antenna gain of 20 dB, and total losses of 6 dB.

11.4 Using the ERP from Problem 11.3, calculate the jammer-to-signal ratio for an escort type jammer with a bandwidth of 100 MHz. The victim radar has a transmit power of 30 kW, an antenna gain of 35 dB, a receiver bandwidth of 10 MHz, and losses of 3 dB. The target has a radar cross section of −10 dBsm and is 30 km from the radar.

11.5 For stand-off type jamming, find the crossover range for a jammer bandwidth of 30 MHz and ERP of 15 dBW. The victim radar has a transmit power of 10 kW, an antenna gain of 25 dB, an average sidelobe level of −30 dB, a receiver bandwidth of 20 MHz, and losses of 6 dB. The target has a radar cross section of 10 dBsm. The range to the jammer is 5 km.

11.6 Using the radar, target, and jammer parameters given in Problem 11.5, calculate the burn through range when the required jammer to signal ratio is −20 dB.

11.7 Describe how space-time adaptive processing is used to mitigate continuous noise jamming and give examples of methods used to reduce the computational complexity.

11.8 How may range gate pull-off and velocity gate pull-off be used in conjunction to be more effective than either method alone?

11.9 Give an example scenario where inverse gain jamming would be more effective than cross-eye jamming.

1.10 Find the number of half-wavelength dipole particles required to produce a chaff cloud with a radar cross section of 30 dBsm. Assume 10% breakage during deployment of the chaff.

1.11 Describe how moving target indication may be used to mitigate chaff countermeasures. Also, discuss the benefit of using multiple delay lines to improve MTI filter performance.

1.12 Explain the advantages of passive deception and give an example scenario where this type of countermeasure technique would be effective.

1.13 Give the major advantages and disadvantages of digital radio frequency memory as a countermeasure technique.

1.14 Describe the recent hardware and software advances used to make DRFM highly effective as a mechanism of jamming a wide variety of radar systems.

References

[1] M. Atrazhev, V. Il'in and N. Mar'in. *Bor'ba s Padioelektronnymi Sredstvami*. Voyenizdat, Moscow, 1972.

[2] *Electronic Warfare and Radar Systems Engineering Handbook*. Naval Air Warfare Center, Point Mugu, CA, 1999.

[3] D. Adamy. *EW 101: A First Course in Electronic Warfare*. Artech House, Norwood, MA, 2001.

[4] M. Skolnik. *Radar Handbook,* 3rd ed. McGraw-Hill Education, 2008.

[5] J. Madewell. Mitigating the effects of chaff in ballistic missile defense. *Proceedings of the 2003 IEEE Radar Conference*, pages 19–22, 2003.

[6] B. R. Mahafza. *Radar Systems Analysis and Design Using MATLAB,* 3rd ed. CRC Press, Boca Raton, FL, 2013.

[7] B. R. Mahafza and A. Elsherbeni. *MATLAB Simulations for Radar Systems Design*. Chapman and Hall/CRC, New York, 2003.

[8] Leonardo, S.p.A. *BriteCloud DRFM (Digital RF Memory) countermeasure*. `https://www.leonardocompany.com/products/britecloud-3`.

[9] Raven Industries. *Raven Military Decoys*. `https://ravenaerostar.com/products/military-decoys`.

[10] Inflatech, S.R.O. *Inflatech Inflatable Decoys*. `http://www.inflatechdecoy.com/`.

[11] M. Maksimov, et al. *Radar Anti-Jamming Techniques*. Artech House, Dedham, MA, 1979.

[12] V. Tikhonov. *Examples and Problems in Statistical Radio Engineering*. Soviet Radio Publishing House, Moscow, 1970.

[13] V. Tikhonov. *Statistical Radio Engineering*. Soviet Radio Publishing House, Moscow, 1966.

[14] Wikipedia Contributors. *Stationary Process*. `https://en.wikipedia.org/wiki/Stationary_process`.

[15] B. Kovit, I. Holahan and P. Dax. *Electronic Countermeasures and Counter-Countermeasure Techniques*. Soviet Radio Publishing House, Moscow, 1966.

[16] V. Druzhinin. *Handbook on Principles of Radar*. Soviet Radio Publishing House, Moscow, 1967.

[17] S. Vakin and L. Shustov. *Principles of Jamming and Electronic Reconnaissance*. Soviet Radio Publishing House, Moscow, 1968.

[18] W. Melvin. A STAP overview. *IEEE Aerospace and Electronic Systems Magazine*, 19(1):19–35, 2004.

[19] J. Guerci. *Space-Time Adaptive Processing for Radar*. Artech House, Norwood, MA, 2003.

[20] D. Dudgeon and R. Mersereau. *Multidimensional Digital Signal Processing*. Prentice-Hall Signal Processing Series, 1984.

[21] H. Van Trees. *Optimum Array Processing*. John Wiley & Sons, New York, 2002.

[22] J. Li and P. Stoica. *MIMO Radar Signal Processing*. John Wiley & Sons, New York, 2009.

[23] X. Tian. Radar deceptive jamming detection based on goodness-of-fit testing. *Journal of Information and Computational Science*, 9(13):3839, 2012.

[24] A. Graham. *Communications, Radar and Electronic Warfare*. John Wiley & Sons, New York, 2011.

[25] T. Tucker and B. Vidger. Cross-eye jamming effectiveness, 2013. Tactical Technologies, Inc.

[26] D. Adamy. *EW 104: EW Against a New Generation of Threats*. Artech House, Norwood, MA, 2015.

[27] S. Spector. A coherent microwave memory using digital storage: The loopless memory loop. *Electronic Defense*, 1975.

[28] J. Strydom and J. Cillers. Advances in the testing and evaluation of airborne radar through realtime simulation of synthetic clutter. *48th Annual AOC International Symposium and Convention*, 2011.

[29] J. Strydom, et al. Advances in the realtime simulation of synthetic clutter for radar testing and evaluation. *47th Annual AOC International Symposium and Convention*, 2010.

[30] Wikipedia Contributors. *Digital radio frequency memory*. `https://en.wikipedia.org/wiki/Digital_radio_frequency_memory`.

[31] Dynetics, Inc. *PhantomRF: Advanced DRFM target generator for radar test and evaluation*. `https://www.dynetics.com/_files/fact-sheets/Dynetics\%20Phantom\%20RF.pdf`.

[32] Mercury Systems, Inc. *DRFM Technology*. `https://www.mrcy.com/drfm-solutions/airborne1177/`.

[33] BAE Systems, Inc. *Digital Electronic Warfare System*. `https://www.baesystems.com/en-us/product/digital-electronic-warfare-system-dews`.

[34] CSIR. *The Council for Scientific and Industrial Research*. `https://www.csir.co.za`.

[35] Kratos Defense and Security Solutions, Inc. *Wideband Digital RF Memory*. `http://www.kratosdefense.com/page/search?q=drfm-wb`.

About the Author

Andy Harrison received a bachelor of science, master of science, and doctor of philosophy in electrical engineering from the University of Mississippi in 1994, 1996, and 1999, respectively. His studies focused on electromagnetic theory, microwave measurements, antenna analysis and design, acoustics, and computational electromagnetics. His graduate research dealt with the use of wavelet-like basis functions in the finite element solution of partial differential equations. He has authored several papers in areas of computational electromagnetics and global optimization techniques and taught short courses in radar system design, simulation, and analysis.

Dr. Harrison served as an associate editor for the Applied Computational Electromagnetics Society and the Southeastern Symposium on System Theory. He is a senior member of the Institute of Electrical and Electronics Engineers and served as vice chair as well as education chair for the Mississippi section. He also served on the Electrical Engineering Industrial Advisory Board. He is a member of the electrical engineering honor society, Eta Kappa Nu (Epsilon Omega Chapter), the scientific research honor society, Sigma Xi, and the Antenna Measurements Techniques Association.

Dr. Harrison received recognition for his outstanding performance while supporting both the Ground-based Midcourse Defense X-Band Radar System Engineering Division and the Sea-Based X-Band Radar Requirements and Analysis team. He was also recognized for his significant contributions to the Missile Defense Agency's Project Hercules.

Index

Recent Titles in the Artech House Radar Series

Dr. Joseph R. Guerci, Series Editor

Adaptive Antennas and Phased Arrays for Radar and Communications, Alan J. Fenn

Advanced Techniques for Digital Receivers, Phillip E. Pace

Advances in Direction-of-Arrival Estimation, Sathish Chandran, editor

Airborne Pulsed Doppler Radar, Second Edition, Guy V. Morris and Linda Harkness, editors

Basic Radar Analysis, Mervin C. Budge, Jr. and Shawn R. German

Basic Radar Tracking, Mervin C. Budge, Jr. and Shawn R. German

Bayesian Multiple Target Tracking, Second Edition , Lawrence D. Stone, Roy L. Streit, Thomas L. Corwin, and Kristine L Bell

Beyond the Kalman Filter: Particle Filters for Tracking Applications, Branko Ristic, Sanjeev Arulampalam, and Neil Gordon

Cognitive Radar: The Knowledge-Aided Fully Adaptive Approach, Joseph R. Guerci

Computer Simulation of Aerial Target Radar Scattering, Recognition, Detection, and Tracking, Yakov D. Shirman, editor

Control Engineering in Development Projects, Olis Rubin

Design and Analysis of Modern Tracking Systems, Samuel Blackman and Robert Popoli

Detecting and Classifying Low Probability of Intercept Radar, Second Edition, Phillip E. Pace

Digital Techniques for Wideband Receivers, Second Edition, James Tsui

Electronic Intelligence: The Analysis of Radar Signals, Second Edition, Richard G. Wiley

Electronic Warfare in the Information Age, D. Curtis Schleher

Electronic Warfare Target Location Methods, Second Edition, Richard A. Poisel

ELINT: The Interception and Analysis of Radar Signals, Richard G. Wiley

EW 101: A First Course in Electronic Warfare, David Adamy

EW 102: A Second Course in Electronic Warfare, David Adamy

EW 103: Tactical Battlefield Communications Electronic Warfare, David Adamy

FMCW Radar Design, M. Jankiraman

Fourier Transforms in Radar and Signal Processing, Second Edition, David Brandwood

Fundamentals of Electronic Warfare, Sergei A. Vakin, Lev N. Shustov, and Robert H. Dunwell

Fundamentals of Short-Range FM Radar, Igor V. Komarov and Sergey M. Smolskiy

Handbook of Computer Simulation in Radio Engineering, Communications, and Radar, Sergey A. Leonov and Alexander I. Leonov

High-Resolution Radar, Second Edition, Donald R. Wehner

Highly Integrated Low-Power Radars, Sergio Saponara, Maria Greco, Egidio Ragonese, Giuseppe Palmisano, and Bruno Neri

Introduction to Electronic Defense Systems, Second Edition, Filippo Neri

Introduction to Electronic Warfare, D. Curtis Schleher

Introduction to Electronic Warfare Modeling and Simulation, David L. Adamy

Introduction to RF Equipment and System Design, Pekka Eskelinen

Introduction to Modern EW Systems, Andrea De Martino

An Introduction to Passive Radar, Hugh D. Griffiths and Christopher J. Baker

Introduction to Radar using Python and MATLAB®, Lee Andrew Harrison

Linear Systems and Signals: A Primer, JC Olivier

Meter-Wave Synthetic Aperture Radar for Concealed Object Detection, Hans Hellsten

The Micro-Doppler Effect in Radar, Second Edition, Victor C. Chen

Microwave Radar: Imaging and Advanced Concepts, Roger J. Sullivan

Millimeter-Wave Radar Targets and Clutter, Gennadiy P. Kulemin

MIMO Radar: Theory and Application, Jamie Bergin and Joseph R. Guerci

Modern Radar Systems, Second Edition, Hamish Meikle

Modern Radar System Analysis, David K. Barton

Modern Radar System Analysis Software and User's Manual, Version 3.0, David K. Barton

Monopulse Principles and Techniques, Second Edition, Samuel M. Sherman and David K. Barton

MTI and Pulsed Doppler Radar with MATLAB®, Second Edition, D. Curtis Schleher

Multitarget-Multisensor Tracking: Applications and Advances Volume III, Yaakov Bar-Shalom and William Dale Blair, editors

Non-Line-of-Sight Radar, Brian C. Watson and Joseph R. Guerci

Precision FMCW Short-Range Radar for Industrial Applications, Boris A. Atayants, Viacheslav M. Davydochkin, Victor V. Ezerskiy, Valery S. Parshin, and Sergey M. Smolskiy

Principles of High-Resolution Radar, August W. Rihaczek

Principles of Radar and Sonar Signal Processing, François Le Chevalier

Radar Cross Section, Second Edition, Eugene F. Knott, et al.

Radar Equations for Modern Radar, David K. Barton

Radar Evaluation Handbook, David K. Barton, et al.

Radar Meteorology, Henri Sauvageot

Radar Reflectivity of Land and Sea, Third Edition, Maurice W. Long

Radar Resolution and Complex-Image Analysis, August W. Rihaczek and Stephen J. Hershkowitz

Radar RF Circuit Design, Nickolas Kingsley and J. R. Guerci

Radar Signal Processing and Adaptive Systems, Ramon Nitzberg

Radar System Analysis, Design, and Simulation, Eyung W. Kang

Radar System Analysis and Modeling, David K. Barton

Radar System Performance Modeling, Second Edition, G. Richard Curry

Radar Technology Encyclopedia, David K. Barton and Sergey A. Leonov, editors

Radio Wave Propagation Fundamentals, Artem Saakian

Range-Doppler Radar Imaging and Motion Compensation, Jae Sok Son, et al.

Robotic Navigation and Mapping with Radar, Martin Adams, John Mullane, Ebi Jose, and Ba-Ngu Vo

Signal Detection and Estimation, Second Edition, Mourad Barkat

Signal Processing in Noise Waveform Radar, Krzysztof Kulpa

Signal Processing for Passive Bistatic Radar, Mateusz Malanowski

Space-Time Adaptive Processing for Radar, Second Edition, Joseph R. Guerci

Special Design Topics in Digital Wideband Receivers, James Tsui

Systems Engineering of Phased Arrays, Rick Sturdivant, Clifton Quan, and Enson Chang

Theory and Practice of Radar Target Identification, August W. Rihaczek and Stephen J. Hershkowitz

Time-Frequency Signal Analysis with Applications, Ljubiša Stanković, Miloš Daković, and Thayananthan Thayaparan

Time-Frequency Transforms for Radar Imaging and Signal Analysis, Victor C. Chen and Hao Ling

Transmit Receive Modules for Radar and Communication Systems, Rick Sturdivant and Mike Harris

For further information on these and other Artech House titles, including previously considered out-of-print books now available through our In-Print-Forever® (IPF®) program, contact:

Artech House
685 Canton Street
Norwood, MA 02062
Phone: 781-769-9750
Fax: 781-769-6334
e-mail: artech@artechhouse.com

Artech House
16 Sussex Street
London SW1V HRW UK
Phone: +44 (0)20 7596-8750
Fax: +44 (0)20 7630-0166
e-mail: artech-uk@artechhouse.com

Find us on the World Wide Web at: www.artechhouse.com